Lecture Notes in Computer Science   9419

Commenced Publication in 1973
Founding and Former Series Editors:
Gerhard Goos, Juris Hartmanis, and Jan van Leeuwen

More information about this series at http://www.springer.com/series/7409

Jianyong Wang · Wojciech Cellary
Dingding Wang · Hua Wang
Shu-Ching Chen · Tao Li
Yanchun Zhang (Eds.)

Web Information Systems Engineering – WISE 2015

16th International Conference
Miami, FL, USA, November 1–3, 2015
Proceedings, Part II

 Springer

Editors

Jianyong Wang
Tsinghua University
Beijing
China

Wojciech Cellary
Poznan University of Economics
Poznan
Poland

Dingding Wang
Florida Atlantic University
Boca Raton, FL
USA

Hua Wang
Victoria University
Melbourne, VIC
Australia

Shu-Ching Chen
Florida International University
Miami, FL
USA

Tao Li
Florida International University
Miami, FL
USA

Yanchun Zhang
Victoria University
Melbourne, VIC
Australia

ISSN 0302-9743   ISSN 1611-3349   (electronic)
Lecture Notes in Computer Science
ISBN 978-3-319-26186-7   ISBN 978-3-319-26187-4   (eBook)
DOI 10.1007/978-3-319-26187-4

Library of Congress Control Number: 2015953784

LNCS Sublibrary: SL3 – Information Systems and Applications, incl. Internet/Web, and HCI

Springer Cham Heidelberg New York Dordrecht London

Printed on acid-free paper

Springer International Publishing AG Switzerland is part of Springer Science+Business Media
(www.springer.com)

Preface

Welcome to the proceedings of the 16th International Conference on Web Information Systems Engineering (WISE 2015), held in Miami, Florida, USA, in November 2015. The series of WISE conferences aims to provide an international forum for researchers, professionals, and industrial practitioners to share their knowledge in the rapidly growing area of Web technologies, methodologies, and applications. The first WISE event took place in Hong Kong, China (2000). Then the trip continued to Kyoto, Japan (2001); Singapore (2002); Rome, Italy (2003); Brisbane, Australia (2004); New York, USA (2005); Wuhan, China (2006); Nancy, France (2007); Auckland, New Zealand (2008); Poznan, Poland (2009); Hong Kong, China (2010); Sydney, Australia (2011); Paphos, Cyprus (2012); Nanjing, China (2013); and Thessaloniki, Greece (2014). This year, for a second time, WISE was held in North America, in Miami, supported by Florida International University (FIU).

WISE 2015 hosted several well-known keynote and invited speakers. Moreover, two tutorials were presented on the topics of building secure Web systems and accessing the Princeton Wordnet.

A total of 171 research papers were submitted to the conference for consideration, and each paper was reviewed by at least two reviewers. Finally, 53 submissions were selected as full papers (with an acceptance rate of 31 % approximately), plus 17 as short papers. The research papers cover the areas of big data techniques and applications, deep/hidden Web, integration of Web and Internet, linked open data, the Semantic Web, social network computing, social Web and applications, social Web models, analysis and mining, Web-based applications, Web-based business processes and Web services, Web data Integration and mashups, Web data models, Web information retrieval, Web privacy and security, Web-based recommendations, and Web search.

In addition to regular and short papers, the WISE 2015 program also featured three special sessions, including a special session on Data Quality and Trust in Big Data (QUAT 2015), a special session on Decentralized Social Networks (DeSN 2015), and an invited session.

QUAT is a qualified forum for presenting and discussing novel ideas and solutions related to the problems of exploring, assessing, monitoring, improving, and maintaining the quality of data and trust for "big data." It provides a forum for researchers in the areas of Web technology, e-services, social networking, big data, data processing, trust, and information systems and GIS to discuss and exchange their recent research findings and achievements. This year, the QUAT 2015 program featured six accepted papers on data cleansing, data quality analytics, reliability assessment, and quality of service for domain applications. QUAT 2015 was organized by Prof. Deren Chen, Prof. William Song, Dr. Xiaolin Zheng, and Dr. Johan Håkansson.

The goal of DeSN 2015 was to serve as a forum for researchers or professionals from both academia and industry to exchange new ideas, discuss new solutions, and share their experience in the design, implementation, analysis, experimentation, or measurement related to decentralized social networks. The DeSN 2015 program included two invited speakers, Dr. Sarunas Girdzijauskas and Dr. Bogdan Carbunar, and three accepted papers. The DESN 2015 co-chairs included Dr. Antoine Boutet, Dr. Sarunas Girdzijauskas, and Dr. Frederique Laforest.

The invited session included five research papers from leading research groups. Each invited paper featured a specific domain, with five papers covering recommender systems, demand trend prediction in cloud computing, deep learning, database security, and social network privacy.

We wish to take this opportunity to thank the honorary co-chairs, Prof. S.S. Iyengar and Prof. Marek Rusinkiewicz; the tutorial and panel co-chairs, Prof. Guandong Xu and Prof. Mitsunori Ogihara; the WISE challenge program co-chairs, Prof. Weining Qian and Qiulin Yu; the workshop co-chairs, Prof. Hill Zhu and Prof. Yicheng Tu; the publication chair, Prof. Hua Wang; the Local Organizing Committee co-chairs, Mr. Carlos Cabrera and Ms. Catherine Hernandez; the publicity co-chairs, Prof. Mark Finlayson, Prof. Giovanni Pilato, and Prof. Yanfang Ye; the registration chair, Mr. Steve Luis; the financial co-chairs, Ms. Lian Zhang and Ms. Donaley Dorsett; and the WISE society representative, Prof. Xiaofang Zhou. The editors and chairs are grateful to the website and social media masters, Mr. Steve Luis and Mr. Bin Xia, for their continuous active support and commitment, and Dr. Rui Zhou and Ms. Sudha Subramani for their effort in preparing the proceedings.

In addition, special thanks are due to the members of the International Program Committee and the external reviewers for a rigorous and robust reviewing process. We are also grateful to the School of Computing and Information Sciences of Florida International University and the International WISE Society for supporting this Conference. The WISE Organizing Committee is also grateful to the special session organizers for their great efforts to help promote Web information system research to broader domains.

We expect that the ideas that have emerged in WISE 2015 will result in the development of further innovations for the benefit of scientific, industrial, and societal communities.

November 2015

Jianyong Wang
Wojciech Cellary
Dingding Wang
Hua Wang
Shu-Ching Chen
Tao Li
Yanchun Zhang

Organization

Honorary Chairs

S.S. Iyengar — Florida International University, USA
Marek Rusinkiewicz — New Jersey Institute of Technology, USA

General Chairs

Shu-Ching Chen — Florida International University, USA
Tao Li — Florida International University, USA
Yanchun Zhang — Victoria University, Australia

Program Co-chairs

Wojciech Cellary — Poznan University of Economics, Poland
Dingding Wang — Florida Atlantic University, USA
Jianyong Wang — Tsinghua University, China

Publication Chair

Hua Wang — Victoria University, Australia

Local Arrangements Chairs

Ivana Rodriguez — Florida International University, USA
Catherine Hernandez — Florida International University, USA

Financial Co-chairs

Carlos Cabrera — Florida International University, USA
Lian Zhang — Florida International University, USA

Web and Social Media Masters

Steve Luis — Florida International University, USA
Bin Xia — Nanjing University of Science and Technology, China

WISE Society Representative

Xiaofang Zhou — University of Queensland, Australia

Tutorial and Panel Co-chairs

Guandong Xu   University of Technology Sydney, Australia
Mitsunori Ogihara   University of Miami, USA

WISE Challenge Program Chairs

Weining Qian   East China Normal University, China
Qiulin Yu   Ping An Technology (Shenzhen) Co., Ltd., China

Workshop Co-chairs

Hill Zhu   Florida Atlantic University, USA
Yicheng Tu   University of South Florida, USA

Publicity Co-chairs

Mark Finlayson   Florida International University, USA
Giovanni Pilato   Italian National Research Council (CNR), Italy
Yanfang Ye   West Virginia University, USA

Program Committee

Karl Aberer   EPFL, Switzerland
Markus Aleksy   ABB, Switzerland
Boualem Benatallah   University of New South Wales, Australia
Azer Bestavros   Boston University, USA
Antonis Bikakis   University College London, UK
David Camacho   Universidad Autonoma de Madrid, Spain
Barbara Catania   University of Genoa, Italy
Tiziana Catarci   Sapienza University of Rome, Italy
Richard Chbeir   LIUPPA Laboratory, France
Fei Chen   HP, USA
Enhong Chen   University of Science and Technology of China
Yueguo Chen   Renmin University of China
Gao Cong   Nanyang Technological University, Singapore
Alexandra Cristea   University of Warwick, UK
Schahram Dustar   Vienna University of Technology, Austria
Damiani Ernesto   University of Milan, Italy
Marie-Christine Fauvet   Joseph Fourier University of Grenoble, France
Bry Francois   Ludwig Maximilian University, Germany
Jun Gao   Peking University, China
Hong Gao   Harbin Institute of Technology, China
Yunjun Gao   Zhejiang University, China
Claude Godart   University Henri Poincare, France

Daniela Grigori   Laboratoire LAMSADE, University Paris-Dauphine,
                         France
Hakim Hacid   Bell Labs, France
Armin Haller   CSIRO ICT Centre, Australia
Xiaofeng He   East China Normal University, China
Yuh-Jong Hu   National Chengchi University, Taiwan
Luke Huan   University of Kansas, USA
Jianbin Huang   Xidian University, China
Marta Indulska   University of Queensland, Australia
Pokorny Jaroslav   Charles University, Czech Republic
Yan Jia   National University of Defense Technology
Lili Jiang   Max Planck Institute for Informatics, Germany
Peiquan Jin   University of Science and Technology of China
Ruoming Jin   Kent State University, USA
Yiping Ke   Nanyang Technological University, Singapore
Fang Li   Shanghai Jiao Tong University, China
Xue Li   School of ITEE, University of Queensland, Australia
Chengkai Li   University of Texas at Arlington, USA
Lei Li   Samsung, USA
Mengchi Liu   Carleton University, Canada
Shuai Ma   Beihang University, China
Xiaofeng Meng   Renmin University of China
Mikolaj Morzy   Poznań University of Technology, Poland
Wilfred Ng   HKUST, China
Kjetil Nørvåg   Norwegian University of Science and Technology,
                         Norway
Mitsunori Ogihara   University of Miami, USA
George Pallis   University of Cyprus, Cyprus
Dasgupta Partha   Arizona State University, USA
Zbigniew Paszkiewicz   PricewaterhouseCoopers, Cyprus
Olivier Pivert   ENSSAT, France
Weining Qian   East China Normal University, China
Tieyun Qian   Wuhan University, China
Jarogniew Rykowski   Poznań University of Economics, Poland
Rizos Sakellariou   University of Manchester, UK
Wei Shen   Nankai University, China
John Shepherd   UNSW, Australia
Dezhao Song   Thomson Reuters, USA
Dandan Song   Beijing Institute of Technology, China
Reima Suomi   University of Turku, Finland
Stefan Tai   KIT, Germany
Dimitri Theodoratos   New Jersey Institute of Technology, USA
Farouk Toumani   Limos, Blaise Pascal University, Clermont-Ferrand,
                         France
Xiaojun Wan   Peking University, China
Guoren Wang   Northeast University, China

Contents – Part II

Contents – Part I

The Influence of Client Platform on Web Page Content: Measurements, Analysis, and Implications

Sean Sanders[✉], Gautam Sanka, Jay Aikat, and Jasleen Kaur

University of North Carolina at Chapel Hill, Chapel Hill, USA
ssanders@cs.unc.edu

Abstract. Modern web users have access to a wide and diverse range of client platforms to browse the web. While it is anecdotally believed that the *same* URL may result in a different web page across different client platforms, the extent to which this occurs is not known. In this work, we systematically study the impact of different client platforms (browsers, operating systems, devices, and vantage points) on the content of base HTML pages. We collect and analyze the base HTML page downloaded for 3876 web pages composed of the top 250 web sites using 32 different client platforms for a period of 30 days — our dataset includes over 3.5 million web page downloads. We find that client platforms have a statistically significant influence on web page downloads in both expected and unexpected ways. We discuss the impact that these results will have in several application domains including web archiving, user experience, social interactions and information sharing, and web content sentiment analysis.

Keywords: Web page measurement · Mobile web · Content analysis

1 Introduction

Users have many choices of *client platforms* — browsers, operating systems, devices, and vantage points — that can be used to request web-based data. Although, it is known that certain client platforms such as device type (e.g., smartphones or laptops) and vantage point can have an influence on the base HTML page that is downloaded [14,26], the extent to which this occurs has not been studied before. Any difference in base HTML pages that is due to client platform can result in data that is incomplete or view-specific. This can present issues for several web-related applications, such as web archival [13,21], document summarization [9,22], and information sharing, because additional care must be taken when (i) designing experiments that yield complete and/or unbiased data and (ii) developing processing scripts that are robust to different web-page designs — the need for understanding these differences has also been recently discussed in [2].

In this paper, we ask the question — *to what extent do different client platforms influence the content of a base HTML web page for the **same** URL*

© Springer International Publishing Switzerland 2015
J. Wang et al. (Eds.): WISE 2015, Part II, LNCS 9419, pp. 1–16, 2015.
DOI: 10.1007/978-3-319-26187-4_1

request? We perform the *first* measurement study that aims to understand this influence. Our methodology includes collecting measurements across different browsers (Opera, Internet Explorer, Google Chrome, Firefox, and Safari), operating systems (Mac OSX, Windows, Linux, iOS, and Android), devices (laptops, tablets, and smartphones), and vantage points (13 planetLab nodes located in 8 different countries) — this includes over 3.5 million measurements obtained from 3876 unique URLs composed of the top 250 web sites collected over a period of 30 days. We extract both HTML tag-based and content-based features from this data and find differences in web pages across different client types that are both practically and statistically significant. Our key findings are:

- *Expected and Unexpected Results:*
    1. As expected, device type (smartphones, tablets, and laptops) has a significant impact on web page content, with smaller devices being returned leaner pages. However, there is no consensus among current web designers and content providers on which type of page should be designed for tablets (i.e., should tablets simply return default laptop pages, mobile optimized pages, or have a special type of page altogether). An unexpected result is that, surprisingly, the manufacturer of a device, say an iPad Tablet or a Galaxy Tablet, may impact the type of page that is downloaded, say a default laptop or mobile optimized page.
    2. The differences that we find across different browsers are largely unexpected. For example, we find that different browsers may provide different default number of comments to be shown in a comment section for news articles and social media sites. We also find that content providers handle outdated browsers in multiple ways including: (1) fail to fulfill the web page request; and (2) fulfill the web page request by sending a similar, but different, web page that is likely more compatible with the user's browser version (e.g., sending a mobile-optimized page to an outdated laptop browser).
    3. As expected, we find that vantage point has a modest influence on web pages. For example, some content providers provide international versions of web pages that is dependent on the country of a user's vantage point while others provide the same content irrespective of vantage point. An unexpected result is that search results are highly influenced by vantage point, even for search queries where vantage point is not an obvious contributing factor to the result set.
- *Implications of Results on Web-related Applications:* Differences in web pages across different client type have implications in several web-related application domains including web archival [23], document summarization [9,22], sentiment analysis [16,22], and web browsing/systems design [14]. Some examples include: (1) the number of default comments and/or product reviews provided on a page is influenced by client platform— this may impact document summarization and sentiment analysis techniques that leverage this information; (2) web page designs may be client platform-specific which influences the type of content that is available and the effectiveness of parsing scripts that

is targeted for a specific page design. This can also be problematic for sharing information on social media because hyperlinks may be client platform specific (hence users may be referring to different content and/or formatting context).

The remainder of this paper is organized as follows. We present our methodology in Sect. 2. The results and implications of our analysis is provided in Sect. 3. Related work is presented in Sect. 4 and a summary of our study along with intended future work is provided in Sect. 5.

2   Methodology

Our methodology consists of two components: (1) Data collection and (2) Statistical analysis. We describe these two aspects in this section.

2.1   Data Collection

Selection of Web Pages to Study: In this study, we target web pages that are comprised from the top 250 web sites of the world according to Alexa [1] — a recent study shows that 99 % of web requests comprise the top 250 web domains [6]. We manually browse each of these 250 web sites to obtain a diverse sample of URLs from each. Our web page sample includes landing pages, video streaming pages, search result pages (e.g., web, image, and news search), mobile web pages, clickable content, audio streaming pages, and social networking. We do this manual browsing for URL collection instead of leveraging a web crawler in order to better control the diversity and representativeness of our dataset. In total, we collect a list of 3876 unique URLs that are used to drive our data collection procedure.

Client Platforms Used: We next select a diverse set of client platforms, that are used to download the web pages we previously identified. As noted before, we intend to study the impact that browsers, operating systems, devices, and vantage points have on base HTML pages. We control for these different client platforms by requesting web pages using an User-Agent string that corresponds to the appropriate client platform of interest. User-Agent strings encapsulate the operating system, browser type, browser version, and even hardware information about client platforms — content providers use this information when responding to web requests [14]. User-Agent strings can be easily set by using scripts (we use python for this) to download base HTML pages. Table 1 lists the 32 User-Agents used for our study[1].

Our definition of "client platform" also includes location (vantage point). Thus, we also download web pages from different vantage points around the world — we use the PlanetLab network for this [8]. The 13 planetLab nodes

---

[1] Please note that each of the User-agents we use in this study were obtained from deep packet inspection of web traffic as generated using known client platforms.

Table 1. Overview of user-agents used for web page requests

Operating system	Browser(s)	Device
Windows 7	Chrome 38.0.2125.122 - Chrome 33.0.1750.154	Laptop
Windows 7	Firefox 33.0 - Firefox 26.0	Laptop
Windows 7	Internet Explorer 11.0 - Internet Explorer 9.0	Laptop
Windows 7	Opera 25.0.1614.68 - Opera 12.16	Laptop
Windows 7	Safari 5.1.7	Laptop
Windows 8	Chrome 39.0.2171.95 - Firefox 32.0	Laptop
Windows 8	Internet Explorer 11.0 - Opera 24.0.1558.61	Laptop
MacOSX 10.6.8	Chrome 39.0.2171.65 -Firefox 33.0	Laptop
MacOSX 10.6.8	Safari 5.1.9-Opera 25.0.1614.71	Laptop
MacOSX 10.9.4	Chrome 38.0.2125.122-Firefox 33.0	Laptop
MacOSX 10.9.4	Safari 7.0.5-Opera 25.0.1614.68	Laptop
Ubuntu	Firefox 34.0	Laptop
Solaris	Firefox 17.0	Laptop
Fedora	Firefox 2.0.0.19	Laptop
Android 4.4.4	Chrome 37.0.2062.117	Motorola Smartphone
Android 4.4.2	Samsung SM-T230NU-Chrome 35.0.1916.141	Samsung GalaxyTablet
Android 4.4.2	Amazon Silk 3.37	Fire Tablet
iOS 7	Safari 8.0 Mobile/12B41	iPhone Smartphone
iOS 7	Safari 7.0 Mobile/11A501	iPad Tablet
iOS 7	Safari 8.0 Mobile/12A405	iPod Touch
iOS 3	Safari 4.0 Mobile/7D11	iPod Touch

that we use are located in Australia, China, Japan, Brazil, Poland, Canada, and the United States (7 nodes — Oregon, Rhode Island, California, Florida, New Mexico, Kentucky, and Ohio).

Repeated Measurements: Modern web content is highly dynamic and may change multiple times a day [11]. We take repeated measurements of each web page across each client platform to eliminate differences in web page content observed across client platform, that are likely simply due to variation over time. Specifically we take 30 repeated measurements over a period between December 18, 2014 and January 18, 2015. Thus, our dataset includes 3,771,348 page downloads.

2.2   Statistical Analysis

Overview of Features: We extract different types of quantitative features from our HTML data to describe the properties of the downloaded web pages. A brief overview of the types of features that we extract are provided below:

HTML Tag-based features are used primarily for the analysis of page formatting — these have commonly been used in other HTML-based analysis [3, 7].

In particular, we count the occurrence of several HTML tags/attributes that are present on a given web page. These tags represent different established categories of HTML information [3][2]. Our feature set includes:

1. *Flow content:* Used within the body of HTML documents (e.g., "table", "form", "option", "text area", and "menu" tags)
2. *Sectioning content:* Used to partition HTML documents (e.g., "area", "article", "body", "div", and "section" tags)
3. *Heading content:* Used for header-level markup (e.g., "header", "title", and "meta" tags)
4. *Phrasing content:* Used for text-level markup (e.g.,"abbr","b", "p", "strong", and "span" tags)
5. *Embedded content:* Used for elements that load external resources into the HTML document (e.g., "script", "image", "audio", "embed", "param", and "iframe" tags).

Count statistics that are derived from tags that represent (i) hyperlink-level information (e.g., "a" and "link" tags) and (ii) the *extensions* of embedded objects that is referenced by a page (e.g., .jpeg, .gif, and .png extensions for embedded image objects) are used for *Object-based features* and analysis. We also derive *Content-based features* from our HTML data. We use a simple bag-of-words model to count the frequency of all the words that are present in a document — bag-of-words models are commonly used in natural language processing, machine learning, and computer vision [25]. A *word* in this model is defined as any sequence of characters that is present in an HTML document that is delimited by $>$, $<$, ", newline, or whitespace characters. This model allows us to derive features that can measure the overall text-related differences between two documents. We derive features such as (i) the number of words that are shared between two documents (i.e., a baseline document and a test document), and (ii) the number of words that are different between two documents to compactly represent these content-related differences. We use these features simply as a measure to flag significant differences in text for further analysis.

While we are able to obtain a lot of information from base HTML files we are unable to collect all of the information that is referenced by a particular web page. This is because modern web pages make significant use of AJAX and scripting technology. It is nontrivial to extract features that are derived from this information using base HTML pages alone. An analysis of the network traffic generated by web page downloads is needed to obtain this data. Such traffic analysis is beyond the scope of this paper.

Statistical Analysis Procedure: In order to determine which of our 134 features differ significantly across web pages downloaded using different client platforms, we use a standard non-parametric statistical test. The use of a non-parametric test allows us to make minimal assumptions about the distribution of these features. In particular, we use the *Kruskal-Wallis* test to determine whether

---

[2] Please refer to [3] for a complete list of these features.

there is a statistically significant difference between the measured web page samples across multiple appropriate groups of client platforms for each feature. The Kruskal-Wallis test yields p-values that represent the statistical significance of each feature for different client platforms. Here, lower p-values correspond to results that have greater statistical significance. We then use these results to dig deeper into our dataset to (i) determine the source of any significant difference and (ii) discuss the practical significance of our findings.

3   Results

Impact of Browser Platform. We first investigate the impact that different browser platforms have on web page content. We initially focus on the influence of different browser platforms installed on the *same* operating system. In particular, we compare the latest versions of the Internet Explorer, Chrome, Firefox, and Opera browsers that correspond to the Windows 7 operating system — refer to Table 1 for more details about these browsers. The Kruskal-Wallis test for this feature group yields 8 features that have p-value $<.05$ across browser platforms — in fact, these p-values are generally less than 10^{-3}. These 8 statistically significant features are: the number of "label" tags, the number of "tr" tags, the number of "table" tags, the number of "td" tags, the number of "style" tags, the number of "legend" tags, javascript length (i.e., the number of characters present in script tags), and the number of different words present. Upon further analysis of our data, we find that these statistically significant features correspond to the following trends:

- *Differences in javascript:* We find that many content providers such as soundcloud.com and bing.com (particularly image search results) use different javascript code that is suited for particular browsers — these javascript related differences were identified by the number of different words feature. We find that different javascript methods are implemented differently across browser platforms and/or have conditional statements that branch for different client browser platforms. For example, soundcloud.com uses conditional statements that takes the client platform into account during javascript execution to determine whether HLS (HTTP Live Streaming) is supported by the client platform. Alternatively, Fig. 1 shows an example where a Youtube.com page has javascript that is browser-specific — here the Chrome javascript for loading a video appears to be HTML5-based while the Firefox javascript appears to be flash-based (This is identified by the "swf" references in Fig. 1). It is known that if different client platforms are not taken into account, rendering differences across browsers can occur when the same source HTML is processed — for example, target.com has differences in rendered tables across browsers despite having rendering the *same source code* that renders that portion of the page.
- *Ads:* We also observe "ads" that attempt to get a user to download a particular browser or app that is browser dependent. For example, yahoo.com

recommends that users update to the latest version of firefox for non-firefox client platforms, whereas target.com recommends that users on the Chrome browser to download their custom app. These ads seem to be attempts to get users to utilize software that is fully supported by the content provider.

– *Reduced comment and recommendation sections:* Our data also shows that cbssports.com and yelp.com do not provide the same number of comments, recommendations, news feeds, or search results for each browser. The limited information provided by certain browser platforms provides inconsistent data for document summarization and sentiment analysis applications [9,16,22] which can yield misleading and/or incomplete results, depending on the specific choice of browser platform. This limited information also impacts user experience because it may require users to take additional actions, such as a click, to view additional content that may be more readily available (already loaded) on a different browser.

```
<!--Chrome Version-->
ytplayer.load = function() {
yt.player.Application.create("player-api",
ytplayer.config);ytplayer.config.loaded = true;};

(function() {if (!!window.yt && yt.player &&
yt.player.Application)
{ytplayer.load();}}());</script>
<div id="watch-queue-mole" class="video-mole mole-
collapsed hid">

<!--Firefox Version-->
swf = swf.replace('__flashvars__',
encoded.join('&'));document.getElementById("player-
api").innerHTML = swf;ytplayer.config.loaded =
true}());</script>
<div id="watch-queue-mole" class="video-mole mole-
collapsed hid">
```

Fig. 1. Example where javascript is different for different browsers (Chrome vs Firefox).

Impact of Browser Version. We next compare the impact that browser *version* may have on base HTML pages. Our statistical test yields 13 statistically significant features. The most notable features that are not also influenced by browser platform, say Safari vs Firefox, are the number of script tags and the number of HTML5 tags. With respect to the number of script tags, we observe similar differences in scripting behavior as we did with the differences in browser platform. With respect to the number of HTML5 tags, we observe that there tends to be more HTML5-related tags for the latest browser versions as compared to the older versions — we believe this to be a compatibility-related issue.

We also observe cases where content providers treat outdated or unsupported browsers in the following 2 ways. First, the content provider can fulfill the web request, but provide a warning to the user that their browser needs to be updated (zillow.com, soundcloud.com) — this may also result in failed web requests. Second, the content provider can fulfill the web request by responding with a web page that is compatible with the user's browser. This is explained in detail next.

We find multiple instances when browser version has an impact on page content. For example, Fig. 2 shows that a Google search result that is rendered

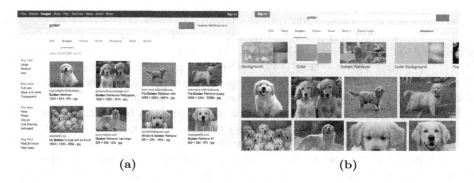

(a)   (b)

Fig. 2. Different HTML pages are returned when an old version of Opera is used (a) in place of a current version of Opera (b).

using an outdated Opera browser (Fig. 2(a)), and an up-to-date Chromium-based Opera browser can be displayed differently (Fig. 2(b)) — though these observed differences are almost purely stylistic with respect to image size and visibility of URLs on images. Figure 3 shows a different example of when a web server responds with a web page for an outdated browser. Here, the web request is for a mobile web page of a product on Amazon.com. Figure 3(a) shows that when a mobile web page is requested using an *up to date mobile device and browser* (an iPhone in particular), the request is satisfied as expected. When we make the same request for a mobile web page using an *outdated Firefox browser on a laptop* we also get the *same* mobile web page — though we do not observe an ad for downloading an app. Figure 3(b) shows that when the same request is made to Amazon using an *up-to-date Firefox browser on a laptop* we get a *different* mobile web page that is clearly representing the same product shown in Fig. 3(a). It is clear that these downloaded web pages are both (i) mobile-optimized web pages and (ii) different, where the version of the page shown in Fig. 3(b) appears to be an older mobile web page design than the page shown in Fig. 3(a). We conclude two things from these observations: (1) mobile web pages may sometimes be used to fulfill web requests to outdated browsers (we observe similar behavior for yahoo.com and att.com); and (2) interesting and unexpected quirks exist for some web page requests that are influenced by browsers[3]. The impact that browser version has on web page downloads is important for web crawling tools because (i) web crawlers may be used for years without receiving any significant upgrades and (ii) content providers may respond to known web-crawler User-Agents in a manner that results in errors or downloading data that is limited (in a manner similar to mobile web pages) [21].

Impact of Operating System. For the purposes of analyzing any implications that operating systems may have on base HTML pages, we compare all laptop-

---

[3] Please note that the significant differences discussed here are primarily true for browser version analysis for Opera and Firefox. This is because we have the largest range in release dates for these two browsers.

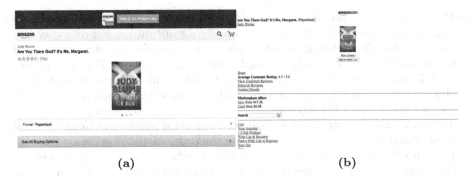

$$(a) \hspace{6cm} (b)$$

Fig. 3. Content providers can return different pages to account for different browser versions — (a) current mobile browser and (b) current desktop browser.

based browsers across each operating system that also has the *same* version of that particular browser. For example, Firefox 33.0 is compared across MacOSX 10.9.4 and Windows 7. We do this for all combinations of browsers where this is valid according to the User-Agents we tested in Table 1. We do not find any statistically significant features that occur for the *same* browser across *different* operating systems. We conclude that browser version and browser type has a much bigger impact on web page downloads than operating systems.

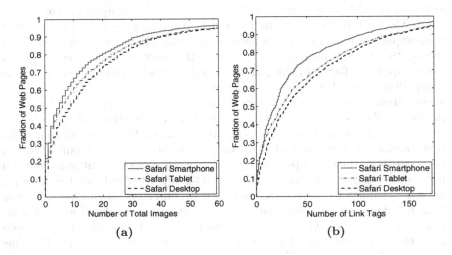

$$(a) \hspace{6cm} (b)$$

Fig. 4. Device type has a significant impact on HTML features.

Impact of Device Type. We next study the impact that different devices have on web page downloads. We start by focusing on comparing the iOS 7 iPhone smartphone, iOS 7 iPad tablet, and the MacOSX 10.9.4 laptop where each device runs a version of Safari. We find that:

1. *Device type has a statistically significant impact on web page downloads:* As can be expected, devices have a statistically significant impact on many features (67 total) by design intent — pages designed for the small screens of mobile devices are likely to have simpler and smaller content. The most prominent features that differ across phones, tablets, and laptops are embedded object-related features such as the number of images, scripts, and CSS references found in an HTML source, content-related features such as the total number of words present on a page, and the total number of links — all of these features have p-values that are on the order of 10^{-10} or less.

2. *Lack of consensus on the design of tablet-specific web pages:* Fig. 4(a) shows the cumulative distributions of the number of images and Fig. 4(b) shows the number of link tags stratified by device type. The smartphone and laptop devices tend to exhibit the fewest and largest number of features respectively. Tablet devices behaves in the middle, where it is similar to a mobile device in some cases, and then slowly transitions to be similar to the laptop device in other cases. This behavior of tablet devices is attributed to the lack of a consensus among content providers on the design of web pages for tablets. Content providers tend to either (i) have a unique web page design for each device type (e.g., 163.com) (ii) leverage the similar web page design for both laptop and tablet devices (e.g., imdb.com), or (iii) leverage the similar web page design for both tablet and smartphone devices (e.g., twitter.com). We also find that different tablet manufacturers may receive different web pages. For example, android devices may receive ads to download android apps where iOS devices will receive ads to downloads apps on the Apple store. More interestingly, we find that the Amazon Fire Tablet will receive a smartphone version of a web page (espn.com) where the iPad Tablet will receive the desktop version of the page — this suggests that screen size is a more important factor in the page that is downloaded than simply referring to the device as a Tablet or smartphone.

3. *Inconsistent redirect behavior that is based on device type across content providers:* We also find that there is a lack of consistency in the device-triggered redirect behavior across content providers. For example, some content providers will redirect mobile web page requests made by laptop clients to its corresponding laptop-based web page, while other content providers will not redirect requests in such a manner. This observed redirect behavior for devices is similar to the redirect behavior we observed for browser versions. This behavior can be problematic for a number of web-related applications. For instance, web crawlers may be redirected from the mobile view of a web page to the laptop view of a web page (in an undesired manner). This has an impact for web page archival because undesirable or even less informative views of a page (mobile or desktop) may be archived instead of the desired page. This also raises concerns for information sharing across social media (e.g., search engines and social networking) because users can be referring to *different* views of information, or, at times, entirely different information altogether, via the *same* hyperlink. For example, if a user shares a link on a social media site, say Facebook.com, and a friend uses a different client plat-

form to view it, the two users could be observing different content (especially comments and recommendations listed on a page). This can be particularly difficult if one user is referring to a particular comment or review on a page that is not immediately viewable by another user.

4. *Different search result sets for web search queries:* Device type is taken into account by web search engines such as bing.com and google.com when returning search results. We find that generally, smartphones tend to have more mobile optimized web pages included in a search result set than tablet and laptop devices — this is because search providers take into account the mobile-friendliness of a web page when providing search results [4]. We also find that the search result set may have different meaning on different devices — this is mainly because search engines are increasingly providing web content to users instead of simply links to pages. For example, the search result set for the "nba standings" search query yields a different order of the basketball team rankings for a smartphone and a laptop (division rankings vs conference standings). This further underscores the impact that device type can have on information sharing and other applications because a user may refer to portions of a page, say the rank of a basketball team, where a friend does not immediately see the same ranking that is being referenced.

Impact of Vantage Point. We next discuss the impact of vantage point on base HTML pages. We find that:

1. Our statistical analysis shows that *none* of the HTML tag-based features are significantly impacted by vantage point. This result shows that web page design and formatting is not significantly influenced by location. This includes locations across different continents, which is surprising given cultural preferences in content layout and appearance.

2. Vantage point has a significant impact on content-related features. Figure 5(a) shows that the average number of *different* words for each vantage point in the U.S is roughly 200–250 words, while Fig. 5(b) shows that the average number of different words for each vantage point that are outside of the U.S is over 500[4] — Fig. 5(a) and (b) both include 95 % confidence interval bars around the average. Most of these differences across all vantage points (both U.S-based and world) correspond to (i) differences in topics of local interests, (ii) differences in search result sets, and (iii) temporal changes (discussed later). We observe a larger difference for vantage points around the world mainly because content providers have international versions of content that is likely to be of interest to the local population (cnn.com and yahoo.com does this). We also find that international web pages may include notes concerning (i) privacy awareness about the use of cookies on web sites and (ii) options to view the U.S version of web pages.

3. Bing search results, whether it is web, news, or image search, may yield different links, ads, and images across different vantage points — please note

---

[4] Please note that while we study the top 250 web sites in the world, many of these sites are served by content providers that are in the U.S.

that we verified that this is not primarily a consequence of time[5]. Some of these differences are obvious due to location-based searches, say when a user is searching for McDonalds, and the search engine returns the address of the nearest McDonalds. Other differences are more complex, such as when more generic and random search queries such as "a hello berry" and "golden" yield different search results. The impact of vantage point on search results is important to note because search engines are a primary tool for various applications including web page scraping [17] and web security [19]. Vantage point driven search results also impact users because location can be misleading for users who access the web via 3G or 4G services — thus, the wrong location can be used to target search results.

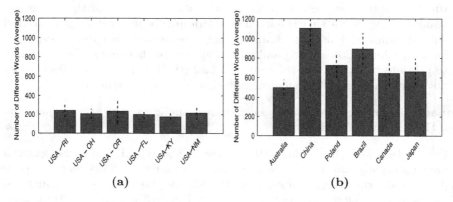

(a)   (b)

Fig. 5. Impact of vantage point on number of different words (U.S. (a) and World(b)). The baseline for comparison located in California.

Impact of Time. Lastly, we investigate the impact that time has on base HTML source files. We perform many univariate Kruskal-Wallis tests between our first measurement (i.e., baseline measurement) and each subsequent measurement. Figure 6(a) shows a time series plot of the p-values for these statistical tests for different tag-based HTML features for the Chrome browser. Figure 6(a) shows that the tag-based features that were statistically significant for some of our prior analysis (e.g., number of images and number of script tags) do not vary significantly (i.e., p-value below .05) over time. In fact, all of the tag-based features that we examine do not change significantly over time. These results imply that the differences across browser and devices that we observe are not significantly influenced by time. This further validates our tag-based findings because it shows that our results are not likely to be due to randomness. We do find rare cases where web page design has changed over time. For example, Fig. 7

---

[5] We discuss results pertaining to bing.com because other search engines such as google.com are blocked in some countries.

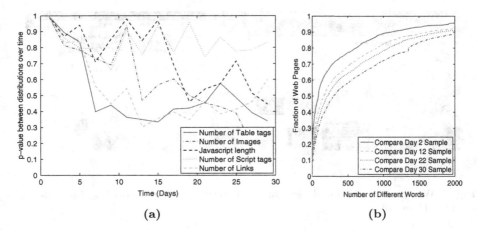

Fig. 6. Impact of time on various HTML features (tag-based features (a) and content-based features (b)).

shows that the format for CNN web pages changed during our data collection procedure. We observe the new format for the CNN page (Fig. 7(b)) for all browsers and devices and conclude that CNN made this format change in order to serve a single web page that adapts to various screen sizes instead of serving multiple web pages to different device types. We also find that Overstock.com will display different versions of a page, one that includes product recommendations and another that does not, at different points in time (we find similar results for zillow.com with respect to content recommendations and imdb.com with respect to ads that completely change the layout of a page). We observe these differences over several browsers and believe that product recommendations are missing at certain instances in time for performance reasons — dynamically generating pages with up-to-date recommendations or ads may be costly. It is important to note these dynamic changes in web page design because it will impact the effectiveness of web page parsing tools that are optimized for a particular page design. This may also impact web crawling procedures because some pages may have links to related/recommended pages while others do not.

Figure 6(b) shows that time has a large influence on content-based features — this is shown by the increasing shift between the CDF plots for the number of different words feature when comparing our day 2, 12, 22, and 30 samples without our initial day 1 sample. This observed difference over time for content-based features is statistically significant, where the pages that are the most heavily influenced tend to correspond news, social networking, homepages in general (e.g., dailymail.com, weather.com, msnmoney.com, and twitter.com) and the pages that are least influenced tend to correspond to business/e-commerce and reference sites (e.g., target.com, dictionary.com, wikipedia.org, and webmd.com).

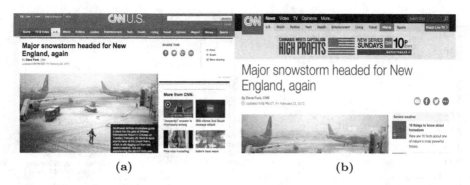

Fig. 7. Web page layout and design can be changed over time — (a) Day 1 sample and (b) Day 11 sample.

4   Related Work

Past work has, to some degree, studied the influence that different type of client platforms may have on mostly performance-related applications. This includes studies that discuss the usability and design trade-offs between mobile web pages and traditional web pages [18,27] and understanding the energy-efficiency and performance-related impact of using mobile browsers and devices for web browsing [15,24]. There has also been recent work that studies (i) the diversity of web page downloads with respect to a single browser [5] and (ii) the impact that different web browsers have on the accuracy of in-browser load time measurements [12]. Time is a factor that is generally accounted for when performing web page measurement studies to ensure that the results are repeatable [5]. [10,11] are examples of measurement studies that thoroughly investigates the influence that time has on the frequency in which web page content changes — these studies also provide insight on the impact that time-related changes have on web crawling. [20] studied the impact that vantage-point has on web page content with respect to price discrimination and found the vantage point has a large influence on the price of goods on many major e-commerce sites.

Our work is different from this prior work because we explicitly study the impact that client platforms have on web page content. In particular, our work (i) investigates the *general* influence that client platforms have on web page *content* without considering performance and (ii) we explicitly consider client platforms that are typically not considered in prior studies including different operating systems, browser types and versions, and tablets.

5   Concluding Remarks

In this paper, we address the question — *to what extent does a client platform influence the content of a base HTML web page for the* **same** *URL request?* We download base HTML-source files in a manner that controls for the influence of over 30 different client platforms. We extract quantitative HTML-based features and perform a comprehensive analysis of the differences that are present across different client platforms. We find differences in web page downloads across client platforms in both expected and unexpected ways. In addition, these observed differences have practical significance in a number of important web-related applications including web archival, mobile web development, document summarization, information sharing, and user experience. While there are many other differences that we find that are due to client platform, such as fonts and colors, we do not discuss them in detail because they have minimal utility in current popular web-related applications. In future work, we intend to (i) study the impact that user personalization (without regard for client platform) has on web page downloads, and (ii) study the influence that client platforms have on the *traffic* generated by web page downloads.

Acknowledgements. This material is based upon work supported by the National Science Foundation Graduate Research Fellowship Program under Grant No. DGE-1144081 as well as by NSF under Grant CNS-1526268.

References

1. Alexa. http://www.alexa.com. Accessed 19 February 2013
2. Future of the web workshop: Introduction and overview. http://netpreserve.org/sites/default/files/resources/OverviewFutureWebWorkshop.pdf. Accessed 29 May 2015
3. Html5 reference: The syntax, vocabulary and apis of html5. http://dev.w3.org/html5/html-author/. Accessed 30 April 2015
4. Make sure your site's ready for mobile-friendly google search results. https://support.google.com/adsense/answer/6196932?hl=en. Accessed May 2015
5. Butkiewicz, M., Madhyastha, H.V., Sekar, V.: Understanding website complexity: measurements, metrics, and implications. In: Proceedings of the 2011 ACM SIGCOMM Conference on Internet Measurement Conference. ACM (2011)
6. Callahan, T., Allman, M., Rabinovich, M.: On modern DNS behavior and properties. ACM SIGCOMM Comput. Commun. Rev. **43**(3), 7–15 (2013)
7. Canali, D., Cova, M., Vigna, G., Kruegel, C.: Prophiler: a fast filter for the large-scale detection of malicious web pages. In: Proceedings of the 20th International Conference on World Wide Web, pp. 197–206. ACM (2011)
8. Chun, B., Culler, D., Roscoe, T., Bavier, A., Peterson, L., Wawrzoniak, M., Bowman, M.: Planetlab: an overlay testbed for broad-coverage services. ACM SIGCOMM Comput. Commun. Rev. **33**(3), 3–12 (2003)
9. de Boer, N., van Leeuwen, M., van Luijk, R., Schouten, K., Frasincar, F., Vandic, D.: Identifying explicit features for sentiment analysis in consumer reviews. In: Benatallah, B., Bestavros, A., Manolopoulos, Y., Vakali, A., Zhang, Y. (eds.) WISE 2014, Part I. LNCS, vol. 8786, pp. 357–371. Springer, Heidelberg (2014)
10. Douglis, F., Feldmann, A., Krishnamurthy, B., Mogul, J.C.: Rate of change and other metrics: a live study of the world wide web. In: USENIX Symposium on Internet Technologies and Systems, vol. 119 (1997)

11. Fetterly, D., Manasse, M., Najork, M., Wiener, J.: A large-scale study of the evolution of web pages. In: Proceedings of the 12th International Conference on World Wide Web, pp. 669–678. ACM (2003)
12. Gavaletz, E., Hamon, D., Kaur, J.: Comparing in-browser methods of measuring resource load times. In: W3C Workshop on Web Performance 8 (2012)
13. He, S., Chan, E.: Surfing notes: an integrated web annotation and archiving tool. In: IEEE/WIC/ACM Web Intelligence and Intelligent Agent Technology (2012)
14. Himmel, M.A.: Customization of web pages based on requester type, US Patent 6,167,441, 26 December 2000
15. Huang, J., Xu, Q., Tiwana, B., Mao, Z.M., Zhang, M., Bahl, P.: Anatomizing application performance differences on smartphones. In: Proceedings of the 8th International Conference on Mobile Systems, Applications, and Services, pp. 165–178. ACM (2010)
16. Iwai, H., Hijikata, Y., Ikeda, K., Nishida, S.: Sentence-based plot classification for online review comments. In: IEEE/WIC/ACM International Joint Conferences on Web Intelligence (WI) and Intelligent Agent Technologies (IAT), 2014, vol. 1, pp. 245–253. IEEE (2014)
17. Jacob, G., Kirda, E., Kruegel, C., Vigna, G.: Pubcrawl: protecting users and businesses from crawlers. Presented as part of the 21st USENIX Security Symposium, Berkeley, CA, pp. 507–522. USENIX (2012)
18. Johnson, T., Seeling, P.: Desktop and mobile web page comparison: characteristics, trends, and implications. IEEE Commun. Mag. **52**(9), 144–151 (2014)
19. Leontiadis, N., Moore, T., Christin, N.: Measuring and analyzing search-redirection attacks in the illicit online prescription drug trade. In: USENIX Security Symposium (2011)
20. Mikians, J., Gyarmati, L., Erramilli, V., Laoutaris, N.: Crowd-assisted search for price discrimination in e-commerce: First results. In: Proceedings of the Ninth ACM Conference on Emerging Networking Experiments and Technologies, pp. 1–6. ACM, December 2013
21. Notess, G.R.: The wayback machine: the web's archive. Online-Weston Then Wilton- **26**(2), 59–61 (2002)
22. Pera, M.S., Qumsiyeh, R., Ng, Y.-K.: An unsupervised sentiment classifier on summarized or full reviews. In: Chen, L., Triantafillou, P., Suel, T. (eds.) WISE 2010. LNCS, vol. 6488, pp. 142–156. Springer, Heidelberg (2010)
23. Roche, X., et al.: Httrack: Website copier. on-line [consulta em 23-12-2008]. Disponível em (2012). http://www.httrack.com
24. Wang, Z., Lin, F.X., Zhong, L., Chishtie, M.: How far can client-only solutions go for mobile browser speed? In: Proceedings of the 21st International Conference on World Wide Web, pp. 31–40. ACM (2012)
25. Weinberger, K., Dasgupta, A., Langford, J., Smola, A., Attenberg, J.: Feature hashing for large scale multitask learning. In: Proceedings of the 26th Annual International Conference on Machine Learning, pp. 1113–1120. ACM (2009)
26. Westfall, J., Augusto, R., Allen, G.: Handling Different Browser Platforms: Beginning Android Web Apps Development, pp. 85–98. Springer (2012)
27. Zhang, D.: Web content adaptation for mobile handheld devices. Commun. ACM **50**(2), 75–79 (2007)

Aspect and Ratings Inference with Aspect Ratings: Supervised Generative Models for Mining Hotel Reviews

Wei Xue[✉], Tao Li, and Naphtali Rishe

Computer Science Department, Florida International University,
11200 Southwest 8th Street, Miami, FL33199, USA
{wxue004,taoli,rishe}@cs.fiu.edu
http://www.cis.fiu.edu

Abstract. Today, a large volume of hotel reviews is available on many websites, such as TripAdvisor (http://www.tripadvisor.com) and Orbitz (http://www.orbitz.com). A typical review contains an overall rating and several aspect ratings along with text. The rating is perceived as an abstraction of reviewers' satisfaction in terms of points. Although the amount of reviews having aspect ratings is growing, there are plenty of reviews including only an overall rating. Extracting aspect-specific opinions hidden in these reviews can help users quickly digest them without actually reading through them. The task mainly consists of two parts: aspect identification and rating inference. Most existing studies cannot utilize aspect ratings which are becoming abundant in the last few years. In this paper, we propose two topic models which explicitly model aspect ratings as observed variables to improve the performance of aspect rating inference over unrated reviews. Specifically, we consider sentiment distributions in the aspect level, which generate sentiment words and aspect ratings. The experiment results show our approaches outperform other existing methods on the data set crawled from TripAdvisor.

Keywords: Sentiment analysis · Information retrieval · Topic model

1 Introduction

The trend that people browse hotel reviews on websites before booking encourages researchers to focus on the analysis of the social media data. Users write down their own experience, and rate hotels with an overall score and/or along with several scores on aspects predefined by websites such as `room`, `service`, and `location`. Overall ratings express a general impression of reviewers which is more abstract than text, but they also hide aspect-specific sentiments. To this end, overall ratings are not informative enough. Although more and more reviews with aspect ratings are available on-line, there is a lot of reviews associated with only an overall rating. Therefore identifying aspect and learning more informative aspect ratings is an attractive topic in opinion mining, which helps users gain more details of each aspect.

© Springer International Publishing Switzerland 2015
J. Wang et al. (Eds.): WISE 2015, Part II, LNCS 9419, pp. 17–31, 2015.
DOI: 10.1007/978-3-319-26187-4_2

Many approaches have been proposed towards simultaneous aspect identification and sentiment inference. A comprehensive survey [13,14] indicated that when using opinion phases, topic model based methods perform better than other bag-of-words based models. Specifically, the vocabulary of a set of reviews is decomposed into two categories: head terms and modifier terms after POS Tagging processing. Each review consists of several pairs of head and modifier. For example, the phrase "nice service" is parsed into a pair of the head term "service" and the modifier term "nice". The words in modifier category can effectively infer the sentiment associated with the aspect implied by the corresponding head terms. While head terms are only responsible for aspect identification, and do not have to express any positive or negative sentiment. Moreover, it is straightforward to consider the dependence between the rating variables generating modifier terms and the topic variables producing head terms. Because reviews usually have different preferences across different aspects.

However, most existing topic models [20,21] cannot gain any benefit from the aspect ratings associated with reviews. For example, given two reviews both of which giving 3 stars overall, it is reasonable to assume on some aspects the reviewer is disappointed. But this information is generally difficult to infer these aspects from text. Even though we use bag-of-phrases and overall ratings, we still cannot tell whether modifier terms are expressing negative or positive views, because the detailed sentiment is mixed into the general overall rating. Motivated by this observation, we propose two new topic models which can simultaneously learn aspects and their ratings of reviews by utilizing aspect ratings and overall ratings. Aspect ratings are now very easy to obtain from websites like TripAdvisor[1] and Orbitz[2] website. TripAdvisor website provides the largest volume of reviews among review host websites. It holds 225 million reviews, most of which are associated with aspect ratings. None of review is without an overall rating. The problem we would like to address is predicting aspect ratings given overall ratings and text. Therefore, our model can be applied to any review data set without aspect ratings. The aspect ratings are only needed for training. Specifically, our model is based on opinion phrases which are pairs of head and modifier terms. The dependences between latent aspects and their ratings are captured by their latent variables. The aspect identification and rating inference is modeled simultaneously. We use Gibbs sampling to estimate the parameters of our models on the training data set, and maximizing a posteriori (MAP) method to predict aspect ratings on unrated reviews.

The rest of paper is organized as follows. Section 2 formulates the problem and notation we use. Section 3 proposes our model and describes the inference methods. Section 4 shows the data, the experiments and discuss experiment results. Finally we draw the conclusion in Sect. 5.

---

[1] http://www.tripadvisor.com.
[2] http://www.orbitz.com.

2   Related Work

The problem of review sentiment mining has been an attractive research topic in recent years. There are several lines of research. The early work focuses on the overall polarity detection, i.e., detecting whether a document expresses positive or negative. The author of [16] found that the standard machine learning techniques outperform human on the sentiment detection. Later, the problem of determining the reviewers sentiment with respect to a multi-point scale (ratings) was proposed in [15], where the problem was transformed into a multi-class text classification problem. Hidden Markov Model (HMM) is specially adapted to identify aspects and their polarity in Topic Sentiment Mixture model (TSM) [12]. Ranking methods are also used to produce numerical aspect scores [17].

In the literature, Latent Dirichlet Allocation (LDA) [3] based methods play a major role, because the ability of topic detection of LDA is very suitable for multi-facet sentiment analysis on reviews. MG-LDA [18,19] (Multi-Grain Latent Dirichlet Allocation) considers a review as a mixture of global topics and local topics. The global topics capture the properties of reviewed entities, while the local topics vary across documents to capture ratable aspects. Each word is generated from one of these topics. In their later work, the authors model the aspect rating as the outputs of linear regressions, and combine them into the model to aggregate relevant words in the corresponding aspect. Joint sentiment/topic model (JST) [9,10] focuses on aspect identification and its ratings prediction without any rating information available. In JST, the words of reviews are determined by the latent variables of topic and sentiment. Aspect and Sentiment Unification model (ASUM) [6] further assumes all the words in one sentence are sampled from one topic and one sentiment. CFACTS model [7] combines HMM with LDA to capture the syntactic dependencies between opinion words on sentence level. Given overall ratings, Latent Aspect Rating Analysis (LARA) [20,21] uses a probabilistic latent regression approach to model the relationship between latent aspect ratings and overall ratings. On the other hand, POS-Tagging technique is also frequently used in the detection of aspect and sentiment. The authors of [11] categorize the words in reviews into head terms and modifier terms with simple POS-Tagging methods and propose a PLSI based model to discover aspects and predict their ratings. Interdependent LDA model [13] captures the bi-direction influence between latent aspects and ratings based on the preprocessing of head terms and modifier terms. Senti-Topic model with Decomposed Prior (STDP) [8] learns different distributions for topic words and sentiment words with the help of basic POS-Tagging. Similar ideas are applied to separate aspects, sentiments, and background words from the text [23].

Our models are based on opinion phrases [11], but overcome the drawback of previous models that cannot take advantage of aspect ratings. We consider the relationship between several factors, such as overall ratings, aspect ratings, head terms and modifier terms.

3   Problem Formulation

In this section, we first introduce the problem and list notations we use in the models.

Formally, we define a data corpus of N review documents, denoted by $\mathcal{D} = \{x_1, x_2, \ldots, x_D\}$. Each review document x_d in the corpus is made of a sequence of tokens. Each review x_d is associated with an overall rating r_d, which takes an integer value from 1 to $S(S = 5)$. An aspect is a frequently commented attribute of a hotel, such as "value", "room", "location" and "service". A review consists of some text paragraphs that express the reviewers' opinions on aspects. For example, the occurrence of word "price" indicates the review comments on aspect "value". Each review is also associated with several integer aspect ratings $\{l_1, l_2, \ldots, l_K\}$, where K is the number of aspects.

Phrase: We assume each review is a set of some opinion phrases f which are pairs of head and modifier terms, i.e., $f = < h, m >$. In most cases, the head term h describes an aspect, and the modifier term m expresses the sentiment of the phrase. The POS-Tagging and basic NLP techniques can be used to extract phrases from raw text for each review.

Aspect: An aspect is a predefined attribute that reviewers may comment on. It also corresponds a probabilistic word distribution in topic models, which can be learned from data.

Rating: Each review contains an overall rating and may contain several aspect ratings. The rating of each review is an integer from 1 to 5. We assume that the overall ratings are available for each review, but the aspect ratings are available only in the reviews used for training. We assume that the rating is equivalent to the sentiment.

Review: A review is represented as a bag of phrases, i.e., $x_d = \{f_1, f_2, \ldots, f_M\}$.

Problem Definition: Given a collection of reviews with overall ratings and aspect ratings, the main problem is to (1) identify aspects of reviews, and (2) infer aspect ratings on the unrated reviews without aspect rating.

4   Models

In this section, we apply two generative models to identify aspects and learn their ratings by incorporating observed aspect ratings. We list the notations of the models in Table 1. We assume reviews are already decomposed into head terms and modifier terms using NLP techniques [13]. We propose two different models incorporating the aspect ratings as observed random variables.

One strong motivation is that existing topic models do not require aspect ratings of reviews during model training and consider it as an advantage. It may be true in the past few years, since there are not many reviews containing aspect ratings. However, more and more review hosts, such as TripAdvisor and Orbitz, let reviewers to rate on predefined attribute as an option. The volume of such reviews is growing rapidly nowadays. It is reasonable to leverage the valuable information to build more precise and accurate models. To our best of knowledge, this study is the first work using aspect ratings.

Table 1. The table of notations

D	the number of reviews
K	the number of aspects
M	the number of opinion phrases
S	the number of distinct integers of ratings
U	the number of head terms
V	the number of modifier terms
z	the aspect/topic switcher
l	the aspect rating
h	the head term
m	the modifier term
r	the overall rating
θ	the topic distribution in a review
π	the aspect rating distribution for each topic
α	the parameter of the Dirichlet distribution for θ
β	the global aspect sentiment distribution
λ	the parameter of the Dirichlet distribution for β
δ	the parameter of the Dirichlet distribution for ϕ and ψ
ϕ	the head term distribution for each topic
ψ	the modifier term distribution for each sentiment

4.1   The Assumptions

We discuss some helpful assumptions for modeling. First, our models presume a flow of generating ratings and text. The reviewer gives an overall rating based on his impression and experience, then rates it on some aspects and writes some paragraphs. In the model of bag-of-phrases, the reviewer chooses a head term for an aspect on which he would like to comment, and a modifier term to express his opinion. This generation process is captured by our models.

Second, there is an interdependency between overall ratings and aspect ratings, and it varies with the numerical value of the overall rating. For example, when a user gives 5 star overall rating, it is extremely unlikely that the user gives low ratings on any of the aspects. On the other hand, however, when a hotel receives a low overall rating, it does not necessarily get low ratings on all aspects. It is possible that the hotel still get positive feedbacks on some aspects. This usually occurs when the traveler is disappointed by a conflict, such as extra charges for unnecessary services. Inspired by this observation, we model this dependency with a multinomial distribution $P(\pi|r)$ and a global aspect sentiment distribution β conditioned on the overall rating in the following models.

Third, aspect ratings imply another interdependency, the one between aspects and sentiments [14]. Basically, it considers that different aspects have different sentiments. We explicitly introduce sentiment variables for modifier terms which are conditioned on aspect variables, so that meaningful aspects and sentiments can be learned from head and modifier terms respectively, but it avoids generating too many non-aspects.

We present two different supervised generative models. They both take aspect ratings as probabilistic variables. The aspect ratings π are merely K scores in the review on K aspects. They are observed in the training data and hence treating them as switchers is quite straightforward. An interesting observation is the distinction between the aspect rating and the phrase sentiment. They are both sentiment switchers and could be conditioned on the overall rating variable r. One is for aspects, the other is for phrases. If we assume they are both necessary and generated from the aspect sentiment distribution β and the overall rating r, then we have ARID model (Aspect and Rating Inference with the Discrimination of aspect sentiment and phrase sentiment) in Fig. 1. The interaction between π and r is through the global aspect sentiment distribution β and the overall rating r. It saves the direct dependency between them. If we assume in given the aspect k, the reviewer holds the same sentiment for all the modifier terms, the discrimination between aspect sentiment and phrase sentiment is redundant. It leads to our second model ARIM (Aspect and Rating Inference with Merging aspect sentiments and phrase sentiments).

4.2   The ARID Model

The ARID model, in Fig. 1, captures the review generation process and the interdependency between aspects and sentiments. Following conventional topic models for review analysis, we use random variables z and l to simulate the generating process of head and modifier terms respectively. The topic selection variable z is governed by a multinomial topic distribution θ. The sentiment variable l for each opinion phrase is also determined by aspect sentiment distribution β, the overall rating r, and the aspect switcher z.

Specifically, in ARID model, the variables π representing aspect ratings are shaded in the graphical representation since they are observed in the training dataset, but become latent variables for prediction over unrated reviews. The latent sentiment variable l is sampled from β_k where k is determined by the value of z. The overall rating variable r is also introduced to serve a switcher for both the aspect rating π and the phrase sentiment l. We would like to estimate

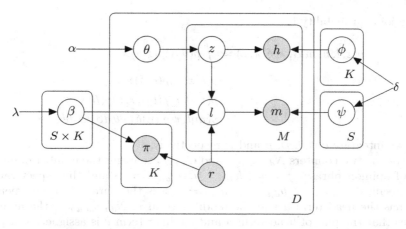

Fig. 1. Graphical Representation of ARID model. The outer box represents D reviews, while the inner box contains M phrases

the aspect rating distribution given the overall aspect sentiment distribution $p(\pi|r)$, and the latent distribution $p(l)$ and $p(z)$.

The formal generative process of our model is as follows:

- For each aspect k and each overall rating value of r
  - Sample the aspect sentiment distribution $\beta_{r,k} \sim \texttt{Dir}(\lambda)$
- For each review x_d,
  - Sample latent topic distribution variable $\theta_d \sim \texttt{Dir}(\alpha)$
  - For each aspect k from 1 to K in the review,
    * Sample aspect rating $\pi_{d,k} \sim \texttt{Mult}(\beta_{r_d,k})$
  - For each phase i from 1 to M in the review,
    * Sample aspect indicator $z_i \sim \texttt{Mult}(\theta_d)$
    * Sample sentiment indicator $l_i \sim \texttt{Mult}(\beta_{r_d,z_i})$
    * Sample head term $h_i \sim \texttt{Mult}(z_i, \phi)$
    * Sample modifier term $m_i \sim \texttt{Mult}(l_i, \psi)$

Estimation. Two parameter estimation methods are widely used for topic models, i.e., Gibbs sampling [4] and variational inference [3]. Since Gibbs sampling updating equations is relatively easy to derive and implement, for this reason, we adopt collapsed Gibbs sampling which integrates out intermediate random variables θ, ϕ, β, and ψ. For prediction, we learn the head term and the modifier term distribution ϕ, ψ, and the global aspect sentiment distribution β from z and l. The Gibbs sampling repeatedly samples latent variables $z_{a,b}$ and $l_{a,b}$ conditioned on all other latent z and l, in document a for phrase b.

The joint probability is

$$p(z,l,h,m|\alpha,\lambda,\delta,\pi,r) = \int p(\theta|\alpha)p(z|\theta)\times$$
$$p(h|z,\phi)p(\phi|\delta)\times \qquad (1)$$
$$p(\pi|\beta,r)p(l|\beta,r,z)p(\beta|\lambda)\times$$
$$p(m|l,\psi)p(\psi|\delta)\,d\theta\,d\beta\,d\phi\,d\psi,$$

where we integrate out θ, ψ, β and ψ respectively.

We define two counters $N_{d,r,k,s,u,v}$ and $C_{d,r,k,s}$ to count the number of occurrence of opinion phrases $f_{d,i} =< h_{d,i} = u,\ m_{d,i} = v >$ and the aspect rating $\pi_{d,k}$. Specifically, $f_{d,i} =< h_{d,i} = u,\ m_{d,i} = v >$ is the phrase i of document d which has the head term u and the modifier term v. $N_{d,r,k,s,u,v}$ is the number of times that the pair of head term u and modifier term v is assigned to aspect k and sentiment s in document d, whose overall rating of the document is r. $C_{d,r,k,s}$ is the indicator of the document d that gives aspect rating s on aspect k when the overall rating of the document is r. Although given document d, its overall rating r_d is determined, we use the overall rating as a subscript for convenience.

$$N_{d,r,k,s,u,v} = \sum_{i=1}^{M} \mathbf{I}[r_d = r,\ z_{d,i} = k,\ l_{d,i} = s,\ h_{d,i} = u,\ m_{d,i} = v], \qquad (2)$$

$$C_{d,r,k,s} = \mathbf{I}[r_d = r,\ \pi_{d,k} = s] \qquad (3)$$

where the function $\mathbf{I}$ is the identify function. Summing out various indices results in the replacement of subscripts of N by $*$. For example,

$$N_{d,r,*,s,u,v} = \sum_{k=1}^{K} N_{d,r,k,s,u,v}. \qquad (4)$$

We sample $z_{a,b}$ and $l_{a,b}$ simultaneously

$$p(z_{a,b}|z_{-(a,b)},\alpha,\delta,\lambda,h,m,r,\pi) \propto (N_{a,r_a,z_{a,b},*,*,*}^{-(a,b)} + \alpha)\times$$
$$\frac{N_{*,*,z_{a,b},*,h_{a,b},*}^{-(a,b)} + \delta}{N_{*,*,z_{a,b},*,*,*}^{-(a,b)} + U\delta}\times$$
$$\frac{N_{*,r_a,z_{a,b},l_{a,b},*,*}^{-(a,b)} + C_{*,r_a,z_{a,b},l_{a,b}} + \lambda}{N_{*,r_a,z_{a,b},*,*,*}^{-(a,b)} + C_{*,r_a,z_{a,b},*} + S\lambda}\times \qquad (5)$$
$$\frac{N_{*,*,*,l_{a,b},*,m_{a,b}}^{-(a,b)} + \delta}{N_{*,*,*,l_{a,b},*,*}^{-(a,b)} + V\delta}.$$

It turns out that the aspect ratings π could be considered as pre-observed phrase sentiment counts for the global aspect sentiment distribution β. We drop the prior parameter λ, and estimate the aspect sentiment distribution β with aspect ratings π and overall ratings r of the training data before Gibbs sampling using Eq. (6).

$$\beta_{r,k,s} = \frac{C_{*,r,k,s}}{C_{*,r,k,*}}. \tag{6}$$

The third term of the right hand of Eq. 5 is replaced by

$$\frac{N^{-(a,b)}_{*,r_d,z_{a,b},l_{a,b},*,*} + \tilde{\lambda}\beta_{r_d,z_{a,b},l_{a,b}}}{N^{-(a,b)}_{*,r_d,z_{a,b},*,*,*} + \tilde{\lambda}}, \tag{7}$$

where $\tilde{\lambda}$ is the scaling factor for β. The parameters of AIRD ψ, ϕ, θ are estimated by

$$\phi_{k,u} = \frac{N_{*,*,k,*,u,*} + \delta}{N_{*,*,k,*,*,*} + U\delta}, \; \psi_{s,v} = \frac{N_{*,*,*,s,*,v} + \delta}{N_{*,*,*,s,*,*} + V\delta}, \; \theta_{d,k} = \frac{N_{d,r_d,k,*,*,*} + \alpha}{N_{d,r_d,*,*,*,*} + K\alpha}. \tag{8}$$

Incorporating Prior Knowledge. We use a small set of seed words to initialize the aspect term distribution ϕ [20]. Learning the head term distribution for each aspect is difficult to converge without any prior knowledge, since each review use similar set of words for commenting on hotels. We consider the seed words as the pseudo-count which means the amount of δ words are added to $\phi_{k,u}$ by before Gibbs sampling.

Prediction. The focus of applying our model is the prediction on the unrated reviews without aspect ratings. Given an opinion phrase $f_{d,i} =< h_{d,i}, \; m_{d,i} >$ and the overall rating r_d in a new document d, we identify which aspect $\hat{z}_{d,i}$ does that phrase belongs to, and predict the aspect rating $\hat{l}_{d,i}$. We drop the two subscripts d and i for simplicity. we first predict $\hat{z}$ by maximizing the posterior probability $p(z|h, m, r, \alpha, \beta, \phi, \psi)$. Using Bayes theorem, it is equivalent to maximize

$$p(z, h, m, r|\alpha, \beta, \phi, \psi) = \int p(\theta|\alpha)p(z|\theta)p(h|z, \phi)p(l|z, r, \beta)p(m|l, \psi) \, d\theta dl, \tag{9}$$

then we predict $\hat{l}$ with

$$\mathbb{E}[p(l|\hat{z}, h, m, r, \beta, \phi, \psi, \alpha)]. \tag{10}$$

The reason to consider the expectation of l is that the aspect rating is actually a numerical value, rather than a discrete category label. The importance of each possible value l is measured by its probability. The aspect weight for a new document could be learned again via Gibbs sampling, but we simply assume θ is a uniform distribution, because a review on hotel should probably comment on all the most concerned aspects. The terms in Eq. (9) we need to compute are $p(h|z, \phi) = \phi_{z,h}$, $p(l|z, r, \beta) = \beta_{r,k,l}$, and $p(m|l, \psi) = \psi_{l,m}$.

4.3 The ARIM Model

In this model, we assume the aspect sentiment is equivalent to the phrase sentiment. In other words, if all the modifier terms are categorized into the same aspect k, they share the same sentiment, i.e., the aspect sentiment,*. Therefore, we could just use only one sentiment indicator for both the aspect and the phrase. ARIM (Aspect and Rating Inference Merging aspect sentiments and phrase sentiments) is illustrated in Fig. 2.

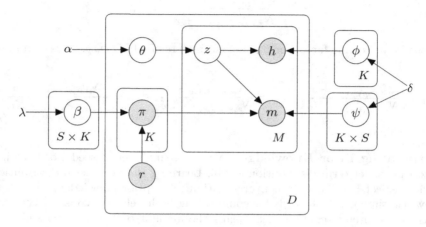

Fig. 2. Graphical Representation of the ARIM model

ARIM models aspect ratings as π like in ARID, but π is also used for phrase sentiment. The aspect ratings are available in the training data, the influence from β to m is blocked by π by d-separation theory [2] of graphical models. Therefore, the modifier term is directly determined by the aspect ratings π instead of β. In the generative procedure of ARIM, the modifier term m_i is sampled from $\psi_{z_i, \pi_{z_i}}$, and π follows a multinomial distribution with parameter β.

We still use Gibbs sampling to estimate z and β. The iterative updating function is

$$p(z_{a,b}|z_{-(a,b)}, \alpha, \delta, \lambda, h, m, r, \pi) \propto (N_{a, r_a, z_{a,b}, *, *, *}^{-(a,b)} + \alpha) \times$$
$$\frac{N_{*, *, z_{a,b}, *, h_{a,b}, *}^{-(a,b)} + \delta}{N_{*, *, z_{a,b}, *, *, *}^{-(a,b)} + U\delta} \times \qquad (11)$$
$$\frac{N_{*, *, z_{a,b}, \pi_{a, z_{a,b}}, *, m_{a,b}}^{-(a,b)} + \delta}{N_{*, *, z_{a,b}, \pi_{a, z_{a,b}}, *, *}^{-(a,b)} + V\delta}$$

The parameters of ARIM model ϕ, θ and β is estimated by Eqs. (8) and (6). But the number of ψ is $K \times S$. It is estimated by

$$\psi_{k,s,v} = \frac{N_{*, *, k, s, *, v} + \delta}{N_{*, *, k, s, *, *} + V\delta}. \qquad (12)$$

When ARIM is applied on the reviews without aspect ratings, we integrate out the latent aspect rating variable π to compute MAP $\hat{z}$ of $p(z|h, m, r, \alpha, \beta, \phi, \psi)$, which equals to

$$p(z, h, m, r|\alpha, \beta, \phi, \psi) = \int p(\theta|\alpha)p(z|\theta)p(h|z, \phi)p(m|z, \psi, \pi)p(\pi|\beta, r) \, d\pi d\theta .$$
(13)

Like Eq. (9), we again assume θ is a uniform distribution, and the terms in Eq. (13) $p(h|z, \phi) = \phi_{z,h}$, $p(m|z, r, \beta, \psi) = \sum_{s=1}^{5} \phi_{z,s,m}\beta_{r,z,s}$ by integrating out π. The estimated aspect rating $\mathbb{E}[p(\pi_k|\hat{z}, h, m, r, \beta, \phi, \psi, \alpha)]$ is computed by all the opinion phrase whose $\hat{z} = k$.

5   Experiments

In this section, we describe the review data we use and evaluate the performance of our models.

5.1   Data

The data set we use for performance evaluation is crawled from TripAdvisor [20]. Each of review in the data set is associated with an overall rating and 7 aspect ratings all within the range from 1 to 5. However some aspects such as *Cleanliness*, *Check in/front desk* are rarely rated. To better train and evaluate methods, we use only four mostly commented aspects, *Value*, *Room*, *Location* and *Service*. We only keep reviews with all four aspect ratings to evaluate and compare different models. We use NLTK [1] to tokenize the review text, remove stop words, remove infrequent words, apply POS-Tagging technique [13] to extract opinion phrases, and filter out short reviews which contains less than 10 phrases. The final data set contains 1,814 hotels and 31,013 reviews. We randomly take 80 % data as the training data set, the rest is the testing data set. The seed words used to initialize the head term distribution ϕ is in Table 2, which form a very small set of words.

5.2   Aspect Identification

In this section, we demonstrate that ARID and AIRM can identify meaningful aspects. In Table 3, we present top 3 frequentest head terms for each aspect

Table 2. Seed words

Aspect	Seed words
Value	value, fee, price, rating
Room	windows, room, bed, bath
Location	transportation, walk, traffic, shop
Service	waiter, breakfast, staff, reservation

Table 3. Frequentest head terms and modifier terms by ARIM

Aspect	Head terms	Modifier terms
Value	deal, price, charge	good, great, reasonable
Room	house, mattress, view	comfortable, clean, nice
Location	parking, street, bus	great, good, short
Service	manager, check-in, frontdesk	friendly, good, great

learned by ARIM. In other words, they have highest values in ϕ_k. We also list top 3 frequentest modifier terms for each aspect. As we can see, ARIM successfully extracted ratable aspects from reviews, and learned aspect-specific sentiment words as well. For example, "comfortable" is frequently used to describe aspect "Room", but not for other aspects. We also observe that people also like to use vague sentiment words for all aspects, such as "good", "great".

5.3 Metric

We use RMSE(Root-mean-square error)[3] to measure the performance of predicting aspect ratings for each hotel in the testing set. Letting the predicted aspect rating for hotel d on aspect k be $\hat{\pi}_{d,k}$ with ground-truth being $\pi_{d,k}$, the RMSE can be represented as Eq. (14).

$$\text{RMSE}(\hat{\pi}_{d,k}, \pi_{d,k}) = \sqrt{\frac{1}{DK} \sum_{d=1}^{D} \sum_{k=1}^{K} (\hat{\pi}_{d,k} - \pi_{d,k})^2} \tag{14}$$

RMSE measure shows how accurate one model could predicate aspect ratings. We also use Pearson correlation to describe the linear relationship between the predicted and the ground-truth aspect ratings, which is Eq. 15. π_d is the vector of the aspect ratings of document d.

$$\rho_{\text{aspect}} = \frac{1}{D} \sum_{d=1}^{D} \rho(\pi_d, \hat{\pi}_d) \tag{15}$$

Since the rating is merely an ordinal variable, whose value does not have the meaning as the numerical value. But its value has a clear ordering. Therefore, we adopt Pearson linear correlation ρ_{aspect} on the aspect ratings within each review to evaluate how a model keeps the aspect order in terms of ratings. For each aspect, it is reasonable to compute the linear correlation across hotels ρ_{hotel} as in Eq. (16). The measure is used to test whether the model could predict the order of hotels in teams of an aspect rating. π_k consists of all the aspect ratings of all the hotels on the aspect k,

$$\rho_{\text{hotel}} = \frac{1}{K} \sum_{k=1}^{K} \rho(\pi_k, \hat{\pi}_k) . \tag{16}$$

---

[3] http://en.wikipedia.org/wiki/RMSE.

5.4   Aspect Rating Prediction

In this section, we present the experiment results on the reviews without any aspect rating in Table 4. We compared three different models and one baseline. The baseline predicts all the aspect ratings of each review with the given overall rating. Since the baseline predicts the aspect ratings of a review with a constant value, $\rho_{aspect} = 0$. From the results, we observe that ARID and ARIM have close performance, but both of them outperform the baseline and LARAM [21]. The main reason is that ARID and ARIM can capture the interdependency between aspects, their ratings and modifier terms, thanks to the aspect ratings in the training data set.

Moreover, ARIM is better than ARID, which confirms our observation. The sentiment of aspect and modifier terms is not so different from each other. Reviewers hold similar attitude with different modifier terms when commenting on one aspect. Therefore, merging aspect sentiment with modifier sentiment does not deteriorate the power of the models. The information learned from the training data in ARID and ARIS is stored in β, ϕ, ψ, which are used to predict the aspect ratings in both models. ARID model has K kinds of modifier term distributions ψ; while ARIS has $K \times S$, since the modifier term m in ARIS is dependent on the aspect switcher z and the sentiment l. ARID estimates a general sentiment distribution across all aspects, but ARIM could learn aspect-specific sentiment distribution by modeling aspect-dependent sentiment. During the inference, although the aspect on which the opinion phrases comment is determined by its head term h, ARID infers the sentiment for each modifier term from a coarse sentiment distribution; while ARIM can obtain more find-grained sentiment using its $K \times S$ modifier term distributions. The ψ in ARIM fine-tunes the predicting results based on β and ϕ. Therefore, in terms of Pearson correlation metric, ARIM has better performance. In terms of ρ_{hotel}, all four approaches have similar scores. On the hotel level, the aspect ratings are averaged across all reviews, while the goals of these four methods are predicting the ratings of each individual review. The difference between each method on predicted aspect ratings for each review is small. Therefore, there is no much difference on the measure ρ_{hotel}.

Table 4. Performance of aspect inference

Measure	Baseline	LARAM	ARID	ARIM
RMSE	0.702	0.632	0.588	0.510
ρ_{aspect}	0.0	0.217	0.176	0.248
ρ_{hotel}	0.755	0.755	0.723	0.758

6   Conclusion

In this paper, we propose two models for aspect and its sentiment inference, ARID and ARIM. Both of them can employ the overall ratings and the aspect ratings in reviews to identify the aspects on which an unrated review comments, and uncover the corresponding latent aspect ratings. The two models are based on topic models, but explicitly consider the interdependency between aspect ratings aspect terms, and sentiment terms. The opinion phrases of head terms and modifier terms are extracted by using simple POS-Tagging techniques. The most important contribution is that the two models incorporate the aspect ratings as observed variables into the models, and significantly improve the prediction performance of aspect ratings. The difference between them is whether the sentiment of modifier terms should be merged with the sentiment of aspects. Gibbs sampling and MAP is used for estimation and inference, respectively. The experiments on large hotel reviews show that ARID and ARIM have better performance in terms of RMSE and Pearson correlation. In the future, we would investigate the methods that can automatically generate ratable aspects from text, not from the predefined seed words. Another interesting research topic is to explore the relation between different aspects [5, 22]. The different aspects in one review may share the similar sentiments.

Acknowledgment. The work is partially supported by National Science Foundation under grants CNS-1126619, IIS-121302, and CNS-1461926 and the U.S. Department of Homeland Security under grant Award Number 2010-ST-062-000039, the U.S. Department of Homeland Security's VACCINE Center under Award Number 2009-ST-061-CI0001.

References

1. Bird, S., Klein, E., Loper, E.: Natural Language Processing with Python. O'Reilly Media (2009)
2. Bishop, C.M.: Pattern Recognition and Machine Learning. Information Science and Statistics. Springer-Verlag New York Inc., Secaucus (2006)
3. Blei, D.M., Ng, A.Y., Jordan, M.I.: Latent dirichlet allocation. J. Mach. Learn. Res. **3**(4–5), 993–1022 (2003)
4. Griffiths, T.L., Steyvers, M.: Finding scientific topics. Proc. Nat. Acad. Sci. U.S.A. **101**(Suppl. 1), 5228–5235 (2004)
5. Guo, Y., Xue, W.: Probabilistic multi-label classification with sparse feature learning, pp. 1373–1379, August 2013
6. Jo, Y., Oh, A.H.: Aspect and sentiment unification model for online review analysis. In: Proceedings of the Forth International Conference on Web Search and Web Data Mining, p. 815. ACM Press, New York (2011)
7. Lakkaraju, H., Bhattacharyya, C.: Exploiting coherence for the simultaneous discovery of latent facets and associated sentiments. In: Proceedings of the 2011 SIAM International Conference on Data Mining, pp. 498–509 (2011)
8. Li, C., Zhang, J., Sun, J.T., Chen, Z.: Sentiment topic model with decomposed prior. In: SIAM International Conference on Data Mining (SDM 2013). Society for Industrial and Applied Mathematics (2013)
9. Lin, C., He, Y.: Joint sentiment/topic model for sentiment analysis. In: Proceedings of the 18th ACM Conference on Information and Knowledge Management, p. 375. ACM Press, New York, November 2009

10. Lin, C., He, Y., Everson, R., Ruger, S.M.: Weakly supervised joint sentiment-topic detection from text. IEEE Trans. Knowl. Data Eng. **24**(6), 1134–1145 (2012)
11. Lu, Y., Zhai, C., Sundaresan, N.: Rated aspect summarization of short comments. In: Proceedings of the 18th International Conference on World Wide Web, p. 131. ACM Press, New York (2009)
12. Mei, Q., Ling, X., Wondra, M., Su, H., Zhai, C.: Topic sentiment mixture: modeling facets and opinions in weblogs. In: Proceedings of the 16th International Conference on World Wide Web, pp. 171–180. ACM (2007)
13. Moghaddam, S.: ILDA: interdependent LDA model for learning latent aspects and their ratings from online product reviews categories and subject descriptors. In: Proceeding of the 34th International ACM SIGIR Conference on Research and Development in Information Retrieval, pp. 665–674 (2011)
14. Moghaddam, S., Ester, M.: On the design of LDA models for aspect-based opinion mining. In: Proceedings of the 21st ACM International Conference on Information and Knowledge Management, pp. 803–812 (2012)
15. Pang, B., Lee, L.: Seeing stars: exploiting class relationships for sentiment categorization with respect to rating scales. In: Proceedings of the 43rd Annual Meeting of the Association for Computational Linguistics. pp. 115–124, June 2005
16. Pang, B., Lee, L., Vaithyanathan, S.: Thumbs up? In: Proceedings of the ACL 2002 Conference on Empirical Methods in Natural Language Processing - EMNLP 2002, vol. 10, pp. 79–86. Association for Computational Linguistics, Morristown, July 2002
17. Snyder, B., Barzilay, R.: Multiple aspect ranking using the good grief algorithm. In: Human Language Technology Conference of the North American Chapter of the Association of Computational Linguistics, pp. 300–307, April 2007
18. Titov, I., McDonald, R.: A joint model of text and aspect ratings for sentiment summarization. In: Proceedings of the 46th Annual Meeting of the Association for Computational Linguistics, pp. 308–316. ACL (2008)
19. Titov, I., McDonald, R.: Modeling online reviews with multi-grain topic models. In: Proceedings of the 17th International Conference on World Wide Web, p. 111. ACM Press, New York (2008)
20. Wang, H., Lu, Y., Zhai, C.: Latent aspect rating analysis on review text data. In: Proceedings of the 16th ACM SIGKDD International Conference on Knowledge Discovery and Data Mining, p. 783. ACM Press, New York (2010)
21. Wang, H., Lu, Y., Zhai, C.: Latent aspect rating analysis without aspect keyword supervision. In: Proceedings of the 17th ACM SIGKDD International Conference on Knowledge Discovery and Data Mining, p. 618. ACM Press, New York (2011)
22. Zeng, C., Li, T., Shwartz, L., Grabarnik, G.Y.: Hierarchical multi-label classification over ticket data using contextual loss. In: 2014 IEEE Network Operations and Management Symposium (NOMS), pp. 1–8. IEEE, May 2014
23. Zhao, W., Jiang, J., Yan, H., Li, X.: Jointly modeling aspects and opinions with a MaxEnt-LDA hybrid. In: Proceedings of the 2010 Conference on Empirical Methods in Natural Language Processing, pp. 56–65, October 2010

A Web-Based Application for Semantic-Driven Food Recommendation with Reference Prescriptions

Devis Bianchini[✉], Valeria De Antonellis, and Michele Melchiori

Department of Information Engineering, University of Brescia, Via Branze, 38,
25123 Brescia, Italy
{devis.bianchini,valeria.deantonellis,michele.melchiori}@unibs.it

Abstract. Food recommendation, as well as searching for health-related information, presents specific characteristics if compared with conventional recommender systems, since it often has educational purposes, to improve behavioural habits of users. In this paper, we discuss the application of Semantic Web technologies in a menu generation system, that uses a recipe dataset and annotations to recommend menus according to user's preferences. Reference prescription schemes are defined to guide our system for suggesting suitable choices. The recommended menus are generated through three steps. First, relevant recipes are selected by content-based filtering, based on comparisons among features used to annotate both users' profiles and recipes. Second, menus are generated using the selected recipes. Third, menus are ranked taking into account also prescription schemes. The system has been developed within a regional project, related to the main topics of the 2015 World Exposition (EXPO2015, Milan, Italy), where the University of Brescia aims at promoting healthy behavioural habits in nutrition.

1 Introduction

Recommender systems find information of interests, properly customized according to the users' own preferences [1]. This is valid also for specific application domains, such as health and nutrition, where any choice made upon automatically provided recommendations might have an impact on users' health and wellness. Several researches on food recommendation and automatic menu generation have been carried on or are currently active (e.g., [2–4]), taking into account different aspects, such as personal and cultural preferences, health and religion constraints, menu composition and recipe co-occurrence. However, the problem within food recommender systems is still how to suggest recipes and menus that not only meet the user's preferences, but also are compliant with best food habits. Let's consider, for example, Jasmine, who is looking for recipe suggestions to have lunch during her working hours. Jasmine is registered to a food recommender system and has an associated profile. She prefers to have pasta and meat during meals. She suffers from long-term diseases, such as diabetes and high-blood-pressure, therefore white meat should be more advisable.

© Springer International Publishing Switzerland 2015
J. Wang et al. (Eds.): WISE 2015, Part II, LNCS 9419, pp. 32–46, 2015.
DOI: 10.1007/978-3-319-26187-4_3

She belongs to the Islamic religion, so recommendations about any food containing alcohol or pork are not acceptable, since this food is prohibited to Muslims. These aspects may be represented through features, for example a feature representing the religion which a recipe is not advisable for, the course type (e.g., first course, appetizer) and many others. The same features can be used to describe available recipes and Jasmine's preferences, associated to her profile. Features may represent either short-term, immediate preferences (e.g., when they are explicitly specified in a request for suggestions issued by the user), or long-term preferences, extracted from the history of past choices made by the user [5]. A food recommender system would be very useful, not only for the high number of available recipes to be suggested[1], but also because it is really difficult to manually check all the constraints (e.g., religion constraints) and preferences to generate proper menus. Feature-based matching between profiles and recipes is the basis for content-based filtering for food recommendation [2,6,7]. However, some Jasmine's preferences (e.g., having pasta and meat during meals, all the days throughout the week) may contrast with best habits, according to up-to-date medical prescriptions. This means that food recommendations should be able to improve the behavioural habits of the users.

Taking the opportunity of the 2015 World Exposition (EXPO2015, Milan, Italy), the University of Brescia is promoting several projects to incentivate healthy habits. Among them, within the Smart BREAK regional project, funded by the Lombardy region, Italy, we are developing *PREFer* (**P**rescriptions for **RE**commending **F**ood), a menu generation system that uses a recipe dataset and reference prescription schemes to suggest suitable menus. The recommended menus are generated through three steps: (i) relevant recipes are selected by content-based filtering, based on comparisons among features used to annotate recipes and to represent users' preferences; (ii) candidate menus are generated using the selected recipes; (iii) candidate menus are ranked also taking into account reference prescription schemes. As the contribution of this paper, we present the application of Semantic Web technologies within a food recommendation scenario, where the recommendation method is *education-oriented*, that is, aims at satisfying both user's preferences and reference prescriptions.

The paper is organized as follows: in Sect. 2 related approaches for the design of food recommender systems are presented; Sect. 3 provides detailed definitions about our ontology-based recommendation model; in Sect. 4 we describe the three steps of the menu generation procedure; Sect. 5 discusses implementation issues and preliminary experimental results; finally, in Sect. 6 we sketch conclusions and future work.

---

[1] The http://allrecipes.com web site lists thousands of recipes; for example, just considering appetizers, we can found more than 7,700 choices (http://allrecipes.com/recipes/appetizers-and-snacks/).

2   Related Work

Literature on recommender systems covers several domains and has been developed in parallel with the Web, to properly suggest movies, books, applications, e-learning materials, recipes, etc. (a survey on recommender systems can be found in [1]). Domain-independent categories of recommender systems hold, based on the filtering algorithm used (e.g., demographic, content-based, collaborative, knowledge- or ontology-based, context-aware, hybrid) and on the employed techniques (e.g., probabilistic approaches, nearest neighbors techniques, fuzzy models, similarity metrics). Nevertheless, given the number of domains where recommender systems have been applied and their specific features, a cross-domain comparison might be difficult and useless. Therefore, in the following we will focus on recent approaches on food recommendation domain.

Some existing approaches for recommending food and health-related information focus on content-based filtering (considering aspects like personal and cultural preferences, health and religion constraints) [2, 6–8]. In [2] recipes are modelled as complex aggregations of different features, extracted from ingredients, categories, preparation directions, nutrition facts, and authors propose a content-driven matrix factorization approach to face the latent dimension of recipes, users and their features. The HealthFinland project [6] is a portal that helps the users to find relevant health information using simple keywords instead of medical vocabularies. Personalized Health Information System (PHIRS) [8] is a recommender system for health information that matches the user's profile against the retrieved health information, also considering culture and religion in the profile. Similarly, food recommendations are provided in [7].

Teng et al. [9] apply collaborative filtering for recipe recommendation: recipes taken from the `allrecipes.com` Web site are suggested on the basis of users' ratings and reviews and on the basis of co-occurrences of ingredients used to prepare them. In the paper, an interesting survey is provided on other approaches that consider ingredients, recipe ratings and cooking directions. The same information are used in [10, 11], where content-based, collaborative and hybrid filtering are compared for recipe recommendation purposes.

Other approaches combine content-based and demographic filtering techniques with ontology-based and knowledge-based tools to enhance recommendation results [12, 13]. Ontologies are used to model personal and cultural preferences, health and religion constraints, but no educational issues are taken into account. CarePlan [3] is a semantic representation framework for healthcare plans, that mixes the patients' health conditions with personal preferences, but ignores other aspects, such as personalization coming from educational health information, user's culture and religion, that impact on the food choice. In [4] an ontology containing fuzzy sets is used to sort recommended recipes according to prices and users' ratings, in combination with attributes like sex, age, weight, physical activity, used to calculate Basal Metabolic Rate (BMR), Activity Factor (AF) and Body Mass Index (BMI). Authors implement a demographic filtering algorithm, thus providing common suggestions to people with common attributes.

This variety of approaches demonstrates that users' profiling, in particular for sectors and domains such as the food and health recommendation, is mainly addressed in an ad-hoc manner, without aiming at providing some educational effect on the users. The papers described in [14,15] highlighted this open issue. In particular, [14] presents preliminary research on how to detect bad and correct food habits by analyzing users' ratings on allrecipes.com, while in [15] authors discovered that online food consumption and production are highly sensitive in time. Although these approaches do not provide a recommender system, their research could be fruitfully exploited for food recommendation purposes. Other works [16–18] explicitly address the issue of promoting healthful choices, by suggesting recipes to users based on their past food selections and nutrition intakes. We will propose a step forward compared to these approaches, promoting healthy behaviour through reference prescriptions, that are based not only on nutrition intakes, but are specifically modelled considering phenotypes, that classify ideal users' nutrition behaviour. A proper domain ontology is used to model such knowledge and is used with content-based filtering for enhanced food recommendation.

3   Recommendation Model

Let's consider the running example introduced above, where Jasmine is looking for a personalized menu for her meals. Some important aspects should be considered here. First, recipes can be combined into different menus, but not all aggregations are suitable. Specific combinations of recipes might be due to particular *menu configurations* (e.g., appetizer, first course, second course, dessert), according to user's preferences. Second, recommendations might be given according to *reference prescriptions*, that should be used as first-class citizens in recommending recipes to users who present particular profiles. Third, although prescriptions can be used to improve the habits of users for what concerns food and nutrition, they cannot be imposed to users, disregarding their own preferences. Prescriptions should *gradually* move users' choices towards more healthy recipes.

In this paper we propose a recommendation model that is based on the ontology shown in Fig. 1. Following the rationale presented in [19], we distinguish between the ontology and the recipe and menu database, that contains data such as the ones shown in Fig. 2 for the running example. The database contains specific instances of recipes, menus and prescriptions, that are annotated with concepts taken from the ontology. The adopted ontology extends the food.owl ontology[2] with the concepts of CookingStyle (e.g., Asian cuisine), Health&CulturalConstraint such as Religion (e.g., Islamic) and Pathology (e.g., diabetes, high-blood pressure), CourseType (e.g., appetizer, first course, second course, fruits, dessert), PrescriptionType and Phenotype, that will be presented in the following. The concepts defined within the food.owl ontology have been considered as specializations of the RecipeType concept. Semantic

---

[2] http://krono.act.uji.es/Links/ontologies/food.owl/view.

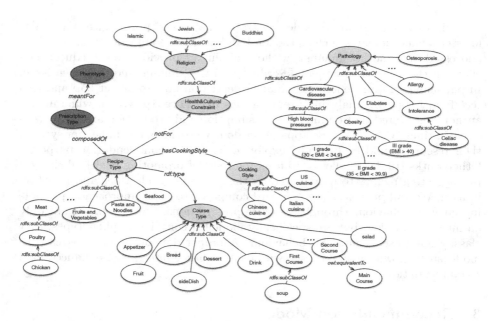

Fig. 1. Main concepts of the ontology adopted for food recommendation.

relationships in the ontology are used to provide food recommendation as discussed in the rest of the paper.

Recipes. Recipes represent the most fine-grained items to be recommended. A recipe is stored in the database as a record $r_i = \langle R_i, n_i, C_i \rangle$ $(\forall i = 1, \ldots N)$, where: R_i is the unique identifier of the recipe (we denote with $\mathcal{R}$ the overall set of N recipes available within the dataset); n_i is the name of the recipe; C_i is a set of concepts taken from the ontology, used to characterize the recipe. In particular, in our approach each recipe can be classified through the CourseType, the CookingStyle, the RecipeType, the Health&CulturalConstraint (for which the recipe is not advised) and their sub-concepts shown in Fig. 1. In Fig. 2 eight different recipes are depicted, with concepts extracted from the ontology.

In our approach, semantic annotation is supported using the semantic disambiguation techniques we applied in other Semantic Web applications [20]. When a new recipe is published, a text field is provided to enter concepts to annotate it. As the user inputs the characters of the concept name he/she wants to use for annotating the recipe, the system provides an automatic completion mechanism based on the set of concepts contained within the ontology. Starting from the name specified by the user, the system queries the ontology, retrieves the concept with the specified names and/or other concepts related to the specified one through semantic relationships, in order to enable the user to explore the ontology and refine the annotation. Other candidate concepts are also provided according to the string distance between concept names and terms contained in the recipe name and descriptions. A thesaurus (WordNet) is also used in this phase to identify candidate concepts for annotation, using lists of synonyms within WordNet synsets, to face polisemy (that is, the same term refers to different concepts) and synonymy (i.e., the same concept is pointed out using different terms).

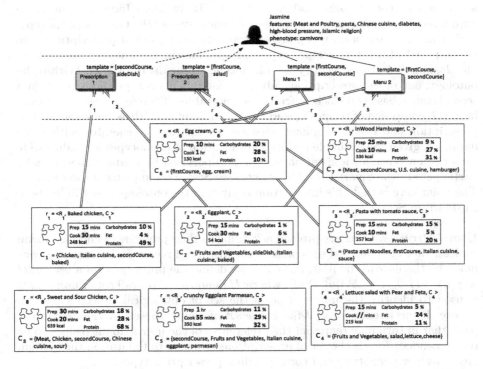

Fig. 2. Recipes to recommend, menus and prescriptions of the running example.

Menus and Prescriptions. Recipes are aggregated to be proposed in a combined way. In the context of our food recommendation approach, we distinguish two kinds of aggregations: (a) available *menus*, that is, combinations of recipes chosen in the past by the users of the system (these menus are used to extract the preferences of the users, exploiting them during the recommendation phase, see for details Sect. 4.1); (b) *prescriptions*, that is, proper combinations of recipes that are advisable for specific kinds of users. Formally, we define an aggregation (either a menu or a prescription) $a_j \in \mathcal{A}$ as $a_j = \langle n_{a_j}, \mathcal{R}[a_j], \tau_{a_j} \rangle$, where: $\mathcal{A}$ denotes the overall set of aggregations; n_{a_j} is the name of the aggregation; $\mathcal{R}[a_j] \subseteq \mathcal{R}$ is the set of recipes aggregated in a_j; τ_{a_j} is the template of the aggregation, expressed in terms of values of a specific concept. In our approach, given an aggregation a_j, τ_{a_j} is identified considering the `CourseType` concept and corresponding sub-concepts (e.g., `Appetizer`, `Fruit`, `sideDish`, etc., see Fig. 1).

Examples of templates may be [Appetizer, FirstCourse, SecondCourse, Dessert] or [FirstCourse, Fruit]. Templates play an important role for the formulation of the request for suggestions (see Sect. 4.1) and to speed up the generation of the recommendation output (see Sect. 4.4). The way prescriptions are associated to users depends on the features used to describe users' profiles. In our food recommendation approach, Food Frequency Questionnaires (FFQ) are issued to collect users' habits and BMI (Body Mass Index), in order to automatically classify users within specific phenotypes [21]. Given a phenotype, one or more prescription types are advisable for it, and each prescription type is composed of a set of recipes types, as specified in the ontology. For example, Jasmine's features identify her phenotype as *carnivore* (Fig. 2). Within the ontology, one of the prescriptions advisable for this phenotype should contain a second course based on chicken, fruits and vegetables. Therefore, prescription1 in Fig. 2, composed of recipes r_1 and r_2, is compliant with these constraints. The prescription and other compliant ones are automatically generated within the database, given the available recipes. Specification of phenotypes and admissible prescription types for a given phenotype is supervised by medical doctors, who participate to the Smart BREAK project (see Sect. 5 on implementation issues). The point here is that this information is given in the ontology and will be used, as shown in Sect. 4.5.

Users' Profiles. Users are profiled according to their preferences and past menu choices, that are collected to represent the history of recipe and menu selections made by the user in the past. Formally, we define the profile $p(u)$ of a user $u \in \mathcal{U}$ as $p(u) = \langle ID_u, \mathcal{C}[u], \mathcal{M}[u], \mathcal{P}[u] \rangle$, where: $\mathcal{U}$ denotes the overall set of users; ID_u is used to identify the user u; $\mathcal{C}[u]$ is the set of ontological concepts used to denote the preferences of u; $\mathcal{M}[u]$ is the set of menus chosen by the user in the past, that in turn may represent the preferences of the user u about recipes to be recommended; $\mathcal{P}[u]$ is the set of prescriptions assigned to the user in the system, given his/her phenotype and corresponding prescription types.

4   Menu Recommendation System

4.1   Formulating a Request for Suggestions

When Jasmine is looking for menu suggestions, she generates a request $r_r(u)$ formulated as $r_r(u) = \langle \mathcal{C}_r, \tau_r \rangle$, where: $\mathcal{C}_r$ is a set of concepts that represent immediate, short-term preferences of Jasmine; τ_r is the menu template Jasmine is searching for. The recommender system takes into account the profile $p(u)$ of the user u (Jasmine), whom the request comes from. To this aim, the request $r_r(u)$ is expanded with the concepts that are present within the Jasmine's profile $p(u)$. We denote with $\widehat{r}_r(u)$ the expanded version of the request, where $\widehat{r}_r(u) = \langle \widehat{\mathcal{C}}_r, \tau_r \rangle$. The set $\widehat{\mathcal{C}}_r$ contains both the concepts specified in $\mathcal{C}_r$ and the concepts within $p(u)$. Concepts used to characterize $p(u)$ represent long-term preferences of the user, that might be collected and updated using traditional techniques from the literature [5]. The set $\mathcal{C}_r$ might also be empty, thus denoting that the system should exclusively rely on the preferences contained within $p(u)$. Each concept $c_r \in \widehat{\mathcal{C}}_r$ is weighted by means of a coefficient $\omega_r \in [0, 1]$ such that:

$$\omega_r = \begin{cases} 1 & if \ c_r \in \mathcal{C}_r \\ freq(c_r) \in [0, 1] & otherwise \end{cases} \tag{1}$$

The value of ω_r means that a concept explicitly specified in the request $r_r(u)$ will be considered the most for identifying candidate recipes. The term $freq(c_r)$ computes the frequency of concept c_r among all the concepts that annotate the recipes contained in the profile $p(u)$. Less frequent concepts will be considered as less important for identifying candidate recipes. If a concept c_r is present both in $\mathcal{C}_r$ and in the profile, then $\omega_r = 1$. If u is a new user, without a history of past choices, then $\widehat{r}_r(u) = r_r(u)$ (no expansion). In this case, prescriptions are used to differentiate the user's choices, as explained in Sect. 4.5. In future versions of the *PREFer* system we aim at integrating here further collaborative filtering and demographic filtering recommendation techniques [1].

Example 1. Let's consider the recipes and Jasmine's profile of the running example (Fig. 2), and the following request, issued to search for menus and recipes containing `baked poultry`, according to [`FirstCourse, SecondCourse`] template: $r_r(u) = \langle \{\texttt{poultry}, \texttt{baked}\}, [\texttt{FirstCourse}, \texttt{SecondCourse}] \rangle$. The following expanded version of the request is generated (frequency values are specified between parenthesis):

$\widehat{\mathcal{C}}_r = \{\texttt{poultry}(1.0), \texttt{meat}(0.5), \texttt{chicken}(0.5), \texttt{SecondCourse}(1.0), \texttt{Chinesecuisine}(0.5),$
$\qquad \texttt{PastaandNoodles}(0.5), \texttt{FirstCourse}(0.5), \texttt{Italiancuisine}(1.0), \texttt{FruitsandVegetables}(0.5)\}$
$\widehat{\mathcal{T}}_r = \{\texttt{baked}(1.0), \texttt{sour}(0.5), \texttt{cream}(0.5), \texttt{egg}(0.5), \texttt{eggplant}(0.5), \texttt{parmesan}(0.5)\}$

As can be noticed, frequencies are computed on a menu basis, since recipes are recommended only within aggregations, represented as menus.

4.2 Menu Recommendation Steps

The approach followed here for food recommendation is articulated over a set of steps, that are summarized in Fig. 3:

– *feature-based recipe filtering* - the overall set of recipes $\mathcal{R}$ is properly pruned taking into account the menu template τ_r, specified in the request (all recipes that do not present a `CourseType` that is included within τ_r are filtered out from the set of recommendation results) and the features, using proper ontology-based similarity metrics; let's denote with $\mathcal{R}' \subseteq \mathcal{R}$ the set of filtered recipes;
– *candidate menu generation* - candidate menus that are compliant with τ_r are generated, only considering the recipes included within $\mathcal{R}'$; let's denote with $\mathcal{A}^*$ the set of generated candidate menus;

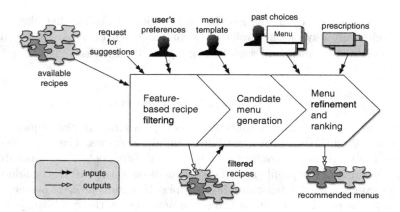

Fig. 3. The three steps of food recommendation approach driven by user's preferences and prescriptions.

– *menu refinement and ranking* - candidate menus contained in $\mathcal{A}^*$ are properly ranked according to their average similarity with the past menu choices made by the user, who is looking for suggestions, and with the prescriptions advertised for that user.

4.3 Feature-Based Recipe Filtering

The input of this step is the set $\mathcal{R}$ of all the available recipes and the request $\widehat{r}_r(u)$. First, τ_r element specified in the request is considered. Recipes such that their `CourseType` is not included within τ_r will not pass the feature-based filtering step. With reference to the running example, only the r_1, r_3, r_5, r_6, r_7 and r_8 recipes will be further considered. To speed up the pre-selection based on `CourseType`, recipes are stored in the underlying dataset indexed with respect to the `CourseType` feature. Another important aspect to be considered is that not all features can be exploited in the same way to filter out not relevant recipes. For instance, let's consider some constraints imposed by the Islamic religion or by some allergies. Recipes that do not respect these constraints must be excluded before any other kind of comparison. These constraints, to keep our model as more general as possible, are defined within the domain ontology and are expressed in terms of other features. For example, the Islamic religion within the Jasmine's profile excludes all recipes that are annotated with `pork` or `alcohol`. Modeling of such constraints must be accurate; this explains why we inserted them within the domain ontology, that is developed in a controlled way.

After τ_r and ontological constraints have been used to pre-select recipes, the filtering based on remaining features is applied, according to the *concept-based relevance*. This metric is computed as follows.

Concept-Based Relevance. The relevance of a recipe $r_i = \langle R_i, n_i, C_i \rangle$ with respect to the request $\widehat{r}_r(u) = \langle \widehat{C}_r, \tau_r \rangle$ taking into account concepts in C_i and $\widehat{C}_r$, denoted with $Sim(\widehat{r}_r, r_i) \in [0, 1]$, is computed as:

$$Sim(\widehat{r}_r, r_i) = \frac{2 \cdot \sum_{c_r, c_i} \omega_r \cdot ConceptSim(c_r, c_i)}{|C_i|} \in [0, 1] \qquad (2)$$

where c_r ranges over the set $\widehat{C}_r$, c_i ranges over the set C_i, $|C_i|$ denotes the number of concepts in the set C_i, ω_r denotes the weight of concept $c_r \in \widehat{C}_r$, as assigned according to Eq. (1), to take into account both short-term and long-term preferences (see Sect. 4.1). $ConceptSim(c_r, c_i)$ represents the *concept similarity* between c_r and c_i:

$$ConceptSim(c_r, c_i) = \frac{2 \cdot |c_r \cap c_i|}{|c_r| + |c_i|} \in [0, 1] \qquad (3)$$

In Eq. (3), we consider the two concepts c_r and c_i as more similar as the number of recipes that have been annotated with both the concepts, denoted with $|c_r \cap c_i|$, increases with respect to the overall number of recipes annotated with c_r, denoted with $|c_r|$, and with c_i, denoted with $|c_i|$. The domain ontology is considered in this case as well: in fact, given two concepts c_i and c_j such that $c_i \sqsubseteq c_j$ (c_i is subclassOf c_j), due to the semantics of the subclassOf relationship, all recipes annotated with c_i are considered as annotated with c_j as well. For example, $|\text{Chicken}| = |\{r_1, r_8\}| = 2$, $|\text{Poultry}| = |\{r_1, r_8\}| = 2$, since Chicken $\sqsubseteq$ Poultry, $|\text{Chicken} \cap \text{Poultry}| = |\{r_1, r_8\}| = 2$, therefore $ConceptSim(\text{Chicken}, \text{Poultry}) = 1.0$.

Pairs of concepts to be considered in the $Sim(\widehat{r}_r, r_i)$ computation are selected according to a maximization function, that relies on the assignment in bipartite graphs and ensures that each concept in C_i participates in at most one pair with one of the concepts in $\widehat{C}_r$ and the pairs are selected in order to maximize the overall $Sim(\widehat{r}_r, r_i)$. The rationale behind Eq. (2) is that the closer $Sim()$ to 1.0, the more concepts in C_i are similar to one of the concepts in $\widehat{C}_r$. In the running example, for computing $Sim(\widehat{r}_r, r_1)$, the pair $\langle \text{Poultry}, \text{Chicken} \rangle$ ($\omega_r = 1.0$) is considered instead of $\langle \text{Chicken}, \text{Chicken} \rangle$ ($\omega_r = 0.5$) in order to maximize the final result, therefore $Sim(\widehat{r}_r, r_1) = (1.0 + 1.0 + 1.0)/3 = 1.0$.

The recipes included in the set $\mathcal{R}' \subseteq \mathcal{R}$, as output of the *feature-based recipe filtering*, are those whose concept-based relevance with respect to the request $\widehat{r}_r(u)$ is equal or greater than a threshold $\gamma \in [0, 1]$ set by the user.

4.4   Candidate Menu Generation

In this step, recipes are aggregated into menus that must be compliant with the template τ_r specified in the request $\widehat{r}_r(u)$. This significantly reduces the number of menu configurations to be generated: in fact, a candidate menu can not contain two recipes r_i and r_j annotated with the same CourseType. If we consider, for example, m `CourseTypes`, with an average number of n candidate recipes for each `CourseType`, the number of possible menu configurations without considering the constraint imposed by the menu template would be equal to $f_1(n, m) = \frac{(n \cdot m!)}{m!(n \cdot m - m)!}$ (since we have $n \cdot m$ elements from which m elements must be selected to be composed, without repetitions). In our approach, the number of possible menu configurations is equal to $f_2(n, m) = n^m$. Moreover, the menu generation in our approach is not performed through a brute force procedure, where all possible n^m configurations are generated and, only after generation, properly ranked. The candidate recipes are already sorted, according to the concept-based similarity $Sim(\widehat{r}_r, r_i)$, therefore the candidate menus are generated as illustrated in Fig. 4 with the running example.

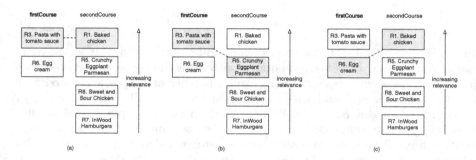

Fig. 4. The menu generation step.

The first candidate menu that is generated is the one where candidate recipes are the best ranked ones for each `CourseType` (Fig. 4(a)). The next candidate menus that are generated are the ones shown in Fig. 4(b–c), where, for instance, $Sim(\widehat{r}_r, r_5) > Sim(\widehat{r}_r, r_6)$. This explains why we choose the combination $r_3 - r_5$, before $r_6 - r_1$. This procedure does not ensure that the list of generated menus will be properly ranked as well. The final ranking of menus is performed in the next step.

4.5 Menu Refinement and Ranking

Menus that have been generated in the previous step are ranked according to their similarity with: (i) past menu choices made by the user u who is issuing the request for suggestions, represented by the set $\mathcal{M}[u]$; (ii) prescriptions prepared for the user u according to his/her profile, represented by the set $\mathcal{P}[u]$. Since both menus and prescriptions are formally defined as sets of recipes, the building block in this step is the similarity measure between items aggregations (*item aggregation similarity*), that is computed as follows:

$$Sim_{agg}(a_i, a_j) = \frac{2 \cdot \sum_{r_i, r_j} Sim(r_i, r_j)}{|\mathcal{R}[a_i]| + |\mathcal{R}[a_j]|} \in [0, 1] \tag{4}$$

where a_i and a_j represent the two compared aggregations (menus or prescriptions), r_i (resp., r_j) is a recipe included within a_i (resp., within a_j), $|\mathcal{R}[a_i]|$ (resp., $|\mathcal{R}[a_j]|$) denotes the number of recipes included within a_i (resp., within a_j). The rationale behind $Sim_{agg}()$ computation is the same as the one of the concept-based relevance: we consider two aggregations as more similar as the number of similar items in the two aggregations increases.

The final ranking of a generated menu $a_k \in A^*$, recommended to the user u who issued a request for suggestions, is performed through a ranking function $\rho : A^* \mapsto [0,1]$, computed as follows:

$$\rho(a_i) = \omega_m \cdot \frac{\sum_{a[u] \in \mathcal{M}[u]} Sim_{agg}(a_i, a[u])}{|\mathcal{M}[u]|} + \omega_s \cdot \frac{\sum_{\widehat{a}[u] \in \mathcal{P}[u]} Sim_{agg}(a_i, \widehat{a}[u])}{|\mathcal{P}[u]|} \quad (5)$$

where $\omega_m, \omega_p \in [0,1]$, $\omega_m + \omega_p = 1.0$, are weights used to balance the impact of past menu choices and prescriptions on the ranking of recommended menus. We have chosen $\omega_m < \omega_p$ (i.e., $\omega_m \cong 0.4$ and $\omega_p \cong 0.6$) in order to stimulate users on improving their food and nutrition habits, without recommending menus and recipes that are too much distant from users' preferences. This is one the most innovative aspects of our approach compared with recent food recommendation literature (see Sect. 2).

5   Implementation and Experimental Issues

We implemented the *PREFer* system as a web application, whose functional architecture is shown in Fig. 5. The *PREFer Web Interface* guides the user through the registration process, the menu recommendation, the publication of new recipes, also supporting semantic annotation (through the *Sense Disambiguation module*), both during the publication of new recipes and the formulation of a request for suggestions, using a wizard similar to the one described in [20]. The Jena reasoner is used to access knowledge stored within the ontology, that is formalised using the Web Ontology Language (OWL). Registration is performed by answering a food frequency survey (FFQ), that is used to collect information about the user in order to compute his/her BMI and identify his/her phenotype [21], in order to prepare suggested prescriptions. This task is executed by medical doctors, who participate to the regional project where PREFer is being developed. The description of this task is out of the scope of this paper. To just give an idea, medical doctors are supported in the identification of phenotypes and have a simple web interface at their disposal (*Prescription Manager*) to prepare and insert prescription types as sets of recipe types. Specific instances of prescriptions, that are compliant with prescription types specified in the ontology, are automatically generated starting from available recipes. These prescriptions are finally assigned to users classified in the phenotype for which prescription types have been built in the ontology. FFQ results are also stored to enable data analysis by doctors for statistical purposes. *Menu recommendation module* implements the recommendation process described in Sect. 4. It supports

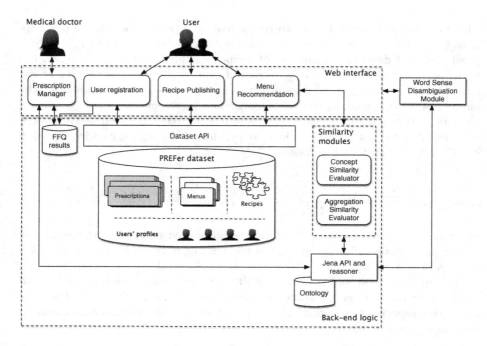

Fig. 5. The functional architecture of the PREFer system for food recommendation.

the user throughout the formulation of the request for suggestions through a proper wizard.

Preliminary Experimentation. Experiments on our food recommendation approach, that are being carried on within the Smart BREAK project, are twofold: (a) to demonstrate the performances of the approach in terms of average precision of the proposed recommendations; (b) to verify the impact of the approach in improving the users' habits concerning food and nutrition. With respect to the former objective, our work has been meant as a complementary approach to recent food recommendation efforts, where content-based filtering techniques based on recipes, ingredients, cultural and contextual features have been implemented. Performance tests are being performed on a dataset that extends an existing one (http://mslab.csie.ntu.edu.tw/~tim/recipe.zip), containing about 220k recipes, randomly aggregated into about 100k menus, where the PREFer system presents comparable average precision with respect to recent approaches. Main experiments in the scope of the Smart BREAK project are being focused on the second objective. They are being performed on a population of about two hundreds students, equally distributed among males and females, with an age included between 18 and 24. Within the population of students, we identified users with pathologies directly related with nutrition (e.g., diabetes, different grades of obesity or various kinds of intolerances) or having bad nutrition behaviour, by submitting Food Frequency Questionnaires. The compliance of users' choices with reference prescriptions, in order to quantify how much the system is able to improve their behaviour, is quantified through the average aggregation similarity between users' choices and reference prescriptions, starting from Eq. (4). Experiments will be carried on until November 2015. Monthly, statistics are generated that, with respect to users' profiles, show the percentage of requests and menu choices that are compliant with or closer to reference prescriptions. Experiments carried on during the first months showed a satisfying increment of closeness between past preferences and reference prescriptions (around 24 % on average, but reaching about 43 % if we consider only users with preferences that are far from the advisable ones, that is, average closeness that is lower than 0.5). These first results are very encouraging and an online community will be created to enabling exchange of food experiences between students who are participating to the experiment.

6 Conclusion and Future Work

In this paper, we presented *PREFer*, a menu generation system that uses a recipe dataset and annotations to recommend menus according to user's preferences. Compared to recent food recommendation efforts, the *PREFer* system takes into account also reference prescriptions schemes, aiming at improving nutritional habits of users. The system has been developed within a regional project, related to the main topics of the 2015 World Exposition (EXPO2015, Milan, Italy), where the University of Brescia aims at promoting healthy behavioural habits in nutrition. The approach will be further extended to refine the recommendation of recipes: (a) by enhancing variety of food in menu preparation; (b) in cases where the violation of health and cultural constraints is due to specific ingredients, by introducing the possibility of suggesting similar recipes, where only the prohibited ingredients are substituted. A semi-automatic functionality for supporting medical doctors in the generation of prescription types will be developed as well. Finally, experimentation is being performed on the approach, but further experiments will be carried on till the end of the Smart BREAK project in order to check the effectiveness of the proposed approach in improving nutritional habits and lifestyles.

References

1. Bobadilla, J., Ortega, F., Hernando, A., Gutièrrez, A.: Recommender systems survey. Knowl.-Based Syst. **46**, 109–132 (2013)
2. Lin, C.-J., Kuo, T.-T., Lin, S.-D.: A content-based matrix factorization model for recipe recommendation. In: Tseng, V.S., Ho, T.B., Zhou, Z.-H., Chen, A.L.P., Kao, H.-Y. (eds.) PAKDD 2014, Part II. LNCS, vol. 8444, pp. 560–571. Springer, Heidelberg (2014)
3. Abidi, S., Chen, H.: Adaptable personalized care planning via a semantic web framework. In: 20th International Conference of the European Federation for Medical Informatics (2006)

4. Maneerat, N., Varakulsiripunth, R., Fudholi, D.: Ontology-based nutrition planning assistance system for health control. Asean Eng. J. **1**(2), 28–41 (2013)
5. Gauch, S., Speretta, M., Chandramouli, A., Micarelli, A.: User profiles for personalized information access. In: Brusilovsky, P., Kobsa, A., Nejdl, W. (eds.) Adaptive Web 2007. LNCS, vol. 4321, pp. 54–89. Springer, Heidelberg (2007)
6. Suominen, O., Hyvonen, E., Viljanen, K., Hukka, E.: HealthFinland - a national semantic publishing network and portal for health information. In: Web Semantics: Science, Services and Agents on the World Wide Web 2009, pp. 287–297 (2009)
7. Dominguez, D., Grasso, F., Miller, T., Serafin, R.: PIPS. An integrated environment for health care delivery and healthy lifestyle support. In: 4th Workshop on Agent applied in Healhcare ECAI 2006 (2006)
8. Wang, Y., Liu, Z.: Personalized health information retrieval system. In: AMIA Annual Symposium Proceedings, p. 1149 (2005)
9. Teng, C., Lin, Y., Adamic, L.A.: Recipe recommendation using ingredient networks. In: Proceedings of the 4th Annual ACM Web Science Conference, pp. 298–307 (2011)
10. Freyne, J., Berkovsky, S.: Recommending food: reasoning on recipes and ingredients. In: De Bra, P., Kobsa, A., Chin, D. (eds.) UMAP 2010. LNCS, vol. 6075, pp. 381–386. Springer, Heidelberg (2010)
11. Sobecki, J., Babiak, E., Słanina, M.: Application of hybrid recommendation in web-based cooking assistant. In: Gabrys, B., Howlett, R.J., Jain, L.C. (eds.) KES 2006. LNCS (LNAI), vol. 4253, pp. 797–804. Springer, Heidelberg (2006)
12. Kim, J., Chung, K.: Ontology-based healthcare context information model to implement ubiquitous environment. Multimedia Tools Appl. **71**, 873–888 (2014)
13. Al Nazer, A., Helmy, T., Al Mulhem, M.: User's profile ontology-based semantic framework for personalized food and nutrition recommendation. Procedia Comput. Sci. **32**, 101–108 (2014)
14. Said, A., Bellogín, A.: You are what you eat! tracking health through recipe interactions. In: Proceedings of the 6th Workshop on Recommender Systems and the Social Web (RSWeb) (2014)
15. Kusmierczyk, T., Trattner, C., Nørvåg, K.: Temporality in online food recipe consumption and production. In: Proceedings of the ACM 2015 International Conference on World Wide Web Companion (WWW 2015) (2015)
16. Geleijnse, G., Natchtigall, P., van Kaam, P., Wijgergangs, L.: A personalized recipe advice system to promote healthful choices. In: IUI, pp. 437–438. ACM (2011)
17. Kamieth, F., Braun, A., Schlehuber, C.: Adaptive implicit interaction for healthy nutrition and food intake supervision. In: Jacko, J.A. (ed.) Human-Computer Interaction, Part III, HCII 2011. LNCS, vol. 6763, pp. 205–212. Springer, Heidelberg (2011)
18. Hsiao, J., Chang, H.: SmartDiet: a personal diet consultant for healthy meal planning. In: Proceedings of IEEE 23rd International Symposium on Computer-Based Medical Systems (CBMS) (2010)
19. Garcia, P., Mena, E., Bermùdez, J.: Some common pitfalls in the design of ontology-driven information systems. In: International Conference on Knowledge Engineering and Ontology Development (KEOD 2009), pp. 468–471 (2009)
20. Bianchini, D., De Antonellis, V., Melchiori, M.: A linked data perspective for effective exploration of web APIs repositories. In: Daniel, F., Dolog, P., Li, Q. (eds.) ICWE 2013. LNCS, vol. 7977, pp. 506–509. Springer, Heidelberg (2013)
21. Rankinen, T., Bouchard, C.: Genetics of food intake and eating behavior phenotypes in humans. Annu. Rev. Nutr. **26**, 413–434 (2006)

Incorporating Cohesiveness into Keyword Search on Linked Data

Ananya Dass[1], Aggeliki Dimitriou[2], Cem Aksoy[1], and Dimitri Theodoratos[1(✉)]

[1] New Jersey Institute of Technology, Newark, USA
dth@njit.edu
[2] National Technical University of Athens, Athens, Greece

Abstract. Keyword search is a popular technique for querying the ever increasing repositories of RDF graph data because it frees the user from knowing a formal query language and the structure of the data. However, the imprecision of keyword queries results in overwhelming numbers of candidate results making the identification of relevant results challenging and hindering the scalability of the query evaluation algorithms.

To address these issues, we introduce cohesive keyword queries on RDF data. Cohesive queries allow the user to flexibly and effortlessly convey her intention using cohesive keyword groups. A cohesive group of keywords in a query indicates that the keywords of the group should form a cohesive unit in the query results. We provide formal semantics of cohesive queries. We design a query evaluation algorithm which relies on the structural summary of the RDF graph to generate pattern graphs that satisfy the cohesiveness constraints. Pattern graphs are structured queries that can be evaluated over the RDF data to compute the query results. Our experiments demonstrate the efficiency of our algorithm and the effectiveness of cohesive keyword queries in improving the result quality and in pruning the space of pattern graphs compared to flat keyword queries. Most importantly, these benefits are achieved while retaining the simplicity and convenience of traditional keyword search.

Keywords: Keyword search · RDF graph · Cohesive query · Pattern graph

1 Introduction

In recent years, there is a constant increase in the number and volume of RDF graph data repositories. Keyword search is by far the most popular technique for querying data on the web including tree and graph data. The reasons of this success is well known and twofold: (a) the user does not need to know a complex query language (e.g., SPARQL) to retrieve information, and (b) the user does not need to know the schema (structure) of the data sources in order to express a query. In fact, the same keyword query can be used to extract information from multiple data sources with different schemas. Despite its convenience and simplicity, keyword search has a drawback: keyword queries are imprecise and

© Springer International Publishing Switzerland 2015
J. Wang et al. (Eds.): WISE 2015, Part II, LNCS 9419, pp. 47–62, 2015.
DOI: 10.1007/978-3-319-26187-4_4

ambiguous. For this reason, they return a very large number of results. This is a typical problem in Information Retrieval (IR). However, in the context of tree and graph data it is exacerbated because the result of a query is not anymore a whole document but a substructure (e.g., a subtree, or a subgraph). This characteristic exponentially increases the number of candidate results. As a consequence, keyword search on graph data faces two major challenges: (a) effectively identifying relevant results among a plethora of candidates, and (b) coping with the performance scalability issue (degradation of the performance when the size of the input data and the number of keywords increase).

In order to identify relevant results, previous algorithms for keyword search over graph data compute candidate results in an approximate way by considering only those which maintain the keyword instances in close proximity [2,6,9,11, 13,14,16,20]. The filtered results are ranked and top-k processed usually by employing IR-style metrics for flat documents (e.g., tf*idf or PageRank) adapted to the structural characteristics of the data [7,10,21,22]. Nevertheless, these probability theory-based metrics cannot disambiguate the results and identify the intent of the user. As a consequence, the produced rankings are, in general, of low quality.

Further, despite the size restriction of the candidate results, these algorithms are still of high complexity and they do not scale satisfactorily. Recognizing the difficulties of keyword search over semi-structured data, recent approaches cluster the results usually by leveraging a structural summary of the data [1,3, 12,18,21,22]. Though promising, these approaches require interaction with the user who has to select clusters of results and possibly navigate in a clustering hierarchy.

Our approach. In this paper, we claim that existing techniques to keyword search on RDF data are not sufficient for producing results of high quality without additional information from the user. Current RDF graphs are very large and integrate data from various application domains. Query keywords on RDF graphs can have not only numerous instances of the same type, but also numerous instances of different types. For example, the keyword job can be the name of different people, the name of an employment agency, a relationship between a person and his occupation, the name of a property of an entity person, the name of a RDF class etc. These multiple instances generate a multitude of result graphs corresponding to different interpretations of the user query.

Example 1. Consider the keyword query $Q = $ (Andrew Job Gordon Network). With this query the user is looking for an article authored by Andrew Job on Gordon Network. Figure 1(a) shows a result graph which corresponds to the user intent. Underlined words in a result graph indicate the instances of the keywords of Q. However, existing algorithms [3,8,15,21–23] will also compute other result graphs like those shown in Fig. 1(b), (c), (d) and (e) which correspond to different undesirable interpretations of the query. These result graphs associate the keywords in a way which is not the one intended by the user. For instance, the result graph of Fig. 1(b) represents a person named Andrew Gordon who has a job of a Network analyst.

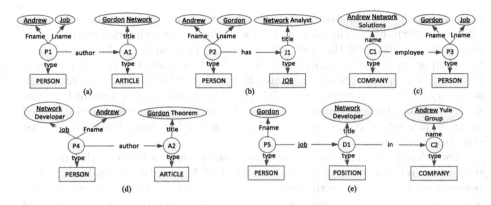

Fig. 1. Result graphs of the query $Q = $ (Andrew Job Gordon Network)

Using scoring functions based on statistical information to rank the results might help but certainly cannot by itself disambiguate the different interpretations and rank on the top position the result graphs that correspond to the user intent. Indeed, the result graph that matches the user intent might be one with low probability which loosely correlates the keyword instances.

To address this problem, we introduce the concept of *cohesive group* of keywords in a query. The keywords of a cohesive group are to be interpreted as forming a cohesive whole. That is, the instances of the keywords outside the group should not "penetrate" the subgraph defined by the instances of the keywords of the group in the result graph. Cohesive groups of keywords are specified naturally and effortlessly by the user while formulating the query. For instance, the previous example query Q can be written as ((Andrew Job) (Gordon Network)), where parentheses are used to delimit the cohesive groups (Andrew Job) and (Gordon Network). This cohesive query excludes the graphs of Fig. 1(b), (c), (d) and (e) from the set of legal result graphs since, as we explain in more detail later, they breach the cohesiveness of the specified cohesive groups. As an example, the keyword instance Job penetrates the subgraph defined by the instances of the keywords Gordon and Network in the result graph of Fig. 1(e). Since result graphs not intended by the user are excluded, the precision of the answer improves. Further, the evaluation time of the query is reduced since the system does not waste time computing unwanted results. These benefits are obtained thanks to the grouping of the keywords provided effortlessly by the user.

Cohesive groups can be nested within other cohesive groups in a query. An example is the keyword query (((Person (Andrew Job)) (Gordon Network)). Further, in contrast to flat keyword queries, cohesive keyword queries can contain repeated keywords provided they occur in different cohesive groups. For instance, the keyword query ((Andrew Job) job (Network Analyst)). These features of cohesive queries increase their expressive power and their capacity to narrow down the search to relevant results by excluding irrelevant ones. It is important to note that the user can naturally specify cohesive groups. In fact, cohesive queries offer more flexibility to the users and allow them to express queries with a clearer meaning. For instance, the user who is looking in a bibliographic data source for a paper authored by John Smith and edited by John Brown can naturally formulate the query ((author (John Smith)) (editor (John Brown))). In summary, cohesive queries are intuitive and as simple as flat keyword queries while retaining both advantages of flat keyword search: the user does not need to know any query language and he does not need to have knowledge of the structure of the data sources.

Note that in IR, flat, document-based search engines like Google allow the user to search for whole phrases (sequence of keywords) by enclosing them between quotes to improve the accuracy of the search. Cohesiveness queries also aim at improving the accuracy and execution time but they are different and more flexible than phrase matching over flat documents since they are designed for data with some form of structure and they do not impose any order on the keywords. The user naturally groups the keywords in a cohesive query based on the associations she wants to express on them and she is not required to know how these keywords are sequenced in the dataset.

Contribution. The main contributions of the paper are the following:

- We formally define cohesive keyword queries which involve cohesive groups of keywords and allow keyword group nesting and keyword replication. Cohesive queries can better express the user intent. They are as simple as flat keyword queries and they can be formulated naturally and effortlessly by the user.
- We provide semantics for cohesive queries on RDF graphs which interprets cohesive keyword groups in a query as cohesive units. This means that the instances of the query keyword occurrences which are not in the cohesive group cannot penetrate and be part of the subgraph of a query result graph representing the cohesive group (Sect. 3).
- We design an algorithm to efficiently evaluate cohesive queries. Our algorithm exploits the structural summary of the RDF graph to compute pattern graphs which are r-radius Steiner graphs with a minimal r. The pattern graphs are structured queries representing alternative interpretations of the cohesive queries and can be evaluated against the RDF data graph to produce the query results. The algorithm constructs pattern graphs incrementally excluding early on graphs under construction which violate the cohesiveness of keyword groups (Sect. 4).
- We ran experiments to assess the effectiveness of cohesive queries and the efficiency of our algorithm. Our results show that the pattern graphs of cohesive queries can be computed by our algorithm much faster than the pattern graphs of flat keyword queries. Cohesive keyword queries considerably improve the quality of the results compared to flat keyword queries, importantly reducing the number of patterns graphs returned to the user (Sect. 5).

2 Data Model and Flat Keyword Queries

Data Model. The Resource Description Framework (RDF) provides a framework for representing information about web resources in a graph form. The RDF vocabulary includes elements that can be broadly classified into Classes, Properties, Entities and Relationships. All the elements are resources. Our data model is an RDF graph defined as follows:

Definition 1 (RDF Graph). An *RDF graph* is a quadruple $G = (V, E, L, l)$:

V is a finite set of vertices, which is the union of three disjoint sets: V_E (representing entities), V_C (representing classes) and V_V (representing values).

E is a finite set of directed edges, which is the union of four disjoint sets: E_R (inter-entity edges called *Relationship* edges which represent entity relationships), E_P (entity to value edges called *Property* edges which represent property assignments), E_T (entity to class edges called *type* edges which represent entity to class membership) and E_S (class to class edges called *subclass* edges which represent class-subclass relationship).

L is a finite set of labels that includes the labels "type" and "subclass".

l is a function from $V_C \cup V_V \cup E_R \cup E_P$ to L. That is, l assigns labels to class and value vertices and to relationship and property edges.

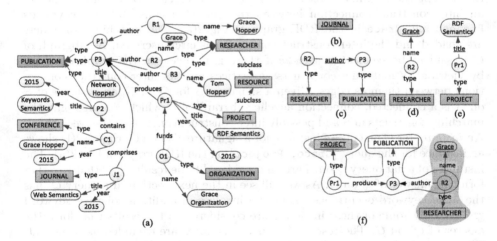

Fig. 2. (a) An RDF Graph, (b), (c), (d) and (e) class, relationship, value and property matching constructs respectively, (f) inter-construct connection and query instance

Entity and class vertex and edge labels are Universal Resource Identifiers (URIs). Vertices are identified by IDs which in the case of entities and classes are URIs. Every entity belongs to a class. Figure 2 shows an example RDF graph which is an excerpt from a large bibliographic RDF database. For simplicity, vertex and edge identifiers are not shown in the figure.

Query Instance. A traditional keyword query Q on an RDF graph G is a set of keywords. An *instance* in G of a keyword k in Q is an occurrence of k (in a vertex or edge label) in G. In order to facilitate the interpretation of the semantics of the keyword instances, every instance of a keyword in Q is matched against a small subgraph of G which involves this keyword instance and the corresponding class vertex or vertices. This subgraph is called *matching construct*. Figure 2(b), (c), (d) and (e) show a class, relationship, value and property matching construct, respectively, for different keyword instances in the RDF graph of Fig. 2(a). Underlined labels in a matching construct denote the keyword instances based on which a matching construct is defined. Note that a matching construct might have more than one underlined keyword instance as it might be a matching construct for more than one keyword instance. Each matching construct provides information about the semantic context of the keyword instance under consideration. For example, the matching construct of Fig. 2(d) shows that Grace is the name of an entity $R2$ of type Researcher.

A *signature* of Q is a function that matches every keyword k in Q to a matching construct for k in G. Given a query signature S, an *inter-construct connection* between two distinct matching constructs C_1 and C_2 in S is a simple path augmented with the class vertices of the intermediate entity vertices in the path (if not already in the path) such that: (a) one of the terminal vertices in the path belongs to C_1 and the other belongs to C_2, and (b) no vertex in the connection except the terminal vertices belong to a construct in S. Figure 2(f) shows an inter-construct connection between the matching constructs for keywords Project and Grace in the RDF graph of Fig. 2(a). The matching constructs are shaded and the inter-construct connection is circumscribed. A subgraph of G is said to be *connection acyclic* if there is no cycle in the graph obtained by viewing its matching constructs as vertices and its inter-construct connections between them as edges. Given a signature S for Q on G, an *instance* of S on G is a connected, connection acyclic subgraph of G which contains only the matching constructs in S and possibly inter-construct connections between them. An *instance* for Q on G is an instance for a signature of Q on G. Figure 2(f) shows an instance for the query {Grace, Project} on the RDF graph of Fig. 2(a). The instances of a flat query Q on G are all considered to be results of Q that together form the answer of Q on G. As we will see in the next section, if a query Q' has the same keyword occurrences as Q and involves in addition cohesive keyword groups, only some of these instances are considered to be results that form the answer of Q' on G. The rest of the query instances are excluded as irrelevant. Therefore, the instances of Q' are its candidate results.

3   Keyword Queries with Cohesive Keyword Groups

We define in this section the syntax and semantics of keyword queries with cohesive keyword groups (called *cohesive keyword queries*).

Syntax. We start by providing a recursive definition of the concept of term which corresponds to a cohesive group of keywords: a *term* is a set of at least two keywords and/or terms.

Definition 2 (Cohesive Query). A cohesive query is: (a) a set of a single keyword, or (b) a term. *Notation*: sets are delimited in a query using parentheses, and elements are separated within sets using spaces.

For instance, Q_1 =(Publication (Grace Hopper) (Project Semantics 2015)) is a cohesive keyword query and (Grace Hopper) and (Project Semantics 2015) are two terms in it. Query Q_2 = (((RDF Project) publication) (author (Tom Hopper))) is another cohesive keyword query where the term (Tom Hopper) is nested within the term (author (Tom Hopper)) and the term (RDF Project) is nested within the term ((RDF Project) Publication).

The same keyword may appear multiple times in a query but, of course, the same keyword or term cannot appear multiple times as an element of a set. For instance, in the cohesive keyword query Q_3 = ((author Grace) Publication (Conference (Grace Hopper))), the keyword Grace occurs twice: once in the term (author Grace) and once in the term (Grace Hopper). In the following unless stated differently, 'query' refers to a cohesive keyword query.

Semantics. The queries are matched against a data graph G. An instance I of a cohesive query Q on G is defined similarly to an instance of the flat query that involves the same keywords. A difference appears only when Q involves multiple occurrences of the same keyword k. In this case, I might contain multiple instances of k and the occurrences of k in Q can be matched to the same or different instances of k in I.

Figure 3(a), (b) and (c) show different instances of the query (Publication (Project (Semantics 2015)) (author (Grace Hopper))) on a bibliography database. This bibliography database encompasses the one of Fig. 2 and is not shown here for the sake of space.

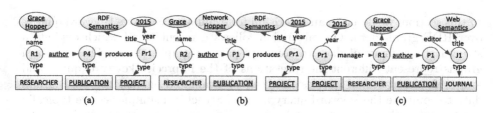

(a)   (b)   (c)

Fig. 3. (a) result graph, (b) and (c) query instances which are not result graphs for the query (Publication (Project (Semantics 2015)) (author (Grace Hopper)))

As mentioned above, a term in a cohesive query contains keywords and/or other terms. A term expresses a cohesiveness relationship on its elements. Intuitively, a term states that the instances of its keyword occurrences in a result of the query should form a *cohesive unit*. That is, in a result graph of a query, they should form a subgraph where the instances of the keyword occurrences which are external to the term do not interfere.

More formally, let I be an instance of a query Q on G, and t be a term in Q. The *instance* I_t *of* t *in* I is a minimal connected subgraph of I that comprises the matching constructs of the keyword occurrences of t in I.

Definition 3 (Query result). A *result* of Q on G is an instance I of Q on G such that for every term t in Q and for every keyword occurrence k in Q which is not in t, the following conditions hold:

(a) The instance of k in I does not occur in I_t unless it is a class vertex label,
(b) The instance of k in I is not the label of a property edge or a value of an entity vertex in I_t unless this entity vertex in I_t is incident to only one non-type edge (that is, only one relationship or property edge).

The *answer* of a query Q on G is the set of the results of Q on G.

Consider again the query and the query instances of Fig. 3. With this query the user is looking for publications authored by Grace Hopper which were produced by a project on Semantics that started in 2015. The query instance of Fig. 3(a) is a result graph for this query as it satisfies the conditions of Definition 3. In contrast, the query instance of Fig. 3(b) is not a result graph for the query. Indeed, the instance of the keyword `author`, which is not in the term (`Grace Hopper`), occurs within the instance of this term in the query instance (condition (a) in Definition 3). Similarly, the query instance of Fig. 3(c) is not a result graph of the query since the instances of keyword `Grace` and `Hopper`, which are not in the term (`Semantics 2015`), are values of an entity vertex (vertex R1) in the instance of this term in the query instance (condition (b) in Definition 3).

4   An Algorithm for Evaluating Cohesive Queries

In this section, we describe an algorithm for evaluating cohesive keyword queries over RDF graph data. Our algorithm follows a recent trend which exploits a structural summary of data to compute pattern graphs [3,21]. These pattern graphs represent different interpretations of the imprecise keyword query and are, in fact, structured queries that can be evaluated against the RDF graph data to compute the keyword query answer. Structural summaries are typically much smaller than the actual RDF data. Therefore, the pattern graphs can be generated efficiently. Moreover, this process scales smoothly when the size of the data increases. Our algorithm computes pattern graphs which are r-radius Steiner graphs and satisfy the cohesiveness of the terms in the keyword query. The algorithm proceeds bottom up in the cohesive query hierarchy to prune the search space of pattern graphs by excluding early on pattern subgraphs that breach the cohesiveness of terms (cohesive keyword groups) in the query.

Structural Summary and Pattern Graphs. Roughly speaking, the structural summary is a graph which summarizes an RDF graph. The details of structural summary construction are omitted in the interest of space. Figure 4(a) shows the structural summary for the RDF graph G of Fig. 2(a). Similarly to matching constructs on the data graph, we define matching constructs on the structural summary. Since the structural summary does not have entity vertices, a matching construct on a structural summary possesses one distinct entity variable vertex, for every class vertex, labeled by a distinct variable.

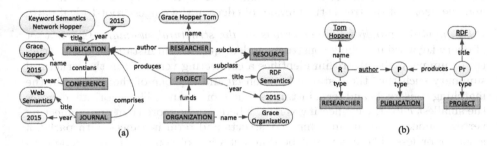

(a)   (b)

Fig. 4. (a) Structural Summary, (b) Query Pattern Graph

Pattern graphs are subgraphs of the structural summary strictly consisting of one matching construct for every keyword in a query Q and the connections between them without these connections forming a cycle.

Definition 4 (Pattern Graph). A *pattern graph* for a keyword query Q is a graph similar to an instance of Q with the following two exceptions:

(a) the labels of the entity vertices in the instance, if any, are replaced by distinct variables in the pattern graph. These variables are called *entity variables* and they range over entity vertex labels in the RDF graph.
(b) The labels of the value vertices are replaced by distinct variables whenever these labels in the query instance do not involve a keyword instance. These variables are called *value variables* and they range over keywords in the value labels of the RDF graph.

Figure 4(b) shows an example of a pattern graph for the query $Q_2 = $ `(((Project RDF) publication) (author (Tom Hopper)))` on the RDF graph of Fig. 4(a).

The pattern graphs of a cohesive query satisfy or violate the cohesiveness of its terms specified in it in the same way the instances of the query do. As mentioned above, pattern graphs are structured queries. Those pattern graphs that satisfy the cohesiveness of the query terms can be used to compute the results of the query. Interestingly, pattern graphs can be expressed as SPARQL queries, and all the machinery of query engines and optimization techniques developed for SPARQL can be leveraged to efficiently compute the results.

The basic components of the Algorithm. Our algorithm proceeds by first parsing the cohesive query. Then, it uses the produced query hierarchy to incrementally construct r-radius Steiner pattern graphs. During the process of pattern graph generation, it checks whether the pattern graph under construction satisfies the cohesiveness constraints.

(a) Parsing the query. The parsing of a cohesiveness query produces a parse tree. The leaf vertices of the parse tree are labeled by the query keyword occurrences. The root represents the query and the internal vertices represent the query terms. The *level* of a vertex in the tree is the number of edges of its path from the root and the *height* of the tree is the number of edges in the longest root-to-leaf path.

(b) Computing r-radius Steiner graphs on the structural summary. Given a set of query keyword matching constructs and/or query term instances on the structural summary, the algorithm identifies a connecting vertex cv in the structural summary such that the distance between cv and any one of the vertices in the matching constructs and term instances is no more than r. The algorithm chooses the smallest r for the connecting vertex. There can be more than one connecting vertex connected to the matching constructs and term instances with paths of length r or less. There can also be different ways of connecting the same connecting vertex with all the matching constructs and term instances with paths of length r or less. All the alternative ways to link the connecting vertices to the matching constructs and term instances define alternative r-radius Steiner graphs. Given a term t in a query, we use the algorithm presented in [3] to compute all the r-radius Steiner graphs with minimal r for the matching constructs and the term instances corresponding to the keywords and nested terms, respectively, of t. This algorithm extends the one in [16], which computes r-radius Steiner graphs on general graphs, to allow for keyword instances on the edges. Once the r-radius Steiner graphs for a term of a query are computed, they can be used for computing the r-radius Steiner graphs of the parent term in the tree.

(c) Checking cohesiveness semantics. Given a set of matching constructs for the keywords and the term instances for the nested terms of a term t, the algorithm checks whether any two elements of the set overlap in a way that breaches the cohesiveness of t (Definition 3). If this is the case, the algorithm discards this set of matching constructs and term instances for term t, and does not use it to construct minimal r-radius Steiner graphs to be propagated to the parent term of t in the query parse tree.

Algorithm description. Our Algorithm, called CohesivePGGen (Cohesive Pattern Graph Generation), is outlined in Algorithm 1. It takes as input a cohesive keyword query Q, and outputs a set of r-radius Steiner pattern graphs which satisfy the cohesiveness semantics. We assume that the structural summary of the RDF data graph is available. Initially, the algorithm computes the matching constructs for all the keywords in Q over the structural summary (lines 1–2), parses the query into the parse tree Δ (line 3), and instantiates a variable l representing the level of a node in Δ to the height of Δ (line 4). The algorithm constructs pattern graphs incrementally, in a bottom up manner over the parse

Algorithm 1. CohesivePGGen

Input: Q: a cohesive keyword query.
Output: a set of pattern graphs.

1: **for** every keyword $k \in Q$ **do**
2:     $I_k \leftarrow$ set of matching constructs of k on the structural summary;
3: $\Delta \leftarrow$ ParseQuery(Q);
4: $l = height(\Delta)$;
5: **while** $l \geq 0$ **do**
6:     **for** every vertex n at level l of Δ **do**
7:        **if** n is a leaf node labeled by keyword k **then**
8:           $I_n = I_k$
9:        **if** n is a term or the root of the tree **then**
10:          $L \leftarrow I_{c_1} \times \ldots \times I_{c_m}$; $\triangleright$ $c_1, \ldots, c_m$ are the children of n.
11:          **if** n contains a term **then**
12:             **for** every combination $L_i \in L$ **do**
13:                **if** $CheckCohessivenessSemantics(L_i)$ = false **then**
14:                   $L \leftarrow L - L_i$;
15:             **for** every combination $L_i \in L$ **do**
16:                $I_n \leftarrow I_n \cup rRadiusSteinerGraphs(L_i)$; $\triangleright$ I_n is the set of instances
                                           of term n on the structural summary
17:     $l \leftarrow l - 1$
18: Return I_n

---

tree Δ, starting with the deepest leaf vertices (lines 5–17). For a vertex n, variable I_n represents the set of matching constructs if n is a leaf (keyword) vertex (lines 7–8), and the set of term instances of n on the structural summary if n is a term or the root of Δ (lines 9–16). For a term vertex n, variable L denotes the Cartesian product of the sets I_{c_i} for the children c_i of n (line 10). Every element of L which violates the cohesiveness of at least one term instance in it is removed from L (lines 11–14). The rest of the elements of L are used to produce r-radius Steiner graphs with minimal r which are instances of the term vertex n (lines 15–16). The process continues until the root vertex is reached. At this point, I_n represents the pattern graphs of Q, which are returned to the user.

5 Experimental Evaluation

We implemented our approach and ran experiments to evaluate: (a) the effectiveness of cohesive keyword queries, and (b) the efficiency of CohesivePGGen.

Datasets and Queries. We used the DBLP and Jamendo[1] real datasets for our experiments. DBLP is a bibliography database of 600MB of size, containing 8.5M triples. Jamendo is a repository of Creative Commons licensed music of 85MB of size, containing 1.1M triples. The extracted structural summaries and the keyword inverted lists for both datasets were stored in a Relational database.

---

[1] http://dbtune.org/jamendo/.

Table 1. Queries on the Jamendo and DBLP datasets

Dataset	Q#	Cohesive keyword query	#MCs	#Sigs	#PGs
Jamendo	1	(document (teenage (text fantasie)))	16	105	162
	2	((lyrics sweet) recorded_as onTimeLine)	18	64	64
	3	((MusicArtist Cicada) performance (track knees))	21	405	488
	4	((Record (date title)) track (Lyrics good))	43	127,008	185,916
	5	((MusicArtist Briareus) (cool (girl Reflections)))	27	1,440	1,950
	6	(time Mako (record (track (down passion))))	39	39,690	49,070
	7	(Kouki (electro (record revolution (track good))))	43	119,070	172,541
	8	(Nuts track (chillout (record spy4)))	28	1,764	2,248
	9	((biography guitarist) (track (title Lemonade)))	25	1,512	2,538
	10	((record (title divergence)) (track obsession))	27	1,323	1,805
DBLP	1	((journal design) creator (person (phdthesis CAD)))	49	69,120	447,086
	2	((name Charles) creator (Proceedings forward))	33	68,200	25,324
	3	((article editor person) creator (inproceedings hybridization))	32	4,320	12,519
	4	((person name) creator (performance 2002))	59	100,800	479,542
	5	(Oliver (Article (Linux year))	21	480	2,960
	6	(inproceedings Tolga (mastersthesis warehouses))	8	5	22
	7	(((compiler cite) Charles) (creator peephole))	34	5,280	20,660
	8	(creator (decentralized IEEE) (coscheduling 2004))	51	25,300	141,531
	9	((Milne 2005) homepage person)	35	1,320	3,788
	10	(((name Yahiko) person) (editor (conceptual springer)))	34	18,900	97,512

The experiments were conducted on a standalone machine with an Intel i5-3210M @2.5 GHz processor and 8 GB memory.

We experimented with a large number of cohesive keyword queries and we report on 10 of them for each dataset. The queries cover a broad range of cases. They involve 4 to 6 keywords and 1 to 3 levels of term nesting. Table 1 shows the queries used and their statistics. #MCs denotes the total number of matching constructs for the keywords of a query, #Sigs denotes the total number of matching constructs combinations for a query (signatures), and #PGs denotes the number of pattern graphs of a query on the structural summary ignoring the cohesiveness semantics.

Effectiveness of the cohesive queries. In order to evaluate the effectiveness of the cohesive queries, we measured: (a) the reduction in the number of pattern graphs of a query, and (b) the improvement in the quality of the results of a query due to the cohesiveness constraints.

(a) Reduction in the space of pattern graphs. We compare, for each cohesive keyword query, the number of pattern graphs generated with the number of pattern graphs of the corresponding flat keyword query (i.e., the flat keyword query obtained by removing cohesiveness constraints and keyword duplicates). Figure 5 reports on the percentage reduction of the number of pattern graphs for the queries of Table 1 on both datasets. As one can see, the cohesiveness constraints reduce substantially the number of pattern graphs from which the user has to choose the relevant ones.

(b) Improvement in the quality of results. The number of pattern graphs of a query can be very large in order to allow an expert user go through them and select the relevant ones. For instance, observe that in Table 1 some queries have hundreds of thousands of pattern graphs. Therefore, we adopt the *path length* and *popularity score* metrics introduced in [21] to rank the pattern graphs of a query.

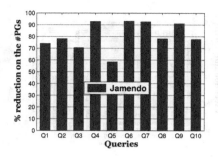
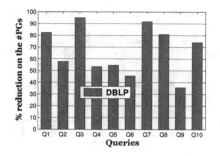

Fig. 5. % reduction on the number of pattern graphs for the queries on the two datasets.

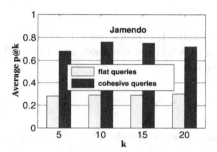
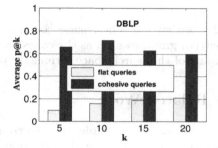

Fig. 6. Average *precision@k* for cohesive queries and their corresponding flat queries varying k on the two datasets.

We then select the top-k pattern graphs and have an expert user identify those of them which are relevant. In order to measure the quality of the results, we use the *precision@k* (*p@k*) metric. The *precision@k* is the ratio of the number of relevant pattern graphs in the first k positions to k. Figure 6 displays the average *p@k* over the queries of Table 1, for different values of k, on the two datasets. For comparison, the figure displays both: the average *p@k* of the cohesive queries and the average *p@k* of their corresponding flat queries. The results show that in all cases, the quality of the cohesive queries is several times higher than that of their corresponding flat queries. This is not surprising, since the cohesive queries benefit from the cohesiveness constraints expressing the user intention.

Efficiency of Algorithm CohesivePGGen. We compare the execution time of our algorithm on cohesive queries with the computation time of the pattern graphs of their corresponding flat keyword queries. Figure 7 shows the execution times of CohesivePGGen for the queries of Table 1 on the two datasets. Note that the Y axis is in logarithmic scale. Algorithm CohesivePGGen on cohesive queries is much faster, in some cases by more than one order of magnitude. In fact, algorithm CohesivePGGen on cohesive queries has to check for the satisfaction of cohesiveness constraints and this incurs additional cost. However, the algorithm does not produce all the pattern graphs of the flat version of the query to check

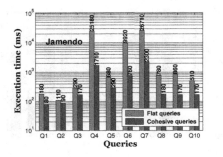 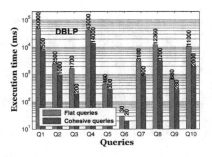

Fig. 7. Execution time of CohesivePGGen on cohesive and flat queries.

if they satisfy the cohesiveness constraints. Instead, it stops the construction of a pattern graph as soon as the cohesiveness of a term is violated, and this early pruning of the search space ultimately pays off.

6   Related Work

Keyword search on graphs tries to compute results which relate the keyword instances in the data graph in some minimal way. Different keyword search algorithms compute results which are Steiner trees [2,11,13]. However trees cannot fully capture the nature of graph data. Other keyword search algorithms compute results which are graphs including r-radius Steiner graphs [16], graphs which minimize the lengths between keyword instance pairs [20], and r-cliques [14]. All the above approaches are proposed for generic graphs, and cannot be used directly for keyword search over RDF graph data. This is because the edges of an RDF graph represent predicates, which can also be matched by the keywords of a keyword query. Traditional keyword search approaches on RDF graphs and their drawbacks are described in the introduction and are not repeated here. Retaining the expressive power of structured queries while incorporating the flexibility of flat keyword queries is an issue of great interest [17,19]. In [17] an entity relationship query language over unstructured document data is introduced. Unlike our cohesive keyword query language which does not follow a strict structure and allows a user to group keywords and terms, the query language in [17] follows a strict structure and allows keywords only to express entity properties and relationships. [19] presents a keyword-based structured query language to be used over structured knowledge bases extracted from the web. The main goal of this work is to extract entities from the knowledge base (possibly in conjunction with the relevant text documents). Although the desire of trading off flexibility and convenience for expressivity is common with our work, the structured queries in [19] are schema dependent: the user needs to characterize some keywords as relations in order to build nested structured queries beyond flat keyword queries. In contrast, in our query language, nesting is incorporated in queries based on the desire of the user to form cohesive groups irrespectively of any schema information. Cohesive queries were also introduced in [5]. However, those queries are for tree-structured data. Therefore, their semantics is different not involving any semantic information.

Algorithm CohesivePGGen exploits the structural summary of the RDF graph to compute pattern graphs. Different variation of this technique have been adopted in recent years to compute the results of keyword search on RDF graphs [3,8,15,21–23]. Pattern graphs cluster result graphs with the same structure. In [3] the user navigates through a clustering hierarchy to select a pattern graph that better meets her intention. Relaxation techniques that allow broadening the result set of a pattern graph are presented in [4].

7 Conclusion

In this paper we claim that without additional information from the user, keyword queries cannot effectively retrieve information from RDF graph data. Therefore, we introduce a novel keyword query language which allows the user to better express her intention by permitting the specification of cohesive keyword groups, keyword group nesting and keyword repetition. We provide formal semantics for cohesive keyword queries, and we design a query evaluation algorithm, called CohesivePGGen, which exploits the structural summary of the RDF graph to produce r-radius Steiner pattern graphs. Our algorithm prunes early on the search space of pattern graphs by retaining only those that satisfy the cohesiveness constraints. Our experiments show that CohesivePGGen largely outperforms the generation of pattern graphs for flat keyword queries. They also show that cohesive queries substantially improve the *precision@k* of flat keyword queries allowing the search for relevant pattern graphs in a much smaller set while retaining the simplicity and convenience of flat keyword queries.

We are currently working on refining the cohesiveness semantics in order to further narrow down the query search space while improving the result quality.

References

1. Aksoy, C., Dass, A., Theodoratos, D., Wu, X.: Clustering query results to support keyword search on tree data. In: Li, F., Li, G., Hwang, S., Yao, B., Zhang, Z. (eds.) WAIM 2014. LNCS, vol. 8485, pp. 213–224. Springer, Heidelberg (2014)
2. Bhalotia, G., Hulgeri, A., Nakhe, C., Chakrabarti, S., Sudarshan, S.: Keyword searching and browsing in databases using BANKS. In: ICDE, pp. 431–440 (2002)
3. Dass, A., Aksoy, C., Dimitriou, A., Theodoratos, D.: Exploiting semantic result clustering to support keyword search on linked data. In: Benatallah, B., Bestavros, A., Manolopoulos, Y., Vakali, A., Zhang, Y. (eds.) WISE 2014, Part I. LNCS, vol. 8786, pp. 448–463. Springer, Heidelberg (2014)
4. Dass, A., Aksoy, C., Dimitriou, A., Theodoratos, D.: Keyword pattern graph relaxation for selective result space expansion on linked data. In: Cimiano, P., Frasincar, F., Houben, G.-J., Schwabe, D. (eds.) ICWE 2015. LNCS, vol. 9114, pp. 287–306. Springer, Heidelberg (2015)

5. Dimitriou, A., Dass, A., Theodoratos, D.: Cohesiveness relationships to empower keyword search on tree data on the web (2015). arXiv preprint arXiv:submit/ 1331603
6. Ding, B., Yu, J.X., Wang, S., Qin, L., Zhang, X., Lin, X.: Finding top-k min-cost connected trees in databases. In: ICDE, pp. 836–845 (2007)
7. Elbassuoni, S., Ramanath, M., Schenkel, R., Weikum, G.: Searching RDF graphs with SPARQL and keywords. IEEE Data Eng. Bull. **33**, 16–24 (2010)
8. Fu, H., Gao, S., Anyanwu, K.: Disambiguating keyword queries on RDF databases using "Deep" segmentation. In: ICSC, pp. 236–243 (2010)
9. Golenberg, K., Kimelfeld, B., Sagiv, Y.: Keyword proximity search in complex data graphs. In: SIGMOD, pp. 927–940 (2008)
10. Guo, L., Shao, F., Botev, C., Shanmugasundaram, J.: XRANK: ranked keyword search over XML documents. In: SIGMOD, pp. 16–27 (2003)
11. He, H., Wang, H., Yang, J., Yu, P.S.: Blinks: ranked keyword searches on graphs. In: SIGMOD, pp. 305–316 (2007)
12. Jiang, M., Chen, Y., Chen, J., Du, X.: Interactive predicate suggestion for keyword search on RDF graphs. In: Tang, J., King, I., Chen, L., Wang, J. (eds.) ADMA 2011, Part II. LNCS, vol. 7121, pp. 96–109. Springer, Heidelberg (2011)
13. Kacholia, V., Pandit, S., Chakrabarti, S., Sudarshan, S., Desai, R., Karambelkar, H.: Bidirectional expansion for keyword search on graph databases. In: VLDB, pp. 505–516 (2005)
14. Kargar, M., An, A.: Keyword search in graphs: finding r-cliques. VLDB **4**, 681–692 (2011)
15. Le, W., Li, F., Kementsietsidis, A., Duan, S.: Scalable keyword search on large RDF data. IEEE Trans. Knowl. Data Eng. **26**(11), 2774–2788 (2014)
16. Li, G., Ooi, B.C., Feng, J., Wang, J., Zhou, L.: Ease: an effective 3-in-1 keyword search method for unstructured, semi-structured and structured data. In: SIG-MOD, pp. 903–914 (2008)
17. Li, X., Li, C., Yu, C.: Entity-relationship queries over Wikipedia. ACM TIST **3**(4), 70 (2012)
18. Liu, X., Wan, C., Chen, L.: Returning clustered results for keyword search on XML documents. IEEE Trans. Knowl. Data Eng. **23**(12), 1811–1825 (2011)
19. Pound, J., Ilyas, I.F., Weddell, G.E.: Expressive and flexible access to web-extracted data: a keyword-based structured query language. In: ACM SIGMOD, pp. 423–434 (2010)
20. Qin, L., Yu, J.X., Chang, L., Tao, Y.: Querying communities in relational databases. In: ICDE, pp. 724–735 (2009)
21. Tran, T., Wang, H., Rudolph, S., Cimiano, P.: Top-k exploration of query candidates for efficient keyword search on graph-shaped (RDF) data. In: ICDE, pp. 405–416 (2009)
22. Wang, H., Zhang, K., Liu, Q., Tran, T., Yu, Y.: Q2Semantic: a lightweight keyword interface to semantic search. In: Bechhofer, S., Hauswirth, M., Hoffmann, J., Koubarakis, M. (eds.) ESWC 2008. LNCS, vol. 5021, pp. 584–598. Springer, Heidelberg (2008)
23. Xu, K., Chen, J., Wang, H., Yu, Y.: Hybrid graph based keyword query interpretation on RDF. In: ISWC (2010)

Privacy-Enhancing Range Query Processing over Encrypted Cloud Databases

Jialin Chi[1,2]([envelope]), Cheng Hong[1], Min Zhang[1], and Zhenfeng Zhang[1]

[1] Trusted Computing and Information Assurance Laboratory, Institute of Software,
Chinese Academy of Sciences, Beijing, China
{chijialin,hongcheng,mzhang,zfzhang}@tca.iscas.ac.cn
[2] University of Chinese Academy of Sciences, Beijing, China

Abstract. The Database-as-a-Service (DAS) model allowing users to outsource data to the clouds has been a promising paradigm. Since users' data may contain private information and the cloud servers may not be fully trusted, it is desirable to encrypt the data before outsourcing and as a result, the functionality and efficiency has to be sacrificed. In this paper, we propose a privacy-enhancing range query processing scheme by utilizing polynomials and kNN technique. We prove that our scheme is secure under the widely adopted honest-but-curious model and the known background model. Since the secure indexes and trapdoors are indistinguishable and unlinkable, the data privacy can be protected even when the cloud server possesses additional information, such as the attribute domain and the distribution of this domain. In addition, results of experiments validating our proposed scheme are also provided.

Keywords: Database-as-a-Service · Cloud database · Range query

1 Introduction

With the rapid developments of networking and Internet technologies, Database-as-a-Service (DAS) [1] has been a promising paradigm, such as Amazon Web Services [2] providing both relational and NoSQL cloud-based database services. In the DAS model, organizations could outsource their data to service providers and retrieve data anytime and anywhere, as long as they have access to the Internet. In other words, the DAS model provides an approach for corporations to share the hardware and software resources as well as the expertise of database professionals, thereby cutting the cost of maintaining their own DBMSs. On the other hand, since data is stored at the third-party server and most organizations view their data as a valuable asset, at least two privacy issues arise. First, data needs to be protected from thefts from outsiders who may break into the providers' websites and scan disks. For instance, in 2014, hackers broke into the computers of Community Health Systems which operates 206 hospitals across the United States and stole data on 4.5 million patients [3]. Second, data also needs to be protected against the service providers, if they cannot

© Springer International Publishing Switzerland 2015
J. Wang et al. (Eds.): WISE 2015, Part II, LNCS 9419, pp. 63–77, 2015.
DOI: 10.1007/978-3-319-26187-4_5

be fully trusted. For example, an engineer in Google's Seattle offices broke into the Gmail and Google Voice accounts of several children in 2010 [4]. To guard data security and privacy, a straightforward solution is to encrypt data before outsourcing. When performing search, all encrypted data is returned and corporations could execute the query at the client after decrypting it. Obviously, this naive approach is impractical for incurring high decryption and network workloads for organizations.

The system model considered in this paper is comprised of three fundamental parties: the *data owner*, the *data user* and the *cloud server*, as illustrated in Fig. 1. The data owner outsources his/her data to the clouds in encrypted form, together with the secure index applied to enable the searching capability of the cloud server. To search over the encrypted data, the data user obtains the secure trapdoor from the data owner through search control mechanisms, e.g., broadcast encryption [8] and then submits it to the cloud server. Upon receiving the trapdoor, the cloud server searches the secure index and returns the matching encrypted data to the data user. Finally, the access control mechanism [5] is used to manage decryption capabilities. However, the search control and access control mechanisms are out of the scope of this paper. For a simple example, the data user can store his/her own data and query his/her own data on the cloud server. In this architecture, the data owner and the data user are trusted while the cloud server hosted by service providers cannot be fully trusted.

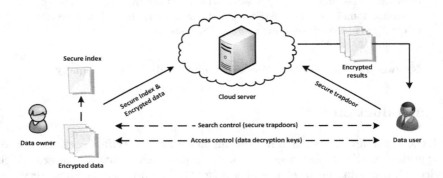

Fig. 1. Architecture of the range query processing over encrypted cloud databases

The problem discussed in this paper is range query which is a major type of database queries. The data in a relational DBMS are represented as tuples (i.e., records or rows) and tuples having the same attributes are organized in a table. For example, the Member data in Table 1 are personal information tuples of a special interest group and all tuples are identified by attributes *ID*, *NAME*, *AGE* and *ADDRESS* where the underlined attribute *ID* denotes the key of this table. For the attribute which can be represented as numerical values (e.g., *AGE*), consider a range query specified by an interval (e.g., *select ∗ from Member where AGE ∈* [20, 30)), the matching results are the tuples whose attribute value falls into the

Table 1. Member

ID	NAME	AGE	ADDRESS
1	John	25	Chicago
2	Eric	21	Miami
3	Thomas	27	New York
4	Franklin	33	San Francisco
5	Aaron	18	Boston

Table 2. Enc_Member

Enc_tuple	I^S_{AGE}
SfahFrwierh3JsejjsdfblklR5fsfWer	α
uDmsmdfOwe05u5jgNfjkgkroeRledlcf	β
mcuj48djhg2EdfslFlsfqjosdjRlslMd	γ
Ijfgkf8djkfsldDiseemJKllfnlfsj8k	η
osdfnsfklJfsnKdfannlK0nlfjalnn2z	π

interval (e.g., $ID \in \{1, 2, 3\}$). In addition, equal query can be viewed as a special kind of range query. Since computers can handle only inherently finite and discrete numbers, the attribute values and the lower and upper bounds of range queries can be assumed as nonnegative integers without loss of generality [25].

To protect the data privacy, there are several granularity choices for encryption, such as encrypting at the level of individual table, attribute, tuple and cell. Encrypting at the level of individual table or attribute implies that the entire table should be returned as the result, although it is more efficient to encrypt and decrypt the data. On the contrary, encrypting at the level of individual cell will incur high encryption and decryption workloads, while providing more efficient query processing. As in previous works, our proposed scheme performs encryption at the tuple level. To provide cloud servers with the ability to check whether one tuple matches the range query, we also associate with each encrypted tuple multiple secure indexes built based on the attribute values to be queried. Hence, each plaintext table can be stored as a table with one attribute for the encrypted tuple and several additional attributes for the secure indexes. More specifically, the plaintext tuple $t(A_1, \ldots, A_n)$ will be mapped onto a new tuple $t^S(Enc(t), I^S_1, \ldots, I^S_m)$ where $m \leqslant n$. The attribute $Enc(t)$ is utilized to store the encrypted tuple and I^S_i corresponds to the secure index over some A_j. In addition, the encryption function $Enc()$ is treated like a black box in this paper. For instance, as shown in Table 2, the indexed table Enc_Member contains the attribute Enc_tuple representing encrypted tuples and I^S_{AGE} representing secure indexes over attribute AGE. Here, the I^S_{AGE} attribute values are denoted as Greek letters.

The main challenge in this context is how to enable efficient range query processing over encrypted cloud databases without sacrificing privacy. In real situations, the service providers may possess background knowledge (e.g. the attribute domain and the distribution of this domain) which can be obtained from similar databases or historical data. So it is desirable that the cloud server can't learn more than the secure indexes, the secure trapdoors and the encrypted results even with such information. However, existing works [17–26] disclose useful statistical properties of data items through indexes or query processing, so the cloud server could learn additional information than allowed such as the exact attribute value of each tuple. Besides, to avoid incurring high decryption workload and network workload, it is better to reduce the interactions between server and client as well as the unmatching data items contained in results.

Our Contribution. In this paper, we propose a privacy-enhancing range query processing scheme. Our approach utilizes canonical ranges and polynomials to build indexes and trapdoors for attribute values and range queries. Then kNN technique is applied to encrypt the indexes and trapdoors. We provide thorough security analysis that the proposed scheme is secure under the honest-but-curious (HBC) model and the known background model. Since the indexes and trapdoors built in our scheme are indistinguishable and unlinkable, security and privacy can be protected even when the cloud servers possess the distribution of the attribute domain. Furthermore, we evaluate the performance of our scheme.

Organization. The rest of the paper is organized as follows. Section 2 presents the previous works related to our proposed scheme. The threat model and security goals are discussed in Sect. 3. In Sect. 4, we describe our privacy-enhancing range query processing in detail. Section 5 shows the security analysis and Sect. 6 gives our experimental results. We conclude the paper in Sect. 7.

2 Related Work

Previous works related to our scheme mainly fall into two categories: range queries, where search conditions are represented as intervals, and keyword queries, where search conditions are denoted as single or multiple keywords. The first keyword search scheme is proposed by Song *et al.* [6]. After this work, many novel schemes have been designed to improve the functionality and efficiency. Instead of scanning every word, [7,8] build secure indexes based on documents and corresponding search protocols to improve efficiency. Subsequent works [9–12] propose schemes to support multi-keyword retrieval, i.e., conjunctive or disjunctive keyword search. Moreover, [13–16] extend the search capability to fuzzy keyword search which can tolerate errors in the query to some extent. However, above works are limited and insufficient in executing range queries over encrypted data.

There are several works designed to enable privacy preserving and yet efficient range query processing. The bucketization technique proposed in [17] partitions the attribute domain into multiple buckets (in an equi-depth or equi-width manner) and each bucket is identified by a unique tag which can be realized by a collision-free hash function. The bucket tag is maintained on the cloud servers as an index together with the encrypted tuples in this bucket. The trapdoor of a range query consists of the tags of buckets that overlap with the search condition and then all the encrypted tuples indexed by these tags will be returned. As mentioned in [18], the security limitation of data partitioning approaches is that the cloud servers can statistically estimate the attribute values of tuples and the lower and upper bounds of queries through domain knowledge and historical results. In addition, since false positives are incurred, i.e., the results may contain tuples that don't satisfy the query, users have to decrypt all the encrypted results and filter the unmatching items. Hence, the bucket size is positively related to security and negatively related to efficiency. In particular, increasing the bucket size can help to improve the data security, but will incur more unmatching results. There are several works focusing on how to trade off the security and efficiency. In [18], Hore et al. claim an optimal bucketization algorithm to maximize the efficiency as well as two measures of privacy, and propose a re-bucketization technique that yields bounded overhead while maximizing the defined notions of privacy. Subsequently, Wang et al. [19] introduce new security and efficiency metrics based on probability distribution variance and overlap ratio respectively, and design a 2-phase local overlap bucket (LOB) algorithm.

In [20], Agrawal et al. describe the first order preserving encryption approach for range queries over encrypted data, followed by [21] which proposes a provable secure scheme to achieve the same functionality. The schemes based on order preserving encryption are deterministic, since ciphertext must keep the same numerical ordering as plaintext. Hence, the cloud servers can obtain the relationship between the attribute values of two tuples directly from their corresponding indexes. In addition, these approaches assume that the distribution of attribute domain remains fixed and the encryption function is conscious of this distribution. The advantage of order preserving encryption schemes is that all results returned are matching tuples, so there is no additional workload incurred to the client.

In [22], damiani et al. design to build encrypted B^+-tree by encrypting the plaintext B^+-tree at node level. As the encrypted tree is not visible, interactions between server and client are required from the root to the leaf, and the number of interactions is equal to the depth of the tree. Obviously, it will increase decryption and network workloads for data users. In [23], Lu constructs the first provably secure logarithmic search mechanism. Pang et al. [24] show that the privacy of Damiani's encrypted B^+-tree can be defeated by monitoring the I/O activities on the server or tracking the sequence of nodes retrieved during range query processing. So they propose a privacy-enhancing PB^+-tree index which groups the nodes in each index level randomly into buckets, but this approach incurs higher I/O cost and computation overheads on the server. Since the ciphertexts are sorted in B^+-tree, the cloud server who possesses the distribution of domain can guess the attribute value of each tuple with high probability.

Besides the schemes based on bucketization technique, order preserving encryption and B^+-tree, there are also several works relying on other mechanisms. Li and Omiecinski [25] adapt a prefix-preserving encryption scheme to create index and this scheme is subject to certain attacks. In [26], Li *et al.* propose the first range query processing scheme that achieves index indistinguishability by utilizing PBtree (where "P" stands for privacy and "B" stands for Bloom filter) which has the property of node indistinguishability and structure indistinguishability. However, since the cloud sever can learn the tuples belonging to each prefix during the query processing, if the distribution of attribute domain is obtained, each plaintext value and the prefixes in trapdoors will be disclosed. In addition, the results in [26] may contain unmatching data items because this scheme utilizes Bloom filter to store each node's prefix family.

3 Problem Formulation

3.1 Threat Model

In this paper, we only consider attacks from the server providers while the data owner and the data user are trusted. We assume that the server is honest-but-curious (HBC) [27]. That means the cloud server will honestly and correctly follow the protocols, but may attempt to infer and analyze more information than allowed. Furthermore, we also consider the known background model, i.e., the cloud server possesses additional information about the data. For example, the attribute domain and its statistical information such as the distribution of this domain which may be obtained from similar databases or historical data.

3.2 Design Goals

Our design bears the following security and performance goals.

Security Goals

Index Confidentiality. The cloud server can't learn the exact attribute value of each tuple from its corresponding secure index. In addition, the secure index generation should be randomized instead of deterministic, i.e., the secure indexes are indistinguishable. For example, when given two secure indexes, the cloud server can't obtain the relationship between their corresponding attribute values.

Trapdoor Confidentiality. The cloud server can't learn the upper and lower bounds as well as the size of range query from its corresponding secure trapdoor. In addition, the secure trapdoor generation should be randomized instead of deterministic, i.e., the secure indexes are unlinkable. For instance, the cloud server can't deduce the relationship of any given trapdoors.

Query Processing Confidentiality. During the query processing, the cloud server can't obtain more than what can be derived from the results, even with background knowledge such as the attribute domain and the distribution of this domain.

Usability and Efficiency. For range query processing over clouds, since client devices mostly have limited storage and computing resources compared with cloud servers, it is better to process as much of the work as possible at the server, without having to decrypt the data. Besides, to avoid high decryption and network workloads, it is desirable to minimize the number of unmatching data items to be returned and interactions between client and server.

4   Privacy-Enhancing Range Query Processing

4.1   Main Idea

To design a privacy-enhancing and well-functioning method for range query, we focus on three important and closely inter-related aspects: (1) data structure utilized to build indexes for tuples and trapdoors for search conditions; (2) effective and efficient search algorithm that can check whether one tuple matches the given search condition; (3) privacy mechanisms that can be integrated with the above two aspects so that the privacy of indexes, trapdoors and query processing can be protected simultaneously. In this subsection, we describe the key ideas to realize data structure and search algorithm without privacy mechanisms. The more detailed scheme will be discussed in the next subsection.

In this paper, we assume the attribute to be queried is A whose value domain is $\mathbb{Z}_{2^n}$. Our proposed scheme first converts the attribute value $t.A$ of tuple t and the range query $\mathcal{Q} = [a, b)$ into canonical range representation which is also applied in [29]. Then use polynomials to construct index $\mathcal{I}$ for value $t.A$ and trapdoor $\mathcal{T}$ for search condition $\mathcal{Q}$. The detailed steps are explained as follows.

Canonical Range Representation of Value/Query. Consider the range query with $\mathbb{Z}_{2^n}$, a canonical range with level $i \in \mathbb{Z}_n$ is $[x2^i, (x+1)2^i)$ for some integer $x \in \mathbb{Z}_{2^{n-i}}$. There are 2^{n-i} canonical ranges in level i and the total number of different canonical ranges is $\sum_{i=0}^{n-1} 2^{n-i} = 2^{n+1} - 2$. Hence, we can identify each canonical range as a unique integer $cr \in \mathbb{N}_{2^{n+1}-2}$, which can be realized by a collision-free hash function $h : (i, x) \to \mathbb{N}_{2^{n+1}-2}$. The corresponding level of cr is denoted as $level(cr)$.

In particular, given the attribute value $t.A \in \mathbb{Z}_{2^n}$, for each level $i \in \mathbb{Z}_n$, compute x_i such that $t.A \in [x_i2^i, (x_i+1)2^i)$. The total number of canonical ranges containing $t.A$ is n and there is only one canonical range for each level. Hence the attribute value $t.A$ can be represented as CRA = $\{cra_0, cra_1, \ldots cra_{n-1}\}$ where $cra_i \in \mathbb{N}_{2^{n+1}-2}$. Given the range query $\mathcal{Q} = [a, b)$, for each level $i \in \mathbb{Z}_n$, find, if any, the minimum y_i such that $[y_i2^i, (y_i+1)2^i) \in [a, b)$ and the maximum z_i such that $[z_i2^i, (z_i+1)2^i) \in [a, b)$. If every element of one canonical range belongs to another canonical range, then the small range should be deleted from the range set. The total number of canonical ranges inside the given query is not fixed since it depends on the interval's upper and lower bounds. Hence the search condition $\mathcal{Q}$ can be represented as CRQ = $\{crq_0, crq_1, \ldots\}$ where $crq_i \in \mathbb{N}_{2^{n+1}-2}$. To check whether t is a matching tuple, i.e., $t.A \in \mathcal{Q}$, we only need to compute

CRA $\wedge$ CRQ. More specifically, If CRA $\wedge$ CRQ $\neq \varnothing$, then $t.A \in \mathcal{Q}$ and t is a matching tuple; otherwise, $t.A \notin \mathcal{Q}$ and t is an unmatching tuple.

For example, assume $n = 4$ and $h : (i, x) \rightarrow \mathbb{N}_{2^5 - 2}$. Given the attribute value $t.A = 9$, the canonical ranges containing $t.A$ are $\{[9, 10), [8, 10), [8, 12), [8, 16)\}$ and CRA $= \{h(0, 9), h(1, 4), h(2, 2), h(3, 1)\}$. Given the range query $\mathcal{Q} = [6, 11)$, the total canonical ranges inside the interval are $\{[6, 7), [10, 11), [6, 8), [8, 10)\}$ and CRQ $= \{h(0, 10), h(1, 3), h(1, 4)\}$ after deleting $h(0, 6)$. Since CRA $\wedge$ CRQ $= \{h(1, 4)\}$, we can obtain that $t.A \in \mathcal{Q}$ and t is a matching tuple.

Polynomial Representation of Index/Trapdoor. In this paper, we convert the canonical range set CRA of attribute value $t.A$ and CRQ of range query $\mathcal{Q} = [a, b)$ into polynomials. Before converting, we first put the n levels into M buckets randomly where M is an factor of n and each bucket has $m = n/M$ levels. Levels in the ith bucket can be denoted as $B_i = \{l_{i,0}, l_{i,1}, \ldots, l_{i,m-1}\}$. Second, we construct a $m + 1$-degree polynomial, whose $m + 1$ roots are denoted as $R = \{a_0, \ldots a_m\}$:

$$\mathbf{P}_R(x) = (x - a_0)(x - a_1) \cdots (x - a_m) = \sum_{i=0}^{m+1} \alpha_i x^i \tag{1}$$

Then produce a m-variable polynomial based on $\mathbf{P}_R(x)$:

$$\mathbf{F}_R(x_0, x_1, \ldots, x_{m-1}) = \mathbf{P}_R(x_0)\mathbf{P}_R(x_1) \cdots \mathbf{P}_R(x_{m-1}) \tag{2}$$

Combine the terms of $\mathbf{F}_R(x_0, x_1, \ldots, x_{m-1})$ only if they have the exact same coefficient. Then we utilize the coefficients U_R to build indexes and the variable parts V_X to generate trapdoors, where $R = \{a_0, \ldots, a_m\}$ and $X = \{x_0, \ldots, x_{m-1}\}$. For instance, assume $m = 2$ and construct a 2-variable polynomial $\mathbf{F}_{(a_0, a_1, a_2)}(x_1, x_2)$ based on the 3-degree polynomial $\mathbf{P}_{(a_0, a_1, a_2)}(x)$. The coefficients and variables of each term in $\mathbf{F}_{(a_0, a_1, a_2)}(x_1, x_2)$ are shown in Table 3.

The detailed protocols to construct indexes and trapdoors are described as follows:

Index. For each bucket $i \in \mathbb{Z}_M$, we build a sub-index $\mathcal{I}_i$ for attribute value $t.A$. First generate a random integer $r_i \notin \mathbb{Z}_{2^{n+1}-1}$. The canonical ranges containing $t.A$ that fall into the ith bucket together with r_i are denoted as $CRA_i = \{cra_{i,0}, \ldots, cra_{i,m-1}, r_i\}$, where $level(cra_{i,j}) \in B_i$ and $B_i = \{l_{i,0}, l_{i,1}, \ldots, l_{i,m-1}\}$ stands for the levels in the ith bucket. Then use the coefficients U_{CRA_i} to represent the sub-index $\mathcal{I}_i$, where the root set $\{a_0, \ldots, a_m\}$ is replaced by $CRA_i = \{cra_{i,0}, \ldots, cra_{i,m-1}, r_i\}$.

Trapdoor. For each bucket $i \in \mathbb{Z}_M$, we build two sub-trapdoors $\mathcal{T}_{i,0}$ and $\mathcal{T}_{i,1}$ for the range query $\mathcal{Q} = [a, b)$. The canonical ranges inside $\mathcal{Q}$ that fall into the ith bucket are $CRQ_i = \{crq_{i,j}\}$ where $level(crq_{i,j}) \in B_i$. Then add zeros to CRQ_i such that $|CRQ_i| = 2m$ and split the set into two sets $CRQ_{i,0}$ and $CRQ_{i,1}$, where $|CRQ_{i,0}| = |CRQ_{i,1}| = m$. Then we use the variable parts $V_{CRQ_{i,0}}$ and $V_{CRQ_{i,1}}$ to construct two sub-trapdoors $\mathcal{T}_{i,0}$ and $\mathcal{T}_{i,1}$ respectively, where $\{x_0, \ldots, x_{m-1}\}$ is replaced by $\{crq_{i,0,0}, \ldots, crq_{i,0,m-1}\}$ and $\{crq_{i,1,0}, \ldots, crq_{i,1,m-1}\}$.

Table 3. The coefficients and variables of $\mathbf{F}_{(a_0,a_1,a_2)}(x_1,x_2)$

Coefficients	Variables
$a_0a_1 + a_0a_2 + a_1a_2$	$x_0^3x_1 + x_0x_1^3$
$(a_0a_1 + a_0a_2 + a_1a_2)^2$	x_0x_1
$-(a_0a_1a_2)$	$x_0^3 + x_1^3$
$(a_0a_1a_2)^2$	1
$-(a_0a_1 + a_0a_2 + a_1a_2)(a_0a_1a_2)$	$x_0 + x_1$
$(a_0 + a_1 + a_2)^2$	$x_0^2x_1^2$
$-(a_0 + a_1 + a_2)$	$x_0^3x_1^2 + x_0^2x_1^3$
1	$x_0^3x_1^3$
$-(a_0 + a_1 + a_2)(a_0a_1 + a_0a_2 + a_1a_2)$	$x_0^2x_1 + x_0x_1^2$
$a_0a_1a_2(a_0 + a_1 + a_2)$	$x_0^2 + x_1^2$

If there exists one canonical range cr satisfying $cr \in \mathrm{CRA}_i \wedge \mathrm{CRQ}_{i,k}$, and we assume $cr = cra_{i,0} = crq_{i,k,0}$. Since cr is a root of $\mathbf{P}_{\mathrm{CRA}_i}(x)$, i.e.,

$$\mathbf{P}_{\mathrm{CRA}_i}(cr) = 0 \tag{3}$$

then

$$\begin{aligned}
\mathcal{I}_i\mathcal{T}_{j,k} &= \mathbf{F}_{\mathrm{CRA}_i}(cr, crq_{i,k,1}, \ldots, crq_{i,k,m-1}) \\
&= \mathbf{P}_{\mathrm{CRA}_i}(cr)\mathbf{P}_{\mathrm{CRA}_i}(crq_{i,k,1})\cdots\mathbf{P}_{\mathrm{CRA}_i}(crq_{i,k,m-1}) \tag{4} \\
&= 0
\end{aligned}$$

The whole index $\mathcal{I}$ can be denoted as $\{\mathcal{I}_0, \ldots, \mathcal{I}_{M-1}\}$ while the whole trapdoor can be represented as $\{\mathcal{T}_{0,0}, \mathcal{T}_{0,1}, \ldots, \mathcal{T}_{M-1,0}, \mathcal{T}_{M-1,1}\}$.

4.2   Scheme Construction

In this subsection, we introduce the detailed process of our proposed scheme and pay more attention to the privacy mechanisms. In particular, we adopt the secure k-nearest neighbor (kNN) scheme proposed by Wong *et al.* [28] to encrypt the index $\mathcal{I}$ and trapdoor $\mathcal{T}$. The steps are explained as follows:

Setup. In this initialization phase, for each bucket $i \in \mathbb{Z}_M$, the data owner takes a security parameter λ_i and outputs the secret key SK_i, which includes: (1) a d-bit randomly generated vector S_i, (2) two invertible random matrices $M_{i,1}, M_{i,2} \in R^{d\times d}$, where d is the length of $\mathcal{I}_i$ and $\mathcal{T}_i$. Hence, SK_i can be denoted as a 3-tuple $\{S_i, M_{i,1}, M_{i,2}\}$.

BuildIndex. For each sub-index $\mathcal{I}_i$, the data owner encrypts the vector using the secret key SK_i. First, split $\mathcal{I}_i$ into two random vectors as $\{\mathcal{I}_i', \mathcal{I}_i''\}$ following the rule: for each element $\mathcal{I}_i[j], 0 \leqslant j \leqslant d - 1$, if the jth bit of S_i is 1, set $\mathcal{I}_i'[j] = \mathcal{I}_i''[j] = \mathcal{I}_i[j]$; if the jth bit of S_i is 0, $\mathcal{I}_i'[j]$ and $\mathcal{I}_i''[j]$ are set to two random numbers so that their sum is equal to $\mathcal{I}_i[j]$. Finally, the encrypted sub-index vector $\mathcal{I}_i^S$ is built as $\{M_{i,1}^T \mathcal{I}_i', M_{i,2}^T \mathcal{I}_i''\}$. The entire secure index $\mathcal{I}^S$ for attribute value $t.A$ is $\{\mathcal{I}_0^S, \ldots, \mathcal{I}_{M-1}^S\}$ and then the data user sends it to the cloud server.

BuildTrapdoor. For each sub-trapdoor $\mathcal{T}_{i,k}, k \in \{0, 1\}$, the data owner encrypts the vector using the secret key SK_i. First, split $\mathcal{T}_{i,k}$ into two random vectors as $\{\mathcal{T}_{i,k}', \mathcal{T}_{i,k}''\}$ following the rule: for each element $\mathcal{T}_{i,k}[j], \leqslant j \leqslant d - 1$, if the jth bit of S_i is 0, set $\mathcal{T}_{i,k}'[j] = \mathcal{T}_{i,k}''[j] = \mathcal{T}_{i,k}[j]$; if the jth bit of S_i is 1, $\mathcal{T}_{i,k}'[j]$ and $\mathcal{T}_{i,k}''[j]$ are set to two random numbers so that their sum is equal to $\mathcal{T}_{i,k}[j]$. Finally, the encrypted sub-trapdoor vector $\mathcal{T}_{i,k}^S$ is built as $\{M_{i,1}^{-1}\mathcal{T}_{i,k}', M_{i,2}^{-1}\mathcal{T}_{i,k}''\}$. The entire secure trapdoor $\mathcal{T}^S$ for the search condition is $\{\mathcal{T}_{0,0}^S, \mathcal{T}_{0,1}^S, \ldots, \mathcal{T}_{M-1,0}^S, \mathcal{T}_{M-1,1}^S\}$ and then the data user sends it to the cloud server.

RangeQuery. With the secure trapdoor $\mathcal{T}^S$, the cloud server computes the inner product of $\mathcal{I}_i^S$ and $\mathcal{T}_{i,k}^S$ for each sub-trapdoor and check whether the result is zero.

$$
\begin{aligned}
\mathcal{I}_i^S \mathcal{T}_{i,k}^S &= \{M_{i,1}^T \mathcal{I}_i', M_{i,2}^T \mathcal{I}_i''\} \cdot \{M_{i,1}^{-1}\mathcal{T}_{i,k}', M_{i,2}^{-1}\mathcal{T}_{i,k}''\} \\
&= \mathcal{I}_i' \cdot \mathcal{T}_{i,k}' + \mathcal{I}_i'' \cdot \mathcal{T}_{i,k}'' \\
&= \mathcal{I}_i \cdot \mathcal{T}_{i,k}
\end{aligned} \tag{5}
$$

If there is some $\mathcal{T}_{i,k}^S$ that satisfies $\mathcal{I}_i^S \cdot \mathcal{T}_{i,k}^S = 0$, then tuple t is a matching tuple and should be returned to the data user; otherwise, t is an unmatching tuple. Then check the next tuple.

Discussion. In above protocols, the number of sub-trapdoors of $\mathcal{T}^S$ is $2M$, so we have to compute $2M$ inner products for each tuple. To improve the performance, if the number of canonical ranges in the jth bucket is less than m, we can add zeros to the set so that $|\text{CRQ}_i| = m$. Then we only need to construct one sub-trapdoor according to bucket j and hence reduce the computation workload.

5   Security Analysis

In this section, we discuss the security issues of our proposed scheme under the HBC model and the known background model.

Index Confidentiality. When constructing the secure index $\mathcal{I}^S$ for attribute value $t.A$, we insert random integer r_i to each canonical range set CRA_i used to generate the sub-index $\mathcal{I}_i$. In addition, each sub-index is also encrypted by using the secret key SK_i. Thus, as long as SK_i is kept private by the data owner, the secure index $\mathcal{I}^S$ is a totally obfuscated vector and even two tuples have the same attribute value, their corresponding secure indexes are different. As a result, the cloud server can only use the secure index to check whether one tuple is matching without directly learning any additional information, such as the exact value and the relationship between two tuples. Then the confidentiality of index can be protected.

Trapdoor Confidentiality. When generating the secure trapdoor T^S for range query $Q = [a, b)$, each sub-trapdoor $T_{i,k}$ is encrypted by the secret key SK_i and then the secure trapdoor is indistinguishable from a random vector. Thus, as long as SK_i is kept private, the cloud server can't obtain the lower and upper bounds as well as the size of the query. In addition, the relationship between two secure trapdoors can't be determined. Then the confidentiality of trapdoor can be protected.

Query Processing Confidentiality. During the query processing, the cloud server only obtains whether a tuple matches the query and which secure sub-trapdoor is matched. Since there are m levels in a bucket, even two tuples both match the same sub-trapdoor, the cloud server can't determine whether they belong to the same canonical ranges. Thus the relationship between two tuples won't be disclosed during the processing. In addition, the more levels in one bucket, the more secure the scheme is. If there is only one level in one bucket, the cloud server can obtain that the matching tuples corresponding to the same sub-trapdoor are close, so m should satisfy $m \geqslant 2$.

Privacy Against Statistics Analysis. In addition to the privacy of index, trapdoor and query processing, we also consider several certain statistics analyses. Since the adversary model in this paper is honest-but-curious, the cloud server can store the search history of each tuple t such as $history_t = \{Tag_{time}, Tag_{subtd}\}$, where Tag_{time} represents the time when the range query is required and Tag_{subtd} denotes which sub-trapdoor is matched. As the history becomes longer, the cloud server may learn that $t.A = t'.A$ with high probability, if $history_t$ and $history_{t'}$ are the same. In addition, if the number of matching tuples corresponding to the secure sub-trapdoor T_i^S is small, the cloud server may infer that there may be only one small canonical range used to build T_i^S and the attribute values of these tuples are close.

To prevent these certain attacks, we can add dummy integers to the canonical range set used to build trapdoor, since we have inserted random numbers to the canonical ranges applied to construct index. In particular, if the canonical range set $\mathrm{CRQ}_{i,k}$ used to build the sub-trapdoor satisfies $\sum_{j=0}^{m-1} |crq_{i,k,j}| < 2^m$ where $|crq_{i,k,j}|$ represents the size of $crq_{i,k,j}$, then we use dummy integers r to replace one zero element in the set $\mathrm{CRQ}_{i,k}$. As a result, the tuple t whose sub-index I_i has been added r will be returned no matter whether its attribute value $t.A$ satisfies $t.A \in Q$. Because of the random numbers added to indexes, the similarity between two histories of tuples t and t' has been broken even they correspond to the same attribute value. The limitation of this method is that the data user has to decrypt all the tuples received from the server and filter the unmatching results. The number of unmatching data items in results depends on the total number of random integers that can be chosen.

6 Performance Evaluation

We implement our proposed scheme in JAVA on desktop PC running Windows 7 Professional with *3.4 GHz Intel(R) Core(TM) i7-3770* processor and 4 GB RAM inside. Each data point is averaged over 10 runs.

6.1 Evaluation of Index Construction

In this paper, the secure index generated for tuple t is $\mathcal{I}^S = \{\mathcal{I}_0^S, \ldots, \mathcal{I}_{M-1}^S\}$, where $M = n/m$ and n is the number of total levels while m is the number of levels in one bucket. The length of each sub-index $|\mathcal{I}_i^S|$ depends on m directly. Thus main impacts that influence the time for generating indexes include m and n. As shown in Fig. 2(a), given $n = 12$, we set $m = 3$ and $m = 4$ respectively. The time required for $m = 4$ is higher because $|\mathcal{I}_i^S| = 126 * 2$ when $m = 4$ while $|\mathcal{I}_i^S| = 35 * 2$ when $m = 3$. Figure 2(b) illustrates that the time is also evidently affected by the number of levels n. Given that each bucket has $m = 3$ levels, the time for constructing indexes increases as n becomes larger, since the number of sub-trapdoors increases. When the number of data items is fixed, the ratio of time required for $n_1 = 15$ to $n_2 = 12$ approximately equals to n_1/n_2. In addition, to ensure the index indistinguishability, our scheme constructs the secure index for each tuple and thus the total time is linearly affected by the total number of data items.

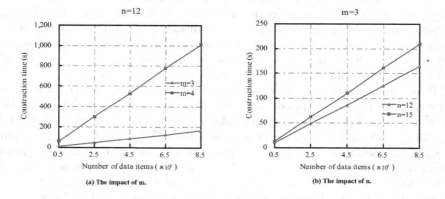

(a) The impact of m.   (b) The impact of n.

Fig. 2. Time for building indexes.

6.2 Evaluation of Query Processing

The time for query processing depends on the number of sub-trapdoors and the length of each sub-trapdoor. Figure 3(a) shows that if we put more levels in one bucket, the total time will increase since the length of each sub-trapdoor becomes longer. As illustrated in Fig. 3(b), as the number of sub-trapdoors in

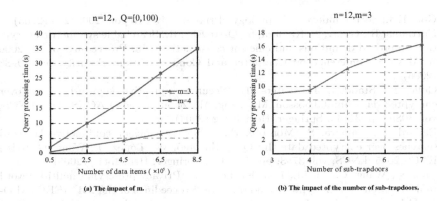

Fig. 3. Time for query processing.

a secure index grows, the total time will increase too. In addition, the time for query processing is also linearly related to the number of data items as shown in Fig. 3(a).

7 Conclusion

In this paper, we propose a privacy-enhancing scheme to realize range query processing over encrypted cloud databases. We first utilize the canonical ranges and polynomials to represent indexes and trapdoors, and then use kNN technique to encrypt them. Our scheme is secure under the widely adopted honest-but-curious model and the known background model. During the query processing, the data privacy can be protected even when the cloud server possesses the distribution of the attribute domain, since the indexes are indistinguishable and the trapdoors are unlinkable. In addition, by evaluating the performance of our scheme, we show that our system is also efficient.

References

1. Hacigumus, H., Iyer, B., Mehrotra, S.: Providing database as a service. In: Proceedings of the 2002 ICDE International Conference on Data Engineering, pp. 29–38 (2002)
2. Amazon Web Services. http://aws.amazon.com/running_databases/?nc2=h_ls
3. Hospital Network Hacked, 4.5 Million Records Stolen. http://money.cnn.com/2014/08/18/technology/security/hospital-chs-hack
4. Google Fires Engineer for Privacy Breach. http://edition.cnn.com/2010/TECH/web/09/15/google.privacy.firing
5. Yu, S., Wang, C., Ren, K., Lou, W.: Achieving secure, scalable, and fine-grained data access control in cloud computing. In: Proceedings of the 2010 INFOCOM International Conference on Computer Communications, pp. 1–9 (2010)
6. Song, D., Wagner, D., Perrig, A.: Practical techniques for searches on encrypted data. In: Proceedings of the 2000 IEEE Symposium on Security and Privacy, pp. 44–55 (2000)

7. Goh, E.J.: Secure indexes. Cryptology ePrint Archive: Report 2003/216 (2003)
8. Curtmola, R., Garay, J., Kamara, S., Ostrovsky, R.: Searchable symmetric encryption: improved definitions and efficient constructions. In: Proceedings of the 2006 ACM CCS Conference on Computer and Communications Security, pp. 79–88 (2006)
9. Golle, P., Staddon, J., Waters, B.: Secure conjunctive keyword search over encrypted data. In: Jakobsson, M., Yung, M., Zhou, J. (eds.) ACNS 2004. LNCS, vol. 3089, pp. 31–45. Springer, Heidelberg (2004)
10. Ballard, L., Kamara, S., Monrose, F.: Achieving efficient conjunctive keyword searches over encrypted data. In: Qing, S., Mao, W., López, J., Wang, G. (eds.) ICICS 2005. LNCS, vol. 3783, pp. 414–426. Springer, Heidelberg (2005)
11. Cao, N., Wang, C., Li, M., Ren, K., Lou, W.: Privacy-preserving multi-keyword ranked search over encrypted cloud data. In: Proceedings of the 2011 IEEE INFOCOM International Conference on Computer Communications, pp. 222–233 (2011)
12. Yu, J., Lu, P., Zhu, Y., Xue, G., Li, M.: Toward secure multi-keyword top-k retrieval over encrypted cloud data. In: Proceedings of the 2013 IEEE TDSC Transactions on Dependable and Secure Computing, pp. 239–250 (2013)
13. Li, J., Wang, Q., Wang, C., Cao, N., Ren, K., Lou, W.: Fuzzy keyword search over encrypted data in cloud computing. In: Proceedings of the 2010 IEEE INFOCOM International Conference on Computer Communications, pp. 1–5 (2010)
14. Chuah, M., Hu, W.: Privacy-aware bed-tree based solution for fuzzy multi-keyword search over encrypted data. In: Proceedings of the 2011 ICDCSW International Conference on Distributed Computing Systems Workshops, pp. 273–281 (2011)
15. Kuzu, M., Islam, M.S., Kantarcioglu, M.: Efficient similarity search over encrypted data. In: Proceedings of the 2012 IEEE ICDE International Conference on Data Engineering, pp. 1156–1167 (2012)
16. Wang, B., Yu, S., Lou, W., Hou, Y.T.: Privacy-preserving multi-keyword fuzzy search over encrypted data in the cloud. In: Proceedings of the 2014 IEEE INFOCOM International Conference on Computer Communications, pp. 2112–2120 (2014)
17. Hacigumus, H., Iyer, B., Li, C., Mehrotra, S.: Executing SQL over encrypted data in the database-service-provider model. In: Proceedings of the 2002 ACM SIGMOD Special Interest Group on Management of Data, pp. 216–227 (2002)
18. Hore, B., Mehrotra, S., Tsudik, G.: A privacy-preserving index for range queries. In: Proceedings of the 2004 VLDB International Conference on Very Large Data Bases, pp. 720–731 (2004)
19. Wang, J., Du, X.: LOB: bucket based index for range queries. In: Proceedings of the 2008 WAIM International Conference on Web-Age Information Management, pp. 86–92 (2008)
20. Agrawal, R., Kiernan, J., Srikant, R., Xu, Y.: Order preserving encryption for numeric data. In: Proceedings of the 2004 ACM SIGMOD Special Interest Group on Management of Data, pp. 563–574 (2004)
21. Boldyreva, A., Chenette, N., Lee, Y., O'Neill, A.: Order-preserving symmetric encryption. In: Joux, A. (ed.) EUROCRYPT 2009. LNCS, vol. 5479, pp. 224–241. Springer, Heidelberg (2009)
22. Damiani, E., Vimercati, S.D.C., Jajodia, S., Paraboschi, S., Samarati, P.: Balancing confidentiality and efficiency in untrusted relational DBMSs. In: Proceedings of the 2003 ACM CCS Conference on Computer and Communications Security, pp. 93–102 (2003)

23. Lu, Y.: Privacy-preserving logarithmic-time search on encrypted data in cloud. In: Proceedings of the 2012 NDSS Symposium Network and Distributed System Security Symposium (2012)

24. Pang, H., Zhang, J., Mouratidis, K.: Enhancing access privacy of range retrievals over B+-trees. IEEE TKDE Trans. Knowl. Data Eng. **25**, 1533–1547 (2013)

25. Li, J., Omiecinski, E.R.: Efficiency and security trade-off in supporting range queries on encrypted databases. In: Jajodia, S., Wijesekera, D. (eds.) Data and Applications Security 2005. LNCS, vol. 3654, pp. 69–83. Springer, Heidelberg (2005)

26. Li, R., Liu, A.X., Wang, A.L., Bruhadeshwar, B.: Fast range query processing with strong privacy protection for cloud computing. In: Proceedings of the 2014 VLDB International Conference on Very Large Data Bases, pp. 1953–1964 (2014)

27. Canetti, R., Feige, U., Goldreich, O., Naor, M.: Adaptively secure multi-party computation. In: Proceedings of the 2001 Annual ACM Symposium on Theory of Computing, pp. 639–648 (2001)

28. Wong, W.K., Cheung, D.W.L., Kao, B., Mamoulis, N.: Secure kNN computation on encrypted databases. In: Proceedings of the 2009 ACM SIGMOD International Conference on Management of data, pp. 139–152 (2009)

29. Pappas, V., Krell, F., Vo, B., Kolesnikov, V., Malkin, T., Choi, S.G., et al.: Blind Seer: a scalable private DBMS. In: Proceedings of the 2014 IEEE Symposium on Security and Privacy (SP), pp. 359–374 (2014)

Multi-Window Based Ensemble Learning for Classification of Imbalanced Streaming Data

Ye Wang[1,2(✉)], Hu Li[2], Hua Wang[1], Bin Zhou[2,3],
and Yanchun Zhang[1]

[1] Centre for Applied Informatics, Victoria University,
Melbourne, VIC, Australia
ye.wang10@live.vu.edu.au,
{hua.wang,yanchun.zhang}@vu.edu.au
[2] College of Computer, National University of Defense Technology,
Changsha 410073, China
{lihu,binzhou}@nudt.edu.cn
[3] State Key Laboratory of High Performance Computing,
National University of Defense Technology, Changsha 410073, China

Abstract. Imbalanced streaming data is widely existed in real world and has attracted much attention in recent years. Most studies focus on either imbalance data or streaming data; however, both imbalance data and streaming data are always accompanied in practice. In this paper, we propose a multi-window based ensemble learning (MWEL as short) method for the classification of imbalanced streaming data. Three types of windows are defined to store the current batch of instances, the latest minority instances and the ensemble classifier. The ensemble classifier consists of a set of latest sub-classifiers, and instances each sub-classifier trained on respectively. All sub-classifiers are weighted before predicting new arriving instance's class labels and new sub-classifiers are trained if a precision is below a threshold. Extensive experiments on synthetic datasets and real world datasets demonstrate that the new approach can efficiently and efficiently classify imbalanced streaming data and outperform existing approaches.

Keywords: Streaming data · Class imbalance · Multi-window · Ensemble learning

1 Introduction

As one of the most important problems in data mining and machine learning area, classification has attracted much attention in recent years. Traditionally, a classifier is supposed to be trained on a static dataset, which means the dataset don't changed over time. Therefore, the dataset is considered to have all the information needed to learn underlying concepts. However, in many real world scenarios, including spam filtering [1], credit card fraud identification [2], intrusion detection [3] and webpage classification [4], datasets are not static. New instances may arrive one by one (incremental) or

© Springer International Publishing Switzerland 2015
J. Wang et al. (Eds.): WISE 2015,Part II,LNCS 9419, pp. 78–92, 2015.
DOI: 10.1007/978-3-319-26187-4_6

batch by batch (batch) in a very high speed. In this case, the incoming instances need to be classified within a finite time and space. No matter new instances arrive incremental or in batch, only data before time step t can be used to trained the classifier and predict the instances arrive at time step $t + 1$. In other words, the classifier can only be trained on part of the information other than all the information.

The problem of class imbalance is very common in both static datasets and streaming datasets. For instance, in spam filtering, the amount of spam is usually much less than the amount of normal mails; the fraud is usually the minority comparing with normal customers in credit card fraud identification, and intrusions are not common compared with normal actions. And thus, class imbalance has become a vital problem that cannot be ignored when deal with real-word problems. Since we are more interested in rare class instances, usually, we take minority instances or rare instances as positive instances and majority instances as negative instances.

In the early stage, classification on streaming data and imbalance problems were studied separately. But in recent years, more and more attentions were paid to tackle these two problems together. However, as mentioned in [5], most researches combined algorithms designed for stream and imbalance data in a simple way, and few works took an insight look into these two problems together. In this paper, we analyze this problem in a novel way and propose a method using multi-window ensemble learning for classification of imbalanced streaming data. Main contributions of this paper are as follows:

- A multi-window framework is proposed to record the current batch of instances, selected positive instances and the ensemble classifier along with corresponding instances each sub-classifier trained on. The framework enables us to accumulate the latest positive instances, enhance the weight of the positive instances accordingly. Furthermore, concept drift in stream can be detected by the error rate together with similarity between current window and history windows each sub-classifier trained on.
- A novel ensemble learning mechanism is designed to classify the incoming instances. The weight of each sub-classifier is determined by the classification error rate and the window similarity. New instances are classified using weighted majority voting rule. Adjusting weight according to error rate can improve the classification accuracy and similarity-based weight adjustment can solve the reoccurring concept drift issue [5] to some extent.
- Extensive experiments on both synthetic datasets and real world datasets in different application domains are carried out, by which we get optimal parameters on each datasets and demonstrate that the proposed approach outperforms other alternatives.

To simplify the model, we only focus on single-label two-class classification problem, which means each instance can only belong to one class and there are only two classes in all. Single-label two-class classification tasks exist in many real-word applications, for example, emails can be classified into spam and normal; credit card user can be labeled as fraud and regular customer and the internet behavior can be divided into intrusion and normal. As to multi-label classification problem, it can be divided into several single-label two-class classification problems [6]. Multi-class problems can also be divided and conquered [7–9]. In addition, we mainly focus on low dimension problems, as high dimension can always be reduced [10].

The rest of the paper is organized as follows. Related works of imbalanced stream data classification are presented in Sect. 2. Section 3 proposes a multi-window based ensemble learning approach and describes it in details. The experiments on both real world and synthetic datasets, and the discussion of the results are detailed in Sect. 4, followed by a conclusion and outlook in Sect. 5.

2 Related Work

Classification on streaming data has attracted much attention for a long period and has been widely studied, especially problems related to concept drift. Meanwhile, many researchers focused on imbalanced data classification problem. In recent years, an increasing number of scholars paid their attentions on problems with both class imbalance and concept drift. We briefly review related works in the following.

Boundary Definition [11] was proposed to build classifier based on boundary instances which are easier to be misclassified. The approach divides the majority instances into correctly classified set and misclassified set. Random under-sampling is done on each set separately to guarantee the distribution consistency. Eleftherios et al. suggested keeping two windows to record the positive and negative classes separately for multi-label stream classification [12]. A learning framework for online imbalanced classification problem, including an imbalance detector, a concept shift detector and an online learner, was proposed in [13]. They also proposed so called OOB (Oversampling-based Online Bagging) and UOB (Undersampling-based Online Bagging) to improve classification accuracy. In their work, they simplified the concept drift to the change of imbalance ratio among classes. Thereafter, the authors proposed SOB (sampling-based Online Bagging) [14], which aim at maximizing the G-mean and balanced classes by adjusting the parameter λ of Poisson distribution in Onling Bagging. Recently, A selective re-train approach based on clustering was proposed in [15]. In this work, a new sub-classifier is trained when new data arrives, and if the overall classification performance is unsatisfied, the worst one of the existing sub-classifier will be replaced. Since each time one new instances arrives, a new classifier should be trained, it is of high computation complex.

3 Method

3.1 Problem Definition

We assume that at time step t the existing data instances forms a data sequence $D = \{(X_0, l_0), (X_1, l_1), \ldots, (X_t, l_t)\}$ in chronological order, where X_i is a d-dimensional feature vector and corresponds to a class label l_i. All class labels constitute a label sequence $L = \{l_1, l_2, \ldots, l_t\}$. In two-class classification task, $l_i \in \{+, -\}$, '+' and '−' represent positive (minority) and negative (majority) class respectively. Our goal is to train a single classifier or a set of classifiers on existing instances so that the classifier(s) can be used to predicate the incoming instance X_{t+1} at time step $t+1$. And once X_{t+1} is predicated, we are aware of the true label of X_{t+1}, i.e. l_{t+1}. In addition, we take the dataset as imbalanced if the ratio of the mount of different class instances, i.e. $\#\{(X_i, l_i)| l_i = `+\'\}/\#\{(X_j, l_j)|l_j = `-\'\}$, $i,j = 0, 1, \ldots, t \ldots$, exceeds a predefined threshold. The task is to build classifier(s) on imbalanced steam data and get acceptable results.

In this paper, we adopt a windowing or batching approach, where the window size is pre-calculated and then fixed. The window move forward once there is window size instances arrived. Here, the window on stream is no-overlapping and is in chronological order. Suppose the window size is M_b, then at time step t, existing instances are divided into $\lceil t/M_b \rceil$ windows or batches. Especially, when $M_b = 1$, the batching approach turn to incremental learning, which means processing one instance at a time.

3.2 Multi-Window Mechanism

In this section, we describe our algorithm of multi-window ensemble learning in detail. First of all, we have to determine the size of sliding window *WB*. A too small window cannot represent the class characters and may result in a classifier with poor generalization. On the other side, too large size will cause much more time and space overhead, which is restricted in practical application. For instance, a window may not be fulfilled with instances in a limited time while the prediction is expected to be done right now. There is scant theoretical guide for determining the optimal window size but through experiments.

Under imbalanced environment, the positive instances are sparse or even there is no positive instance in some sliding windows at all, and the classifiers trained on this kind of windows may not be able to represent the positive class. Therefore, we use a minority window *WM* to store those newly incoming minority or positive instances. The minority window was also utilized in [11] to keep all arrived minority instances. But it is space and time consuming, moreover, long-ago instances maybe meaningless in concept drift environment. We adopt a strategy similar to [16] which fix the minority window size and add minority instances that near to current positive instances into minority window. However, it is time consuming to select the nearest instances when the window size is very large and the nearest one may not be of same concept. Consequently, we choose a simple substitution policy here with fixed size, but add minority instances into *WM* before reach its upper size limit; otherwise, the oldest one will be replaced by the newest one. Thus, the minority window always represents the up to date positive instances over time.

Moreover, we use a classifier window *WC* to preserve a certain number of new trained sub-classifiers and its corresponding weights WC_{weight}, as well as windows of instances WC_{ins} on which each sub-classifier trained on. The size of *WC* is determined in advance through experiments and fixed in upcoming stream instances. New trained sub-classifiers are added to *WC* before reach the predefined size; otherwise, the oldest one will be replaced.

3.3 Prerequisite to Update *WC*

As accuracy is not a reliable metric for imbalanced data, for example, if there are 99 normal emails and 1 spam in the dataset, the classifier can treats them all as normal emails and get an accuracy of 99 % easily, which misses the minority class at all. So, it is necessary to evaluate the classification result on each class at same time in imbalances environment. In this paper, we use the precision both on minority and majority class to evaluate the classification performance. If both the precision of majority class $Precision_{maj}$ and precision of minority class $Precision_{min}$ of the new trained classifier are greater than 0.5, which means better than random guess, we then add it into *WC*. In addition, the corresponding weight WC_{weight} is set as current accuracy and current window on which new sub-classifier trained is saved as WC_{ins}. Otherwise, the classifier is discarded.

3.4 Algorithm

The main procedure of Multi-Window Ensemble Learning (MWEL) is shown in Algorithm 1. When the first window arriving, *WC* and *WM* are initialized empty and the first classifier is trained on current window. Then, it is decided whether to add the classifier to the classifier window *WC* or not by metric described in Sect. 3.3. At the time, we update *WM* according to rules described in Sect. 3.2.

When the next window arrives, we need to update the weights of sub-classifiers to fit the new instances. We firstly compute and normalize the similarity between current window and each of the existing windows which corresponds to existing sub-classifiers to modify the weights based on the following observation.

OBSERVATION 1: The greater the similarity between window *W*1 and window *W*2 is, the closer the concept between *W*1 and *W*2. Therefore, those sub-classifiers trained on more similar windows to current window should be given bigger weights.

$$WC_{weight \cdot i} = WC_{weight \cdot i}/(1 - sim(WB, WC_{ins \cdot i}))i = 1, 2, \ldots, M_c \qquad (1)$$

Where $WC_{weight \cdot i}$ is the weight of the ith sub-classifier in WC_i. *WB* is the current sliding window. $WC_{ins \cdot i}$ is the window of instances corresponding to the ith sub-classifier in WC_i. $sim(WB, WC_{ins \cdot i})$ is the similarity between two windows of instances. Moreover, it can be used to deal with the reoccurring concept drift [5]. Because when the same concept reoccurred over time, the corresponding existing classifier apparently can helps to get better result.

For each instance in *WB*, we use a majority weighted voting method to predicate the class label.

$$l'_{newIns} = \sum_{i=1}^{|WC|} WC_{weight \cdot i} \cdot WC_i(newIns) \qquad (2)$$

$WC_i(newIns)$ denotes classifying instance *newIns* with the ith sub-classifier in *WC*.

Algorithm 1. Multi-Window Ensemble Learning

Input: ordered instances $D = \{(X_0, l_0), (X_1, l_1), ..., (X_t, l_t)\}$

        Real class labels $L = \{l_0, l_1, ..., l_t\}$

        Maximum size of batch M_b

        Maximum size of classifier ensemble M_c

        Maximum size of minority window M_m

Output: Window of Minority instances WM

        Window of Classifiers WC

        Window of Weight of Classifiers WC_{weight}

        Window of instances each classifier trained on WC_{ins}

        Predicted labels $L' = \{l'_0, l'_1, ..., l'_t\}$

Initialize: Window of batch of instances $WB = \{\}$

        Window of Minority instances $WM = \{\}$

        Window of Classifiers $WC = \{\}$

        $WC_{weight \cdot i} = 1,\ i = 0, 1, ..., M_c$

1: FOR each ordered instance in D

2:    $WB \leftarrow getBatchOfInstances(M_b)$

3:    IF $|WC| = 0$ THEN

4:       $c \leftarrow trainClassifier(WB)$

5:       $\{precision_{min}, precision_{maj}, errorRate\} \leftarrow classify(WB, c)$

6:       IF $precision_{min} > 0.5\ \&\&\ precision_{maj} > 0.5$ THEN

7:           $updateClassifierWindow(WC, c, WB, errorRate)$

8:    ELSE

9:          $WB_s \leftarrow sample(WB)$

10:         $WB_s \leftarrow WB_s \cup WM$

11:         $c' \leftarrow trainClassifier(WB_s)$

12:         $\{precision_{min}, precision_{maj}, errorRate\} \leftarrow classify(WB, c')$

13:         IF $precision_{min} > 0.5\ \&\&\ precision_{maj} > 0.5$ THEN

14:            $updateClassifierWindow(WC, c', WB, errorRate)$

15:         END IF

16:    END IF

17:     ELSE
18:         FOR each window instances in WC_{ins}
19:             $sim \leftarrow getSimScore(WB, WC_{ins \cdot i})$
20:             $WC_{weight \cdot i} \leftarrow WC_{weight \cdot i} / (1 - sim)$
21:         END FOR
22:         $\{precision_{min}, precision_{maj}, errorRate\} \leftarrow classify(WB, WC)$
23:         IF $precision_{min} < 0.5 \parallel precision_{maj} < 0.5$ THEN
24:             $WB_s \leftarrow sample(WB)$
25:             $WB_s \leftarrow WB_s \cup WM$
26:             $c' \leftarrow trainClassifier(WB_s)$
27:             $\{precision_{min}, precision_{maj}, errorRate\} \leftarrow classify(WB, c')$
28:             IF $precision_{min} > 0.5 \,\&\&\, precision_{maj} > 0.5$ THEN
29:                 $updateClassifierWindow(WC, c')$
30:             END IF
31:         END IF
32:     END IF
33:     $WM \leftarrow updateMinorityWindow(WM, WB)$
34:END FOR

Then we compare the predicated label with the true label of each instance in current window. If one or both of the $Precision_{maj}$ and $Precision_{min}$ is less than 0.5, instances in current sliding window will be resampled. The sampling procedure is showed in Algorithm 2. Those rightly classified positive instances remain unchanged because they are not likely to help much in increasing the precision, while wrongly classified positive instances should be oversampled to increase their weights when learned by new sub-classifier. Here, SMOTE [17] is used to oversample minority instances and random under sample method is applied on correctly classified majority instances to reduce the size of majority. After several iterations, the imbalance ratio will gradually decrease to the predefined threshold. Finally, correctly classified minority instances, wrongly classified majority instances, oversampled wrongly classified minority instances and under sampled correctly classified majority instances are combined together as the new training set.

Algorithm 2. sample

Input: batch of instances WB
        Imbalance rate IR

Output: sampled instances WB_s

1:   $\{correctClassifiedMinority, wrongClassifiedMinority,$
     $correctClassifiedMajority, wrongClassifiedMajority\} \leftarrow getSubset(WB)$

2: WHILE $|Minority| / |Majority| < IR$

3:      $sampledMinority \leftarrow SMOTE(wrongClassifiedMinority)$

4:      $sampledMajority \leftarrow randomUnderSample(correctClassifiedMajority)$

5: END WHILE

6:   $WB_s = correctClassifiedMinority \cup sampledMinority \cup$
     $wrongClassifiedMajority \cup sampledMajority$

Notice that the new classifier will get the greatest weight 1 by default on the basis of observation below.

OBSERVATION 2: Considering the influence of time factor, the latest classifier is more likely to accord with the concept of current window, which deserve a higher weight.

In general, the factors of time, similarity and sub-classifiers' precision are considered overall by our methods, which should be more reasonable and comprehensive.

As for ensemble classifier, it is very important to increase its diversity, since it is essential for its robustness and performance. However, if there is no or little concept drift occurring in current window, new sub-classifier build on it cannot bring diversity but result in a huge computational burden. Therefore, we believe there is no need to build new sub-classifier on each sliding window. Instead, our approach builds new sub-classifiers only when the precision of majority or minority is less than 0.5, which assures the diversity, as well as reduces the computation load.

4 Experiment

We evaluate our approach on both real world and synthetic datasets.

4.1 Datasets

As shown in Table 1, we conducted experiments on four kinds of real world datasets, the first three are public available on MOA datasets[1]:

---

[1] http://moa.cs.waikato.ac.nz/datasets/.

Table 1. Dataset statistics

ID	Name	#All instances	#Positive instances	#Negative instances	#Attributes	Positive index	Negative index	Imbalance rate
ele	Elec	45,312	19,237	26,075	8	1	2	1:1.35
for	Forest	286,048	2747	283,301	54	4	2	1:103
air	Airlines	539,383	240,264	299,119	7	2	1	1:1.24
th1	Thyroid1	6,832	166	6,666	21	1	3	1:40
th2	Thyroid2	7,034	368	6,666	21	2	3	1:18
gcd	GCD	100,000	24,652	75,348	20	2	1	1:3
scd	SCD	100,000	25,178	74,822	20	2	1	1:3
rcd	RCD	100,000	24,280	75,720	20	2	1	1:3

- ElecNormNew (*ele*) was collected from the Australian New South Wales Electricity Market to predicate the prices rise or fall. In which, prices are affected by demand and supply of the market and are set every five minutes.
- Forest Coverage Type (*for*) contains the forest cover type for 30*30 meter cells obtained from US Forest Service and 7 classes in original dataset. We extract the class 4 and 2 as minority and majority.
- Airlines (*air*) was used to predict whether the flight will delay or not.
- Thyroid Disease datasets[2] (*th1*, *th2*) was used to identify whether the patient have thyroid disease. To obtain imbalanced dataset, we select class 1 and class 3 to form the first dataset (*th1*), class 2 and class 3 to form the second dataset (*th2*).

Moreover, three types of synthetic datasets are generated by algorithm provided in MOA[3]:

- Gradual Concept Drift (*gcd*), in which concept drift starts from the 30,000[th] instance to 100,000[th] instance gradually.
- Sudden Concept Drift (*scd*) takes place in the 50,000[th] instance suddenly and the dataset keep the new concept until the 100,000[th] instance.
- Reoccuring Concept Drift (*rcd*) happened from the 30,000[th] instance and begins to shift back to the original concept around the 50,000[th] instance and last until the 100,000[th] instance (Fig. 1).

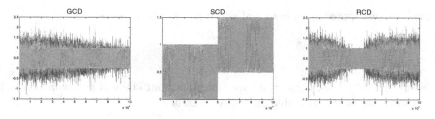

Fig. 1. Distribution of synthetic datasets

---

[2] https://archive.ics.uci.edu/ml/datasets/Thyroid+Disease.

[3] http://moa.cms.waikato.ac.nz/.

During the instance generation, we randomly remove some instances from one class to form imbalanced datasets.

4.2 Evaluation Criterion

We use accuracy, precision, recall, F1 and G-mean to evaluate our method in this paper. As explained in Sect. 3.3, accuracy may skew to negative class. While $G - mean = \sqrt{Recall_{min} \times Recall_{maj}}$ considers both classes' recall together and can only be high when both minority and majority have high recalls, so it is a better choice in imbalanced circumstance.

When evaluate the classification performance on stream data, we adopt the strategy mentioned in [11], which evaluate the classifier using the average performance over all of the batches in the data stream.

$$F = \frac{1}{\lceil t/M_b \rceil} \sum_{i=1}^{\lceil t/M_b \rceil} f_i \qquad \begin{array}{l} f, F \in \{Accuracy, Precision_{min}, Precision_{maj}, \\ Recall_{min}, Recall_{maj}, F1_{min}, F1_{maj}, G - mean\} \end{array} \qquad (3)$$

In which f is the indicator of each sliding windows and F is the average indicator of the data stream. We compared all the indicators in our experiments, but limited to the paper length, only results on accuracy, G-mean, recall of positive class and run time are show in the paper. In many real world applications, the result is expected in finite time, at least before next batch arriving. Therefore, the algorithm on stream needs a tradeoff between efficiency and effectiveness.

4.3 Sliding Window Size Setup

As discussed in Sect. 3.2, there is no widely accepted standard to set up the sliding window size with regard to different types of datasets. The larger window size means fewer amounts of windows to a specific length dataset, which result in reducing the frequency of classifier training, and contrarily, the smaller window size means more windows to a specific length dataset, increasing the frequency of classifier training. Moreover, smaller window size also means less training time of each classifier. Therefore, in this paper, experiments are conducted to find the optimal size for different datasets.

Figure 2 shows how sliding window size affects the classification results. It can be seen there is great difference on optimal windows size of different dataset with various indicators. For instance, in Fig. 2(a), datasets *for*, *th1* and *th2* are of quite high accuracy when the window size is 500 and remain stable relatively on datasets *for* and *th1*, but fall slowly on *th2* when the size growing.

However, in Fig. 2(b), maximum G-mean of dataset *for* is at window size 1300, while 100 and 200 of *th1* and *th2*. By comparing the recall of minority in Fig. 2(c), we realize that as the window size increasing, the recall of minority of *th1* and *th2*, which lack of positive instances, declined sharply which lead to decrease on G-mean. The

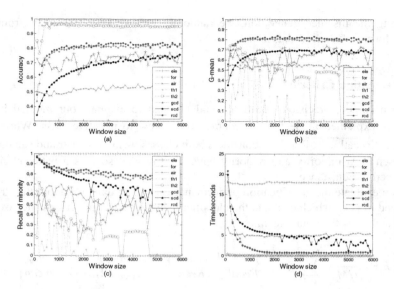

Fig. 2. Classification performance affected by sliding window size

bigger the window, the sparser the density of the minority instance. Furthermore, we notice that a bigger window cause a shorter training and classifying time firstly, but after the size larger than 1500, the time cost shows the trend of slow growth in Fig. 2(d) on most datasets except *for* and *scd*. In case of fixed data stream length, running time is the product of two parts. One is the training and classifying time of each window, and another is the number of windows. The early reduction is due to the decrease of window number, however, longer processing time on each window results in the general rise in later stages.

4.4 Minority Window Size Setup

We examine the influence of minority window size on classification performance and the result is show in Fig. 3. We observe that besides *th1*, *th2*, *for* and *ele,* the accuracy of other datasets decrease when the window size of minority increases in Fig. 3(a). While in Fig. 3(b), G-mean in other datasets rise quickly and reach the highest value around 400 except *for* and *ele*. Since the minority window is designed to improve the probability of identifying positive class in imbalanced context, the results in Fig. 3(c) demonstrate that it is helpful to raise the recall of minority. However, the *for* and *ele* keep nearly unchanged all the time, because minority instances within these two datasets are distributed more evenly than other datasets. In addition, *ele* dataset is of low imbalance rate (1:1.35) and thus with subtle influence.

From Fig. 3(d), we can find while the minority window size increase, the running time increases inevitably. The running time of *th1* and *th2* remain nearly unchanged because the total amounts of minority instances within these two datasets are only 166 and 368.

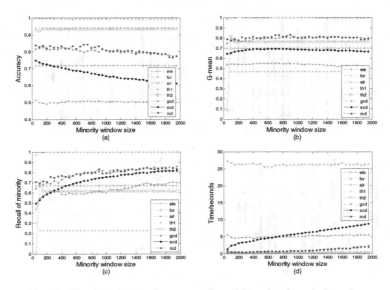

Fig. 3. Classification performance affected by minority window size

4.5 Comparing with Existing Methods

This subsection compares the proposed methods with two existing approaches, named BD [11] and CS [15]. The authors all claimed that their approaches are better than others, but these two methods have not be compared directly with each other under the same evaluation metrics. In this paper, we compare the performance of our method MWEL (MW) to BD and CS, and the result is shown in Fig. 4.

The comparison of accuracy in Fig. 4(a) indicates that on most of datasets, MW is better than CS but comparable to BD. Since MW pay more attention on minority instances, and thus impact the precision of majority, finally affect the accuracy. At the same time, we observe in Fig. 4(b) that the G-mean of MW is close to BD and outperforms CS on all datasets. When analysis G-mean and recall of minority together, we notice that on *th1* and *th2*, BD has obvious advantages in recall of minority than MW while with lower G-mean. The reason is that BD accumulates all positive instances to build successor sub-classifiers and thus get better recall of minority. However, too much minority instances may overwhelm the majority and thus suffer from a low G-mean value. For CS, though the accuracy on some datasets is very high, the G-mean and minority recall of some corresponding datasets present the opposite trend, since many positive instances were misclassified. In addition, too much minority instances also means more running time as showed in Fig. 4(d). MW is much more efficient than BD and CS except for dataset *ele* which is nearly balanced. Actually, run time of BD on air is more than twenty hours but we set it to 250 s for display.

Besides the visual diagrams comparison in Fig. 4, we also use Wilcoxon signed rank test to compare the statistical differences among MW, BD and CS, and the corresponding p-value is shown in Table 2. There is not statistically significant difference among three methods in accuracy with a significance level $\alpha = 0.05$. However,

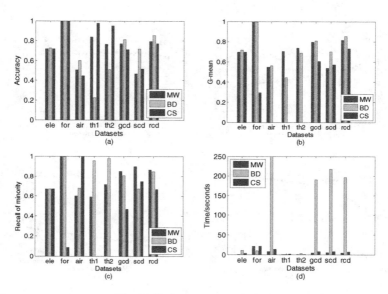

Fig. 4. Performance comparison

Table 2. Wilcoxon signed rank test statistics

	p-value			
	Accuracy	G-mean	Recall	Time
MW vs. BD	0.74218750	0.74218750	0.84375000	0.07812500
MW vs. CS	0.64062500	0.03906250	0.07812500	0.01562500
BD vs. CS	0.74218750	0.00781250	0.14843750	0.07812500

for G-mean and recall of minority, MW is better than CS. Although BD achieves higher G-mean and recall of minority than MW in some datasets, the test results show that there is no significant with $\alpha = 0.05$. As for running time, MW performs better than BD at significance level $\alpha = 0.05$ and when $\alpha = 0.10$, the running time of MW outperforms both BD and CS.

5   Conclusions

For the problem of classification on imbalanced stream data, we propose a multi-window based ensemble learning (MWEL) framework to predicate the class label of new arrival instances. We utilize multiple windows to preserve the current data batch, selected positive instances and set of latest sub-classifiers as well as the corresponding sets of instances each sub-classifier trained on. Moreover, before predicting the label of incoming instances, we update the weight of each sub-classifier by calculating the similarity between current window and previous windows each sub-classifier trained on. And then, a weighed majority voting strategy is used to predict the class label.

New sub-classifier will be trained only when current ensemble classifier is of low precision on both classes. When imbalance problem exists, we oversample minority instances and under sample majority instances at the same time. Extensive experiments on both real datasets and synthetic datasets demonstrate that our method can deal with imbalanced stream data efficiently and effectively and outperform existed methods in some extent, especially in running time. Considering the problem complexity, we only involve two classes in the paper. However, many applications in real world are of multi-class or multi-label, so imbalanced multi-class and multi-label stream classification is our future direction.

Acknowledgements. This work was supported by ARC DP project (DP 130101327), 973 Program (Grant No. 2013CB329601, 2013CB329602, 2013CB329604), NSFC (Grant No. 60933005, 91124002), 863 Program (Grant No. 2012AA01A401, 2012AA01A402), National Key Technology R&D Program (Grant No. 2012BAH38B04, 2012BAH38B06).

References

1. Delany, S.J., Cunningham, P., Tsymbal, A., Coyle, L.: A case-based technique for tracking concept drift in spam filtering. In: Macintosh, A., Ellis, R., Allen, T. (eds.) Applications and Innovations in Intelligent Systems XII, pp. 3–16. Springer, London (2005)
2. Wang, H., Fan, W., Yu, P.S., Han, J.: Mining concept-drifting data streams using ensemble classifiers. In: Proceedings of the Ninth ACM SIGKDD International Conference on Knowledge Discovery and Data Mining, pp. 226–235. ACM, New York, NY, USA (2003)
3. Parveen, P., Weger, Z.R., Thuraisingham, B., Hamlen, K., Khan, L.: Supervised learning for insider threat detection using stream mining. In: Proceedings of the 2011 IEEE 23rd International Conference on Tools with Artificial Intelligence, pp. 1032–1039. IEEE Computer Society, Washington, DC, USA (2011)
4. Wang, X., Jia, Y., Chen, R., Fan, H., Zhou, B.: Improving text categorization with semantic knowledge in Wikipedia. IEICE Trans. Inf. Syst. **E96-D**, 2786–2794 (2013)
5. Hoens, T.R., Polikar, R., Chawla, N.V.: Learning from streaming data with concept drift and imbalance: an overview. Prog. Artif. Intell. **1**, 89–101 (2012)
6. Shen, X., Boutell, M., Luo, J., Brown, C.: Multilabel machine learning and its application to semantic scene classification. Presented at the storage and retrieval methods and applications for multimedia 2004, 1 December 2003
7. Liu, W., Wang, L., Yi, M.: Simple-random-sampling-based multiclass text classification algorithm. Sci. World J. **2014**, 1–7 (2014)
8. Rifkin, R., Klautau, A.: In defense of one-vs-all classification. J. Mach. Learn. Res. **5**, 101–141 (2004)
9. Domingos, P., Hulten, G.: Mining high-speed data streams. In: Proceedings of the Sixth ACM SIGKDD International Conference on Knowledge Discovery and Data Mining, pp. 71–80. ACM, New York, NY, USA (2000)
10. Shi, J., Luo, Z.: Nonlinear dimensionality reduction of gene expression data for visualization and clustering analysis of cancer tissue samples. Comput. Biol. Med. **40**, 723–732 (2010)
11. Lichtenwalter, R.N., Chawla, N.V.: Adaptive methods for classification in arbitrarily imbalanced and drifting data streams. In: Theeramunkong, T., Nattee, C., Adeodato, P.J.L., Chawla, N., Christen, P., Lenca, P., Poon, J., Williams, G. (eds.) New Frontiers in Applied Data Mining. LNCS, vol. 5669, pp. 53–75. Springer, Heidelberg (2010)

12. Xioufis, E.S., Spiliopoulou, M., Tsoumakas, G., Vlahavas, I.: Dealing with concept drift and class imbalance in multi-label stream classification. In: Proceedings of the Twenty-Second International Joint Conference on Artificial Intelligence, vol. 2, pp. 1583–1588. AAAI Press, Barcelona, Catalonia, Spain (2011)

13. Wang, S., Minku, L.L., Yao, X.: A learning framework for online class imbalance learning. In: 2013 IEEE Symposium on Computational Intelligence and Ensemble Learning (CIEL), pp. 36–45 (2013)

14. Wang, S., Minku, L.L., Yao, X.: Online class imbalance learning and its applications in fault detection. Int. J. Comput. Intell. Appl. 12, 1340001 (2013)

15. Zhang, D., Shen, H., Hui, T., Li, Y., Wu, J., Sang, Y.: A selectively re-train approach based on clustering to classify concept-drifting data streams with skewed distribution. In: Tseng, V.S., Ho, T.B., Zhou, Z.-H., Chen, A.L.P., Kao, H.-Y. (eds.) PAKDD 2014, Part II. LNCS, vol. 8444, pp. 413–424. Springer, Heidelberg (2014)

16. Chen, S., He, H.: Towards incremental learning of nonstationary imbalanced data stream: a multiple selectively recursive approach. Evol. Syst. 2, 35–50 (2010)

17. Chawla, N.V., Bowyer, K.W., Hall, L.O., Kegelmeyer, W.P.: SMOTE: synthetic minority over-sampling technique. J. Artif. Int. Res. 16, 321–357 (2002)

A Dynamic Key Length Based Approach for Real-Time Security Verification of Big Sensing Data Stream

Deepak Puthal[1($\boxtimes$)], Surya Nepal[2], Rajiv Ranjan[2], and Jinjun Chen[1]

[1] Faculty of Engineering and IT,
University of Technology Sydney, Ultimo, Australia
{deepak.puthal,jinjun.chen}@gmail.com
[2] Digital Productivity Flagship, CSIRO, Canberra, Australia
{Surya.Nepal,Rajiv.Ranjan}@csiro.au

Abstract. The near real-time processing of continuous data flows in large scale sensor networks has many applications in risk-critical domains ranging from emergency management to industrial control systems. The problem is how to ensure end-to-end security (e.g., integrity, and authenticity) of such data stream for risk-critical applications. We refer this as an *online security verification* problem. Existing security techniques cannot deal with this problem because they were not designed to deal with high volume, high velocity data in real-time. Furthermore, they are inefficient as they introduce a significant buffering delay during security verification, resulting in a requirement of large buffer size for the stream processing server. To address this problem, we propose a Dynamic Key Length Based Security Framework (DLSeF) based on the shared key derived from synchronized prime numbers; the key is dynamically updated in short intervals to thwart Man in the Middle and other Network attacks. Theoretical analyses and experimental results of DLSeF framework show that it can significantly improve the efficiency of processing stream data by reducing the security verification time without compromising the security.

Keywords: Security · Sensor networks · Big data stream · Key exchange

1 Introduction

A variety of applications, such as emergency management, SCADA, remote health monitoring, telecommunication fraud detection, and large scale sensor networks, require real-time processing of data stream, where the traditional store-and-process method falls short of addressing the challenge [1]. These applications have been characterized to produce high speed, real-time, sensitive and large volume data input, and therefore require a new paradigm of data processing. The data in these applications falls in the big data category, as its size is beyond the ability of typical database software tools and applications to capture, store, manage and analyze in real time while ensuring end-to-end security [6]. More formally, the characteristics of big data are defined by "4Vs" [10, 11] such as Volume, Velocity, Variety, and Veracity. The

© Springer International Publishing Switzerland 2015
J. Wang et al. (Eds.): WISE 2015,Part II,LNCS 9419, pp. 93–108, 2015.
DOI: 10.1007/978-3-319-26187-4_7

streaming data from sensing source meets these characteristics. Our focus in this paper is thus on secure processing of high volume, high velocity data stream.

Big data stream is continuous in nature and it is critical to perform real-time analysis as: (i) the life time of the data is often very short (i.e., the data can be accessed only once) [2, 3] and (ii) the data is utilized for detecting events (e.g., flooding of highways, collapse of railway bridge, etc.) in real-time in many risk-critical applications (e.g., emergency management). Since big data stream has high volume and velocity, it is not economically viable to store and then process (as done in the batch computing model). Hence, stream processing engines (e.g. Spark, Storm, S4) have evolved in the recent past that have capability to undertake real-time big data processing. Stream processing engines offer two significant advantages: firstly, they circumvent the need to store large volume of data and secondly, they enable real-time computation over data as needed by emerging applications such as emergency management and industrial control systems. Further, integration of stream processing engines with elastic cloud computing resources has further revolutionized the big data stream computation as the stream processing engines can now be easily scaled [2, 5] in response to changing volume and velocity.

Though, the stream data processing has been studied in the past several years within the database research community, the focus has been on query processing [13], distribution [14] and data integration. Data security related issues, however, have been largely ignored. Since many emerging risk-critical applications, as discussed above, need to process big streaming data while ensuring end-to-end security. For example, consider emergency management applications that collect soil, weather, and water data through field sensors. Data from these sensors are processed in real-time to detect emergency events such as sudden flooding and landslides on railways and highways. In these applications, compromised data can lead to wrong decision making and in some cases even loss of lives and critical public infrastructure. Hence, the problem is how to ensure end-to-end security (i.e., integrity, and authenticity) of such data stream in near real-time processing. We refer this as an *online security verification* problem.

The problem in processing big data becomes extremely challenging when millions of small sensors in self-organizing wireless networks are streaming data through intermediaries to the data stream manager. In these cases, intermediaries as well as the sensors are prone to different kinds of security attacks such as Man in the Middle Attack. In addition, these sensors have limited processing power, storage, and energy; hence, there is a requirement to develop light-weight security verification schemes. Furthermore, data streams need to be processed on-the-fly in a correct sequence. In this paper, we address these issues by designing an efficient approach for online security verification of big data streams.

The most common approach for ensuring data security is to apply the cryptographic methods. In the literature, there are two most common types of cryptographic encryption methods: asymmetric and symmetric key encryption. Asymmetric key encryption methods (e.g., RSA, ElGamal, DSS, YAK, Rabin, etc.) perform a number of exponential operations over a large finite field. Therefore, they are 1000 times slower than the symmetric key cryptography [15, 16]. Hence, efficiency become an issue if asymmetric key such as the Public Key Infrastructure PKI [18] is applied to end-to-end security of big data streams. Thus, the symmetric key encryption is the most efficient cryptographic solution for such applications. However, existing symmetric key methods (e.g., DES, AES, IDEA, RC$_4$, etc.) fail to meet the requirements of real time security verification. Hence, there is a need to develop an efficient and scalable approach for performing the online security verification of big data stream. The main contributions of the paper can be summarized as follows:

- We have designed and developed a Dynamic Key Length Based Secure Framework (DLSeF) to provide end-to-end security for big data stream processing. Our approach is based on a common shared key that is generated by exploiting synchronize prime number. The proposed method avoids excessive communication between data sources and Data Stream Manager (DSM) for the rekey process. Hence, this leads to reduction in the overall communication overhead. Due to the reduced communication overhead, our approach is able to do security verification on-the-fly (with minimum delay) with minimal computational overhead.
- Our proposed approach adopts dynamic key length from the set of 128-bit, 64-bit, and 32-bit. This enables faster security verification at DSM without compromising the security. Hence, our approach is suitable to process high volume of data without any delay.
- We compare our proposed approach with the standard symmetric key solution (AES) in order to evaluate the relative computational efficiency. The results show that our approach performs better than the standard AES method.

The rest of this paper is organized as follows. Section 2 gives the background and defines the problem space. Section 3 describes our proposed solution, DLSeF. Section 4 presents the formal security analysis of our approach. Section 5 evaluates the performance and efficiency of the approach through extensive experiments. Section 6 concludes our work and points out to potential future directions.

2 Background and the Problem Definition

Data stream management system has been studied in the past several years with the main focus on query processing, data distribution, and data integrity. The security aspects have been largely overlooked. Nehme et al. [17] initially highlighted the need of security framework in streaming data. They divided the security problem into two: data security problem (also known as data security punctuation) and query security problem (also known as query security punctuation). Data security punctuation deals with the data security, whereas query security punctuation deals with the security and access control during the query processing. They extensively work on access control by focusing on both data security and query security punctuation in their papers [7, 17]. For example, FENCE, a continuous access control framework in dynamic data stream environments, deals with both data and query security restrictions [7]. It gives low overhead which is suitable for data stream environments. Similarly, ASSIST, an application system based effective and efficient access control framework, is proposed to protect streaming data from unauthorized access [8]. They have implemented ASSIST on top of StreamInsight, a commercial stream processing engine. In this paper, we are focusing on the data security punctuation, where our approach is to protect the data efficiently from potential attacks from/on untrusted intermediaries before the data reaches to the DSM.

Figure 1 shows an overall architecture for big data stream process from source sensing devices to the data processing center including our proposed security framework. We refer to [4] for further information on stream data processing in datacenter cloud. We refer [19] for cloud infrastructure and data processing in cloud. In sensor networks, data packets from the source are transmitted to the sink (data collector) through multiple intermediaries hops (e.g. routers and gateways).Collected data at sink node is forward to the DSM as data stream which may also pass through many untrusted intermediaries. The number of hops and intermediaries depend on the network architecture designed for a particular application. The intermediaries in the network may behave as a malicious attacker by modifying and/or dropping the data packets. Hence, the traditional communication security techniques [9, 12] are not sufficient to provide end-to-end security. In our framework both the queries and data security related techniques are handled by DSM in coordination with the on-field deployed sensors. It is important to note that the security verification of streaming data has to be performed before the query processing phase and in near real time (with minimal delay) with a fixed (small) buffer size. The processed data is stored in the big data storage system supported by cloud infrastructure. Queries used in DSM are defined as "continuous" since they are continuously applied to the streaming data. Results (e.g. significant events) are pushed to the application user each time the streaming data satisfies a predefined query predicate.

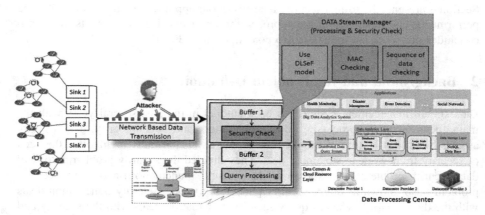

Fig. 1. High level of architecture from source sensing device to big data processing center.

The discussion in the above architecture clearly identifies the following most important features for security verification for big data stream processing. In summary, they include: (a) security verification needs to be performed in real time (on-the-fly), (b) framework has to deal with high volume of data at high velocity, (c) data items should be read once in the prescribed sequence, and (d) original data is not available for comparisons which is widely available in store-and-process batch processing paradigm.

These features need to be enabled by the big data stream processing framework in addition to meeting the end-to-end data security requirements.

Based on the above features of big data stream processing, we categorize existing data security methods into two classes: communication security [9, 12] and server side data security [26, 27]. Communication security deals with the data security between two nodes when it is in motion, and does not deals with the end-to-end security, server side data security approaches focus on ensuring the security of data when it is at rest in repository. They are not suitable to use in the big data stream. Furthermore, symmetric cryptographic based security solutions are either static shared key or centralized dynamic key. In static shared key, we need to have a long key to defend from potential attackers. It is well known that length of the key is always proportional to the security verification time and hence longer keys leading elevated computation time are not suitable for applications that need to do real-time processing over high volume, high velocity data. For the dynamic key, centralize processor rekey and distribute keys to all the sources; this is a time consuming process. Moreover, big data stream is always continuous in nature and impossible to put data in halt for rekeying process. To address this problem, we propose a distributed and scalable approach for big data stream security verification.

Our proposed approach works as follows: we use a common shared key for both sensors and DSM. The key is updated dynamically by generating synchronize relative prime numbers without having further communication between them after handshaking. This procedure reduces the communication overhead and increases the efficiency of the solution, without compromising the security. Due to the reduced communication overhead, our approach performs the security verification with minimum delay. Based on the shared key properties, individual source sensor updates their dynamic key and key length independently.

3 Proposed Approach

Our approach is motivated by the concept of moving target defense. The basic idea is that if we keep on moving the keys in spatial (dynamic key size) and temporal (same key size, but different key) dimensions, we can achieve the required efficiency without compromising the security. Our proposed approach, Dynamic Key Length Based Security Framework (DLSeF), provides the robust security by changing both key and key length dynamically. In our approach, if an intruder/attacker eventually hacks the key, he/she cannot predict the key or its length for the next session. We argue that it is very difficult for an intruder to guess the appropriate key and its length as our approach dynamically changes the both across the sessions. Similar to any secret key based symmetric key cryptography, our DLSeF approach consists of 4 independent components and related processes: system setup, handshaking, rekeying, and security verification. As stream processing is expected to be performed in near real time processing, we assume that data packets should not take more than few hours to reach DSM, as the end-to-end delay is an important QoS parameter to measure the performance of sensor networks [28]. Table 1 provides the notations used in modelling our approach. We next describe the approach.

Table 1. Notations used in our approach

Acronym	Description
S_i	i^{th} sensor's ID
K_i	i^{th} sensor's secret key
K_{si}	i^{th} sensor's session key
kl	Key length
$K_1/K_2/K_3/K_4$	Initial keys for authentication
K_{SH}	Secret shared key calculated by the sensor and DSM
K_{SH^-}	Previous secret shared key maintain at DSM
$P_1/P_2/P_3/P_4$	Communicated format during authentication
r	Random number generated by the sensors
t	Interval time to generate the prime number
P_i	Random prime number
K_d	Secret key of the DSM
I_D	Encrypted data for integrity check
A_D	Secret key for authenticity check
$E()$	Encryption function
$H()$	One-way hash function
$Prime(P_i)$	Random prime number generation function
$KeyGen$	Key generation procedure
Key-Length ()	Key length selection procedure
$\oplus$	Bitwise X-OR operation
$\parallel$	Concatenation operation
$DATA$	Fresh data at sensor before encryption

3.1 DLSeF System Setup

We have made a number of realistic and practical assumptions while designing and modelling our approach. First, we assume that DSM has all deployed sensor's identities (IDs) and secret keys because the network is untrusted. Sensors and DSM implement some common primitives such as hash function (H()), and common key ($K1$), which are executed during the initial identification and system setup steps.

The proposed authentication process includes five steps. The first three steps are for the sensors and DSM where they authenticate with each other and the next two steps are for the generating shared key. The shared key is utilized during the handshaking process.

Step 1: A sensor (S_i) generates a pseudorandom number r and encrypts it along with its own secret key K_i. The encryption process uses the common shared key ($K1$), which is initialized during the deployment. The output of encryption ($E_{K1}(r \parallel K_i)$) is denoted as $P1$. The output is then sent to DSM: $S_i \rightarrow$ **DSM:** $P1$

Step 2: Upon receiving the message, the DSM decrypts $P1$ (i.e. $D_{K1}(P1)$) and retrieves the corresponding source ID ($S_i \leftarrow retrieveKey(K_i)$). If the source sensor's ID matches with its own database, then the DSM computes the hash of the key to generate another key for encryption $K2 \leftarrow H(K1)$. The DSM then encrypts the pseudorandom number (r) with the newly generated key as $P_2 \leftarrow E_{K2}(r)$ and sends it to the sensor for DSM authentication as follows: $S_i \leftarrow$ **DSM:** P_2

Step 3:   Corresponding sensor receives the encrypted pseudorandom number and decrypts it to authenticate the DSM, i.e. $r' \leftarrow D_{K2}(P_2)$. It calculates the current secret shared key using the hash of existing shared key i.e. $K2 \leftarrow H(K1)$. If the received random number is the same as the sensor had before (i.e. $r = r'$), the sensor sends an acknowledgement (ACK) to DSM. The ACK is encrypted with the new key, which is computed using hash of the current key ($K3 \leftarrow H(K2)$). The encrypted ACK is denoted as $P_3 \leftarrow E_{K3}(ACK)$, and sends to DSM as follows: $S_i \rightarrow$ **DSM:** P_3

Step 4:   The DSM decrypts the ACK ($ACK \leftarrow D_{K3}(P_3)$) to confirm that the sensor is now ready to establish the session. The current secret key is updated using the hash of existing secret key i.e. $K3 \leftarrow H(K2)$. After the confirmation of ACK, the DSM generates a random session key i.e. $K_{si} \leftarrow randomKey()$ for handshaking. The generated session key (K_{si}) is encrypted with the hash of the current key e.g. ($K4 \leftarrow H(K3)$) and then sent to individual sensors as $S_i \rightarrow$ **DSM:** $\{P_4\}$, where $P_4 \leftarrow E_{K4}(K_{si})$.

Step 5:   The sensor decrypts P_4 and extracts the session key for handshaking ($K_{si} \leftarrow D_{K4}(P4)$). It follows the same procedure as before, i.e., the current shared key is updated with the hash value of existing shared key ($K4 \leftarrow H(K3)$). We update the shared key in every transaction to ensure the strength of security for handshaking.

3.2   DLSeF Handshaking

In the handshaking process, the DSM sends the key generation and synchronization properties to sensors based on their individual session key (K_{si}) established earlier. Generally, a larger prime number is used to strengthen security process. However, a larger prime number requires greater computation time. In order to make the rekeying process efficient (lighter and faster), we recommend reducing the prime number size. The challenge is how to maintain the security while avoiding large prime number size. We achieve this by dynamically changing the key size as described next.

The dynamic prime number generation function is defined in Algorithm II. We calculate the prime number and shared key on both sensing sources and DSM ends to reduce communication overhead and minimize the chances of disclosing the shared key. The computed shared keys have of multiple lengths (32 bit, 64 bit, and 128 bit) which are varied across the sessions. Initial key length is set to 64 bit and is dynamically updated as per the logic depicted in Algorithm I. After a certain time interval, the next shared key is generated by applying Algorithm II where the size is determined by Algorithm I as follows.

$Prime(P_i)$ periodically computes the relative prime number at both the sensor and DSM ends after a time interval t, which are updated based on function $KeyLength()$. The shared secret key (K_{SH}) generation process needs K_d, and P_i. In the handshaking process, DSM transmits all properties required to generate shared key to sensors $(K_d, t, P_i, Prime(), KeyLength(), K_{SH}, KeyGen)$ as follows: $S_i \leftarrow \textbf{DSM} : \{K_{si}(K_d, t, P_i, Prime(), KeyLength(), K_{SH}, KeyGen)\}$.

All of these above transferred information are stored in the trusted part of source for future rekeying process (e.g., TPM) [25]).

3.3   DLSeF Rekeying

Our proposed approach not only calculates the dynamic prime number to update the shared key without further communication after handshaking, but also proposes a novel way of dynamically changing key length at source and DSM following the steps described in Algorithm I. We change the key periodically in DLSeF Rekeying process to ensure that the protocol remains secured. If there are any types of key or data compromise at a source, the corresponding sensor is desynchronized with DSM instantly. Following that the source sensor need to reinitialize and synchronize with DSM as described above. We assume that the secret information is stored in the trusted part of the sensor (e.g. TPM) and it is sent by the sensor to DSM for synchronization. In some cases, data packet can arrive at DSM after shared key is updated. Such data packets are encrypted using previous shared key. We add a time stamp field to individual data packet to identify the encrypted shared key. If the data is encrypted using previous key then the DSM uses K_{SH^-} key for the security verification; otherwise, it follows the normal process.

The above defined *DLSeF Handshaking* process makes sensors aware about the *Prime (Pi)*, *KeyLength*, and *KeyGen*. We now describe the complete secure data transmission and verification process using those functions and keys. As mentioned above, our approach uses the synchronized dynamic prime number generation Prime *(Pi)* on both sides, i.e., sensors and DSM as shown in Fig. 1. At the end of the handshaking process, sensors have their own secret keys, initial prime number and initial shared key generated by the DSM. The next prime generation process is based on the current prime number and the time interval as described in Algorithm II. The prime number generation process (Algorithm II) always calls Algorithm I to fetch shared key length information and associated time interval. Sensors generate the shared key $K_{SH} = (E(P_i, K_d))$ using the prime number P_i, and DSM's secret key $(E(P_i, K_d))$. We use the secret key of DSM to improve the robustness of the security verification process. Each data block is associated with the authentication and integration tag and contains two different parts. One is encrypted DATA based on shared key K_{SH} for integrity checking (i.e., $I_D = DATA \oplus K_{SH}$), and the other part is for the authenticity checking (i.e., $A_D = S_i \oplus K_{SH}$). The resulting data block $((DATA \oplus K_{SH}) \| (S_i \oplus K_{SH}))$ is sent to DSM as follows: $Si \rightarrow \textbf{DSM:} \{(I_D\|A_D)\}$.

3.4   DLSeF Security Verification

In this step, the DSM first checks the authenticity in each individual data block A_D and then the integrity with randomly selected data blocks I_D. The random value is calculated based on the corresponding prime number *i.e.* $j = P_i\% 5$, when the key length is 32; $j = P_i\% 9$ when the key length is 64; and there is no integrity verification when the key length is 128. DSM also checks the time stamp of each individual data block to find the shared key used for encryption. For the authenticity check, the DSM decrypts A_D with shared key $S_i = A_D \oplus K_{SH}$. Once Si is obtained, the DSM checks its source database and extracts the corresponding secret key K_i ($K_i \leftarrow retrieveKey(S_i)$). In the integrity check process, the DSM decrypts the selected data such as $DATA = I_D \oplus K_{SH}$ to get the original data and checks MAC for the data integrity.

4 Security Analysis of DLSeF

In this section, we provide a theoretical analysis on our approach. We made the following assumptions: (a) any participant in our scheme cannot decrypt the data that was encrypted by DLSeF algorithm unless it has the shared key which was used to encrypt the data; (b) as DSM is located at the big data processing system side, we assume that DSM is fully trusted and no one can attack it; and (c) sensors' secret key, Prime (P_i) and secret key calculation procedures reside inside the trusted part of the sensor (such as the TPM) so that they are not accessible to the intruders.

Similar to most security analyses of communication protocols, we now define the attack models for the purpose of verifying the authenticity and integrity.

Definition 1 (attack on authentication). A malicious attacker M_a can attack on the authenticity if it is capable to monitor, intercept, and introduce itself as an authenticated source node to send data in the data stream.

Definition 2 (attack on integrity). A malicious attacker M_i can attack on the integrity of the data if it is an adversary capable to monitor the data stream regularly and try to access and modify the data blocks before it reaches to DSM.

Theorem 1: The data security of data streams is not compromised by changing the size of shared key (K_{SH}).

Proof: The dynamic prime number generation generates and updates the key on both source and DSM. The dynamic shared key length is either 32 bit or 64 bit or 128 bit. The ECRYPT II recommendations on key length say that a 128-bit symmetric key provides the same strength of protection as a 3,248-bit asymmetric key [16]. Even smaller symmetric key provides more security as it is never shared publicly. Advanced processor (*Intel i7 Processor*) took about 1.7 ns to try out one key from one block. With this speed it would take about *1.3 × 10^{12} × the age of the universe* to check all the keys from the possible key set [16]. By reducing the size of the prime number, we vary the key length to confuse the adversary, but achieve faster security verification at DSM using the data reported in Table 2. Further, Table 2 shows that 128 bit symmetric key takes 3136e +19 ns (more than a month), 64 bit symmetric key takes 3136e +19 ns (more than a week), and so on. We fixed the time interval (t) to generate prime number and updated the shared key as follows: $t = 720\ h$ for 128-bit key length, $t = 168\ h$ for 64-bit length, and $t = 20\ h$ for 32 bit length key (see Algorithm I). Dynamic shared key is computed based on the calculated prime number and associated properties initialized accordingly (See Algorithm II). Based on these calculation, we conclude that an attacker cannot intercept within the interval time t. The key has been already changed four times before an attacker knows the key and this knowledge is not known to the attackers. Data blocks arrived after 20 h are discarded as they might be compromised.

ALGORITHM I. SYNCHRONIZATION OF DYNAMIC KEY LENGTH GENERATION

Key-Length (x_{n-1})

1:   $x_{n-1} \leftarrow 64$ (For First Iteration)
2:   $x_n \leftarrow x_{n-1} + x_{n-1} \cos x_{n-1}$
3:   $i \leftarrow x_n \% 3$
4:   If $i = 0$ then
5:      Set $kl \leftarrow 128$
6:           $t \leftarrow 720$ hours (1 month)
7:           $j \leftarrow$ no checking
8:   Else If $i = 1$ then
9:      Set $kl \leftarrow 64$
10:          $t \leftarrow 168$ hours (1 week)
11:          $j \leftarrow P_i \% 9$
12: Else
13:     Set $kl \leftarrow 32$
14:          $t \leftarrow 20$ hours (1 day)
15:          $j \leftarrow P_i \% 5$
16: End If
17: End If
18: Return (x_n) // use to initialize x_{n-1} for next iteration.

Table 2. Time taken by symmetric key (AES) algorithm to get all possible keys using the most advanced Intel i7 processor.

Key Length	8	16	32	64	128
Key domain size	256	65536	4.295e + 09	1.845e + 19	3.4028e + 38
Time (in nanoseconds)	1435.2	1e + 05	7.301e + 09	3136e +19	5.7848e + 35

Theorem 2: Relative prime number P_i is calculated in Algorithm II is always synchronized between the source sensors (S_i) and DSM.

Proof: The normal method to check the prime number is $6k + 1$, $\forall k \in N^+$ (an integer). Here, we first initialize the value of k based on this primary test formula stated above. Our prime number generation method is based on the nth prime number generation concept and from the extended idea of [24]. In our approach, the input P_i is the currently used prime number (initialized by DSM) and the return P_i is the calculated new prime number. Intially P_i is intianized by DSM at DLSeF Handshaking process and the interval time is t (see Algorithm I).

By applying the *Algorithm II*, we calculate the new prime number P_i based on the previous one P_{i-1}. The complete process of the prime number calculation and generation is based on the value of *m, where m* is initialized from k. The value of k is kept constant at source because it is calculated from the current prime number. This is initialized during *DLSeF Handshaking*. Since k is constant the procedure *Prime* (P_i) returns identical values at both seniors and DSM. In Algorithm II, the value of $S(x)$ is computed as follows, if the computed value is 1 then x is a prime; otherwise it is not a prime.

ALGORITHM II. DYNAMIC PRIME NUMBER GENERATION

Prime (P_i)

1:   $P_{i-1} = P_i$
2:   Set $k := \left\lceil \frac{P_{i-1}}{6} \right\rceil$
3:   Set $m := 6k + 1$
4:   If $m \geq 10^7$ then
5:      $k := k/10^5$
6:      GO TO: 14
7:   If $S(m) = 1$ then
8:      GO TO: 13
9:   Set $m := 6k + 5$
10: If $S(m) = 1$ then
11:     GO TO: 13
12: $k := \lfloor k^3 + \sqrt{k} \rfloor \bmod 17 + k$
13: GO TO: 3
14: $P_i = m$
15: Return (P_i) // calculated new prime number

$$S_1(x) = \frac{(-1)}{\left\lfloor \frac{\sqrt{x}}{6} \right\rfloor + 1} \sum_{k=1}^{\left\lfloor \frac{\sqrt{x}}{6} \right\rfloor + 1} \left\lfloor \left\lfloor \frac{x}{6k+1} \right\rfloor - \frac{x}{6k+1} \right\rfloor,$$

$$S_2(x) = \frac{(-1)}{\left\lfloor \frac{\sqrt{x}}{6} \right\rfloor + 1} \sum_{k=1}^{\left\lfloor \frac{\sqrt{x}}{6} \right\rfloor + 1} \left\lfloor \left\lfloor \frac{x}{6k-1} \right\rfloor - \frac{x}{6k-1} \right\rfloor$$

$$S(x) = \frac{S_1(x) + S_2(x)}{2}$$

If $S(x) = 1$ then x is prime, otherwise x is not a prime.
$x \not\equiv 0 \bmod i \forall 1 \leq i \leq x - 1$, if x is prime.
Put the value of x as a prime number, then derivations as follows:

$$\Rightarrow \left\lfloor \left\lfloor \frac{x}{6k+1} \right\rfloor - \frac{x}{6k+1} \right\rfloor = -1$$

Same as $\left\lfloor \left\lfloor \frac{x}{6k-1} \right\rfloor - \frac{x}{6k-1} \right\rfloor = -1$

$\forall k$ within the specified range i.e. 10^7, then

$$S_1(x) = \frac{(-1)}{\left\lfloor \frac{\sqrt{x}}{6} \right\rfloor + 1} \sum_{k=1}^{\left\lfloor \frac{\sqrt{x}}{6} \right\rfloor + 1} (-1) = 1$$

Same $S_2(x)$ is also 1 and then $S(x) = \frac{S_1(x) + S_2(x)}{2} = 1$

Hence, the property of $S(x)$ is proved.

Theorem 3: An *attacker* M_a cannot read the secret information from a sensor node (S_i) or introduce itself as an authenticated node in DLSeF.

Proof: Following *Definition 1* and considering the computational hardness of secure module (such as TPM), we know that M_a cannot get the secret information for P_i generation, K_i and *KeyGen*. So there are no possibilities for the malicious node to trap sensor, but M_a can introduce him/herself as an authenticated node to send its information. In our approach, a sensor (S_i) sends $((I_D) \parallel (A_D))$, where the second part of the data block $(S_i \oplus K_{SH})$ is used for authentication check. DSM decrypts this part of the data block for authentication check. DSM retrieves S_i after decryption and matches corresponding S_i within its database. If the calculated S_i matches with the DSM database, it accepts; otherwise it rejects the node as source and it is not an authenticated sensor node. Hence, we conclude that an *attacker* M_a cannot attack on big data stream.

Theorem 4: An *attacker* M_i cannot read the shared key K_{SH} within the time interval t in DLSeF model.

Proof: Following *Definition 2*, we know that an attacker M_i has full access to the network to read the shared key K_{SH}, but M_i cannot get correct secret information such as K_{SH}. Considering the method described in *Theorem 1*, we know that M_i cannot get the currently used K_{SH} within the time interval t (see Table 2), because our proposed approach calculates P_i randomly after time t and then uses the value P_i to generate K_{SH} as described in *Theorems 1 and 2*.

5   Experimental Evaluation

The proposed DLSeF security approach though deployed in big sensor data stream in this paper is a generic approach and can be used in other application domains. In order to evaluate the efficiency and effectiveness of the proposed architecture and protocol, even under the adverse conditions, we experimented with two different approaches in two different simulation environments. We first verify the security approach using Scyther [22], and then measure the efficiency of the approach using JCE (Java Cryptographic Environment) [23].

5.1   Security Verification

The protocols in our proposed approach is written in Scyther simulation environment using Security Protocol Description Language (.spdl). According to the features of Scyther, we define the role of S and D, where S is the sender (i.e., sensor nodes) and D is the recipient (i.e., DSM). In our scenario, S and D have all the required information that are exchanged during the handshake process. This enables S and D to update their own shared key. S sends the data packets to D and D performs the security verification. In our simulation, we introduce two types of attacks by adversaries. In the first type of attacks, a malicious attacker change the data while it is being transmitted from S to D through intermediaries (integrity attack). In the second type of attacks (authentication attack), an adversary acquires the property of S and sends the data packets to D pretending that it is from S. We experimented with 100 runs for each claim and found out no attacks at D as shown in Fig. 2(a).

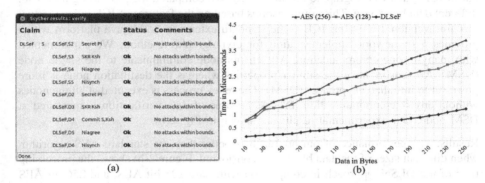

(a)   (b)

Fig. 2. (a) Scyther simulation result of successful security verification at DSM. (b) Performance of our approach compared in efficiency to 128 bit AES and 256 bit AES.

Experiment model: In practice, attacks may be more sophisticated and efficient than brute force attacks. However, this does not affect the validity of the proposed DLSeF approach as we are interested in efficient security verification without periodic key exchanges and successful attacks. Here, we model the process as described in the previous section and vary the key size between 32 bits, 64 bits, and 128 bits (see Table 2). We used Scyther, an automatic security protocols verification tool to verify our proposed model.

Results: We did our simulation using a different number of data blocks in each run. Our experiment ranged from 10 to 100 instances with 10 intervals. We checked authentication for each data block, whereas the integrity check is performed on the selected data blocks. As the key generation process is saved in the trusted part of the sensors, no one can get access to those information except the corresponding sensor. Hence, we did not find any authentication attacks. For integrity attacks, it is hard to get shared key (K_{SH}), as we are frequently changing the shared key (K_{SH}) and its length based on the dynamic prime number P_i on both source sensor (S_i) and DSM. In the experiment, we did not encounter any integrity attacks. Figure 2(a) shows the result of security verification experiments in Scyther environment. This shows that our approach is secured from integrity and authentication attacks.

5.2 Performance Comparison

Experiment model: It is clear that the actual efficiency improvement brought by our approach highly depends on the size of the key and rekeying without further communication between sensor and DSM. We have performed experiments with different size of data blocks. The results of our experiments are given below.

We compare the performance of our proposed approach DLSeF with advanced encryption standard (AES), the standard symmetric key encryption algorithm [20, 21]. Our approach is efficient compared with two standard symmetric key algorithm such as 128-bit AES and 256-bit AES. This performance comparison experiment was carried out in JCE (Java Cryptographic Environment). We compared the processing time with different data block size. This comparison is based on the features of JCE in java virtual machine version 1.6 64 bit. JCE is the standard extension to the java platform which provides a framework implementation for cryptographic method. We experimented with many-to-one communication. All sensors node communicate to the single node (DSM). All sensors have the similar properties whereas the destination node is more powerful to initialize the process (DSM). The rekey process is executed at all the nodes without any intercommunication. Processing time of data verification is measured at DSM node. Our experimental results are shown in Fig. 2(b).

Results: The performance of our approach is better than the standard AES algorithm when different sizes of the data blocks are considered. Figure 2(b) shows the processing time of the DLSeF approach in comparison with base 128-bit AES, and 256-bit AES for different size of the data blocks. The performance comparison shows that our proposed approach is efficient and faster than the baseline AES protocols.

From the above two experiments, we conclude that our proposed DLSeF approach is secured (from both authenticity and integrity attacks), and efficient (compare to standard symmetric algorithms such as 128-bit AES and 256-bit AES).

6 Conclusion and Future Works

In this paper, we have proposed a novel authenticated key exchange approach, namely Dynamic Key Length Based Security Framework (DLSeF), which aims to provide real-time security verification approach for big sensing data stream. Our approach has been designed based on the symmetric key cryptography, dynamic key length and dynamic prime number generation. We proved our proposed DLSeF approach significantly improve the security verification time without compromising the security by theoretical analysis and experimental evaluation. In this proof we shown that our approach prevent malicious attacks on authenticity and integrity. In our approach, dynamic key initialization at both sensor and DSM reduce the communication and computation overhead. We plan to pursue a number of research avenues in future to perform importance for better security of big data stream. We will further investigate the technique to develop a moving target defense strategy for the Internet of Things.

Acknowledgment. This research is funded by Australia India Strategic Research Grant titled "Innovative Solutions for Big Data and Disaster Management Applications on Clouds (AISRF - 08140)" from the Department of Industry, Australia.

References

1. Stonebraker, M., Çetintemel, U., Zdonik, S.: The 8 requirements of real-time stream processing. ACM SIGMOD Rec. **34**(4), 42–47 (2005)
2. Bifet, A.: Mining big data in real time. Informatica (Slovenia) **37**(1), 15–20 (2013)
3. Dayarathna, M., Suzumura, T.: Automatic optimization of stream programs via source program operator graph transformations. Distrib. Parallel Databases **31**(4), 543–599 (2013)
4. Ranjan, R.: Streaming big data processing in datacenter clouds. IEEE Cloud Comput. **1**(1), 78–83 (2014)
5. Tien, J.M.: Big data: unleashing information. J. Syst. Sci. Syst. Eng. **22**(2), 127–151 (2013)
6. Manyika, J., Chui, M., Brown, B., Bughin, J., Dobbs, R., Roxburgh, C., Byers, A.: Big data: the next frontier for innovation, competition, and productivity (2011)
7. Nehme, R.V., Lim, H.S., Bertino, E.: FENCE: continuous access control enforcement in dynamic data stream environments. In: Proceedings of the Third ACM Conference on Data and Application Security and Privacy, pp. 243–254 (2013)
8. Cao, J., Kister, T., Xiang, S., Malhotra, B., Tan, W.-J., Tan, K.-L., Bressan, S.: ASSIST: access controlled ship identification streams. In: Hameurlain, A., Küng, J., Wagner, R., Amann, B., Lamarre, P. (eds.) TLDKS XI. LNCS, vol. 8290, pp. 1–25. Springer, Heidelberg (2013)
9. Puthal, D.: Secure data collection and critical data transmission technique in mobile sink wireless sensor networks. M. Tech thesis, National Institute of Technology, Rourkela (2012)
10. Bahrami, M., Singhal, M.: The role of cloud computing architecture in big data. In: pedrycz, W., Chen, S.-M. (eds.) Information Granularity, Big Data, and Computational Intelligence, pp. 275–295. Springer, Switzerland (2015)
11. McAfee, A., Brynjolfsson, E.: Big data: the management revolution. Harvard Bus. Rev. **90**, 59–69 (2012)
12. Puthal, D., Sahoo, B.: Secure Data Collection and Critical Data Transmission in Mobile Sink WSN: Secure and Energy Efficient Data Collection Technique. LAP Lambert Academic Publishing, Germany (2012)
13. Deshpande, A., Ives, Z., Raman, V.: Adaptive query processing. Found. Trends Databases **1**(1), 1–140 (2007)
14. Sutherland, T.M., Liu, B., Jbantova, M., Rundensteiner, E.A.: D-cape: distributed and self-tuned continuous query processing. In: Proceedings of the 14th ACM International Conference on Information and Knowledge Management, pp. 217–218. ACM (2005)
15. Burke, J., McDonald, J., Austin, T.: Architectural support for fast symmetric-key cryptography. ACM SIGOPS Operating Syst. Rev. **34**(5), 178–189 (2000)
16. www.cloudflare.com. Accessed 04 August 2014
17. Nehme, R.V., Lim, H.S., Bertino, E., Rundensteiner, E.A.: StreamShield: a stream-centric approach towards security and privacy in data stream environments. In: Proceedings of the ACM SIGMOD International Conference on Management of data, pp. 1027–1030. ACM (2009)
18. Park, K.W., Lim, S.S., Park, K.H.: Computationally efficient PKI-based single sign-on protocol, PKASSO for mobile devices. IEEE Trans. Comput. **57**(6), 821–834 (2008)

19. Puthal, D., Sahoo, B.P.S., Mishra, S., Swain, S.: Cloud computing features, issues, and challenges: a big picture. In: International Conference on Computational Intelligence and Networks (CINE), pp. 116–123. IEEE (2015)
20. Pub, N.F.: 197: advanced encryption standard (AES). Fed. Inf. Process. Stand. Publ. **197**, 441-0311 (2001)
21. Heron, S.: Advanced encryption standard (AES). Netw. Secur. **2009**(12), 8–12 (2009)
22. Scyther. http://www.cs.ox.ac.uk/people/cas.cremers/scyther/
23. Pistoia, M., Nagaratnam, N., Koved, L., Nadalin, A.: Enterprise Java 2 Security: Building Secure and Robust J2EE Applications. Addison Wesley Longman Publishing, Inc., Chicago (2004)
24. Kaddoura, I., Abdul-Nabi, S.: On formula to compute primes and the nth prime. Appl. Math. Sci. **6**(76), 3751–3757 (2012)
25. Nepal, S., Zic, J., Liu, D., Jang, J.: A mobile and portable trusted computing platform. EURASIP J. Wirel. Commun. Netw. **2011**(1), 1–19 (2011)
26. Zissis, D., Lekkas, D.: Addressing cloud computing security issues. Future Gener. Comput. Syst. **28**(3), 583–592 (2012)
27. Liu, C., Zhang, X., Yang, C., Chen, J.: CCBKE—session key negotiation for fast and secure scheduling of scientific applications in cloud computing. Future Gener. Comput. Syst. **29**(5), 1300–1308 (2013)
28. Akkaya, K., Younis, M.: An energy-aware QoS routing protocol for wireless sensor networks. In: 23rd International Conference on Distributed Computing Systems, pp. 710–715 (2003)

Detecting Internet Hidden Paid Posters
Based on Group and Individual Characteristics

Xiang Wang[1]([✉]), Bin Zhou[1,2], Yan Jia[1,2], and Shasha Li[1]

[1] School of Computer, National University of Defense Technology, Changsha, China
{xiangwangcn,binzhou,yanjia,shashali}@nudt.edu.cn
[2] State Key Laboratory of High Performance Computing,
National University of Defense Technology, Changsha 410073, China

Abstract. Online social networks are popular communication tools for billions of users. Unfortunately, they are also effective tools for hidden paid posters (or Internet water army in some literatures) to propagate spam or mendacious messages. Paid posters are typically organized in groups to post with specific purposes and have flooded the communities of microblogging websites. Typical traditional methods only utilize individual characteristics in detecting them. In this paper, we study the group characteristics of paid posters and find that group characteristics are also very important in detecting them comparing to individual characteristics. We construct a classifier based on both the individual and group characteristics to detect paid posters. Extensive experiments show that our method is better than existing methods.

Keywords: Paid posters · Internet water army · Microblogging · Social network

1 Introduction

Nowadays, social networks like Twitter, SINA Weibo and Facebook are becoming popular information sources for billions of people. Due to the ease of forwarding messages, information can disseminate to a large number of interested people via their social network. For example, if a user posts a tweet in Twitter, all its followers can read the tweet immediately. Some users like famous people even have millions of followers. As one of the major social networks, microblogging differs from a traditional blog and allows users to exchange small elements of content such as short sentences, individual images, or video links. There are several famous microblogging platforms like Twitter, SINA Weibo and Yammer. Users can post about topics ranging from the daily chats to the thematic like national policies. The microblogging platforms have significantly influenced people's daily life and brought considerable opportunities to business.

Paid posters [8] are typical employed to promulgate spam or mendacious information to increase normal users' awareness of their targets in a campaign. If there are large number of mendacious messages, normal users can not know the

© Springer International Publishing Switzerland 2015
J. Wang et al. (Eds.): WISE 2015, Part II, LNCS 9419, pp. 109–123, 2015.
DOI: 10.1007/978-3-319-26187-4_8

actual state of affairs. This maybe lead some bad consequences if large number of normal users are misled. For example, the CEO of Smartisan Technology corporation named Luo Yong-hao announced that their new product "Smartisan T1" was attacked by paid posters and publicly offered a reword of CNY 200,000 to find the paid posters[1]. There are many slanderous comments for their new product and the sales of the product are decreased. To reduce the negative effect, it's crucial for us to detect paid posters.

Paid posters are different from traditional spammers. First, paid posters are typical a group of users with group characteristics while spammers (except opinion spam) are usually considered to be individuals. Second, some paid poster groups go far away than posting spam message, the behaviors of them sometimes can hurt others. Third, paid posters are either controlled by a program through platform API or human beings. They are different from Twitter bot [9] which is a program used to produce automated posts or to automatically follow Twitter users. As they can also be human beings which are more covert and complex than Twitter bot. Fourth, paid posters are more covert than spammers. They are normal users at ordinary times, but they become paid posters when they try to promote a campaign. Even Some famous users with high influence can be paid to be paid posters temporarily when they are needed in a promoting campaign. Opinion spam is a kind of paid posters [16,20], but existing researches focused on detect them in electronic-commerce websites like Amason.com and hotel booking website TripAdvisor, rather than social network platforms like microblogging websites.

Spammers have been appearing in many applications like blogs [17,24], email [2,6], Web search engine [12] and videos websites [3,5]. There are a large amount of methods which have been proposed to detect them [11,15] in these platforms. They mainly employ individual statistical characteristics for detecting spam. Our study finds that group characteristics are also important for detecting paid posters. Traditional methods which only utilize individual statistical characteristics are not good enough for detecting paid posters. For example, in a promotional campaign to promote an URL, many paid posters retweet the advertising tweet to their communities to make large number of users see it. Typically most of them do not follow the author of the advertising tweet, so it is important to use group characteristic "retweeting without following" to detect paid posters.

In this paper, we study six group characteristics for detecting paid posters. The group characteristics are discussed in Sect. 3. Some individual characteristics used in traditional spam detecting methods are also utilized in our method. User influence which is calculated by users' multi-relational networks [10] is employed to detect paid posters. We employ the SVM model to combine the individual characteristics and group characteristics to detect paid posters. Experimental results on three real datasets show that our method is better than existing methods.

The main contributions of this paper can be summarized as follows:

---

[1] http://digi.163.com/14/0919/15/A6H2KS8H00162OUT.html.

- We study several useful group characteristics for detecting paid posters and find that group characteristics are also very important in detecting paid posters comparing to traditional individual features.
- We propose a method named "IGCSVM" combining both user's individual and group characteristics for detecting paid posters.
- Extensive experiments have been done on three real datasets of SINA Weibo. Experimental results show that our IGCSVM method is more effective than existing methods in detecting paid posters.

The rest of this paper is organized as follows: Sect. 2 discusses some important related works. Section 3 introduces our method for detecting paid posters. Experimental results are shown in Sect. 4. Finally, conclusion and future work are provided in Sect. 5.

2   Related Works

Spammers have been appearing in a lot of applications, such as blogs [17,24], email [2,6], Web search engine [12] and videos [3,5]. And there are a large amount of methods which have been proposed to detect them [11,15]. Zhang et al. [26] analyze the characteristics of the spam users in two campaigns in Twitter. They explored the mention network to find the characteristics of outdegree and indegree, neighborhood connectivity and burstiness in order to find their relationships with spam users. They also analysis the online social network to get the features of followers/friends and response time. They try to find useful features for spam detection. They also investigate the benefit-cost analysis of spammers based on epidemic model. Yang et al. [27] presented a case study of analyzing inner social relationships of criminal users and proposed a new algorithm named Mr.SPA to detect users that have close relationship with criminal users. They also designed an algorithm named CIA to detect more criminal users based on a seed set by analyzing the social and semantic relationships among users. Gao et al. [13] proposed a method to detect malicious users and posts based on URL and text clustering. They also analysis the characteristics of the malicious users and posts. Thomas et al. [23] characterized the behaviors of 1.1 million spammers on Twitter by analyzing the text of the tweets sent by the suspended users. They also found there was a market providing spam users services. They also explored five spam campaigns and find the tools employed by spammers and the approaches they used in spam activities. Lee et al. [18] analyzed the profile features of spammers and developed a classifier to classify spam users to different categories: promoters, legitimate users and so on. Grier et al. [14] studied spam on Twitter and found that clickthrough rate of spam URLs was much lower than email. The analyze also showed that 84 % spam users are organized by few number of controllers. M. McCord and M. Chuah [19] studied user based and content based features and find that they are different between spammers and legitimate users. They also utilize the features for detecting spammers. Chu et al. [9] build a classifier to determine an account to be a human, bot or cyborg.

There are also some researches about paid posters. Opinion spam is a kind of paid poster. Jindal and Liu [16] found that opinion spam is widespread and in electronic-commerce websites. They trained their models using features like review text, reviewer and product to detect duplicate opinions in Amazon.com. Ott et al. [20] proposed a n-gram based text categorization to detect deceptive opinion spam in hotel booking website TripAdvisor. Chen et al. [8] investigated the behavioral pattern of paid posters and designed a detection mechanism to identify potential paid posters based on user comments in social network. We utilize not only user comments but also user posts, user social friendships and group characteristics for detecting paid posters in this paper. Wang et al. [25] studied five features for detecting paid posters. Zeng et al. [28] investigated the behavior patterns of paid posters in online forums.

3   Detecting Organized Posters

3.1   Typical Organization Structure

To promote a campaign, the organizers of the campaign will typically employ three teams working for them: resource team, poster team and observation and evaluation team. The typical organization structure for paid posters is shown in Fig. 1. The organizers ask the resource team to prepare content of tweets for posting. The content can be not only text content, but also image, audio and even video. There are writers, graphic designers, video makers and so on in the resource team. Poster team is responsible for publishing the content manufactured by the resource team in popular websites like SINA Weibo. The observation and evaluation team is responsible for observing and evaluating the effect of the whole promoting activities and competitors' activities.

Fig. 1. Typical structure for the paid posters

The poster team mainly comes from two sources. First, some companies and organizations control large number of paid posters directly. These paid posters either controlled through open API of the platforms such as SINA Weibo Open Platform or employees in the company or organization. Second, some paid posters come from temporary recruitment. There are some platforms for hiring part-time posters, such as Shuijunwang.com and 51shuijun.net. A company or organization can quickly employ a large number of paid posters from these platforms. The paid posters can be hired to attract public attention to their targets, enhance the strength of their viewpoints to something or even perturb public perspective. Some messages we see sometimes can not be trustworthy due to many rumors posted by them.

3.2   Framework for Detecting Organized Posters

Our framework for detecting paid posters is shown in Fig. 2 based on the individual and group characteristics using SVM model (IGCSVM). Given a user, we first study its individual statistical characteristics and group characteristics. The four individual characteristics and six group characteristics form a 10-dimensional vector. The features in the 10-dimensional vector are normalized to be between 0 and 1. Then we build a classification model from the training dataset to classify a user to be a paid poster or a legitimate user. A record in the training data is represented as the 10-dimensional vector and a class label (1 or −1). Class label 1 represents user u to be a paid poster and −1 represents it to be a legitimate user.

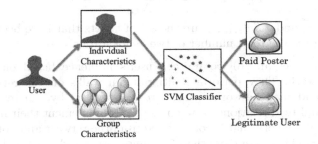

Fig. 2. Framework for detecting paid posters

Individual Statistical Characteristics. The four individual statistical characteristics are discussed in this section.

The Ratio of Friends to Followers. Some paid posters are not likely to be followed by normal users since they always do not post high quality contents. So they can not get many followers. The ratio of friends to followers (RFF) of a paid poster is always larger than normal users. We define the ratio of friends to followers P_{RFF} in Eq. 1.

$$P_{RFF} = \frac{N_{FR}}{N_{FR} + N_{FO}} \tag{1}$$

where N_{FR} is the number of friends of a user and N_{FO} is the number of followers of the user.

The Ratio of Tweets that Contain URLs to User's All tweets. Since the length of a tweet is not allowed to exceed 140 characters in microblogging websites like Twitter and SINA Weibo, there is always an URL in paid posters' tweets to promote a campaign. The ratio of tweets that contain URLs to user's all tweets (URL) for paid posters is probably higher than normal users. Equation 2 is defined to be the ratio of tweets that contain URLs to all tweets P_{URL}.

$$P_{URL} = \frac{N_{URL}}{N_{All}} \tag{2}$$

where N_{URL} is the number of tweets that contain URLs and N_{All} is the total number of tweets of a user.

The Ratio of Replied/Retweeted Tweets to User's All Tweets. Paid posters' tweets are less likely to be replied or retweeted than normal users' tweets. The first reason is that paid posters tend to post low quality tweets. The second one is that there are probably fewer normal users following them. Then the ratio of replied/retweeted tweets to user's all tweets (RRE) can be used distinguish paid posters and normal users. Equation 3 shows how to calculate the ratio of replied/retweeted tweets to user's all tweets P_{RRE}.

$$P_{RRE} = \frac{|TSet_{reply} \cup TSet_{retweet}|}{N_{All}} \tag{3}$$

where $TSet_{reply}$ and $TSet_{retweet}$ are the set of tweets that have been replied or retweeted. N_{All} is the total number of tweets for a user.

Influence. Ding et al. [10] compute a user's influence based on the multi-relational network. They perform multi random walks on the retweet, reply, reintroduce, and read networks which are constructed by the retweet, reply, reintroduce, and read relations between users. We implement their method on a multi-relational network that is constructed from the retweet and notify (@user-name) relations. There are more than 30 million users and a parallel distributed framework MapReduce[2] is used to compute the influence of users on a Hadoop[3] cluster which contains 32 nodes. The influence of a user (IN) $P_{IN}(0 \leq P_{IN} \leq 1)$ is defined to be a feature for detecting paid posters.

Group Characteristics. The six group characteristics are discussed in this section.

Original Tweet Posting. Paid posters tend to post copied tweets (sometimes changing few words) from the resource team which is described in Sect. 3.1 We call this feature "original tweet copying" (OTCopy). This observation has been widely studied in some existing researches [13, 27] for detecting spam. To find the copied tweets, we first segment tweets to process Chinese words using ICTCLAS which is

---

[2] MapReduce: http://en.wikipedia.org/wiki/MapReduce.
[3] Apache Hadoop: http://en.wikipedia.org/wiki/Apache_Hadoop.

developed by Institute of Computing Technology, Chinese Academy of Sciences[4]. Then stopwords are removed and TF-IDF weighting schema is used to calculate weights of words. Finally we use vector space model (VSM) [21] to compute the similarity of two tweets. The threshold in our experiment is set to be 0.85, which is an empirical value, to determine whether two original posts (not a retweet) are the same. For a tweet $tweet_i$, we think it is copied from $tweet_j$ if the similarity between $tweet_i$ and $tweet_j$ is beyond the threshold and the posting time of $tweet_i$ is after $tweet_j$. We compared all tweets in our experiments to find copied tweets. Suppose a user u posts a total of N_{OT} tweets in a campaign, there are N_{OTCopy} tweets that are copied from others in a campaign. Then group characteristics "original tweets posting" P_{OT} for building classification model is obtained from the ratio of N_{OTCopy} and N_{OT} as shown in Eq. 4.

$$P_{OTCopy} = \frac{N_{OTCopy}}{N_{OT}}. \tag{4}$$

Retweeting. A retweet is a reposting of someone else's tweet. It is common to retweet its friends' tweets which can be seen in its timeline in SINA Weibo and add some comments on them. But for paid posters, they always retweet from someone who they do not follow and add the same comments that come from the resource team as other paid posters. Suppose a user u retweets a total of N_{RT} tweets, there are $N_{RTNonFriends}$ tweets that are retweeted from users who are not its friends, then the feature $P_{RTNonFriends}$ of group characteristic "retweeting without following (RTNonFriends)" for building classification model is obtained from the ratio of $N_{RTNonFriends}$ and N_{RT} as shown in Eq. 5.

$$P_{RTNonFriends} = \frac{N_{RTNonFriends}}{N_{RT}} \tag{5}$$

Suppose there are N_{RTCopy} tweets that have the same comments with others, then the feature "retweeting copy (RTCopy)" P_{RTCopy} for building classification model is obtained from the ratio of N_{RTCopy} and N_{RT} as shown in Eq. 6. The VSM model is used to measure if two comments are the same one like what has been done in measuring if two original tweets are the same ones.

$$P_{RTCopy} = \frac{N_{RTCopy}}{N_{RT}} \tag{6}$$

Replying. Everyone can reply tweets in SINA Weibo. Like posting a new tweet, paid posters tend to get the comments from the resource team and they post the same comments (sometimes changing few words) on the target tweets. VSM model is also used to measure the similarity between two comments in a dataset. Paid posters are more likely to comment on users' tweets and the users are not their friends (non-friends). Given a user u who replies N_{RE} times in all tweets of a special campaign, there are $N_{RENonFriends}$ comments replied to non-friends'

---

[4] ICTCLAS: http://ictclas.org/index.html.

tweets, then the feature $P_{RENonFriends}$ of group characteristic "replying without following (RENonFriends)" for building classification model is obtained from the ratio of $N_{RENonFriends}$ and N_{RE} as shown in Eq. 7.

$$P_{RENonFriends} = \frac{N_{\text{RENonFriends}}}{N_{RE}} \tag{7}$$

If there are N_{RECopy} comments are the same as others, the feature P_{RECopy} of group characteristic "replying copy (RECopy)" is obtained from the ratio of N_{RECopy} and N_{RE} as shown in Eq. 8.

$$P_{RECopy} = \frac{N_{RECopy}}{N_{RE}} \tag{8}$$

Mentioning. Mentioning someone enables the mentioned user to receive a notification. It's also a convenient way for normal users to communicate with friends, but paid posters utilize the way to spread messages to the users they want. It's an usual way for paid posters to make others to see their tweets. This feature is also used to detect spammers in many studies [14,26,27]. If a user posts, retweets, replies the same tweet and mentions someone in its tweet, but the mentioned users are neither talked in the tweet nor followed by the poster, then it will be considered to be an abnormal action. Posting, retweeting and replying the same tweet has been studied in this section, we only consider the retweeting action with no comments but mentioning un-followed and un-related users in this paper. Given a user u who mentions un-followed and un-related users $N_{NoFollow}$ times in all N_{ME} tweets of a campaign and we call this feature" mentioning without following (NoFollow)", then the feature "mentioning without following (NoFollow)" P_{ME} can be obtained from the ratio of $N_{NoFollow}$ and N_{ME} as shown in Eq. 9.

$$P_{ME} = \frac{N_{NoFollow}}{N_{ME}}. \tag{9}$$

4   Experiments and Evaluation

4.1   Dataset

SINA Weibo[5], which is a microblogging website like Twitter, is one of the most popular websites in China with over 500 million registered users [1]. We collected public tweets via API in Sina Weibo. We obtained three datasets which are "Sina Campaign", "The Continent" and "Sangfor Tournament". The "Sina Campaign" dataset is conducted to promote a campaign in SINA Weibo. We collected all tweets about "Sina Campaign". To protect privacy, we do not show details in this dataset. We also collect two open public datasets "The Continent" and "Sangfor Tournament". We show the details about how we collected the two datasets. We extracted tweets that contain hashtag "#The Continent#"

---

[5] SINA Weibo: http://www.weibo.com/.

for dataset "The Continent". We collected 79,075 tweets from 72,064 users and 42,325 comments for the tweets between June 25 and July 25, 2014. Dataset for topic "Sangfor Tournament" was collected from tweets that contain keyword "Sangfor Tournament" from Jun 27 to Aug 27, 2014. There are 57,474 tweets from 16,364 users and 1,021 comments in the dataset. The follower/friend relationship and the most recent 200 tweets of all users in the three datasets were crawled.

Since it is hard to know who is exactly a paid poster or a legitimate user, to construct test datasets from topic 'The Continent" and "Sangfor Tournament", we randomly selected 450 users from each dataset and estimated them manually by three volunteers. They were asked to carefully check the content, the client, content of comments, retweeters of the top-100 posts of each user to evaluate whether a user was a paid poster or not. We also asked them to check other features like the user influence, the ratio of friends to followers, the ratio of replied/retweeted tweets to user's all tweets, the ratio of tweets that contain urls to user's all tweets and so on. For example, a user posts a tweet and the content of the tweet is the same as others (We set the number of persons to be 3 in our evaluation), and the client for posting the tweet is not coming from a sharing source like news website. Furthermore, the influence of the user, the ratio of friends to followers, the ratio of replied/retweeted tweets to user's all tweets are low, and the ratio of tweets that contain urls to user's all tweets is very high, then the user is probably a paid poster. If two or all of the three volunteers think the user is a paid posters, then it is. Otherwise, it is a legitimate user. There are 171 paid posters and 279 legitimate users in the "The Continent" dataset, comparing to 351 paid posters and 99 legitimate users in the "Sangfor Tournament" dataset.

For dataset "Sina Campaign", we totally control the dataset and know who are the paid posters. We also randomly select 450 users like the datasets "The Continent" and "Sangfor Tournament". There are 294 paid posters and 156 legitimate accounts.

4.2   Experiments

To evaluate the performance of our methods for detecting paid posters, we compare them with two baseline methods: SpamSVM method [4,18] and Chen2013 method [8]. 10-fold cross-validation is performed to analyze the performance of these methods in all experiments. Details of these methods are described below:

IGCSVM Method. Our IGCSVM method is based on both the individual statistical characteristics and group characteristics discussed in Sect. 3.2. Support Vector Machine (SVM) with a linear kernel was used to learn the classification model from the 10 features in Sect. 3.2. The values of the 10 features are computed by the equations in Sect. 3 like Eq. 1 and so on.

Individual method. Individual method is like the IGCSVM method, but it is only based on the four individual statistical characteristics of paid posters in Sect. 3.2.

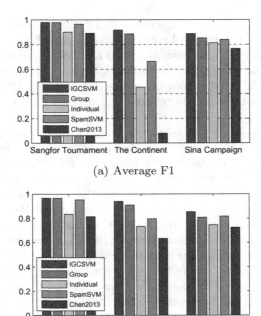

(a) Average F1

(b) Average Accuracy

Fig. 3. Performance of the five methods

Group Method. Group method is like the IGCSVM method, but it is only based on the six group characteristics of paid posters in Sect. 3.2.

SpamSVM Method. Methods for detecting spammers can also be used to detect paid posters. Some researches [4,18] employ profile-based features and user's tweets to build an effective supervised learning model. A classifier is used to learn the model. And then the model is applied on unseen data to filter social spammers. In our experiments, profile-based features which are statistical features in Sect. 3.2 and semantic features which are original tweet copying and replying copy in Sect. 3.2 are employed.

Chen2013 Method. Chen et al. [8] proposed a method to detect paid posters using users' comments. Their method is based on users' comments rather than user's posts. The features they use in their method are ratio of replies, average interval time of posts, active days, the number of news reports and replying copy. LIBSVM [7] is also used in our experiments.

Support Vector Machine (SVM) with a linear kernel is used in all our experiments to learn classification models as it can get state of the art results [22]. SVM is a supervised learning model for classification and regression analysis. An open source implementation of SVM named LIBSVM [7] was used in all our experiments. LIBSVM is an integrated software for support vector classification and the main features of LIBSVM include different SVM formulations,

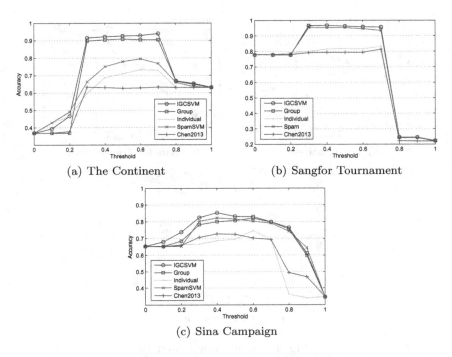

(a) The Continent   (b) Sangfor Tournament

(c) Sina Campaign

Fig. 4. Accuracy comparison with the change of the threshold

efficient multi-class classification, cross validation for model selection, various kernels (including precomputed kernel matrix) and so on.

We compare the five methods in the datasets "The Continent", "Sangfor Tournament" and "Sina Campaign" with accuracy, and F1 score. Figure 3(a) and (b) show the performance results of the five methods in the three datasets. We can find that our ISCSVM method achieves the best performance on F1 score and accuracy in all the three datasets. It's significantly better than traditional spam detection method SpamSVM on F1 score and accuracy. The Group method is also better than traditional spam detection method SpamSVM and Individual method on F1 score and accuracy in all the three datasets. It shows that group features are more discriminative than traditional individual features for detecting spam in detecting paid posters. Chen2013 method is not good enough partly because there are only 1021 comments in the dataset "Sangfor Tournament".

We compare the accuracy of the five methods with the change of the threshold value which is used to distinguish ranges of values for detecting paid poster. If the value of the model predicting the probability of a user to be a paid poster is below the threshold, then it will considered to be a paid posters. Otherwise, it will be considered to be a legitimate user. The results on the three datasets are shown in Fig. 4. We can find that IGCSVM method gets the best performance when the threshold is between 0.3 and 0.7.

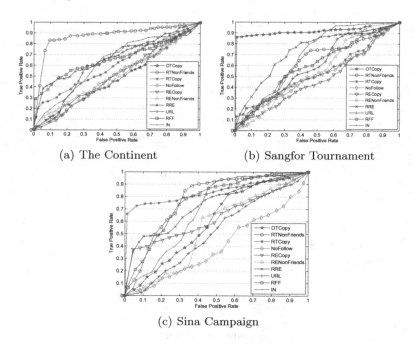

(a) The Continent

(b) Sangfor Tournament

(c) Sina Campaign

Fig. 5. Features comparison

A Receiver Operating Characteristics (ROC) curve is constructed to measure the discrimination power of individual and group characteristics shown in Sect. 3. ROC curve is plotting true positive rate to false positive rate with the change of different threshold value. The four individual characteristics which are "RFF", "RRE", "URL" and "IN" and six group characteristics which are "OTCopy", "RTCopy", "RTNonFriends", "RECopy", "RENonFriends" and "NoFollow" are compared. Figure 5 shows the discrimination power of the ten features.

For the "The Continent" dataset shown in Fig. 5(a), we can find that "RTNon-Friends" is the most discriminative feature in detecting paid posters. Features "NoFollow", "RECopy", "RENonFriends" and "OTCopy" are the least discriminative features. In the dataset "Sangfor Tournament" shown in Fig. 5(b), group feature "OTCopy" and individual feature "RRE" and "IN" are the most discriminative features in detecting paid posters. For the "Sina Campaign" dataset shown in Fig. 5(c), we can find that group feature "RTCopy", "RTNonFriends" and individual feature "RFF", "RRE" are the most discriminative feature in detecting paid posters. It shows that group features and individual features are both important to detect paid posters in dataset "Sina Campaign". It is the reason that our IGCSVM using both group and individual features gets better performance than Group method and Individual method which is based on only group or individual features.

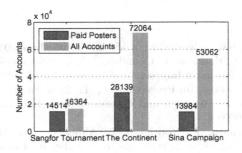

Fig. 6. Number of paid posters detected

We detect paid posters in the three datasets using IGCSVM method which gets the best accuracy and F1 score. The number of paid posters detected by IGCSVM method is shown in Fig. 6. IGCSVM method detects 14,514 paid posters in dataset "The Continent" which contains 16,364 users totally. It is 88.69 % of all users. It finds 28,139 paid posters in dataset "Sangfor Tournament", which is 39.05 % of all users. In "Sina Campaign" dataset, IGCSVM method detects 13,984 paid posters of totally 53,062 users, which is 26.35 % of all users.

5   Conclusion and Future Work

In this paper, we study a special type of online users named paid posters who are organized to post for purposes like advertising and so on in SINA Weibo. Our study is main related to online spam detection in social network. Our method utilizes the group characteristics of paid posters to detect them. Traditional individual statistics characteristics for detecting spam are also used to improve the performance. Our experimental results on the three datasets "Sangfor Tournament", "The Continent" and "Sina Campaign" show that group characteristics are also important in detecting paid posters comparing to traditional individual features. Our IGCSVM method which combines the two types of characteristics is effective in detecting paid posters and better than exiting approaches.

Our method in choosing features for detecting paid posters is empirical. It's better to learn effective features automatically to adapt to the change of paid posters. We will also try to improve the efficiency of our methods in future. For example, our methods based on the bag of words model have to compare all tweets in a dataset, it is not efficient enough. In future, we will try fingerprint based method and construct an index like B-tree to reduce the computational complexity.

Acknowledgments. This work was supported by 973 Program of China (Grant No. 2013CB329601, 2013CB329602, 2013CB329604), NSFC of China (Grant No. 60933005, 91124002), 863 Program of China (Grant No. 2012AA01A401, 2012AA01A402), National Key Technology RD Program of China (Grant No. 2012BAH38B04, 2012BAH38B06).

References

1. Sina weibo. http://en.wikipedia.org/wiki/sina_weibo, June 2014
2. Androutsopoulos, I., Koutsias, J., Chandrinos, K.V., Spyropoulos, C.D.: An experimental comparison of naive Bayesian and keyword-based anti-spam filtering with personal e-mail messages. In: Proceedings of the 23rd Annual International ACM SIGIR Conference on Research and Development in Information Retrieval, pp. 160–167. ACM (2000)
3. Benevenuto, F., Duarte, F., Rodrigues, T., Almeida, V.A., Almeida, J.M., Ross, K.W.: Understanding video interactions in youtube. In: Proceedings of the 16th ACM international conference on Multimedia, pp. 761–764. ACM (2008)
4. Benevenuto, F., Magno, G., Rodrigues, T., Almeida, V.: Detecting spammers on twitter. In: Collaboration, Electronic Messaging, Anti-Abuse and Spam Conference (CEAS), vol. 6, p. 12 (2010)
5. Benevenuto, F., Rodrigues, T., Almeida, V., Almeida, J., Zhang, C., Ross, K.: Identifying video spammers in online social networks. In: Proceedings of the 4th International Workshop on Adversarial Information Retrieval on the Web, pp. 45–52. ACM (2008)
6. Blanzieri, E., Bryl, A.: A survey of learning-based techniques of email spam filtering. Artif. Intell. Rev. **29**(1), 63–92 (2008)
7. Chang, C.-C., Lin, C.-J.: LIBSVM: a library for support vector machines. ACM Trans. Intell. Syst. Technol. **2**(3), 27:1–27:27 (2011)
8. Chen, C., Wu, K., Srinivasan, V., Zhang, X.: Battling the internet water army: detection of hidden paid posters. In: Proceedings of the 2013 IEEE/ACM International Conference on Advances in Social Networks Analysis and Mining, pp. 116–120. ACM (2013)
9. Chu, Z., Gianvecchio, S., Wang, H., Jajodia, S.: Who is tweeting on twitter: human, bot, or cyborg? In: Proceedings of the 26th Annual Computer Security Applications Conference, pp. 21–30. ACM (2010)
10. Ding, Z., Jia, Y., Zhou, B., Han, Y.: Mining topical influencers based on the multi-relational network in micro-blogging sites. China Commun. **10**(1), 93–104 (2013)
11. Drucker, H., Wu, S., Vapnik, V.N.: Support vector machines for spam categorization. IEEE Trans. Neural Netw. **10**(5), 1048–1054 (1999)
12. Fetterly, D., Manasse, M., Najork, M.: Spam, damn spam, and statistics: using statistical analysis to locate spam web pages. In: Proceedings of the 7th International Workshop on the Web and Databases: Colocated with ACM SIGMOD/PODS 2004, pp. 1–6. ACM (2004)
13. Gao, H., Hu, J., Wilson, C., Li, Z., Chen, Y., Zhao, B.Y.: Detecting and characterizing social spam campaigns. In: Proceedings of the 10th ACM SIGCOMM Conference on Internet Measurement, pp. 35–47. ACM (2010)
14. Grier, C., Thomas, K., Paxson, V., Zhang, M.: @ spam: the underground on 140 characters or less. In: Proceedings of the 17th ACM Conference on Computer and Communications Security, pp. 27–37. ACM (2010)
15. Gyöngyi, Z., Garcia-Molina, H., Pedersen, J.: Combating web spam with trustrank. In: Proceedings of the Thirtieth International Conference on Very Large Data Bases, VLDB Endowment, vol. 30, pp. 576–587 (2004)
16. Jindal, N., Liu, B.: Opinion spam and analysis. In: Proceedings of the 2008 International Conference on Web Search and Data Mining, pp. 219–230. ACM (2008)

17. Kolari, P., Java, A., Finin, T., Oates, T., Joshi, A.: Detecting spam blogs: a machine learning approach. In: Proceedings of the National Conference on Artificial Intelligence, vol. 21, pp. 1351. AAAI Press, Menlo Park, CA (1999), MIT Press, Cambridge, London, MA (2006)
18. Lee, K., Caverlee, J., Webb, S.: Uncovering social spammers: social honeypots+ machine learning. In: Proceedings of the 33rd International ACM SIGIR Conference on Research and Development in Information Retrieval, pp. 435–442. ACM (2010)
19. McCord, M., Chuah, M.: Spam detection on twitter using traditional classifiers. In: Alcaraz Calero, J.M., Yang, L.T., Mármol, F.G., Villalba, L.J.G., Li, A.X., Wang, Y. (eds.) ATC 2011. LNCS, vol. 6906, pp. 175–186. Springer, Heidelberg (2011)
20. Ott, M., Choi, Y., Cardie, C., Hancock, J.T.: Finding deceptive opinion spam by any stretch of the imagination. In: Proceedings of the 49th Annual Meeting of the Association for Computational Linguistics: Human Language Technologies, vol. 1, pp. 309–319. Association for Computational Linguistics (2011)
21. Salton, G., Wong, A., Yang, C.-S.: A vector space model for automatic indexing. Commun. ACM **18**(11), 613–620 (1975)
22. Sebastiani, F.: Machine learning in automated text categorization. ACM Comput. Surv. **34**(1), 1–47 (2002)
23. Thomas, K., Grier, C., Song, D., Paxson, V.: Suspended accounts in retrospect: an analysis of twitter spam. In: Proceedings of the 2011 ACM SIGCOMM Conference on Internet Measurement Conference, pp. 243–258. ACM (2011)
24. Thomason, A.: Blog spam: a review. In: CEAS (2007)
25. Wang, K., Xiao, Y., Xiao, Z.: Detection of internet water army in social network. In: 2014 International Conference on Computer, Communications and Information Technology (CCIT 2014). Atlantis Press (2014)
26. Zhang, Y., Ruan, X., Wang, H., Wang, H.: What scale of audience a campaign can reach in what price. In: 2014 IEEE International Conference on Computer Communications (InfoCOM 2014) (2014)
27. Yang, C., Harkreader, R., Zhang, J., Shin, S., Gu,G.: Analyzing spammers' social networks for fun and profit: a case study of cyber criminal ecosystem on twitter. In: Proceedings of the 21st International Conference on World Wide Web, pp. 71–80. ACM (2012)
28. Zeng, K., Wang, X., Zhang, Q., Zhang, X., Wang, F.-Y.: Behavior modeling of internet water army in online forums. World Congr. **19**, 9858–9863 (2014)

A Soft Subspace Clustering Method for Text Data Using a Probability Based Feature Weighting Scheme

Abdul Wahid$^{(\boxtimes)}$, Xiaoying Gao, and Peter Andreae

School of Engineering and Computer Science, Victoria University of Wellington,
19 Kelburn Parade, 6012 Wellington, New Zealand
{abdul.wahid,xgao,pondy}@ecs.vuw.ac.nz
http://ecs.victoria.ac.nz

Abstract. Clustering methods aim to find clusters or groups of similar objects in a given set of data. Common soft subspace clustering methods for text data find different clusters in subspaces using a weighted distance measure. The weighting scheme heavily affects the clustering performance and requires special consideration. Since text data has semantic information along with syntactic information, a weighting scheme, which uses semantic information, is more likely to generate a better clustering solution.

This paper introduces a novel soft subspace clustering method that uses a probabilistic model to extract semantic information from documents for weighting features. We created a feature weight matrix from the probability distribution of terms in subspaces and developed a weighted distance measure for finding similar documents in relevant subspaces. Our experiment results on synthetic and real-world datasets show that our newly developed method outperforms other state-of-the-art soft subspace clustering methods.

Keywords: Clustering algorithms · Soft subspace clustering · Latent dirichlet allocation

1 Introduction

Clustering methods try to find similar documents and group them together in clusters. Documents are generally represented in a Vector Space Model, where each distinct term is treated as a feature. Hence the feature space becomes very large. Traditional clustering methods such as k-means, consider all features at the same time to cluster the data and are only suitable for data with a small number of features.

Subspace clustering methods are widely applied when the number of features is very large. They try to group similar objects using a subset of features (i.e. subspace) instead of all features. In subspace clustering, each cluster represents a set of objects clustered according to a subspace of features. The problem

© Springer International Publishing Switzerland 2015
J. Wang et al. (Eds.): WISE 2015, Part II, LNCS 9419, pp. 124–138, 2015.
DOI: 10.1007/978-3-319-26187-4_9

of subspace clustering is often divided into two sub-problems: determining the subspaces and clustering the data. Based on how these problems are addressed, there are two main categories of subspace clustering methods: *hard subspace clustering* and *soft subspace clustering*. In hard subspace clustering, a feature in a subspace is either present or not present (1 or 0), whereas in soft subspace clustering, a feature in a subspace is determined by its degree of presence (i.e. a weight between 0–1). A feature is considered relevant (i.e. present) if its weight is high in a subspace and considered irrelevant if its weight is low in a subspace.

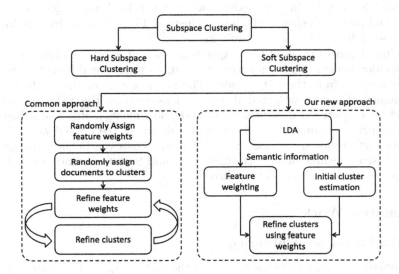

Fig. 1. Differences between of subspace clustering approaches and our new approach

In text datasets, some features can be considered to be partially presented in subspaces. Therefore, soft subspace clustering methods, which assign weights to features instead of determining the exact presence of features in a subspace, are becoming more popular in text clustering.

The most popular soft subspace clustering methods are FWKM [20], EWKM [19] and FGKM [9]. These methods use modified version of k-means to cluster the data in different subspaces according to feature weights. These methods mainly differ in terms of how they compute the feature weights. The main issue with these methods is that they ignore the semantic information of the documents, which might be helpful in improving the clustering process.

Latent Dirichlet Allocation (LDA) is a popular topic modeling method which can be used to extract semantic information from a collection of documents. LDA is based on a generative model, where a document is assumed to be generated from the distribution of terms which form a special theme or topic. The main idea of our method is to treat topics generated from the LDA model as subspaces because each topic specifies a soft subset of related terms (features). Subspaces generated by the LDA were utilized in initializing the clusters in our method.

We use LDA model to compute a probability that a term is relevant in a subspace (topic/subset of terms). These probabilities can represent the semantic information and is used as term or feature weightings in our soft subspace clustering to improve the clustering process. Figure 1 shows the difference between existing clustering methods and our new method. The common existing soft subspace clustering methods use a random approach to initialize weightings and randomly assign objects to clusters. Then the feature weightings and clusters are refined iteratively. In our method, we first use LDA to assign the feature weights and assign objects to the initial clusters. Then we iteratively refine the clusters according to the feature weights.

The main contribution of this paper is a new soft subspace clustering algorithm for documents using semantically weighted terms for different subspaces that are derived from the LDA model. The main novelty of the method is the development of a new weighted distance measure from the LDA probability matrices to compute the distances between the documents in different subspaces.

The paper is organized as follows: Sect. 2 discusses the related work; Sect. 3 describes our proposed method; Sect. 4 explains the experimental design and Sect. 5 presents results along with discussion; and Sect. 6 provides a conclusion of the paper along with the future directions.

2   Related Work

2.1   Hard Subspace Clustering

Hard subspace clustering methods divide the feature space into different subspaces where each feature is either present or absent in a subspace. Hard subspace clustering methods can be further categorized by their search approaches i.e. bottom-up and top-down. The examples of bottom-up hard subspace clustering methods are CLIQUE [3], ENCLUS [10], MAFIA [18] and FINDIT [29]. The examples of top-down hard subspace clustering methods are PROCLUS [1], ORCLUS [2] and δ-Clusters [30]. Our method differs from these methods because it belongs to soft subspace clustering methods.

2.2   Soft Subspace Clustering

In soft subspace clustering, each feature is assigned different weights for different subspaces. Hence some proportion of a feature is present in all subspaces. In clustering process, the features that have higher weight values in a subspace contribute more to form a cluster than the features that have lower weights. Generally the soft subspace clustering methods employ variable weighting scheme and iteratively update the feature weights in the clustering process.

Variable weighting schemes are widely applied in data mining [11–13, 21, 22]. Some of the variable weighting methods can be extended, especially k-means type variable weighting, to develop soft subspace clustering algorithms [7, 14–17, 20].

Recent approaches such as FWKM [20], EWKM [19] and FGKM [8,9] use k-means type variable weighting algorithms and formulate a minimization problem for data clustering. FWKM uses Lagrange multiplier and forms a polynomial weighting formula to compute the feature weights and iteratively refines the clusters using the following objective function.

$$\min J(U, W, \Lambda) = \sum_{i=1}^{k} \sum_{j=1}^{n} u_{ij} \sum_{t=1}^{m} \lambda_{it}[(\mu_{it} - d_{jt})^2 + \sigma] \tag{1}$$

where

- u is a $k \times n$ binary matrix representing the assignment of objects to clusters. $u_{ij} = 1$ iff object j is in cluster i, $u_{ij} = 0$ otherwise.
- λ is $k \times m$ feature weight matrix. It represents k subspaces in rows and m features in columns. The value in a cell is a weight of the feature to its corresponding subspace and the value ranges from 0–1. The sum of the weights of all features in a subspace is 1. i.e. $\sum_{t=1}^{m} \lambda_{it} = 1, 1 \leq i \leq k, 0 < \lambda_{it} < 1$
- μ is a $k \times m$ matrix representing the mean value of a feature in a cluster.
- d_{jt} represents a feature t of the jth object[1].
- σ is an average spread/variance of all the features in a dataset.

EWKM clusters the data in a similar fashion but uses the exponential weighting formula to compute the feature weights. Its objective function is similar to Eq. 1, but instead of using σ, it uses Shanon entropy to control the weights. FGKM has a slightly different approach, it not only uses the individual feature weightings but also uses the feature group weightings scheme. The feature group weightings is computed by combining features into different groups and then assigning weights to those groups.

The above soft subspace clustering methods ignore the semantic information of the documents in a clustering process. The main motivation of our research work is to investigate the use of semantic information (e.g. topics) of documents in soft subspace clustering process.

2.3 Latent Dirichlet Allocation

Latent Dirichlet Allocation (LDA) [6] extracts topics/themes from documents, which have semantic information. It is widely used in other domains such as topic modeling [5] and Entity Resolution [4]. The topics generated by LDA can be considered as subspaces and for each subspace, LDA facilitates to compute a term weight. Our soft subspace clustering method is related to FWKM and EWKM, however our method uses LDA based weighting scheme to utilize the semantic information of the documents.

LDA is a probabilistic model with an assumption that a document is a random mixture over latent topics and each topic is a distribution over terms. The two main parameters in this model are topic-document distributions θ and topic-term distributions ϕ.

---

[1] For clustering a collection of documents, d_{jt} is often the term-frequency of a term in a document.

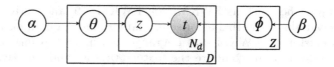

Fig. 2. A common LDA graphical model using plate notation.

Figure 2[2] represents a graphical model for LDA. Arrows represent conditional dependencies between two variables and plates/rectangles represent loop or repetition of the variable mentioned in the corner of the plate. The shaded circle represents the observed variable while unshaded represent unobserved variables. Hyperparameter α is a prior on topic distribution. High value of α favors topic distributions with more topics and low value (<1) of α favors topic distribution with a few topics. Hyperparameter β is a prior on term distribution in every topic, which controls the number of times terms are sampled from a topic. The LDA model infers three latent variables θ, ϕ and z (topics) while observing t (terms) in a document set D.

In Fig. 2, the inner plate (z and t) denotes the continuous sampling of topics and terms until N_d terms are created from document d. The out plate (which is surrounding θ) denotes the continuous sampling of a topic distribution for each document d in a document set D. The plate surrounding ϕ denotes the continuous sampling of a term distribution over each topic z until a total of Z topics are generated. More details of LDA can be found in [5].

To the best of our knowledge, our research work is the first attempt that applies LDA to assign weights and use it in text soft subspace clustering.

3   Our LDA Weighted K-Means Model

This section presents our new subspace clustering method which builds on LDA for document clustering[3]. Figure 3 shows the overall design of our method. The documents are pre-processed by implementing stop words filtration, low frequency words filtration and WordNet lemmatization. Then we use LDA based on Gibbs sampling to generate two matrices: topic-document matrix θ and topic-term matrix ϕ. θ is then used for initializing the clusters and ϕ is used as feature weights for refining the clusters.

3.1   Gibbs Sampling

We implemented LDA model in an unsupervised way (without using training datasets) using Gibbs sampling algorithm explained in [24]. The Gibbs sampling

---

[2] This figure is created by the author. However, similar figures are commonly used in literature to describe LDA.

[3] The code of our method was implemented using lingpipe toolkit (http://alias-i.com/lingpipe/).

iteratively computes the conditional probability of assigning an occurrence of a term (token of a term) to each topic. The common Gibbs sampling method provides the estimates of the posterior distribution over z (topics) but does not provides θ and ϕ. However, we can use the Gibbs sampling technique to approximate θ and ϕ from posterior estimates of z.

For each token i (an occurrence of a term), let v_i, d_i, z_i denote the term for the token, the document for the token and the topic of the token respectively in a document collection. The Gibbs sampling iteratively processes each term token in the document collection and estimates the conditional probability of assigning the current term token to an individual topic, based on the topic assignments to all other term tokens. The conditional distribution is formalized as:

$$Prb(z_i = r | \mathbf{z}_{-i}, ...) \tag{2}$$

where $z_i = r$ is the assignment of ith token to topic r. $\mathbf{z}_{-i}$ denotes the topic assignment of all the tokens excluding the ith token. Other variables for Eq. 2 represented by (...) are v_i, d_i, $\mathbf{v}_{-i}$, $\mathbf{d}_{-i}$, α and β. $\mathbf{v}_{-i}$ represents all terms tokens except the ith term token and $\mathbf{d}_{-i}$ represents document tokens except the ith document token. Griffiths and Steyvers [24] provided a simple way to compute Eq. 2 as:

$$Prb(z_i = r | \mathbf{z}_{-i}, ...) \propto \frac{\mathcal{C}_{rv_i}^{(1)} + \beta}{\sum_{l=1}^{m} \mathcal{C}_{rl}^{(1)} + m\beta} \frac{\mathcal{C}_{rd_i}^{(2)} + \alpha}{\sum_{z=1}^{Z} \mathcal{C}_{zd_i}^{(2)} + Z\alpha} \tag{3}$$

where $\mathcal{C}^{(1)}$ and $\mathcal{C}^{(2)}$ are $Z \times m$ and $Z \times D$ matrices respectively and Z, m, D are the number of topics, terms and documents respectively. The cell values of these matrices represent the frequency of the term/document for the corresponding topics. $\mathcal{C}_{rv_i}^{(1)}$ denotes the number of times the term v_i is assigned to the topic r excluding the ith instance and $\mathcal{C}_{rd_i}^{(2)}$ denotes the number of times a term token in document d is assigned to the topic r excluding the ith instance.

3.2   Generating θ and ϕ

After applying the Gibbs sampling algorithm, we create two matrices: (1) ϕ topic-term matrix and (2) θ topic-document matrix. These matrices are generated from the two count matrices $\mathcal{C}^{(1)}$ and $\mathcal{C}^{(2)}$ according to [24] as follows:

$$\phi_{rt} = \frac{\mathcal{C}_{rt}^{(1)} + \beta}{\sum_{l=1}^{m} \mathcal{C}_{rl}^{(1)} + m\beta}, \; \theta_{rj} = \frac{\mathcal{C}_{rj}^{(2)} + \alpha}{\sum_{z=1}^{Z} \mathcal{C}_{zj}^{(2)} + Z\alpha} \tag{4}$$

ϕ corresponds to the probability that a term t is assigned to topic r and θ corresponds to the probability that a document j is assigned to topic r.

The rows of topic-document matrix θ represent topics and the columns represent documents. The cells of θ represent the probability that a document has the corresponding topic. We use this matrix to form the initial clusters. One should note that LDA naturally provides a simple way for clustering the documents. However, this clustering is not soft subspace clustering. Following is a way to improve the clusters generated from LDA by utilizing the information from LDA and forming soft subspace clustering method.

In LDA model, each term is a feature and each topic corresponds to a subspace, therefore topic-term matrix ϕ can be considered of a feature weight matrix for different subspaces where each feature or term has a degree of presence in all subspaces or topics. We used the values of topic-term matrix ϕ for determining relevant subspaces and developed a new weighted distance measure, which finds similar documents in relevant subspaces.

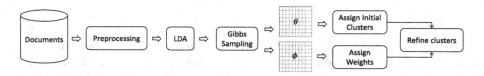

Fig. 3. System diagram of our new method. θ and ϕ are the topic-document and topic-term matrices respectively.

3.3   Objective Function

We perform clustering by formulating the clustering as a minimization problem and our objective is to minimize the sum of squared distances between documents and the nearest cluster centers weighted by different subspaces. The objective function is similar to the objective functions (Eq. 1) of the FWKM or EWKM, however, we do not include σ or Shanon entropy because we are already controlling the feature weighting using two hyper parameters of LDA model (α and β). Moreover, the objective function uses previously computed LDA based feature weights instead of computing the feature weights in iterative manner.

Let $D = \{d_1, d_2, d_3, ..., d_n\}$ be a set of n documents and $T = \{t_1, t_2, t_3, ..., t_m\}$ represents m terms in the documents. Then the objective function for clustering the n documents into k clusters can be defined as:

$$\sum_{i=1}^{k} \left(\sum_{j=1}^{n} \sum_{t=1}^{m} \delta_{ij} \phi_{it} (\mu_{it} - d_{jt})^2 \right) \tag{5}$$

where

- δ is a $k \times n$ binary matrix representing the assignment of documents to clusters. $\delta_{ij} = 1$ iff document j is in cluster i, $\delta_{ij} = 0$ otherwise.
- ϕ is $k \times m$ topic-term matrix generated from LDA model. It represents k subspaces in rows and m terms in columns. The value in a cell is a weight of the term to its corresponding subspace and the value ranges from 0-1. The sum of the weights of all terms in a subspace is 1. i.e. $\sum_{t=1}^{m} \phi_{it} = 1, 1 \leq i \leq k, 0 < \phi_{it} < 1$

- μ is a $k \times m$ matrix representing the mean value of a term in a cluster. It is calculated as:

$$\mu_{it} = \frac{\sum_{j=1}^{n} \delta_{ij} d_{jt}}{\sum_{j=1}^{n} \delta_{ij}} \tag{6}$$

- d_{jt} represents a term t (a feature) of the jth document, which is the term-frequency of the term in the document.

We iteratively assign documents to their nearest cluster centers until the algorithm converges. We minimize the objective function by updating δ using the following:

$$\delta = \begin{cases} \delta_{ij} = 1, & \text{if } i = \text{argmin}_x \, dist(\mu_x, d_j) \\ \delta_{ij} = 0, & \text{otherwise} \end{cases} \tag{7}$$

where $dist(\mu_x, d_j)$ is defined as

$$dist(\mu_x, d_j) = \sum_{t=1}^{m} \phi_{xt} (\mu_{xt} - d_{jt})^2 \tag{8}$$

Equation 8 defines our distance measure. Unlike k-means, our distance measure computes the distance of a document from the cluster centers by using a LDA parameter ϕ, which provides a semantic based feature weighting to different subspaces. Higher value of the probability that a term is assigned to a topic indicates that the term has a higher degree of presence in a subspace. Therefore the difference between a term in the document and the mean value of the term in the cluster for that particular term is more important. The use of LDA differentiates our method from other soft subspace clustering methods.

3.4 Our Algorithm: DWKM

Our Dirichlet Weighted K-mean algorithm is a modified version of k-means algorithm. The details are shown in Algorithm 1.

Algorithm 1. DWKM

Input: document set D and number of clusters k
Output: Clustering solution $\mathcal{C}$

1: Preprocess document set D
2: Initialize the LDA model and assign all term tokens to Z Topics according to Eqs. 2 and 3
3: Perform Gibbs sampling and generate θ and ϕ from LDA model using Eq. 4
4: Initialize δ using θ. $\delta_{ij} = 1$, if $i = \text{argmax}_x \, \theta_x$
5: **repeat**
6:     Update clusters means according to Eq. 6
7:     Assign documents to δ according to Eq. 7
8: **until Convergence**

Algorithm 1 takes two arguments: a document set and the number of clusters and outputs the clustering solution. The algorithm performs preprocessing step on the documents, which includes stop word removal, lemmatization and tokenization of words. Then the algorithm randomly assigns all term tokens to Z topics and performs Gibbs sampling. Once ϕ and θ matrices are generated, line 4 of the algorithm groups documents to different clusters according to their highest probability using θ. The algorithm then, fine tunes the clusters by repeating the update and assignment steps according to Eqs. 6 and 7 until convergence criteria is met. The convergence criterion terminates the loop if there are no more documents to relocate to any clusters or the total number of specified iterations exceeds the predefined limit.

4   Experimental Setup

Our experiments are designed based on two recent papers [9,19]. Our method DWKM was evaluated on four synthetic and six real world datasets, and compared with five clustering methods using different cluster quality measures. Four synthetic datasets were generated by following the same process described in [9] and six real-world datasets were generated as described in [19].

4.1   Datasets

The synthetic datasets SD1, SD2, SD3, SD4 were generated according to [9]. Each consists of 6000 objects, 200 features, three subspaces and three clusters. The noise level in SD1, SD2, SD3 and SD4 are 0, 0.2, 0 and 0.2 respectively (as described in [9]). The percentage of missing values in DS1, DS2, DS3 and DS4 are 0, 0, 0.12, 0.12 respectively. Detailed information about how to reproduce the synthetic datasets can be found in [9].

The six real-word datasets with two or more clusters from 20-Newsgroup[4] are the same as [19]. Table 1 shows the details of these six datasets. The dataset D1, D2 and D3 are easier than datasets D4, D5 and D6. D1 and D2 have semantically different clusters whereas D4 and D5 have semantically related clusters. D3 and D6 have unbalanced clusters (as shown in Table 1).

4.2   Evaluation Measures

In order to compare our method with other methods, we used two evaluation measures: Cluster Accuracy [23] and F-measure [19,25–27] for synthetic dataset and three evaluation measures: F-measure, Normal Mutual Information(NMI) [32] and Entropy [31] for the real-world datasets. These measures are chosen based on [19] and [9] The lower entropy value of a clustering solution indicates the clustering solution has a better quality, whereas higher values of all other evaluation measures indicate a better cluster quality.

---

[4] http://qwone.com/~jason/20Newsgroups/.

Table 1. Six real world datasets created from 20-Newsgroup dataset

Dataset	Clusters	# of docs	Dataset	Clusters	# of docs
D1	alt.atheism	100	D4	talk.politics.mideast	100
	comp.graphics	100		talk.politics.misc	100
D2	comp.graphics	100	D5	comp.graphics	100
	rec.sport.baseball	100		comp.os.ms-windows	100
	sci.space	100		rec.autos	100
	talk.politics.mideast	100		sci.electronics	100
D3	comp.graphics	120	D6	comp.graphics	120
	rec.sport.baseball	100		comp.os.ms-windows	100
	sci.space	59		rec.autos	59
	talk.politics.mideast	20		sci.electronics	20

The evaluation measures can be computed as follows:

$$Cluster\,Accuracy = \frac{\sum_{i=1}^{k} d_i}{n} \qquad (9)$$

$$\text{F-measure} = \sum_{i=1}^{k} \frac{n_i}{n} \cdot \max_{1 \le j \le k} \left\{ \frac{2 \cdot \frac{n_{ij}}{n_i} \cdot \frac{n_{ij}}{n_j}}{\frac{n_{ij}}{n_i} + \frac{n_{ij}}{n_j}} \right\} \qquad (10)$$

$$\text{NMI} = \frac{\sum_{i=1,j=1}^{k} n_{ij} \log\left(\frac{n \cdot n_{ij}}{n_i \cdot n_j}\right)}{\sqrt{(\sum_{i=1}^{k} n_i \log \frac{n_i}{n})(\sum_{j=1}^{k} n_j \log \frac{n_j}{n})}} \qquad (11)$$

$$\text{Entropy} = \sum_{j=1}^{k} \frac{n_j}{n} \left(-\frac{1}{\log k} \sum_{i=1}^{k} \frac{n_{ij}}{n_j} \cdot \log \frac{n_{ij}}{n_j} \right) \qquad (12)$$

where d_i is correctly identified documents in cluster i, k is total number of clusters and n is the total number of documents in a dataset. n_i and n_j represent the number of documents in class i of the original dataset and cluster j in our computed clustering solution respectively, n_{ij} represents the number of documents that are common in both class i and cluster j.

5   Results

We compared our method DWKM with k-means, LDA based simple clustering, FWKM [20], EWKM [19] and FGKM [9].

Table 2. Comparison of clustering methods on synthetic dataset using Accuracy (AC) and F-measure (FM). The values on left are the mean values of 100 runs and the values in parenthesis are standard deviation of 100 runs.

Datasets	Metric	k-means	LDA	FWKM	EWKM	FGKM	DWKM
SD1	AC	0.65 (0.09)	0.66 (0.11)	0.77 (0.14)	0.69 (0.10)	0.82 (0.16)	**0.87 (0.15)**
	FM	0.63 (0.13)	0.65 (0.09)	0.73 (0.19)	0.59 (0.13)	0.75 (0.22)	**0.81 (0.20)**
SD2	AC	0.63 (0.04)	0.68 (0.06)	0.76 (0.10)	0.72 (0.13)	0.87 (0.16)	**0.92 (0.15)**
	FM	0.64 (0.05)	0.69 (0.09)	0.75 (0.12)	0.63 (0.17)	0.82 (0.22)	**0.88 (0.21)**
SD3	AC	0.62 (0.04)	0.64 (0.07)	0.67 (0.07)	0.70 (0.09)	**0.94 (0.13)**	**0.94 (0.12)**
	FM	0.62 (0.06)	0.63 (0.13)	0.64 (0.11)	0.59 (0.11)	0.91 (0.18)	**0.92 (0.17)**
SD4	AC	0.60 (0.04)	0.61 (0.15)	0.61 (0.06)	0.69 (0.08)	0.91 (0.13)	**0.93 (0.13)**
	FM	0.59 (0.05)	0.60 (0.16)	0.60 (0.07)	0.58 (0.11)	0.88 (0.18)	**0.90 (0.19)**

5.1   Comparison

K-means and LDA based simple clustering algorithm were implemented in ling-pipe. We provided predefined number of clusters as a parameter for both algorithms. The simple LDA clustering algorithm uses the same initial steps described in our method without the cluster refinement step. We treated initial clusters as final clusters and skipped the loop which refines the cluster using feature weights. The parameters for LDA are *number of topics = number of clusters in ground truth, number of clusters = number of clusters in ground truth*, $\alpha = 0.1$ and $\beta = 0.01$. We tuned the parameter α and β for the best performance. FWKM, EWKM and FGKM clustering algorithm were implemented in Weka[5] and we used standard parameters as described by the authors.

The performance of all six clustering algorithms for synthetic dataset is shown in Table 2 and for real-world dataset is shown in Table 3.

Table 2 shows the comparison of clustering methods in terms of Accuracy and F-measure on four synthetic datasets. The values in bold represent the best results. In general, DWKM performs better than other clustering methods in terms of both Accuracy and F-measure on the synthetic datasets. The Accuracy and F-measure values on datasets SD1 and SD2 for DWKM and FGKM have large gaps, whereas the differences of the values on datasets SD3 and SD4 are relatively smaller. The LDA based simple clustering performed better than standard k-means, but performed worse than soft subspace clustering algorithms.

Table 3 shows the mean values of F-measure, NMI and Entropy for k-means, FWKM, FGKM and DWKM clustering methods on six real-world datasets. In general, on the six real-world data set DWKM performed better than other clustering methods in terms of F-measure, NMI and Entropy values. The D1 dataset is the easiest dataset. K-means, EWKM, FGKM and DWKM have the same F-measure value **0.96** on D1 dataset, which means these clustering methods produced equally good clustering solutions. However, if we consider the NMI and Entropy values

---

[5] The code for FWKM, EWKM and FGKM was provided by the authors.

Table 3. A comparison of clustering methods in terms of F-measure, NMI and Entropy on six real-world datasets created from 20-Newsgroup dataset. The values listed in the table are the mean values of 100 runs of five clustering methods on six real-world datasets

Datasets	Metric	k-means	LDA	FWKM	EWKM	FGKM	DWKM
D1	F-measure	**0.96**	**0.96**	0.95	**0.96**	**0.96**	**0.96**
	NMI	0.78	0.78	0.79	0.83	0.85	**0.86**
	Entropy	0.21	0.21	0.20	0.16	0.15	**0.13**
D2	F-measure	0.93	0.92	0.90	0.91	0.94	**0.96**
	NMI	0.80	0.78	0.75	0.76	0.78	**0.80**
	Entropy	0.19	0.24	0.25	0.23	0.17	**0.15**
D3	F-measure	0.89	0.90	0.95	0.95	0.95	**0.96**
	NMI	0.71	0.72	0.84	0.86	0.87	**0.88**
	Entropy	0.28	0.20	0.15	0.11	0.10	**0.08**
D4	F-measure	0.88	0.90	0.90	0.94	0.95	**0.96**
	NMI	0.47	0.55	0.60	0.72	0.75	**0.78**
	Entropy	0.52	0.30	0.40	0.28	0.27	**0.20**
D5	F-measure	0.70	0.75	0.86	0.89	0.90	**0.92**
	NMI	0.38	0.48	0.64	0.68	0.70	**0.73**
	Entropy	0.61	0.41	0.35	0.31	0.30	**0.29**
D6	F-measure	0.65	0.81	0.92	0.92	0.93	**0.94**
	NMI	0.37	0.68	0.73	0.75	0.76	**0.78**
	Entropy	0.53	0.28	0.23	0.23	0.22	**0.19**

Table 4. Percentage improvement of DWKM over FGKM in terms of Accuracy(AC) and F-measure (FM) on synthetic datasets

	AC % (IMP)	FM % (IMP)
SD1	5.75	7.41
SD2	5.43	6.82
SD3	0.00	1.09
SD4	2.15	2.22

Table 5. Percentage improvement of DWKM over FGKM in terms of F-measure (FM), NMI and Entropy (EN) on real datasets

	FM % (IMP)	NMI % (IMP)	EN % (IMP)
D1	0.000	1.163	2.299
D2	2.083	2.500	2.353
D3	1.042	1.136	2.174
D4	1.042	3.846	8.750
D5	4.255	4.110	1.408
D6	2.105	2.564	3.704

along with F-measure value of the D1 dataset, we can see that DWKM performed slightly better than other clustering methods. The LDA based simple clustering followed the same trend as in synthetic datasets and performed better than standard k-means, but worse than soft subspace clustering algorithms.

Table 6. P-values of unpaired ttest of DWKM and FGKM on synthetic datasets

SD1		SD2		SD3		SD4	
Accuracy	F-measure	Accuracy	F-measure	Accuracy	F-measure	Accuracy	F-measure
0.0237	**0.0449**	**0.0237**	**0.0449**	1	0.6867	0.278	0.4457

It was also observed from the results that DWKM performed well on data with different level of difficulties (data without noise, with noise, with balanced clusters and with unbalanced clusters). This shows that our semantic weighting of subspaces derived from LDA is reasonably effective for finding clusters in different types of data. Moreover the LDA based simple clustering algorithm performed much better than k-means algorithm when datasets had semantically related clusters (results of D4 and D5). It was also noted that the use cluster refinement step based on feature weighting of LDA model boosted the performance of clustering solution. The DWKM algorithm without the cluster refinement step, performed better than k-means algorithm and slightly worse than other clustering methods.

Tables 4 and 5 provide percentage improvement of DWKM over FGKM on synthetic datasets and real datasets respectively. The results in all tables suggest that DWKM is a better clustering method. We further investigate the performance of all clustering methods by conducting a statistical analysis.

5.2 Statistical Analysis

We performed two types of statistical tests: (1) unpaired t-test and (2) paired Wilcoxon statistical significance test [28] by considering DWKM as the control group. The unpaired ttest was performed using the standard deviation and mean values of evaluation measures listed in Table 2. In general the results from unpaired ttest showed that DWKM achieved statistically significant improvement over three methods k-means, FWKM and EWKM on all synthetic datasets with p-value less then **0.05**. The p-values of unpaired ttest computed for FGKM on SD1 and SD2 synthetic datasets are less than **0.05**, which indicates that our method DWKM has statistical significant improvement on SD1 and SD2 over FGKM. The performance of our method on other SD3 and SD4 synthetic dataset was found to be comparable over FGKM.

For the six real-world dataset we used paired Wilcoxon statistical significance test. The p-values of F-measure, NMI and Entropy values for FGKM were **0.0305**, **0.0028** and **0.0228** respectively. In general the p-values for all five clustering methods were found to be less than **0.05**, which suggested that our method DWKM shows a better performance and significant improvement over five clustering methods (Table 6).

6 Conclusion

In this paper, we introduced a new soft subspace clustering method which uses LDA model to weight the features in the subspaces for clustering documents.

The LDA model was implemented using a standard Gibbs sampling algorithm, and it generated two matrices: topic-term and topic-documents. We used the topic-term matrix to develop a new weighted distance measure, where topics are used as subspaces. We developed a k-mean based soft subspace clustering method based on our new weighted distance measure. The algorithm is initialized using the topic-document matrix, where topics are considered as initial clusters.

Our new method DWKM, was found to achieve a statistically significant improvement over recently developed soft subspace clustering methods on synthetic and real-world datasets.

Currently the method requires users to input the number of topics to initialize the LDA model. In future we will remedy this by investigating non-parametric LDA models and will try to reduce the computational complexity of the overall method. Another direction for the future work is to investigate the use of LDA to generate different candidate clustering solutions for clustering ensemble methods.

References

1. Aggarwal, C.C, Wolf, J.L., Yu, P.S., Procopiuc, C., Park, J.S.: Fast algorithms for projected clustering. In: ACM SIGMOD Record, vol. 28, pp. 61–72. ACM (1999)
2. Aggarwal, C.C., Yu, P.S: Finding generalized projected clusters in high dimensional spaces, vol. 29. ACM (2000)
3. Agrawal, R., Gehrke, J, Gunopulos, D., Raghavan, P.: Automatic subspace clustering of high dimensional data for data mining applications, vol. 27. ACM (1998)
4. Bhattacharya, I., Getoor, L.: A latent dirichlet model for unsupervised entity resolution. In: SDM, vol. 5, p. 59. SIAM (2006)
5. Blei, D.M., Lafferty, J.D.: Topic models. Text Min.: Classif., Clustering, Appl. **10**, 71 (2009)
6. Blei, D.M., Ng, A.Y., Jordan, M.I.: Latent dirichlet allocation. J. Mach. Learn. Res. **3**, 993–1022 (2003)
7. Chan, E.Y., Ching, W.K., Ng, M.K., Huang, J.Z.: An optimization algorithm for clustering using weighted dissimilarity measures. Pattern Recogn. **37**(5), 943–952 (2004)
8. Chen, X., Xu, X., Huang, J.Z., Ye, Y.: Tw-(k)-means: automated two-level variable weighting clustering algorithm for multiview data. IEEE Trans. Knowl. Data Eng. **25**(4), 932–944 (2013)
9. Chen, X., Ye, Y., Xu, X., Huang, J.Z.: A feature group weighting method for subspace clustering of high-dimensional data. Pattern Recogn. **45**(1), 434–446 (2012)
10. Cheng, C.-H., Fu, A.W., Zhang, Y.: Entropy-based subspace clustering for mining numerical data. In: Proceedings of the Fifth ACM SIGKDD International Conference on Knowledge Discovery and Data Mining, pp. 84–93. ACM (1999)
11. De Soete, G.: Optimal variable weighting for ultrametric and additive tree clustering. Qual. Quant. **20**(2–3), 169–180 (1986)
12. De Soete, G.: Ovwtre: a program for optimal variable weighting for ultrametric and additive tree fitting. J. Classif. **5**(1), 101–104 (1988)
13. DeSarbo, W.S., Carroll, J.D., Clark, L.A., Green, P.E.: Synthesized clustering: a method for amalgamating alternative clustering bases with differential weighting of variables. Psychometrika **49**(1), 57–78 (1984)

14. Domeniconi, C., Papadopoulos, D., Gunopulos, D., Ma, S.: Subspace clustering of high dimensional data. In: SDM, vol. 73, p. 93. SIAM (2004)
15. Friedman, J.H., Meulman, J.J.: Clustering objects on subsets of attributes (with discussion). J. R. Stat. Soc.: Ser. B (Stat. Methodol.) **66**(4), 815–849 (2004)
16. Frigui, H., Nasraoui, O.: Simultaneous clustering and dynamic keyword weighting for text documents. In: Berry, M.W. (ed.) Survey of Text Mining, pp. 45–72. Springer, New York (2004)
17. Frigui, H., Nasraoui, O.: Unsupervised learning of prototypes and attribute weights. Pattern Recogn. **37**(3), 567–581 (2004)
18. Goil, S., Nagesh, H., Choudhary, A.: Mafia: efficient and scalable subspace clustering for very large data sets. In: Proceedings of the 5th ACM SIGKDD International Conference on Knowledge Discovery and Data Mining, pp. 443–452 (1999)
19. Jing, L., Ng, M.K., Huang, J.Z.: An entropy weighting k-means algorithm for subspace clustering of high-dimensional sparse data. IEEE Trans. Knowl. Data Eng. **19**(8), 1026–1041 (2007)
20. Jing, L., Ng, M.K., Xu, J., Huang, J.Z.: Subspace clustering of text documents with feature weighting K-means algorithm. In: Ho, T.-B., Cheung, D., Liu, H. (eds.) PAKDD 2005. LNCS (LNAI), vol. 3518, pp. 802–812. Springer, Heidelberg (2005)
21. Makarenkov, V., Legendre, P.: Optimal variable weighting for ultrametric and additive trees and k-means partitioning: methods and software. J. Classif. **18**(2), 245–271 (2001)
22. Modha, D.S., Spangler, W.S.: Feature weighting in k-means clustering. Mach. Learn. **52**(3), 217–237 (2003)
23. Nguyen, N., Caruana, R.: Consensus clusterings. In: Seventh IEEE International Conference on Data Mining, ICDM 2007, pp. 607–612. IEEE (2007)
24. Steyvers, M., Griffiths, T.: Probabilistic topic models. Handb. Latent Semant. Anal. **427**(7), 424–440 (2007)
25. Wahid, A., Gao, X., Andreae, P.: Exploiting user queries for search result clustering. In: Lin, X., Manolopoulos, Y., Srivastava, D., Huang, G. (eds.) WISE 2013, Part I. LNCS, vol. 8180, pp. 111–120. Springer, Heidelberg (2013)
26. Wahid, A., Gao, X., Andreae, P.: Multi-view clustering of web documents using multi-objective genetic algorithm. In: 2014 IEEE Congress on Evolutionary Computation (CEC), pp. 2625–2632. IEEE (2014)
27. Wahid, A., Gao, X., Andreae, P.: Multi-objective multi-view clustering ensemble based on evolutionary approach. In: IEEE Congress on to Appear in Evolutionary Computation, CEC 2015. IEEE (2015)
28. Wilcoxon, F.: Individual comparisons by ranking methods. Biom. Bull. **1**, 80–83 (1945)
29. Woo, K.-G., Lee, J.-H., Kim, M.-H., Lee, Y.-J.: Findit: a fast and intelligent subspace clustering algorithm using dimension voting. Inf. Softw. Technol. **46**(4), 255–271 (2004)
30. Yang, J., Wang, W., Wang, H., Yu, P.: δ-clusters: csubspace correlation in a large data set. In: Proceedings of the 18th International Conference on Data Engineering, pp. 517–528. IEEE (2002)
31. Zhao, Y., Karypis, G.: Comparison of agglomerative and partitional document clustering algorithms. Technical report, DTIC Document (2002)
32. Zhong, S., Ghosh, J.: A comparative study of generative models for document clustering. In: Proceedings of the Workshop on Clustering High Dimensional Data and Its Applications in SIAM Data Mining Conference (2003)

Using Web Collaboration to Create New Physical Learning Objects

André Peres[1]([✉]), Evandro Manara Miletto[1], Fabiana Lorenzi[2],
Elidiane Zayaeskoski[1], Gianfranco Meneguz[1], and Ramon Costa da Silva[1]

[1] Instituto Federal de Educação Ciência e Tecnologia do Rio Grande do Sul - IFRS,
Campus Porto Alegre, Porto Alegre, RS, Brazil
{andre.peres,evandro.miletto}@poa.ifrs.edu.br,
{ezayaeskoski,gmeneguz,ramondasilv}@gmail.com
[2] Universidade Luterana do Brasil - ULBRA, Canoas, RS, Brazil
fabilorenzi@gmail.com

Abstract. This paper describes the construction of a technological and collaborative infrastructure for the creation of physical learning objects. The solution uses a social network, digital fabrication lab and wiki page as an way to propose, build and publish educational objects. The idea is to create an innovative solution to get together a multidisciplinary team of students, teachers, professors and researchers that can propose, define, coordinate, build and publish these objects. The structure is presented and also the first objects and results.

Keywords: Digital fabrication · Social networks · Learning objects

1 Introduction

The reduction of the costs and consequent popularization of electronic equipments allows the creation of new technological solutions outside of the mainstream development industries. This new scenario creates a new generation of people motivated by the do-it-yourself culture, also known as the maker generation.

The expansion of the maker community results in the creation of collaborative spaces named: makerspaces - to develop any kind of object, technological or not; hackerspaces - with a more technological approach; and fablabs - a global network of collaborative digital fabrication labs created by the Center of Bits and Atoms of MIT [1].

Collaborative spaces for digital fabrication has a fundamental role in the creation of new objects because, despise of the reduced cost of the fabrication equipment (like 3D printers and laser cutters), the makers use the collaboration in order to increase the knowledge needed to make something. In these spaces, the users share specialized knowledge and techniques like design, software modeling, machine use, electronics, etc., in order to build new complete solutions. The users normally depend on each other and they grow in knowledge with the collaborative process.

© Springer International Publishing Switzerland 2015
J. Wang et al. (Eds.): WISE 2015, Part II, LNCS 9419, pp. 139–148, 2015.
DOI: 10.1007/978-3-319-26187-4_10

This paper presents the description of a collaborative web based technological infrastructure capable of produce a collection of educational tools in a digital fabrication lab. This collection is composed by physical objects coupled with sensor, actuators and network capabilities to be used in educational experiments. The propose and definition of the objects are made through a collaborative virtual social network and, after the definition process, the fabrication is made in the lab. At the end of the process, the final object is published in the web to be reproduced by anyone.

With this structure, we can use the object in classroom experiments, allowing the students to construct the relation between theory and practice, experimenting collaboration through the fabrication of the object and creativity to use and modify the object.

The authors intend to share the objects so that they can be recreated, used and modified in collaborative educational spaces by students, teachers, researchers and professors in educational and creative activities.

This project aims to build the needed infrastructure so that the collaborative learning process can occur, in a multidisciplinary fashion, through the creation of new physical learning objects.

2   Digital Fabrication and Education

The use of creation/fabrication spaces and the concept of "learning by doing" is aligned with the constructivist theories of Piaget. To Piaget, the "...use of active methods which give broad scope to the spontaneous research of the child or adolescent and requires that every new truth to be learned, be rediscovered, or at least reconstructed by the student and not simply imported to him." [2] apud [3]. To Piaget, teachers at the university and secondary levels should known their subjects and also make an interdisciplinary approach.

To researcher Seymour Papert: "In our image of a school computation laboratory, an important role is played by numerous controller ports which allow any student to plug any device into the computer... The laboratory will have a supply of motors, solenoids, relays, sense devices of various kids, etc. Using them, the students will be able to invent and build an endless variety of cybernetic systems." [2] apud [4].

The use of new technologies in education environments is not something new. The use of low costs electronics, sensors and controllers started with Papert, Michel Resnick and Fred Martin in the 90's using the Lego Mindstorm Kit [5]. Starting from these initial experiments, the increase in the availability of new low cost technologies normally brings new studies in how to use them in learning activities.

The specification and design of new objects normally takes place inside educational institutions, with the teacher acting as the starter of the process. The students engage in the fabrication and experimentation in local labs. Our goal is to expand this process making the specification step more embracing and allowing the share of the creations.

In order to achieve this goal the implemented infrastructure should allow that anyone with basic technological resources (internet connection, some form of fabrication and basic electronics) can study and build the available learning objects. Anyone should be able to add, recreate and modify the objects.

3   Creation and Publication of the Objects

The creation process follows three phases which are presented in Fig. 1.

In the first phase we have the proposer of the object, which describes the objectives of some learning experiment. He/she describes these objectives, publish this information for a community of teachers, students and researchers and invites some other specialist users.

All the people involved in this first definition phase create a micro-community inside our infrastructure. This micro-community collaborates through an "object space" created by the proposer, inside a virtual social network environment. They use this space to post comments, files and ideas in order to make the specification of the proposed object.

The objectives of the new learning object should consider: the theory involved; some initial ideas about the design; if electronic components should be used; and some initial ideas about the educational experiments that can be performed with this object.

Starting from this initial description, the micro-community collaborate through posts and files, discussing and presenting ideas about the object. In this phase, the micro-community should define the design attributes that the new object should have such as [6]:

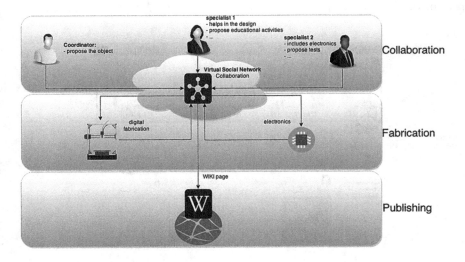

Fig. 1. Creation and publishing process

- **Affordances** - the shape of the object regarding the way that it should be used. The right shape leads the user to the right use of the object;
- **Signifiers** - signals in the object to lead the user like icons, buttons, colors, text, etc.;
- **Discoverability and Feedback** - the user should figure it out how to use the object and have the right feedback to each action performed.

The same process happens with the electronic parts needed in the object. The electronic specialists should interfere in the design process, indicating the restrictions and considerations regarding the electronic aspects of the object.

The construction of the educational aspects and the global/abstract development of the new object happens through collaboration inside this object space. All the creative process will be available and public to the community, serving as a new case for other object spaces.

We used the HumHub system as our virtual social network [7]. This system has a familiar user interface (similar to other social networks, like facebook) and allows the creation of discussion "spaces" among users. Each user can create a new space, publish it and invite other users to participate in that space. Each new object is proposed in a new space and the users collaborate in the space, forming the micro-community.

The Fig. 2 presents a object space in the social network interface. The users can publish posts and files, create wiki pages (through HumHub plugins), create collaborative text files, etc. In Fig. 2 the object space defines an eolic turbine to be used in environmental classes.

The object proposer is responsible for the coordination of the object space, guidance and allowing the collaborative process among the micro-community. Whenever necessary, the proposer can schedule meetings (virtual or face meetings).

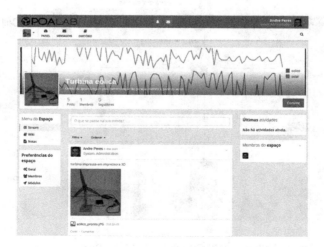

Fig. 2. Object space in the project social network

When the object is defined in its design, functionality, use and electronic requirements, we start the second phase: the fabrication.

In the fabrication phase, the team responsible for the fabrication of the object meets up at the digital fabrication lab. At this point, the collaboration is made in the lab and all the work is documented in the object space at the social network. Also, all the files generated in the softwares used to the design of the object (vector files, 3d files, computer code, etc.) are also stored in the object space.

The fabrication lab that we used has: three simple 3D printers (RepRap based); 1 vinyl cutter; 1 laser cutter; 1 cnc precision mill; arduino boards, sensors and other electronic equipment.

The object space is now used to post questions about the fabrication process, register use cases and to post proposition about modifications in the object.

When the object presents a positive history of use within the community, the final layout and files are published in an external repository in the internet. This publication is the third phase of the project.

For the third phase, we choose to use a wiki page to store the description, specifications, instructions and files of the objects that presents a positive history in the community. We used the MediaWiki system [8].

The wiki page serves as a repository for the files, and also as a way for users outside of our social network to interact with the objects designs. Anyone can edit the wiki, modify the pages and even create new pages (with new objects).

Each object should have published: its description and objectives; the list of electronic components needed; the electronic circuit description and schematics; images/photos; digital fabrication files; code; and examples of use in educational activities.

The Fig. 3 presents a wiki page that describes how to create the eolic turbine. In the page there is the description of: all the parts needed; how to change a computer fan in order to make a turbine; the 3D printer files to make the tower and helices of the turbine; code files to connect the turbine to an Arduino board [9]; and some PHP code to present the generated energy to an end user through the web.

4   First Results

The fabricated objects serve as a base for educational activities in different fields of knowledge. We present the first five objects in the next subsections.

4.1   Electromagnetic Sensor

The Electromagnetic Sensor object was proposed by one of our students based on the similar project published in [10]. The definition and design was made by 2 students in the social network.

The object goal was to be used in physics educational experiments, detecting the intensity of electromagnetic emissions and displaying this information in the computer screen.

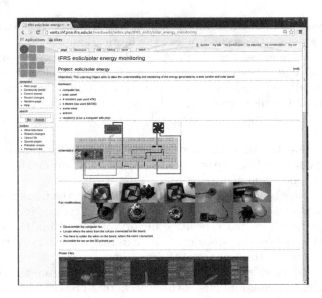

Fig. 3. Project wiki page

The electromagnetic interference is caused by the electron propagation in a conductor. The electric current in this conductor generates the electromagnetic field around it.

The sensor node quantifies the intensity of the electromagnetic field received by an antenna and plot this value in a graphic interface. In order to do so, the antenna is connected to an Arduino Uno board and the data received is send to a computer through the Arduino's USB interface. The computer generates the graphic interface in real time.

This object can be used to demonstrate the concepts of induction, interference and electromagnetism. The Fig. 4 presents the object, circuit layout and graphic interface.

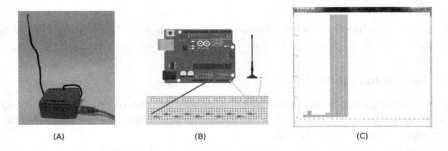

(A)   (B)   (C)

Fig. 4. (A) Electromagnetic measurement object (B) Circuit and (C) User interface

4.2   Water Conductivity Sensor

Similar to the Electromagnetic Sensor, the Water Conductivity Sensor also was proposed by 2 students based on [10]. After proposing the object, the students contact researchers from the Environmental courses in our institution. They contribute to the development of the object and in the lab tests.

The electrical conductivity is used to measure a material's ability to conduct an electric current. In aquatic environments it can be used to identify the presence of extra volume of ions (in polluted water with inorganic matter, for instance).

The authors of this paper has already worked with the Environmental researchers in a water quality project [11]. This new sensor aims to reduce costs in this type of environmental monitoring.

In this object we used an Arduino Uno board and connected two cables in 2 analog ports. We measure the conductivity in water sending 5 V to one of the cables and analyzing the current received by the other cable. With this values we can determine the resistance of the water.

In the same way as the Electromagnetic Sensor, we connected the Arduino board to a computer and generate a real time graphic with the values obtained by the sensor. The Fig. 5 presents the object, circuit layout and graphic interface.

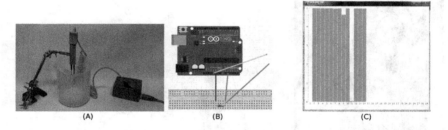

Fig. 5. (A) Water conductivity sensor object (B) Circuit and (C) User Interface

4.3   Greenhouse Monitoring and Control

Getting together researchers from Computer, Biotechnology and Environmental area, the Greenhouse monitoring and control is a web system developed to control a small greenhouse in the institution.

The object was proposed by one computer science student as its final project and evaluated by one computer science and one environmental professors.

This object controls the humidity, light and temperature of the greenhouse and has the ability to change these parameters by turning on led lights, water pump and fans.

The humidity, luminosity and temperature influence in every stage in the agriculture production. The construction of greenhouses helps in the control of these parameters and, the use of automatic resources increases the efficiency of the production.

It was developed a computer system capable of monitoring and controlling these factors using sensors and actuators connected to an Arduino board. The Arduino has an ethernet shield that allows the remote communication. The board collects data from sensors and send the data to a remote database through the Internet. The system has a defined policy regarding the expected humidity, light and temperature based on the specific plant needs.

The user can remotely monitor and interact with the board and the data is kept in the database for historic proposes.

The greenhouse is used by the environmental researchers in order to monitor the soil and make experiments with it, and by the biotechnology researchers to study plants reaction to specific chemicals. The automation of the greenhouse allows a better control over the plants which reflects in the quality in the results (the assurance that the results are not compromised by mistakes in the culture of the plants).

The Fig. 6 presents the arduino board, the sensors and the graphic monitoring interface.

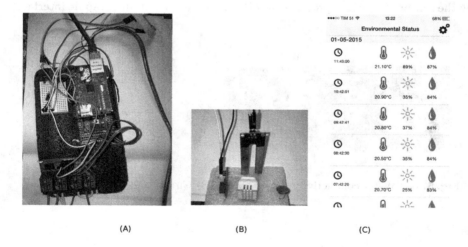

(A)   (B)   (C)

Fig. 6. (A) Arduino controller (B) Object sensors and (C) User interface

4.4   Air Quality Monitorining

Another final project proposed by one of the computer science students was the construction of a node for air quality monitoring. This object was evaluated by one computer science and one environmental professor.

It was created an object that is able to obtain data about the air quality using sensors and publishing it in order to be used in educational activities.

The object uses the National Council of Environment (CONAMA - Conselho Nacional do Meio Ambiente) air quality reference data in order to indicate the air quality.

The collected data is stored in a database and published as: raw data for scientific use; geographic data presenting a map with the sensor location and the air quality digest information; and in an educational interface, displaying the physiological effects of concentrations of obtained compounds data in humans.

The collected compounds are: Methane Gas Sensor - MQ-4, Carbon Monoxide Sensor - MQ-7, Hydrogen Gas Sensor - MQ-8, LPG Gas Sensor - MQ-6 and Optical Dust Sensor.

4.5   Eolic Turbine and Solar Panel

An eolic turbine was proposed by one computer science professor together with one environmental professor. The environmental course has an electric generation lab that is used by its students in order to understand the energy generation.

The lab has some energy generation kits capable of demonstrate how battery and solar panels works. Other objects were created by students using alternative materials.

Using the lab objects as a starting point, the professors proposed the construction of an eolic turbine made with a computer fan, connected to an arduino board. The computer fan was modified in order to generate energy. The arduino board monitors the amount of generated energy and sends this information to an raspberry-pi board through the USB port. The raspberry-pi has an LAMP (Linux, Apache, Mysql and PHP) environment and plots the received data in to a graph.

It was also connected to the arduino board one solar panel. The graph plotted by the raspbberry-pi shows both graphs - the eolic turbine and solar panel.

The eolic turbine tower and solar panel base was fabricated using an 3D printer. The Fig. 7 presents the object, the circuit and the web interface generated by the raspberry-pi.

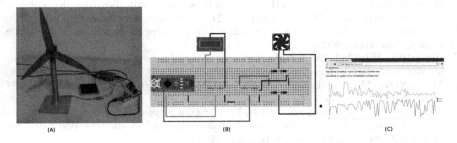

Fig. 7. (A) Eolic turbine and solar panel object (B) Circuit (C) User interface

5   Conclusions

This project put together a set of technological solutions in order to allow a better interaction and collaboration among students, professors, teachers and researchers in the development of new objects that can be used in educational activities.

We created this infrastructure with: social networking as a platform for proposition, definition, coordination and collaboration in the construction of innovative educational objects; a digital fabrication lab to construct the proposed object and be a place to meet and talk about the objects; and a publication platform in a wiki page, that can publish all the objects created together with their files, photos and description, allowing anyone to build, modify, increment or adapt the objects.

We consider that the collaboration process was successful during the proposing and fabrication of the first objects. Also, the authors consider that this first objects can serve as a base for the expansion of the infrastructure through the inclusion of new collaboration mechanisms in the social network and the increase in the number of its users.

Acknowledgment. The authors would like to thank CNP-q and CAPES/LIFE for sponsoring this project.

References

1. Makezine: Is it a hackerspace, makerspace, techshop, or fablab? (2015). http://makezine.com/2013/05/22/the-difference-between-hackerspaces-makerspaces-tech shops-and-fablabs/
2. Martinez, S.L., Stager, G.: Invent To Learn: Making, Tinkering, and Engineering in the Classroom. Constructing Modern Knowledge Press, Torrance (2013)
3. Piaget, J.: To Understand IsTo Invent: The Future of Education. Penguim Books, New York (1976)
4. Papert, S., Solomon, C.: Twenty Things to do with a computer - Artificial Intelligence Memo #248. Massachusetts Institute of Technology, Cambridge (1971)
5. Blikstein, P.: Digital fabrication and 'making' in education: The democratization of invention (2013). https://tltl.stanford.edu/sites/default/files/files/documents/publications/Blikstein-2013-Making_The_Democratization_of_Invention.pdf
6. Norman, D.A.: The Design of Everyday Things. Reprint paperback edn. Basic Books, New York (2002)
7. HumnHub: Hum hub the flexible open source social network kit home page (2015). https://www.humhub.org/en/overview
8. MediaWiki: Mediawiki home page (2014). https://www.mediawiki.org/wiki/MediaWiki
9. Arduino: Arduino home page (2013). http://www.arduino.cc
10. Gertz, E., Justo, P.D.: Environmental Monitoring with Arduino - Watching our World with Sensors. O'Reilly, Sebastopol (2012)
11. Peres, A.P., Miletto, E.M., Kapusta, S., Ojeda, T., Lacasse, A., Gagnon, J.: Waits - an it structure for environmental informatio via open knowledge, dynamic dashboards and social web of things. In: Proceedings of the IADIS International Conference WWW/Internet 2013, vol. 1 (2013)

A Custom Browser Architecture to Execute Web Navigation Sequences

José Losada[✉], Juan Raposo, Alberto Pan, Paula Montoto,
and Manuel Álvarez

Facultad de Informática, Information and Communications
Technology Department, University of A Coruña,
Campus de Elviña, s/n, 15071 A Coruña, Spain
{jlosada,jrs,apan,pmontoto,mad}@udc.es

Abstract. Web automation applications are widely used for different purposes such as B2B integration and automated testing of web applications. Most current systems build the automatic web navigation component by using the APIs of conventional browsers. This approach suffers performance problems for intensive web automation tasks which require real time responses and/or a high degree of parallelism. Other systems use the approach of creating custom browsers to avoid some of the tasks of conventional browsers, but they work like them, when building the internal representation of the web pages. In this paper, we present a complete architecture for a custom browser able to efficiently execute web navigation sequences. The proposed architecture supports some novel automatic optimization techniques that can be applied when loading and building the internal representation of the pages. The tests performed using real web sources show that the reference implementation of the proposed architecture runs significantly faster than other navigation components.

Keywords: Web automation · Optimization · Browser architecture

1 Introduction

Most today's web sources do not provide suitable interfaces for software programs. That is why a growing interest has arisen in so-called web automation applications that are able to automatically navigate through websites simulating the behavior of a human user. Web automation applications are widely used for different purposes such as B2B integration, web mashups, automated testing of web applications, Internet meta-search or business watch. For example, a technology watch application can use web automation to automatically search in the different websites and daily retrieve new patents and articles of a predefined area of knowledge.

A crucial part of web automation technologies is the ability to execute automatic web navigation sequences. An automatic web navigation sequence consists in a sequence of steps representing the actions to be performed by a human user over a web browser to reach a target web page. Figure 1 illustrates an example of a web navigation sequence that retrieves the list of patents matching the search term *"World Wide Web"* in the European Patent Office website (www.epo.org).

© Springer International Publishing Switzerland 2015
J. Wang et al. (Eds.): WISE 2015, Part II, LNCS 9419, pp. 149–163, 2015.
DOI: 10.1007/978-3-319-26187-4_11

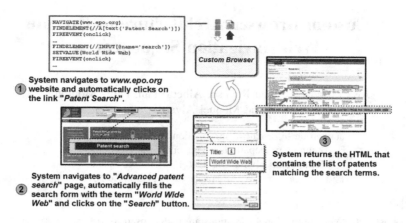

Fig. 1. Navigation sequence example.

The approach followed by most of the current web automation systems [2, 9, 11, 15–17] consists in using the APIs of conventional web browsers to automate the execution of navigation sequences. This approach does not require developing a custom navigation component, and guarantees that the accessed pages will behave the same as when they are accessed by a human user. While this approach is adequate to some web automation applications, it presents performance problems for intensive web automation tasks which require real time responses (because web browsers are client-side applications and they consume a significant amount of resources).

There exist other systems which use the approach of creating custom browsers to execute web navigation sequences [5, 8, 10]. Since they are not oriented to be used by humans, they can avoid some of the tasks of conventional browsers (e.g. page rendering). Nevertheless, they work like conventional browsers when building the internal representation of the web pages. Since this is the most important part in terms of the use of computational resources, their performance enhancements are not very significant.

In this work, we present a custom browser architecture oriented to the efficient execution of web navigation sequences. This architecture is influenced by a set of optimizations that we have designed to be automatically applied during the process of loading and building the internal representation of the web pages. Some of these optimizations are based on the fact that, in the web automation systems, navigation sequences are defined 'a priori' and executed multiple times. Using this peculiarity, the navigation component can extract some useful information during the first execution of the sequence (at definition time) and use that information in the next executions of the same sequence, to minimize the use of resources (CPU, memory, bandwidth and execution time). To support these optimizations, the proposed architecture includes some novel components not present in any other web navigation systems.

The rest of the paper is organized as follows. Section 2 briefly describes the models our approach relies on. Section 3 presents an overview of the architecture and functioning of the conventional and custom browsers, and introduces a set of automatic optimizations that can be applied in custom browsers. Section 4 explains in detail the proposed architecture. Section 5 describes the experimental evaluation of the approach. Section 6 discusses related work. Finally, Sect. 7 summarizes our conclusions.

2   Background

2.1   Document Object Model

The main model we rely on is the Document Object Model (DOM) [4]. This model describes how browsers internally represent the HTML web page currently loaded and how they respond to user performed actions on it. An HTML page is modelled as a tree, where each HTML element is represented by an appropriate type of node. An important type of nodes are the script nodes, used to execute a script code typically written in a scripting language such as JavaScript.

In addition, every node in the tree can receive events produced (directly or indirectly) by the user actions. Event types exist for actions such as clicking on an element (*click*), or moving the mouse cursor over it (*mouseover*), to name but a few. Each node can register a set of listeners for different types of events. An event listener executes arbitrary script code that has the entire page DOM tree accessible and can perform actions such as modifying existing nodes, creating new ones or even launching new events.

2.2   Dependencies Between Nodes

In our previous work [12], we introduced the concept of dependency between nodes of the DOM. This is a key concept in the custom browser architecture proposed in this work. We can summarize the idea with the following definitions:

Definition 1. We say the node *n1* depends on node *n2* when *n2* is necessary for the correct execution of *n1*. We say that *n2* is a dependency of *n1* and denote it as $n1 \rightarrow n2$. The following rules define this type of dependencies:

1. If the script code of a node *s1* uses an element (e.g. a function or a variable) declared or modified in a previous script node *s2*, then $s1 \rightarrow s2$. Rationale: to be able to execute the script code of *s1*, the node *s2* must be executed previously.
2. If the script code of a node *s* uses a node *n*, then $s \rightarrow n$. Rationale: to be able to execute the script code of *s*, the node *n* must be loaded previously, e.g., if *s* obtains a reference to an anchor node (e.g. using the function *getElementById*) and navigates to the URL specified by its *href* attribute, then it will not be possible to execute *s* unless the anchor node is loaded.
3. If the script code of a node *s* makes a modification in a node *n*, then $n \rightarrow s$. Rationale: the action performed by *s* may be needed to allow *n* to be used later, e.g., if *s* modifies the *action* attribute of a form node to set the target URL, then it will not be possible to submit the form unless *s* is executed previously.

Definition 2. We say that there exists a dependency conditioned to the event *e* being fired over the node *n*, between two nodes *n1* and *n2*, when the node *n2* is necessary for the correct execution of the node *n1*, when the event *e* is fired over the node *n*. We denote this as $n1 \rightarrow^{e|n} n2$. For example, suppose *n* is a node with a listener for the *onMouseOver* event. The listener uses a function defined in *s*. Then $n \rightarrow^{onMouseOver|n} s$. Analogous rules to the ones explained before define this type of dependencies, which, in this case, involve nodes containing event listeners.

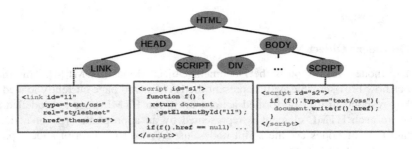

Fig. 2. Example of dependencies between nodes.

Figure 2 illustrates an example of dependencies between nodes. In the example, the script *s1* defines the function *f*. This function access the link node *l1* (using the function *getElementById*), so *s1* → *l1*. In addition, the script *s2* uses the function *f*, so *s2* → *s1*.

3   Overview

This section presents an overview about the architecture and functioning of the conventional browsers (Sect. 3.1), custom browsers (Sect. 3.2), and introduces a set of automatic optimizations that can be applied in custom browsers (Sect. 3.3).

3.1   Conventional Web Browsers

A web browser is a software application used for retrieving and presenting resources downloaded from the WWW. The architecture of the modern web browsers (Fig. 3a) [6] includes the high level components: Graphical User Interface, Browser Engine and Rendering Engine; and the auxiliary subsystems: JavaScript Interpreter, Networking, Display Backend, HTML Parser and Data Persistence.

1. The Graphical User Interface includes the browser display area except the main window where the response page is rendered (address bar, toolbars, main menu, etc.).
2. The Browser Engine is a high level interface for querying and manipulating the rendering engine. It provides methods for high level browser actions, e.g., initiate the loading of a URL, go back to the previous page, etc.
3. The Rendering Engine represents the core of the browser. It is the responsible for processing and painting the HTML contents. The page loading process fires a set of events in cascade and most of them are processed sequentially by this component.

Due to the semantics of JavaScript, web browsers execute scripts in a sequential form. Nevertheless, there are some special cases where they can execute JavaScript in parallel. First, the scripts containing the attribute *async* can be executed asynchronously with the rest of the page loading. This feature has been introduced in HTML5 [18]. The other scenario where JavaScript can be executed in parallel is using Web Workers (also introduced in HTML5). A Web Worker can execute JavaScript in background but have the major limitation that the code cannot access the DOM tree objects.

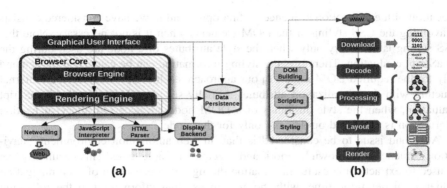

Fig. 3. Web browsers reference architecture and rendering engine (Color figure online).

Figure 3b shows the processing steps of the rendering engine in the web browsers:

1. Download and Decode: the HTML contents are downloaded and decompressed.
2. Processing: the DOM tree is built. For efficiency purposes, this is an incremental process in most of the browsers. When new resources are discovered, they are downloaded and processed (style sheets, scripts, etc.). Style sheets contain presentation information, used to build the page layout. Script nodes contain scripting code.
3. Layout and Rendering: the layout tree contains rectangles with visual attributes like dimensions and colors (this structure is different from the DOM tree). The rendering process paints the layout on the browser window using the display backend layer.

3.2 Custom Browsers

Custom browsers are navigation components, used in web automation systems, specialized in the execution of navigation sequences. Custom browsers usually simulate the behavior of a real browser and they are designed with two main goals: the perfect emulation of a conventional browser (if the custom browser does not behave just the same as a real browser, the sequence execution could lead to wrong web pages) and efficiency in the execution of the navigation sequences.

Custom browsers are not human-oriented and the visualization of the pages is not necessary. This will increase the efficiency because there is no need for building and render the page layout. In the custom browsers architecture, the Browser Engine is also the entry point for accessing the Rendering Engine, but it does not receive commands from the user interface. Instead, the Browser Engine receives the list of commands of the navigation sequence to be executed. These commands will represent events produced by a human user in a real web browser, e.g., navigations to URLs or user events over the DOM elements of the loaded page.

3.3 Automated Optimizations in Custom Browsers

As we have commented, in custom browsers, the rendering of the page layout is not required, because the visualization of the web page is not necessary for the correct execution of the navigation sequence. A first optimization we have considered consists in avoiding the CSS styling of the DOM elements when it is not necessary. Note that CSS styling is necessary only when the style attributes of a node are used during the JavaScript evaluation. Therefore, the styling information can be calculated on-demand only for the required DOM nodes. In our approach, each node will contain an internal structure with the visualization attributes, initially set to null. During the JavaScript evaluation, when the style attributes of a DOM node are accessed, the visualization information is generated on-demand only for that node.

A second issue to be considered is that, in web automation environments, navigation sequences are known 'a priori' and executed multiple times. This peculiarity can be used to extract some useful information during a first execution of each navigation sequence, at definition time, with the goal to use that information in the following executions and improve its efficiency. We will focus in two points:

1. Load minimized DOM trees. As described in our previous work [12], there are a lot of fragments of the web pages that are not necessary for the correct execution of the navigation sequences. For example, if the navigation sequence fills a form and fires a *click* event on the submit button, in most of the cases, many fragments of the page will not be involved in this sequence execution (for example, ads, banners, *iframes*, menus, etc.). These irrelevant fragments can be ignored (not added to the DOM tree), without affecting to the correct execution of the navigation sequence.
2. Parallelize the execution of script nodes. Web browsers execute the scripts contained in the web pages sequentially (except some particular exceptions explained in Sect. 3.1), even when scripts have no dependencies between them. Script elements that are not dependent could be executed in parallel without affecting to the correct execution of the navigation sequence.

To achieve these two objectives, in our approach, the custom browser will work in two phases: optimization and execution. The optimization phase requires one execution of the navigation sequence. In this execution, the navigation component automatically calculates some optimization information and saves it. More in detail, it calculates:

1. Which nodes of the DOM tree are necessary for the correct execution of the sequence, and which ones can be discarded (irrelevant nodes). To do so, the script evaluation is monitored to collect all dependencies between the nodes in the DOM tree (following the rules cited in the Sect. 2.2). Using this dependencies, the irrelevant nodes are identified and represented using XPath-like [19] expressions. This process is deeply described in [12].
2. A script dependency graph, which contains, for each script S in the page, the list of other scripts that must be executed before, because they contain dependencies necessary for the correct execution of the script S. The script dependency graph is also calculated using the dependencies between the nodes obtained during the JavaScript evaluation. The scripts contained in this graph are also represented using XPath-like expressions. This process is deeply described in [13].

The execution phase involves the next executions of the same navigation sequence. In this phase, the rendering engine uses the information previously generated to execute the sequence more efficiently. When each page is loaded, a reduced DOM tree is built, discarding the irrelevant nodes, and the scripts of the page are evaluated in parallel according to the script dependencies graph.

4 Proposed Architecture

In this section, we describe the proposed architecture for a custom browser able to support the previously described optimization techniques.

1. To support the on-demand CSS simulation technique, the custom browser architecture should take into account that:
   (a) The visualization information will be stored in the DOM nodes directly.
   (b) The CSS Subsystem will be accessed only from the JavaScript Engine.
2. To support the minimized DOM optimization technique, the designed architecture should include the following elements:
   (a) A module able to interact with the JavaScript Engine during the optimization phase to collect the dependencies between the nodes.
   (b) A module able to interact with the HTML Subsystem during the execution phase, to detect and discard the irrelevant nodes previously identified.
   (c) The Data Persistent Layer should be extended to provide a mechanism for saving and retrieving the irrelevant nodes of each web page.
3. To support the parallel JavaScript execution technique, the custom browser architecture should include the following elements:
   (a) A module able to calculate the graph with the dependencies between scripts.
   (b) The Data Persistent Layer should be extended to provide a mechanism for saving and retrieving the script dependency graph.
   (c) A pool of reusable threads to execute scripts in parallel.
   (d) A component able to detect available scripts and execute them in parallel using the pool of reusable threads.

4.1 Architecture Core Components

Figure 4 shows the components of the custom browser architecture: Browser Engine, Rendering Engine, Data Persistence Layer and Browser Core Objects; the Rendering Engine subsystems: Main Thread, Event Queue, Dispatcher Thread, Thread Pool (for the parallel script execution); and the auxiliary subsystems: HTML Engine, JavaScript Engine, CSS Subsystem, Networking Layer, and Optimizer.

The **Browser Engine** receives the list of commands from the navigation sequence and translates them into events that are placed in the **Event Queue**. The Event Queue contains the sorted list of events pending for its execution. Each event execution can produce new events (child events) that are also placed in this queue.

The **Dispatcher Thread** is the responsible for assigning events to execution threads. When the custom browser is executed as a regular browser (without using the optimization information), all events are executed in the **Main Thread** one by one, until the queue is empty. When the custom browser is executed using the optimization information previously collected during the optimization phase, the Dispatcher Thread

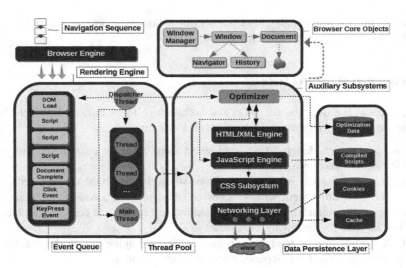

Fig. 4. Browser architecture core components.

analyzes the event queue, looking for scripts ready for its execution that could be evaluated in parallel. A script is parallelizable when all other scripts that depends on, have already finished its execution. The **Thread Pool** (containing reusable threads) is used to evaluate these scripts in parallel. Other events are executed in the Main Thread. If all threads of the pool are busy evaluating scripts, the Main Thread can also execute scripts.

During the optimization phase, the **Optimizer** is the responsible for the calculation of the dependencies between the nodes, the set of irrelevant nodes (not required for the correct execution of the sequence) and the script dependency graph. It is also the responsible for saving the optimization information using the **Data Persistence Layer**. During this phase, when the scripts are evaluated, the **JavaScript Engine** invokes the Optimizer to collect the dependencies between the nodes. Then, after the page loading, when all scripts finished its execution, the Optimizer analyzes these dependencies and generates the optimization information using XPath-like expressions to represent the nodes. In the execution phase, the Dispatcher also uses the Optimizer to detect the scripts that could be executed in parallel. When a script finishes its execution, the Optimizer updates the script dependency graph and the Dispatcher is notified. If there are new scripts ready for execution that could be evaluated in parallel, the Dispatcher places them in the available threads of the pool.

The **CSS Subsystem** parses and stores the CSS snippets. The JavaScript Engine uses the CSS Subsystem to dynamically calculate the CSS style attributes on-demand (during the script evaluation). If the JavaScript code does not reference the style attributes, these calculations can be omitted, saving the corresponding processing time. Figure 5 illustrates an example of on-demand CSS styling and also outlines the pseudo-code of the algorithm that calculates the structure with the CSS properties. In the example, the CSS attributes are calculated only for two nodes (*html* and *body*).

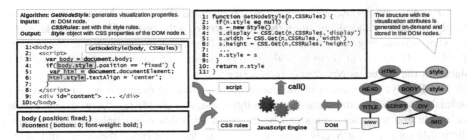

Fig. 5. On-demand CSS styling.

The **HTML Engine** is responsible for parsing the HTML and XML streams. It uses the Optimizer (during the DOM building stage) to identify the irrelevant fragments, and build a minimized version of the DOM tree containing only the relevant nodes.

The **Networking Layer** is responsible for the execution of HTTP requests. Multiple downloads can be executed in parallel. A cache of downloaded files is provided to increase the performance and prevent unnecessary downloads.

All the subsystems can access to the **Browser Core Objects**, including the windows and the documents with the DOM tree of each loaded page. Windows and frames can be accessed thought the window manager object. Each window contains the currently loaded document object (and also the history with previously loaded documents). Each node can contain an additional structure with the visualization information. This structure is generated only if the style attributes are required during the script evaluation.

The **Data Persistence Layer** provides a mechanism for accessing the persistent information, including the cache of downloaded JavaScript and CSS files, the cache of compiled scripts, cookies, optimization information (irrelevant nodes and the script dependency graph) and browser configuration parameters.

4.2   Event Execution Model

The event execution model considers different types of events (e.g. DOM load, script execution, user actions, etc.) and each event stores information about its execution state. Figure 6 shows the supported event states and the transitions between them.

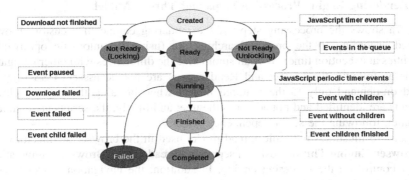

Fig. 6. Event transition states.

Created: initial state, before adding the event to the queue. Most events immediately switch to Ready state when they are inserted in the queue. If the event depends on other actions (e.g. download a file), it will be placed in the queue as *Not Ready (Locking)*. If the event requires a delay (e.g. using *setTimeout* function), it will be placed in the queue as *Not Ready (Unlocking)*.

Not Ready: the event is in the queue but it cannot be executed because there are unfinished pending actions associated to the event. *Not Ready (Locking)* events block the queue and no other events can be executed until this event finishes (except parallelizable scripts). *Not Ready (Unlocking)* events does not block the queue and other events can be executed in the meantime.

Ready: the event is ready to be executed. If it is a script event and it is parallelizable, then it can be executed, in the Main Thread or in a thread of the pool, when all its dependencies (according to the script dependency graph) are in *Completed* state. In other case, it will be executed in the Main Thread following the queue order.

Running: the event is out of the queue and it is being executed. This running event can generate new events that must be executed immediately (even before finishing its own execution). In that case, this running event pauses its execution and returns to the queue. For example, if a style sheet is discovered during the HTML parsing, a CSS event is created and placed in the queue; the HTML parsing stops its execution, returns to the queue (in a position higher than the CSS event), and it will continue just after finishing the CSS processing.

If the event is a periodic timer event (e.g. using *setInterval* function), it will be placed in the queue again, just after finishing its execution.

Finished: the event finished its execution but has unfinished child events. This state is used to correctly detect when a script has completely finished (necessary to evaluate the scripts in parallel). If a script *S*, produces JavaScript child events, other scripts in the page detected as dependent of *S* cannot be executed until the child events of *S* end its execution (at that moment, *S* switch to *Completed* state and the script dependency graph is updated).

Completed: the event and its child events have finished (this is a recursive process).

Failed: the event (or one associated preload action) has finished with errors.

4.3   Rendering Engine Processing Steps and Thread Model

Figure 7a shows the processing steps of the rendering engine of a custom browser designed according to the proposed architecture (using its automatic optimization capabilities at execution time). We can summarize the differences with other navigation systems as follows: the Layout and Render steps are not necessary; CSS rules are applied on-demand only to the required nodes; when building the page, irrelevant fragments are identified and not added to the tree; and finally, the scripts are evaluated in parallel, following the script dependency graph.

Figure 7b illustrates the multi-thread model used in the browser architecture.

Browser Engine Thread: using a separated thread for the Browser Engine allows a better control on the rendering engine. In addition, the navigation sequence commands can be placed in the queue asynchronously.

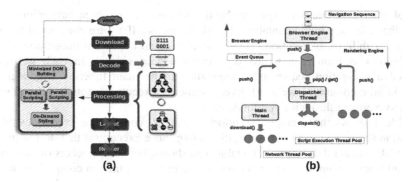

Fig. 7. Rendering engine processing steps and thread model.

Dispatcher Thread: responsible for selecting events from the queue and assigning these events to the execution threads. Events can be selected in queue order (*pop* method) to be executed in the Main Thread (note that, the *pop* method does not always return the first event in the queue, due to the state *Not Ready (Unlocking)*), or not following the queue order (*get* method) if they are script events that can be executed in a thread of the pool.

Main Thread: executes all kind of events. This thread can access to the event queue to place new events generated during the execution of the running event.

Parallel Scripts Thread Pool: execute JavaScript events in parallel. These threads can also place new events in the queue.

Network Thread Pool: executes HTTP requests in parallel.

The inter-thread communication is managed through the Event Queue and the events inserted in it. The Browser Engine will keep a reference to the events added to the queue (when the navigation sequence commands are translated to browser events). Using this reference, the Browser Engine will be able to know the execution progress because each event stores information about its current state, previous transitions between states, child events generated, etc. The Browser Engine will place in the queue special types of events for actions such as stop the execution after a predefined timeout, etc.

5   Evaluation

To evaluate the validity of the proposed architecture, we developed a reference implementation that emulates Microsoft Internet Explorer. This navigation component was implemented in Java using open source libraries. In the experiments, we selected websites from different domains and different countries, included in the top 500 sites on the web according to Alexa [1]. The test machine was a quad-core with 16 GB of RAM. The thread pool size (for parallel script evaluation) was limited to a maximum of 3.

In the first experiment, we tested the architecture performance comparing the execution time of our custom browser using its automatic optimization capabilities with the custom browser without using the optimization techniques, and also with other representative navigation components. On one hand, we used a navigation component based on HtmlUnit [8] because it is a popular open source project with JavaScript and CSS support. On the other hand, we used a navigation component developed using the

APIs of Microsoft Internet Explorer (MSIE from now). The three navigation components (the reference implementation, HtmlUnit and MSIE) were configured to use its caching capabilities. In addition, MSIE was configured to prevent image downloading and plugin execution (e.g. to avoid showing banner videos). In each website we recorded a navigation sequence representative of its main function (e.g. a product search in an e-commerce website). Every sequence executed events to fill and submit forms, to navigate through hyperlinks, etc.

Table 1 shows the average execution time of 30 consecutive executions of each navigation sequence used in the tests, discarding those executions that do not fit in the range of the standard deviation. The table also shows, between brackets (in the second, third and fourth columns), the percentage of the execution time in comparison with the custom browser using the automatic optimization capabilities (first column).

Table 1. Execution times.

	OPTIMIZED (MS)	NOT OPTIMIZED (MS)	HTMLUNIT (MS)	MSIE (MS)
360.CN	2613	5274 (219%)	6116 (234%)	7974 (305%)
ALIBABA.COM	2759	6249 (226%)	13350 (483%)	12025 (435%)
ALLEGRO.PL	1027	5142 (500%)	9372 (912%)	11185 (1089%)
AMAZONWS.COM	2093	3455 (165%)	9802 (468%)	9722 (464%)
BBC.COM	1220	4194 (343%)	7901 (647%)	6898 (565%)
BET365.COM	1092	1787 (163%)	3195 (292%)	10922 (1000%)
BILD.DE	3450	18003 (521%)	21207 (614%)	15161 (439%)
BLOGGER.COM	1004	4033 (401%)	4409 (439%)	7750 (771%)
BLOOMBERG.COM	2880	5738 (199%)	11149 (387%)	11874 (412%)
BOOKING.COM	5105	6731 (131%)	11378 (222%)	12649 (247%)
CNET.COM	1298	2586 (199%)	3360 (258%)	11586 (892%)
ENGADGET.COM	496	1890 (381%)	5843 (1178%)	9198 (1854%)
FORBES.COM	754	3420 (453%)	3846 (510%)	6706 (889%)
GITHUB.COM	1183	3587 (303%)	3001 (253%)	7254 (613%)
GIZMODO.COM	607	1752 (288%)	2124 (349%)	9109 (1500%)
GSMARENA.COM	1388	8172 (588%)	9084 (654%)	10706 (771%)
IGN.COM	2269	4768 (210%)	4834 (213%)	9135 (402%)
IKEA.COM	199	1105 (555%)	1494 (750%)	5470 (2748%)
IMGUR.COM	2048	14556 (710%)	17402 (849%)	13343 (651%)
INDIATIMES.COM	1517	7732 (509%)	6512 (429%)	7930 (522%)
INSTAGRAM.COM	1367	2419 (176%)	1969 (144%)	7867 (575%)
LEMONDE.FR	405	1950 (481%)	7996 (1974%)	8679 (2142%)
LIBERO.IT	852	2605 (305%)	1930 (226%)	4386 (514%)
LIFEHACKER.COM	1162	1862 (160%)	4298 (369%)	8702 (748%)
LINKEDIN.COM	1507	4405 (292%)	6685 (443%)	5754 (381%)
LIVEJOURNAL.COM	1350	10655 (789%)	19849 (1470%)	17942 (1329%)
MARCA.COM	899	8007 (890%)	10026 (1115%)	9741 (1083%)
MASHABLE.COM	666	1879 (282%)	3089 (463%)	6742 (1012%)
MEDIAFIRE.COM	4271	5935 (125%)	6409 (135%)	7832 (165%)
PETFLOW.COM	1283	5433 (423%)	5853 (456%)	6673 (520%)
PINTEREST.COM	5263	6877 (130%)	6463 (122%)	8310 (157%)
REDIFF.COM	1799	5024 (279%)	5865 (326%)	7531 (418%)
REUTERS.COM	7021	18125 (258%)	20031 (285%)	16620 (236%)
RT.COM	2566	7064 (275%)	12033 (468%)	9913 (386%)
SCRIPBD.COM	8005	9673 (120%)	11923 (148%)	14817 (185%)
SOFTONIC.COM	933	3524 (377%)	4821 (516%)	6403 (686%)
SOURCEFORGE.NET	4868	10593 (217%)	12828 (263%)	18260 (375%)
SPEEDTEST.NET	2139	4932 (230%)	6386 (298%)	10823 (505%)
STACKEXCHANGE.COM	2097	5008 (238%)	5541 (264%)	9312 (444%)
TAOBAO.COM	1588	2472 (155%)	8051 (506%)	11610 (731%)
TARINGA.NET	5249	10886 (207%)	8734 (166%)	9695 (184%)
TECHCRUNCH.COM	604	2095 (346%)	5868 (971%)	8551 (1415%)
THEFREEDICTIONARY.COM	915	6307 (689%)	6539 (714%)	7112 (777%)
TIME.COM	4092	6243 (152%)	12103 (295%)	11676 (285%)
TRIPADVISOR.COM	1050	2281 (217%)	6561 (624%)	6997 (666%)
TUMBLR.COM	3469	5054 (145%)	5805 (167%)	7858 (226%)
UPLOADED.NET	1496	3321 (221%)	3682 (246%)	9558 (638%)
UPS.COM	2232	4016 (179%)	3057 (136%)	5846 (261%)
USATODAY.COM	335	1280 (382%)	2161 (645%)	4478 (1336%)
WARRIORFORUM.COM	1540	3091 (200%)	3396 (220%)	7913 (513%)
WEATHER.COM	2045	4457 (217%)	11260 (550%)	10407 (508%)
WIX.COM	1655	3097 (186%)	4024 (241%)	4908 (294%)
WORDPRESS.COM	1975	2770 (140%)	2848 (144%)	10793 (546%)
WORDREFERENCE.COM	832	5086 (611%)	7778 (934%)	6507 (782%)
XDA-DEVELOPERS.COM	3175	5966 (187%)	9793 (308%)	10585 (333%)
YAHOO.COM	3680	5483 (148%)	6590 (179%)	8734 (237%)
YOUTUBE.COM	664	1872 (281%)	2730 (411%)	6334 (953%)
ZIPPYSHARE.COM	1049	2680 (255%)	2228 (212%)	5867 (559%)
AVERAGE		310%	470%	684%
AVERAGE ± STDEV		240%	349%	544%
MEDIAN		246%	378%	534%

The execution of the custom browser using its optimization capabilities always got best results (first column). Calculating the average of the percentages, the execution time of the custom browser without using its automatic optimization capabilities is 3.1 times slower (310 %). Discarding the results that do not fit in the range of the average ± standard deviation it is 2.4 times slower, and the median value indicates that it is 2.46 times slower. Regarding the other two browsers, HtmlUnit is the one that got better results. It is, in average, 4.7 times slower (470 %). Discarding the results that do not fit in the range of the average ± standard deviation it is 3.49 times slower, and the median value indicates that it is 3.78 times slower. The navigation component based on MSIE is, in average, 6.84 times slower (684 %). Discarding the results that do not fit in the range of the average ± standard deviation it is 5.44 times slower, and the median value indicates that it is 5.34 times slower.

In the second experiment, we executed a load test benchmark using multiple browsers executing the same navigation sequence in parallel. This experiment was not executed using the real websites because most of them do not allow the level of concurrency required for the parallel load testing. Instead, we simulated the real web site saving the contents of the downloaded pages (including the JavaScript files, CSS files, etc.) in a local web server, and modifying HTML contents and JavaScript files to emulate the form submission and the AJAX requests (this simulation forbade HTTP requests outside the local web server). In this experiment, 30 different instances of the same type of browser (e.g. 30 custom browser instances, 30 MSIE instances and 30 HtmlUnit instances) executed the same navigation sequence in parallel during 5 min.

Table 2 shows the number of finished executions of the navigation sequence using the custom browser (with and without using its optimization capabilities), and also using HtmlUnit and MSIE. The custom browser when uses its optimization capabilities always got best results (first column). Compared with the custom browser without using optimization capabilities, it completed, in average, 4.89 times more executions (489 %). Discarding the results that do not fit in the range of the average ± standard deviation, it completed 3.5 times executions, and the median value indicates that it completed 3.6 times more executions. Compared with HtmlUnit, the custom browser using its automatic optimization capabilities completed, in average, 14.25 times more executions (1425 %). Discarding the results that do not fit in the range of the average ± standard deviation, it completed 9.29 times more executions and the median value indicates that it completed 9.68 times more executions. Compared with MSIE, it completed, in average,

Table 2. Load tests benchmark.

	OPTIMIZED	NOT OPTIMIZED	HTMLUNIT	MSIE
AMAZON.COM	16520	1728 (956%)	552 (2992%)	345 (4788%)
APPLE.COM	6570	3986 (164%)	654 (1004%)	858 (765%)
EBAY.COM	6171	3504 (176%)	1026 (601%)	492 (1254%)
FLICKR.COM	34792	6244 (557%)	909 (3827%)	588 (5917%)
GOOGLE.COM	19719	1737 (1135%)	2413 (817%)	1823 (1081%)
IMDB.COM	6048	2007 (301%)	519 (1165%)	633 (955%)
LINKEDIN.COM	28364	11483 (247%)	4495 (631%)	2110 (1344%)
WALMART.COM	3342	870 (384%)	240 (1392%)	189 (1768%)
WIKIPEDIA.COM	12722	3779 (336%)	1430 (889%)	669 (1901%)
WSJ.COM	4146	651 (636%)	444 (933%)	516 (803%)
AVERAGE		489%	1425%	2057%
AVERAGE ± STDEV		350%	929%	1233%
MEDIAN		360%	968%	1299%

20.57 times more executions (2057 %). Discarding the results that do not fit in the range of the average ± standard deviation, it completed 12.33 times more executions, and the median value indicates that it completed 12.99 times more executions.

6 Related Work

Most of the current web automation systems (Smart Bookmarks [9], Wargo [15], Selenium [17], Kapow [11], WebVCR [2], WebMacros [16]) use the APIs of conventional browsers to automate the execution of navigation sequences. This approach has two important advantages: it does not require to develop a new browser (which is costly), and it is guaranteed that the page will behave in the same way as when a human user access it with her browser. Nevertheless, it presents performance problems for intensive web automation tasks which require real time responses. This is because web browsers are designed to be client-side applications and they consume a significant amount of resources.

Other systems use the approach of creating simplified custom browsers. For example, Jaunt [10] lacks the ability to execute JavaScript. HtmlUnit [8] and EnvJS [5] use their own custom browser with support for advanced JavaScript features. They are more efficient than conventional web browsers, because they are not oriented to be used by humans and can avoid some tasks (e.g. rendering). Nevertheless, they work like conventional browsers when building the internal representation of the web pages. Since this is the most important part in terms of the use of computational resources, their performance enhancements are smaller than the ones achieved with our approach.

Traditional web browsers (Firefox, Chrome, etc.) implement some optimizations, (e.g., Mozilla Firefox uses the speculative parsing [13] to early discover resources and start preload actions), but they always calculate the CSS visualization information of all the DOM nodes, evaluate scripts in a sequential form, and load the pages completely.

Other browsers exploit different levels of optimization and parallelism. For example, ZOOMM [3] is a parallel browser engine that exploits HTML pre-scanning with resource prefetching, concurrent CSS styling and parallel script compilation, and Adrenaline [7] speeds up page processing by splitting the original page in mini-pages, rendering each of these mini-pages in a separate process.

7 Conclusions

In this paper we presented a complete architecture for a headless custom browser specialized in the execution of web navigation sequences. The architecture supports a set of novel automatic optimization techniques not implemented in any other navigation component and includes some elements not present in any other navigation system.

This architecture design, exploits some peculiarities of web automation environments. First, custom browsers do not require some operations that are unconditionally executed in conventional browsers (e.g. build page layout and rendering), and second, the fact that, in the web automation systems, the same navigation sequences are executed multiple times. This peculiarity is used to extract some useful information during a first execution of each navigation sequence, with the goal to use that information in the following executions and improve the efficiency.

To evaluate the validity of the proposed architecture, we developed a reference implementation following the architecture principles. In the experiments, we analyzed the performance of the architecture comparing our custom browser with other navigation components. The reference implementation, using optimization techniques, got the best results, followed by the same reference implementation without using those optimization capabilities, and, at a greater distance, by the other navigation components.

We can conclude that a custom browser built according to the proposed architecture is able to execute the navigation sequences faster, consuming fewer resources than other existing navigation components.

References

1. Alexa. The Web Information Company. http://www.alexa.com
2. Anupam, V., Freire, J., Kumar, B., Lieuwen, D.: Automating web navigation with the WebVCR. Comput. Netw. **33**(1–6), 503–517 (2000)
3. Cascaval, C., Fowler, S., Montesinos-Ortego, P., Piekarski, W., Reshadi, M., Robatmili, B., Weber, M., Bhavsar, V.: ZOOMM: a parallel web browser engine for multicore mobile devices. In: Proceedings of the 18th ACM SIGPLAN symposium on Principles and Practice of Parallel Programming (PPoPP 2013). ACM, New York, NY, USA, pp. 271–280 (2003)
4. Document Object Model (DOM). http://www.w3.org/DOM/
5. EnvJS. http://www.envjs.com/
6. Grosskurth, A., Godfrey, M.W.: A reference architecture for web browsers. In: ICSM 2005: Proceedings of the 21st IEEE International Conference on Software Maintenance (ICSM 2005). pp. 661–664 (September 2005)
7. Mai, H., Tang, S., King, S.T., Cascaval, C., Montesinos, P.: A case for parallelizing web pages. In: Proceedings of the 4th USENIX Conference on Hot Topics in Parallelism, HotPar 2012, Berkeley, CA, USA. USENIX Association (June 2012)
8. HtmlUnit. http://htmlunit.sourceforge.net/
9. Hupp, D., Miller, R.C.: Smart Bookmarks: automatic retroactive macro recording on the web. In: Proceedings of the 20th Annual ACM Symposium on User Interface Software and Technology, pp. 81–90. ACM New York, Newport (2007)
10. Jaunt. Java Web Scraping and Automation. http://jaunt-api.com
11. Kapow. http://kapowsoftware.com/
12. Losada, J., Raposo, J., Pan, A., Montoto, P.: Efficient execution of web navigation sequences. World Wide Web J. doi:10.1007/s11280-013-0259-8. ISSN 1386-145X
13. Losada, J., Raposo, J., Pan, A., Montoto, P., Álvarez, M.: Optimization techniques to speed up the page loading in custom web browsers. Manuscript accepted for publication in ICEBE 2015. Beijing, China (23–25 October 2015)
14. Mozilla HTML5 Parser. https://developer.mozilla.org/en-US/docs/Web/Guide/HTML/HTML5/HTML5_Parser
15. Pan, A., Raposo, J., Álvarez, M., Hidalgo, J., Viña, A.: Semiautomatic wrapper generation for commercial web sources. In: IFIP WG8.1 Working Conference on Engineering Information Systems in the Internet Context, pp. 265–283. Kluwer, B.V. Deventer, Japan (2002)
16. Safonov, A., Konstan, J., Carlis, J.: Beyond hard-to-reach pages: interactive, parametric web macros. In: 7th Conference on Human Factors and the Web. Madison (2001)
17. Selenium. http://seleniumhq.org
18. HTML5. https://html.spec.whatwg.org
19. XML Path Language (XPath). http://www.w3.org/TR/xpath

File Relation Graph Based Malware Detection Using Label Propagation

Ming Ni[1]($\boxtimes$), Qianmu Li[1], Hong Zhang[1], Tao Li[2], and Jun Hou[3]

[1] School of Computer Science and Engineering, Nanjing University of Science and Technology, Nanjing 210094, China
nq1027@gmail.com
[2] School of Computer Science and Technology, School of Software, Nanjing University of Posts and Telecommunications, Nanjing 210046, China
[3] School of Humanities and Social Sciences, Nanjing University of Science and Technology, Nanjing 210094, China

Abstract. The rapid development of malicious software programs has posed severe threats to Computer and Internet security. Therefore, it motivates anti-malware industry to develop novel methods which are capable of protecting users against new threats. Existing malware detectors mostly treat the file samples separately using supervised learning algorithms. However, ignoring of relationship among file samples limits the capability of malware detectors. In this paper, we present a new malware detection method based on file relation graph to detect newly developed malware samples. When constructing file relation graph, k-nearest neighbors are chosen as adjacent nodes for each file node. Files are connected with edges which represent the similarity between the corresponding nodes. Label propagation algorithm, which propagates label information from labeled file samples to unlabeled files, is used to learn the probability that one unknown file is classified as malicious or benign. We evaluate the effectiveness of our proposed method on a real and large dataset. Experimental results demonstrate that the accuracy of our method outperforms other existing detection approaches in classifying file samples.

Keywords: Malware detection · File relation graph · kNN · Label propagation

1 Introduction

With the rapid development of Computer and Internet technology, computer security becomes more and more prevalent over past decades. Malware (short for **mali**cious soft**ware**), including Viruses, Backdoors, Spyware, Trojans, Worms and Botnets, is software that spread and infect computers for malicious intent of an attacker [5]. In the form of executable code, scripts, active content, and other softwares, malware samples can be used to disrupt computer operation, gather sensitive information, or gain access to private computer systems, and may cause

J. Wang et al. (Eds.): WISE 2015, Part II, LNCS 9419, pp. 164–176, 2015.
DOI: 10.1007/978-3-319-26187-4_12

serious damages and financial losses to computers and users. Malware detection is thus becoming more and more important due to its damage to the security and the economic loss of people.

Currently, the main approach of protecting against malware is signature-based method which is widely adopted by most anti-malware companies [6,7]. Signature is a particular piece of code which is obtained after being analyzed manually by computer security experts and expressed in the form of byte or instruction sequences and is unique for each known malware [8]. However, due to the rapid development of malware techniques, a huge number of malware samples are being generated or mutated every day. Meanwhile, malware writers have employed advanced development toolkit, including encryption, polymorphism, and metamorphism to make malware samples be immune to signature-based detection. It poses a big threat to signature-based detection. Human experts cannot analyze each new file manually, and the required responding time is limited. This issue has motivated anti-malware industry to redesign their security systems for detecting malware samples. Recently, many research efforts have been conducted on malware detection using data mining techniques. Researchers have shifted from traditional signature-based method to file-content-analysis based approaches to detect and classify malware with static or dynamic features [1,2,10–13,19,21,22]. These techniques applied data mining algorithms for malware detection based on content features, such as instructions, control flow extracted from binary codes and API call sequences tracked from runtime environments.

In this paper, instead of using the content information of file samples, we investigate how file relations can be used to detect malware samples and employ a Label Propagation method for classifying file samples based on the constructed file relation graphs. A real and large scale file relation dataset from an anti-malware industry company is used in the experiments. The scale of this dataset is representative including 69,165 file samples (3,095 malware, 22,583 benign files, and 43,487 unknown files) on 3,793 clients.

This paper makes the following contributions:

1. Unlike classic classifiers based on only the file content information, we make use of the relationships among file samples and apply graph mining algorithm for malware detection. Relations with other known files are used to identify the unknown file samples.
2. We use the k-nearest neighbors of each file sample to construct a file relation graph for inferring each file's probability of being malicious or benign.
3. A label propagation algorithm is used to propagate the label information from labeled files to unlabeled files.
4. The empirical evaluation on a real and large data collection from an anti-malware industry company is performed and demonstrates the performance of our method.

The remainder of this paper is organized as follows. Section 2 presents the background and discuss the related work. Details of the dataset is described in Sect. 3. We discuss how file relations and the Label Propagation algorithm can be used to perform malware detection in Sect. 4. Experiments are conducted to evaluate the effectiveness and efficiency of our proposed method by comparing with the baselines in Sect. 5. Finally, we state the conclusions and future studies in Sect. 6.

2　Background and Related Work

In recent years, an increasing number of studies have been conducted on developing efficient algorithms to detect malware by data mining and machine learning techniques [1, 8, 9, 13, 16–19, 21, 22]. In [8], Jeffrey et al. developed a statistical method to extract virus signatures automatically, it is the first major work applying data mining techniques to detect malware. Schultz et al. [12] used DLL information, strings and n-grams to train RIPPER, Naive Bayes and Multi Naive Bayes to classify malware. Assaleh et al. [1] created class profiles of various lengths according to the number of most frequent n-grams within the class with different n-gram sizes. Kolter et al. [9] selected the most relevant n-grams on 1971 benign and 1651 malicious file samples, then different classification methods including Naive Bayes, Support Vector Machine (SVM), and Decision Tree (DT), were compared based on these n-grams for malware detection. In [21], Ye et al. developed an Intelligent Malware Detection System (IMDS) which uses Objective-Oriented Association classification based on Windows API call sequences. An OOA Fast FP-Growth algorithm was developed. Their experiments showed that OOA-based method outperforms the Apriori algorithm for association rule generation.

The aforementioned methods are all based on the file contents, including Application Programming Interface calls and program code strings. Besides file contents, relations among file samples can also be used to extract invaluable information about the properties of file samples. In recent years, some research efforts have been conducted on detecting malware based on file relation graphs [3, 4, 14, 15, 20]. Chau et al. [3] presented a novel method based on Belief Propagation algorithm to infer file reputation using file-machine relations. In [20], Ye et al. built a semi-parametric classifier model that combines file-to-file relationship with file contents information for malware detection. Tamersoy et al. proposed AESOP, a scalable algorithm, which leverages locality-sensitive hashing to measure similarity between files and employs a tuned BP algorithm on the file-bucket graph based on LSH [14].

3　Data Description

In this section, we describe the dataset used in our work. We obtain the dataset from an anti-malware industry which contains 69,165 file samples (3,095 malware, 22,583 benign files, and 43,487 unknown files) and relations between these file samples [20]. Figure 1 shows the structure of the file relation database including 8 fields: file id, file label ("1" is for benign file, "−1" denotes malicious file, and "0" represents unknown file), file name, number of malware that the file co-exists, malware ids that the file co-exists, the number of benign files that the file co-exists, benign file ids that the file co-exists, number of clients in which the file exists.

id	file_sort	file_md5crc	ref_black_count	ref_black_ids	ref_white_cour	ref_white_ids	ref_file_count
1	-1	58414817dbd783...	10	19821:1,19822:1,19837:1,...	14	138:1,140:1,141:1,14535:1,3177:1,32...	00000000002
2	-1	c3baaf5afa8cba8...	9	1:1,13980:1,18575:1,1857...	313	10198:1,10927:1,10930:1,11:1,11276...	00000000010
3	1	6b967b59d4d6a4...	441	1002:2,1003:1,10243:1,10...	6047	10:121,1000:4,1001:1,10029:3,10031...	00000000351
4	1	b786825902bd49...	78	13939:1,1481l:1,16171:1,...	456	10:15,10183:1,10198:1,10268:1,1142...	00000000034
5	1	a8dc6cc4115c0d5...	47	11906:1,14340:1,15381:1,...	538	10:9,10055:1,10198:1,1028:1,10282:...	00000000027
6	1	41d5501224adae...	594	1002:1,10064:1,10189:1,1...	6420	10:159,1000:1,10022:1,10024:1,1002...	00000000394
7	1	0043cbcf44106b3...	302	10505:2,10634:3,10635:1,...	3666	10:55,10022:1,10025:1,10056:1,1018...	00000000141
8	1	6929f9ff15a8f3b0...	1069	1002:1,10033:1,10062:1,1...	10644	10:382,1000:6,10020:7,10021:1,1002...	00000001276
9	1	90b16c00d94e7f7...	581	1002:1,10062:1,10063:1,1...	6790	10:196,1000:4,10020:2,10023:2,1002...	00000000644
10	1	7763b669cda651...	913	1002:1,1003:1,10064:1,10...	8671	1000:5,1001:1,10020:2,10022:1,1002...	00000000897
11	1	d0cb8f4df3fa8b1...	64	11029:1,12305:1,14573:1,...	1359	10:16,10023:1,10024:1,10027:1,1003...	00000000040
12	1	0020553f141db51...	285	10505:2,10634:3,10635:1,...	3428	10:59,10022:1,10056:1,10186:2,1023...	00000000134
13	1	a8c7a7c0bca0e16...	29	11029:1,12728:1,15471:1,...	695	10:24,10029:1,10183:1,10241:1,1024...	00000000030

Fig. 1. Sample File Relation Database

Usually, the number of benign files is much larger than that of malware samples. It leads to the imbalanced data distribution which can be seen from the dataset. Both the number and the relations are imbalanced. Figures 2 and 3 show the distribution of the co-occurrence between malware samples and the co-occurrence between malware samples and benign files respectively.

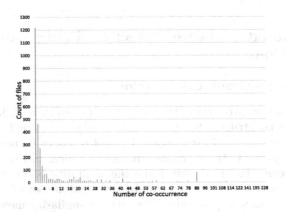

Fig. 2. Co-occurrence between Malware Samples

Note that the file lists were collected from users' clients. It is unnecessary and unpractical for the clients to collect all the file samples from users' machines, the clients only submit the suspicious file samples to the server for further analysis. So that, only the associations with labeled file samples are recorded in the database, the relationship between unknown file samples is missing. We will use the co-occurrence information for each unknown file sample to calculate the similarity between unknown file samples, which will be described in next section.

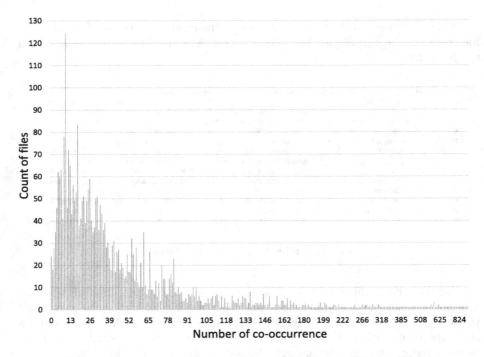

Fig. 3. Co-occurrence between Malware Samples and Benign Files

4   Graph-Based Malware Detection Using Label Propagation

4.1   File Relation Graph Construction

Based on the structure property of the dataset described in Sect. 3, an undirected weighted graph is constructed to represent the relations among file samples. The graph is defined as $G = (V, E, W)$, where V is set of nodes corresponding to the file samples, E represents the relations among the nodes, and W corresponds to the weights of each edge. Here, we define the similarity between file f_i and f_j as the co-occurrence strength. Let C_i and C_j denote the set of clients in which the file f_i and f_j exists respectively. The Jaccard similarity measure is used to calculate the co-occurrence strength as follows.

$$sim(f_i, f_j) = \frac{|C_i \cap C_j|}{|C_i \cup C_j|}, \tag{1}$$

where $|C|$ is the size of set C. The value of this measure is between 0 and 1; "0" indicates no co-occurrence relationship, "1" indicates a full co-occurrence relationship.

As mentioned earlier, the relationship between unknown file samples is missing. Here, we use the relationship with labeled file samples of each unknown file to measure the similarity between unknown file samples. Let M_i represent the set of file samples which co-exist with unknown file sample f_i. The similarity between two unknown file samples f_i and f_j is:

$$sim(f_i, f_j) = \sum_{m \in M_i \cap M_j} sim(m, i) * sim(m, j). \tag{2}$$

In order to filter out the noisy data, we choose the k-nearest neighbors of each file by applying the kNN based method. If file f_i is in k-nearest neighbor of file f_j, then there is an edge between them. The weight of the edge is the similarity between file f_i and f_j.

To further illustrate, a file relation dataset sample is given as Table 1, in which 6(3) means that the file co-exists with file No.6 in three clients. Based on the given relations, an undirected weighted graph is constructed as shown in Fig. 4. The figure shows the relations among the files in the case of $k = 3$, and the number on each edge indicates the weight. The key idea of our problem can be described as: An unlabeled file sample can be labeled as malware or benign based on their co-occurrence with labeled files.

Table 1. File Relation Dataset Sample

ID	Label	Co-exists with Malware	Co-exists with Benign File	Count of Clients
1	0	7(1),8(1),10(1)	6(1)	2
2	0	4(1)	3(2),6(2)	2
3	1	4(1),8(1)	5(2),6(3)	4
4	−1	8(1),10(1)	3(1),6(1)	2
5	1	8(1)	3(2),6(1)	2
6	1	4(1),8(1)	3(3),5(1)	4
7	−1	10(1)	——	1
8	−1	4(1),10(1)	3(1),5(1),6(1)	3
9	0	——	3(1),5(1),6(1)	1
10	−1	4(1),7(1),8(1)	——	2

4.2   Label Propagation

Label Propagation is a graph-based semi-supervised learning method, which lets every labeled data spread its label information to the whole graph until all unlabeled data have a stable label states [23].

Let $(x_1, y_1)...(x_l, y_l)$ denote labeled data, where $y_1...y_l$ are class labels, and $(x_{l+1}, y_{l+1})...(x_{l+u}, y_{l+u})$ be unlabeled data. The key idea of label propagation is that data points with high similarity tend to have the same labels. Label

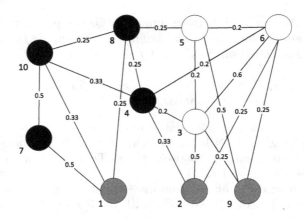

Fig. 4. Constructed Graph based on Table 1

information of labeled nodes need to be propagated to all nodes through the edges.

Define Y as a $(l + u) \times C$ label matrix, where Y_{ij} represents the probability of node x_i being labeled as y_j and C is the number of classes. In other words, Y represents the label probability distribution of each node. T is a probabilistic transition matrix defined as

$$T_{ij} = P(j \rightarrow i) = \frac{w_{ij}}{\sum_{k=1}^{l+u} w_{ij}}, \tag{3}$$

where T_{ij} denotes the probability of jumping from node j to i. Algorithm 1 presents the method proposed by Zhu in [23].

Algorithm 1. Label Propagation

---

1. Initialization. Set Y be the initial labels attached to each node, where $Y_{ij} = 1$ if x_i is labeled as y_j.

repeat

   (1). Propagate labels of any node to its neighbors by $Y \leftarrow \overline{T}Y$, where $\overline{T}$ is row-normalized matrix of T, i.e. $\overline{T}_{ij} = T_{ij}/\sum_k T_{ik}$.

   (2). Clamp the labeled data.

until Y converges

2. Assign x_i with a label using $y_i = \arg\max_j Y_{ij}$.

Define Y_L as the top l rows of Y which are labeled data and Y_U as the remaining u rows standing for unlabeled data. Due to the clamping operation, Y_L never changes, so we only need to focus on Y_U. It is shown that the algorithm converges to a unique fixed point and the solution is $Y_U = (I - \overline{T}_{uu})^{-1}\overline{T}_{ul}Y_L$. Here, $\overline{T}_{uu}$ and $\overline{T}_{ul}$ are sub-matrices obtained by splitting $\overline{T}$ after the l-th row and the l-th column.

4.3   Malware Detection Using Label Propagation

Based on the constructed file relation graph as well as the label propagation algorithm described in previous subsection, the whole process of our proposed method is described in Algorithm 2.

Algorithm 2. Algorithm for malware detection

Input: Raw file lists data
Output: Class label of each file sample
 1. Calculate the similarity for each pair of associated files;
 2. Calculate the similarity for each pair of unlabeled files based on their co-occurrence;
 3. Choose k nearest neighbors for each file as neighbors in the graph;
 4. Initialize graph $G = (V, E, W)$;
 5. Perform the Label Propagation algorithm described in Algorithm 1;
 6. Assign labels(i.e., malicious or benign) to unlabeled file samples.

---

5   Experiment

In this section, we conduct two sets of experiments: (1) In the first set of experiments, we evaluate the effectiveness of kNN method applied for neighbor selection and choose the best value of k for the rest experiments. (2) In the second set of experiments, we evaluate the effectiveness of our proposed method for malware detection by comparing with baseline methods. All algorithms are evaluated with the dataset described in Sect. 3.

5.1   Experiments Setting

All algorithms in following subsections are implemented on a laptop of Windows 8 OS with Intel Core i7 2.7 GHz Duo CPU and 8 GB RAM using JAVA 1.7. The evaluation metrics are described below.

- **True Positive(TP)**: Number of samples labeled as malicious correctly.
- **True Negative(TN)**: Number of samples labeled as benign correctly.
- **False Positive(FP)**: Number of samples labeled as malicious incorrectly.
- **False Negative(FN)**: Number of samples labeled as benign incorrectly.
- **TP Rate(TPR)**: $\frac{TP}{TP+FN}$
- **FP Rate(FPR)**: $\frac{FP}{TN+FP}$
- **Accuracy(ACC)**: $\frac{TP+TN}{TP+TN+FP+FN}$

5.2 Performance Evaluation of Neighbor Selection Using kNN

When constructing the file relation graph, we choose the k-nearest neighbors of each file samples to filter out the noisy data and keep the most similar neighbors. In this section, we evaluate the effectiveness of applying kNN method. We run the algorithm without applying kNN method, and 5 times with $k = 10$, $k = 30$, $k = 50$, $k = 70$ and $k = 100$ respectively. From Table 2 and Fig. 5, we saw an improvement of about 13 % on TP (True Positive) and 10 % on TN (True Negative) after applying kNN by setting $k = 50$.

Table 2. Effectiveness of Applying kNN

Method	TP	FP	TN	FN	ACC
Non-kNN	110	301	3,396	179	0.8796
$k = 10$	109	224	3,473	180	0.8986
$k = 30$	113	179	3,518	176	0.9109
$k = 50$	155	67	3,630	134	0.9496
$k = 70$	126	139	3,558	163	0.9242
$k = 100$	122	199	3,498	167	0.9082

5.3 Comparisons of Label Propagation with Other Methods

In this subsection, we compare the effectiveness of our proposed method with other methods including both graph-based and content-based classification approaches. Four baseline methods were compared: AESOP in [14], Malware Distributor Detector (MDD) in [15], Support Vector Machine (SVM), and Random Forest (RF).

Cross Validation: We use 10-fold cross validation scheme to evaluate the performance of our proposed method. At each round, we set the labels of files in the test set to 0 and the probabilities of being malicious and benign both to 0.5. For each fold, we run our proposed algorithm with baseline methods and record the ACC (Accuracy). Quantitative results on the 10-fold validation are shown in Fig. 6. The notch marks the 95 % confidence interval for the medians. The figure demonstrates that our proposed method makes an significant improvement on accuracy compared to the best performance among the baseline methods, with $k = 50$.

Prediction: 3,986 files with ground truth (includes 289 malware, 3,697 benign files) were selected at random as the test data for evaluation, and the rest data were used as the training data. Here we set $k = 50$. Table 3 and Fig. 7 present the

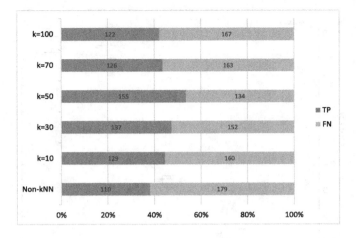

(a) TP Versus FN

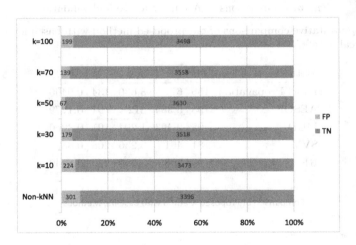

(b) TN Versus FP

Fig. 5. Effectiveness of Applying kNN

results of our proposed method along with the four baseline methods. The comparison results illustrate that our proposed algorithm outperforms other methods in malware detection on the large and real datasets.

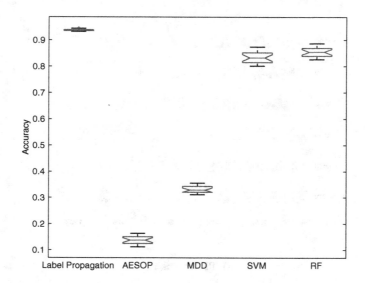

Fig. 6. Comparisons of Accuracy for 10-fold validation

Table 3. Quantitative comparisons of our proposed method with baseline methods on large and real data

Method	TP	FP	TN	FN	ACC
Label Propagation	155	67	3,630	134	0.9496
AESOP	198	3,385	312	91	0.1279
MDD	98	2,493	1,204	191	0.3266
SVM	81	461	3,236	208	0.8321
RF	69	398	3,299	220	0.8449

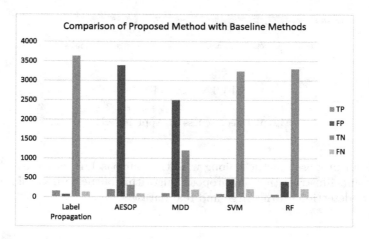

Fig. 7. Comparisons of Proposed Method with Baseline Methods

6   Conclusion and Future Work

In this paper, we study how to use file relations for malware detection. The associations between file samples are used to compute the file similarity values to construct file-relation graph by kNN method. A Label Propagation algorithm is applied for classifying file samples based on the constructed file-relation graph. We use a real and large dataset consisting of file co-occurrence records from users' clients. Comprehensive experiments are performed to compare our proposed method with other existing malware detection approaches. The experimental results demonstrate that the accuracy of our proposed method outperform other malware detection methods using data mining techniques. For the future work, we plan to further explore the combination of file relation information and file contents to reduce the false positive and negative rates.

Acknowledgment. This paper is supported by the National Natural Science Foundation of China under Grant No. 61272419, the Future Network Science Foundation of Jiangsu under Grant No. BY2013095-3-02, the Prospective research project in Jiangsu Province under Grant No. BY2014089, and the Lianyungang Scientific and Technological Project under Grant No. CH1304.

References

1. Abou-Assaleh, T., Cercone, N., Keselj, V., Sweidan, R.: N-gram-based detection of new malicious code. In: Proceedings of the 28th Annual International Computer Software and Applications Conference, COMPSAC 2004, vol. 2, pp. 41–42. IEEE (2004)
2. Bailey, M., Oberheide, J., Andersen, J., Mao, Z.M., Jahanian, F., Nazario, J.: Automated classification and analysis of internet malware. In: Kruegel, C., Lippmann, R., Clark, A. (eds.) RAID 2007. LNCS, vol. 4637, pp. 178–197. Springer, Heidelberg (2007)
3. Chau, D.H., Nachenberg, C., Wilhelm, J., Wright, A., Faloutsos, C.: Polonium: tera-scale graph mining and inference for malware detection. In: SIAM International Conference on Data Mining, vol. 2 (2011)
4. Chen, l., Li, T., Abdulhayoglu, M., Ye, Y.: Intelligent malware detection based on file relation graphs. In: 2015 IEEE International Conference on Semantic Computing (ICSC), pp. 85–92. IEEE (2015)
5. Egele, M., Scholte, T., Kirda, E., Kruegel, C.: A survey on automated dynamic malware-analysis techniques and tools. ACM Comput. Surv. (CSUR) **44**(2), 6 (2012)
6. Filiol, E.: Malware pattern scanning schemes secure against black-box analysis. J. Comput. Virol. **2**(1), 35–50 (2006)
7. Filiol, E., Jacob, G., Le Liard, M.: Evaluation methodology and theoretical model for antiviral behavioural detection strategies. J. Comput. Virol. **3**(1), 23–37 (2007)
8. Kephart, J.O., Arnold, W.C.: Automatic extraction of computer virus signatures. In: 4th Virus Bulletin International Conference, pp. 178–184 (1994)
9. Kolter, J.Z., Maloof, M.A.: Learning to detect malicious executables in the wild. In: Proceedings of the Tenth ACM SIGKDD International Conference on Knowledge Discovery and Data Mining, pp. 470–478. ACM (2004)

10. Masud, M.M., Al-Khateeb, T.M., Hamlen, K.W., Gao, J., Khan, L., Han, J., Thuraisingham, B.: Cloud-based malware detection for evolving data streams. ACM Trans. Manag. Inf. Syst. (TMIS) **2**(3), 16 (2011)
11. Reddy, D.K.S., Pujari, A.K.: N-gram analysis for computer virus detection. J. Comput. Virol. **2**(3), 231–239 (2006)
12. Schultz, M.G., Eskin, E., Zadok, E., Stolfo, S.J.: Data mining methods for detection of new malicious executables. In: Proceedings of the 2001 IEEE Symposium on Security and Privacy, S&P 2001, pp 38–49. IEEE (2001)
13. Siddiqui, M., Wang, M.C., Lee, J.: A survey of data mining techniques for malware detection using file features. In: Proceedings of the 46th Annual Southeast Regional Conference on XX, pp. 509–510. ACM (2008)
14. Tamersoy, A., Roundy, K., Chau, D.H.: Guilt by association: large scale malware detection by mining file-relation graphs. In: Proceedings of the 20th ACM SIGKDD International Conference on Knowledge Discovery and Data Mining, pp. 1524–1533. ACM (2014)
15. Venzhega, A., Zhinalieva, P., Suboch, N.: Graph-based malware distributors detection. In: Proceedings of the 22nd International Conference on World Wide Web Companion, pp. 1141–1144. International World Wide Web Conferences Steering Committee (2013)
16. Ye, Y., Chen, L., Wang, D., Li, T., Jiang, Q., Zhao, M.: SBMDS: an interpretable string based malware detection system using SVM ensemble with bagging. J. Comput. Virol. **5**(4), 283–293 (2009)
17. Ye, Y., Li, T., Huang, K., Jiang, Q., Chen, Y.: Hierarchical associative classifier (HAC) for malware detection from the large and imbalanced gray list. J. Intell. Inf. Syst. **35**(1), 1–20 (2010)
18. Ye, Y., Li, T., Jiang, Q., Han, Z., Wan, L.: Intelligent file scoring system for malware detection from the gray list. In: Proceedings of the 15th ACM SIGKDD International Conference on Knowledge Discovery and Data Mining, pp. 1385–1394. ACM (2009)
19. Ye, Y., Li, T., Jiang, Q., Wang, Y.: CIMDS: adapting postprocessing techniques of associative classification for malware detection. IEEE Trans. Syst. Man Cybern. Part C: Appl. Rev. **40**(3), 298–307 (2010)
20. Ye, Y., Li, T., Zhu, S., Zhuang, W., Tas, E., Gupta, U., Abdulhayoglu, M.: Combining file content and file relations for cloud based malware detection. In: Proceedings of the 17th ACM SIGKDD International Conference on Knowledge Discovery and Data Mining, pp. 222–230. ACM (2011)
21. Ye, Y., Wang, D., Li, T., Ye, D.: IMDS: Intelligent malware detection system. In: Proceedings of the 13th ACM SIGKDD International Conference on Knowledge Discovery and Data Mining, pp. 1043–1047. ACM (2007)
22. Ye, Y., Wang, D., Li, T., Ye, D., Jiang, Q.: An intelligent pe-malware detection system based on association mining. J. Comput. Virol. **4**(4), 323–334 (2008)
23. Zhu, X., Ghahramani, Z.: Learning from labeled and unlabeled data with label propagation. Technical report. Citeseer (2002)

Cross-Domain Collaborative Recommendation by Transfer Learning of Heterogeneous Feedbacks

Jun Wang[1], Shijun Li[1(✉)], Sha Yang[1,2], Yonggang Ding[1,3], and Wei Yu[1]

[1] School of Computer, Wuhan University, Wuhan, China
[2] School of Computer Science and Technology, Hankou University, Wuhan, China
[3] School of Education, Hubei University, Wuhan, China
{wjwj,shjli,yuwei}@whu.edu.cn

Abstract. With the rapid development of information society, the era of big data is coming. Various recommendation systems are developed to make recommendations by mining useful knowledge from massive data. The big data is often multi-source and heterogeneous, which challenges the recommendation seriously. Collaborative filtering is the widely used recommendation method, but the data sparseness is its major bottleneck. Transfer learning can overcome this problem by transferring the learned knowledge from the auxiliary data to the target data for cross-domain recommendation. Many traditional transfer learning models for cross-domain collaborative recommendation assume that multiple domains share a latent common rating pattern which may lead to the negative transfer, and only apply to the homogeneous feedbacks. To address such problems, we propose a new transfer learning model. We do the collective factorization to rating matrices of the target data and its auxiliary data to transfer the rating information among heterogeneous feedbacks, and get the initial latent factors of users and items, based on which we construct the similarity graphs. Further, we predict the missing ratings by the twin bridge transfer learning of latent factors and similarity graphs. Experiments show that our proposed model outperforms the state-of-the-art models for cross-domain recommendation.

Keywords: Cross-domain · Transfer learning · Collaborative filtering · Sparseness · Heterogeneous feedbacks

1 Introduction

With the rapid development of computer and network technologies, especially the mobile internet, people can easily acquire all kinds of information from the internet. Under the big data environment of web, massive information often makes people dazzling and unable to get valuable information rapidly and effectively. In order to solve this problem of information overload, personalized recommendation systems are developed. For example, the famous personalized

J. Wang et al. (Eds.): WISE 2015, Part II, LNCS 9419, pp. 177–190, 2015.
DOI: 10.1007/978-3-319-26187-4_13

recommendation platforms of Amazon, Movielens, Alibaba and so on. They help users obtain the required service quickly and accurately and make huge benefits for the relevant manufacturers. Collaborative Filtering (CF) is an excellent recommendation algorithm which is widely used at present. CF in recommender systems is designed to predict the missing ratings for a user or an item based on the collected ratings from like-minded users or similar items [1,2]. CF is simple, which doesn't need the configuration information of users and has no special requirements to the recommendation objects. CF is effective, which can make multiple recommendations. Although CF has such advantageous properties, the recommendation performance would be degraded seriously when the observed data is very sparse.

To address the sparseness problem, many improved CF methods based on the single domain have been proposed. But these methods are subject to the data quality of the target domain. They may be invalid when the target data is extremely sparse. In reality, especially in the current era of big data, we often have many data in the associated domains of the target domain. Why not try to make use of them? So we research the cross-domain recommendation which combines relevant data from different domains with the original target data to improve the recommendation. Transfer learning [3] can be used for cross-domain CF recommendation particularly. During the transfer learning, useful knowledge will be learned from the auxiliary data and transferred to the target data so that the sparseness problem of the target data can be addressed effectively. However, traditional transfer learning models for cross-domain CF recommendation have some issues as follows:

- They are often limited to the transfer of homogeneous user feedbacks. However, the heterogeneous user feedbacks are common in reality, especially in the current big data era.
- They usually suppose that different domains share a common latent rating pattern based on the user-item co-clustering. In fact, however, the associated domains do not necessarily share such a common latent rating pattern, and the diversity among associated domains may outweigh the advantages of this common latent rating pattern [4], which may degrade the recommendation performance.
- Since the target data is extremely sparse, it is expected that more useful common knowledge is transferred from the auxiliary data to the target data. Only using the latent factors extracted from the auxiliary data may result in that the positive information transferred to the target data is insufficient.

To solve these problems, we propose a new model of cross-domain CF recommendation based on the twin bridge transfer learning of heterogeneous user feedbacks. Our contributions are summarized as follows:

- To transfer the rating information of heterogeneous user feedbacks, the initial data are preprocessed to be homogenous. Then we do the collective factorization to rating matrices of the target data and its auxiliary data, and get the initial latent factors of uses and items respectively, when we consider both the common and domain-specific latent rating patterns.

- Based on the initial latent factors, the similarity graphs are constructed. The model can be formulated as an optimization problem based on the graph regularized weighted nonnegative matrix tri-factorization [5]. In the process of optimization, latent factors and similarity graphs are regarded as a implicit bridge and a explicit bridge for transfer respectively to learn more useful knowledge.
- An efficient gradient descent method is executed to optimize the objective function with convergence guarantee. Extensive experiments on several real-world data sets suggest that our proposed model outperforms the state-of-the-art models for the cross-domain recommendation.

2   Related Work

CF is widely used due to its simpleness and high-efficiency. However, since CF method fully depends on the observed rating data, the sparseness issue has become its major bottleneck [6]. In real life, we may easily find some related CF domains with the similar recommendation as the target domain. A question was then asked in [7]: Can we establish a bridge between related CF domains and transfer useful knowledge from one another to improve the performance?, which is an emerging research topic about cross-domain CF [8].

Transfer learning is used for cross-domain CF recommendation in particular. Liu bin [9] gives a brief survey of the pilot studies on cross-domain CF in CF domains and knowledge transfer styles. Chungyi Li et al. [10] try to match users and items across domains for transfer learning to improve the recommendation quality. Weiqing Wang et al. [11] research cross-domain CF by tag transfer learning. Zhongqi Lu et al. [12] explore selective transfer learning for cross-domain recommendation. Pan et al. [13–15] propose the models to transform knowledge from domains which have heterogeneous forms of user feedbacks.

The majority of the existing transfer learning models for cross-domain recommendation assumes that the target domain and its auxiliary domains are related but doesnt suggest ways to compute the relatedness across multiple domains. The usual way of the existing models is to exploit the common latent structure shared among multiple domains as the information bridge to transfer the useful knowledge. For example, Shi, Y. et al. [16] propose a generalized cross domain CF model by tag transfer learning. They use user-generated tags as the common features to connect multiple domains together and perform transfer learning across different domains for knowledge transfer. Traditional cross-domain recommendation models assume that all the domains share the common latent rating pattern, which is inconsistent with the reality and may cause the recommendation performance of a hard decline. Gao et al. [4] propose a cluster-level latent factor model, which can not only learn the common rating pattern shared across domains with the flexibility in controlling the optimal level of sharing, but also learn the domain-specific rating patterns of users in each domain that involve the discriminative information propitious to performance improvement. This model is referred to as GAO model by us.

Liu bin et al. propose Codebook Based Transfer (CBT) model [7] and Rating Matrix Generative model [17] to transfer the cluster-level codebook to the target data, which are novel and influential. However, because the dimension of codebook is limited, the codebook cannot transfer enough useful knowledge when the observed data are quite sparse. To overcome this problem, Transfer by Collective Factorization (TCF) [15,18], Coordinate System Transfer (CST) [19], and Transfer by Integrative Factorization (TIF) [12] extract both latent tastes of users and latent features of items in forms of latent factor matrices, and transfer the useful knowledge they contain from the auxiliary data to the sparse target data. However, these models do not take full account of the negative transfer. In addition, they only employ a single information bridge to transfer the knowledge. In Graph Regularized Weighted Nonnegative Matrix Factorization (GMF) [5] model, the neighborhood information is integrated into the factorization. The associated information among users or items can be utilized with the help of the similarity of user tastes or item features. But it requires dense ratings to calculate the neighborhood structure. When the data are very sparse, the neighborhood structure may be rather inaccurate so that the recommendation can't be performed effectively. Different from these methods, Shi et al. [20] explore the twin bridge of latent factors and their similarity graphs for the purpose of transferring more useful knowledge to the target data, which is referred to as SHI model by us. SHI model can enhance efficient transfer by transferring more knowledge, while alleviate negative transfer by regularizing the learning model with latent factors and similarity graphs, which can naturally filter out the negative information contained in the latent factors.

3   Problem Definition

Suppose that multiple domain-related rating matrices are given. Let η be the domain index, $R_\eta \in R^{M_\eta \times N_\eta} (\eta \in [1, t], t \in N^+)$ is the rating matrix of the η-th domain, where M_η and N_η represent the numbers of users and items, respectively. A binary weighting matrix Z_η with the same size as R_η is used to mark the missing ratings, where $[Z_\eta]_{ij} = 1$ if $[R_\eta]_{ij}$ is observed and $[Z_\eta]_{ij} = 0$ otherwise. The rating matrix in which missing ratings are to be predicted is deemed as the target data, and other rating matrices related to the target data are deemed as the auxiliary data. The goal is to predict missing ratings in the target data by transferring useful knowledge from the auxiliary data.

Without loss of generality, we prepare to solve a concrete problem which is the same as SHI model. Suppose that $R \in R^{M \times N}$ is the rating matrix of the target data, $Z \in \{0, 1\}$ is the indicator matrix, $Z_{ij} = 1$ if user i has rated item j and $Z_{ij} = 0$ otherwise. R_1 and R_2 are rating matrices of two auxiliary data sets respectively. R_1 shares the common set of users with R, while R_2 shares the common set of items with R. We try to predict the missing ratings of R by transfer learning from R_1 and R_2. Of course, when neither the users nor the items in the ratings matrices across multiple domains such as R, R_1 and R_2 are overlapping, our proposed model will be still effective.

4   Related Models

4.1   GAO Model [4]

In GAO model, the cluster level structures hidden across domains are extracted to learn the rating pattern of user groups on the item clusters for knowledge transfer, and to clearly demonstrate the co-clusters of users and items. The co-clustering of the data matrix in domain η can be performed by the orthogonal non-negative matrix tri-factorization, and the integrated objective function is

$$\min_{U_\eta, S_0, S_\eta, V_\eta \geq 0} f = \sum_\eta ||[R_\eta - U_\eta[S_0, S_\eta]V_\eta^T] \odot Z_\eta||^2 \tag{1}$$

where $\odot$ is the entry-wise product, U_η/V_η denotes the user/item latent factor matrix, S_0 denotes the share rating pattern matrix, S_η denotes the specific rating pattern matrix of domain η. In order to make the factorization more accurate, some prior knowledge can be imposed on the latent factors during the optimization, such as the L1 norm constraint: $U_\eta 1 = 1$ and $V_\eta 1 = 1$.

4.2   SHI Model [20]

Shi et al. extract latent factors U_0 and V_0 from R_1 and R_2 by GMF [5] and construct similarity graphs W_U and W_V based on U_0 and U_0, respectively. W_U and W_V are defined as follows:

If $u_{i_*}^0 \in N_p(u_{j_*}^0)$ or $u_{j_*}^0 \in N_p(u_{i_*}^0)$, $(W_U)_{ij} = 1$; Otherwise $(W_U)_{ij} = 0$
If $v_{i_*}^0 \in N_p(v_{j_*}^0)$ or $v_{j_*}^0 \in N_p(v_{i_*}^0)$, $(W_V)_{ij} = 1$; Otherwise $(W_V)_{ij} = 0$

Where $u_{i_*}^0$ is the i th row of U_0 denoting the latent taste of user i, $v_{i_*}^0$ is the i th row of V_0 denoting the latent feature of item i, $N_p(u_{j_*}^0)$ and $N_p(v_{j_*}^0)$ are the sets of p-nearest neighbors of $u_{i_*}^0$ and $v_{i_*}^0$ respectively. Latent factors and similarity graphs are integrated into a unified optimization framework for twin bridge transfer learning:

$$\min_{U,V,B>=0} O = ||Z \odot (R - UBV^T)||_F^2 + \lambda_U ||U - U_0||_F^2 +$$
$$\lambda_V ||V - V_0||_F^2 + \gamma_U G_U + \gamma_V G_V \tag{2}$$

where $G_U = tr(U^T L_U U)$, $G_V = tr(V^T L_V V)$ (tr refers to the trace of matrix), L_U and L_V are the graph Laplacian matrices for the similarity graphs W_U and W_V respectively. λ_U and λ_V are regularization parameters indicating the confidence on the latent factors. γ_U and γ_V are regularization parameters indicating the confidence on the similarity graphs, U/V denotes the user/item latent factor matrix, B denotes the latent pattern matrix of ratings. $L_U = D_U - W_U$, $L_V = D_V - W_V$, $D_u = diag(\sum_j (W_U)_{ij}$, $D_v = diag(\sum_j (W_V)_{ij}$.

5   HFT Model

Inspired from the related models above, we propose a novel model of cross-domain CF recommendation based on twin bridge transfer learning of heterogeneous user feedbacks named HFT. At first the data is preprocessed for our proposed HFT model.

In a user-item rating matrix, the observed ratings such as the 5-star ratings are often extremely sparse, so over-fitting may easily happen when we predict the missing ratings. However, we observe that some auxiliary data in the form of like/dislike may be more easily obtained. For example, we can easily collect the favored/disfavored data in Moviepilot, the love/ban data in Last.fm and the Want to see/Not Interested data in Flixster, which are implicit feedbacks and heterogeneous to the numerical ratings [15]. Moreover, the implicit feedbacks can be collected from the user behaviors and formalized to be binary ratings. For example, if the user browses, forwards or collects some information, we set the rating as 1, and 0 otherwise. It is more frequently for users to express such implicit tastes than to mark numerical ratings. We can make use of implicit feedbacks to alleviate the sparseness problem in explicit feedbacks. For the explicit feedbacks, such as the numerical rating matrix $R = (R_{ui}), R_{ui} \in \{1, 2, 3, 4, 5\}$, let $R' = (R'_{ui})$, $R'_{ui} = \frac{R_{ui}-1}{4}$. R' is the preprocessed rating matrix of R. R_{ui} can be restored by $R_{ui} = 4R'_{ui} + 1$ at the end of prediction. To alleviate the data heterogeneity between different kinds of feedbacks, such as $\{0, 1\}$ and $\frac{\{1,2,3,4,5\}-1}{4}$, let $\sigma(x) = \frac{1}{1+e^{-(x-0.5)}}$, $x \in \{0, 1\}$ is the implicit feedback, $\sigma(x)$ is a logistic link function to revise the implicit feedbacks. So the heterogeneous rating data are normalized to be homogeneous for the collective factorization. Then HFT model is mainly divided into the following three steps.

5.1   Extraction of Latent Factors

To extract latent factors, we do the flowing two collective factorizations respectively. Each factorization is learning from the GAO model.

$$\min_{U_0, S_0, S_1, S, V, V_1 \geq 0} f_1 = ||[R - U_0[S_0, S]V^T] \odot Z||^2 + \tag{3}$$
$$||[R_1 - U_0[S_0, S_1]V_1^T] \odot Z_1||^2$$

$$\min_{U_2, U, S'_0, S_2, S', V_0 \geq 0} f_2 = ||[R - U[S'_0, S']V_0^T] \odot Z||^2 + \tag{4}$$
$$||[R_2 - U_2[S'_0, S_2]V_0^T] \odot Z_2||^2$$

We solve Eqs. (3) and (4) by the gradient descent method, and get the latent factors U_0 and V_0, respectively.

Solving Equation (3)

Let $V = [V_{00}^T, V_{01}^T] \in R^{N \times L}$, $V_1 = [V_{10}^T, V_{11}^T] \in R^{N \times L_1}$, where $V_{00}^T = V(:, 1 : D)$, $V_{01}^T = V(:, (D + 1) : L)$, $V_{10}^T = V_1(:, 1 : D)$, $V_{11}^T = V_1(:, (D + 1) : L_1)$, D is the dimension of shared common rating pattern. Here $L - D$ is the dimension of domain-specific rating pattern of R, $L_1 - D$ is the dimension of domain-specific rating pattern of R_1, $S_0 \in R^{K \times D}$, $S \in R^{K \times (L-D)}$, $S_1 \in R^{K \times (L_1-D)}$, $U_0 \in R^{M \times K}$, $Z \in R^{M \times N}$. $U_0^T U_0 = I$, $V_1^T V_1 = I$, $V^T V = I$.

Taking the learning of latent factor S as an example, we will show how to optimize S by deriving its updating rule while fixing other latent factors. For this purpose we can rewrite the objective function in Eq. (3) as follows:

$$\min_S f_1(S) = ||R - U_0 S_0 V_{00} - U_0 S V_{01}] \odot Z||^2 +$$

$$||[R_1 - U_0 S_0 V_{10} - U_0 S_1 V_{11}] \odot Z_1||^2 \tag{5}$$

The derivative of $f_1(S)$ with respect to S is

$$\frac{\partial f_1(S)}{\partial S} = 2(U_0{}^T([U_0 S_0 V_{00}] \odot Z)V_{01}^T - U_0{}^T(R \odot Z)V_{01}^T) + 2U_0{}^T([U_0 S_1 V_{01}] \odot Z)V_{01}^T.$$

We use the Karush-Kuhn-Tucker complementary condition for the nonnegativity of S and let $\frac{\partial f_1(S)}{\partial S} = 0$, then can get the following updating rule for learning S:

$$S \leftarrow S \sqrt{\frac{U_0^T(R \odot Z)V_{01}^T}{U_0^T([U_0 S_0 V_{00}] \odot Z)V_{01}^T + U_0^T([U_0 S V_{01}] \odot Z)V_{01}^T}} \tag{6}$$

Similarly, we can get the updating rules for learning other latent factors as follows:

$$S_1 \leftarrow S_1 \sqrt{\frac{U_0^T(R_1 \odot Z_1)V_{11}^T}{U_0^T([U_0 S_0 V_{10}] \odot Z_1)V_{11}^T + U_0^T([U_0 S_1 V_{11}] \odot Z_1)V_{11}^T}} \tag{7}$$

$$U_0 \leftarrow U_0 \sqrt{\frac{(R \odot Z)V[S_0, S]^T}{([U_0[S_0, S]V^T \odot Z) \odot V[S_0, S]^T}} \tag{8}$$

$$V \leftarrow V \sqrt{\frac{[S_0, S]^T U_0{}^T(R \odot Z)}{[S_0, S]^T U_0{}^T([U_0[S_0, S]V^T] \odot Z)}} \tag{9}$$

$$V_1 \leftarrow V_1 \sqrt{\frac{[S_0, S_1]^T U_0{}^T(R_1 \odot Z_1)}{[S_0, S_1]^T U_0{}^T([U_0[S_0, S_1]V_1{}^T] \odot Z_1)}} \tag{10}$$

$$S_0 \leftarrow S_0 \sqrt{\frac{U_0^T(R \odot Z)V_{00}^T + U_0^T(R_1 \odot Z_1)V_{10}^T}{P + Q}} \tag{11}$$

where

$$P = U_0^T([U_0 S_0 V_{00}] \odot Z)V_{00}^T + U_0^T([U_0 S V_{01}] \odot Z)V_{00}^T$$

$$Q = U_0^T([U_0 S_0 V_{10}] \odot Z_1)V_{10}^T + U_0^T([U_0 S_1 V_{11}] \odot Z_1)V_{10}^T$$

Based on the above updating rules for learning different latent factors, it can be proved that the objective function in Eq. (3) will decrease monotonically and the learning algorithm demonstrated above is convergent [4]. At last, we can extract latent factor U_0 by enough iterations. In the same way, we can extract latent factor V_0 by solving Eq. (4).

5.2   Similarity Graph Construction

When the observed user-item rating data is very sparse, two users may rate the common item with a fairly low probability though they have the same taste, so the neighborhood information of users or items can not be effectively utilized. To overcome this problem, the similarity graphs from dense auxiliary data are constructed since the auxiliary data is closely related to the target data. Because U_0 and V_0 are denser than R_1 and R_2, they can better represent the neighborhood information. We construct the user-side similarity graph W_U and item-side similarity graph W_V based on the extracted U_0 and V_0, respectively [20]. In SHI model, the distance is simply defined by the concept of p-nearest neighbors with binary values of 0 and 1, so W_U and W_V may be inaccurate to be the weight matrices for the similarity graphs. Here we adopt the PCC (Pearson Correlation Coefficient) to measure the similarity:

$$(W_U)_{ij} = \rho(u_{i*}^0, u_{j*}^0) = \frac{\sum (u_{i*}^0 - \overline{u_{i*}^0})(u_{j*}^0 - \overline{u_{j*}^0})}{\sqrt{\sum (u_{i*}^0 - \overline{u_{i*}^0})^2} \sqrt{\sum (u_{j*}^0 - \overline{u_{j*}^0})^2}}$$

$$(W_V)_{ij} = \rho(v_{i*}^0, v_{j*}^0) = \frac{\sum (v_{i*}^0 - \overline{v_{i*}^0})(v_{j*}^0 - \overline{v_{j*}^0})}{\sqrt{\sum (v_{i*}^0 - \overline{v_{i*}^0})^2} \sqrt{\sum (v_{j*}^0 - \overline{v_{j*}^0})^2}}$$

where u_{i*}^0 is the i th row of U_0 denoting the latent taste of user i, v_{i*}^0 is the i th row of V_0 denoting the latent feature of item i, $\overline{u_{i*}^0}/\overline{v_{i*}^0}$ is the mean of u_{i*}^0/v_{i*}^0.

5.3   Predicting of Missing Ratings

We predict missing ratings by the graph regularized weighted nonnegative matrix tri-factorization. The optimization function is

$$\min_{U,V,B>=0} O = ||Z \odot (R - UBV^T)||_F^2 + \gamma_U G_U + \gamma_V G_V \qquad (12)$$

where $G_U = tr(U^T L_U U)$, $G_V = tr(V^T L_V V)$, L_U and L_V are the graph Laplacian matrices for the similarity graphs W_U and W_V respectively. γ_U and γ_V are regularization parameters indicating our confidence on the similarity graphs.

We solve the optimization problem in Eq. (12) by the gradient descent method [20]. The derivative of O with respect to U is

$$\frac{\partial O}{\partial U} = -2Z \odot RVB^T + 2Z \odot (UBV^T)VB^T + 2\gamma_U L_U U.$$

We use the Karush-Kuhn-Tucker (KKT) complementary condition for the nonnegativity of U and let $\frac{\partial O}{\partial U} = 0$, then get

$$[-Z \odot RVB^T + Z \odot (UBV^T)VB^T + \gamma_U L_U U] \odot U = 0$$

Because L_U may take any signs, we decompose it as $L_U = L_U^+ - L_U^-$, where $L_U^+ = \frac{1}{2}(|L_U|+L_U)$, $L_U^- = \frac{1}{2}(|L_U|-L_U)$. L_U^+ and L_U^- are positive-valued matrices. We get the following updating rule for learning U:

$$U \leftarrow U \odot \sqrt{\frac{Z \odot RVB^T + \gamma_U L_U^- U}{Z \odot (UBV^T)VB^T + \gamma_U L_U^+ U}} \qquad (13)$$

Similarly, we can obtain the updating rules for learning V and B as follows:

$$V \leftarrow V \odot \sqrt{\frac{(Z \odot R)^T UB + \gamma_V L_V^- U}{(Z \odot (UBV^T))^T UB + \gamma_V L_V^+ V}} \qquad (14)$$

$$B \leftarrow B \odot \sqrt{\frac{U^T(Z \odot R)V}{U^T(Z \odot (UBV^T))V}} \qquad (15)$$

According to [5], the objective function in Eq. (12) will monotonically decrease until convergence when updating U, V and B sequentially and iteratively by Eqs. (13–15). R can be calculated by UBV^T, so the missing ratings can be predicted.

6 Experiments

6.1 Data Sets

MovieLens10M[1]. It contains 10000054 ratings and 95580 tags applied to 10681 movies by 71567 users of the online movie recommender service MovieLens. The preference of the user for a movie is rated on a 5-star scale, with half-star increments. If the movie is not rated by any user, the rating is marked as 0.
Epinions[2]. It contains 2,372,198 ratings given by 44,157 users to 50,682 products. Users can mark products with integer ratings ranging from 1 to 5.
Book-Crossing[3]. It contains 278,858 users providing 1,149,780 ratings expressed on a scale from 0 to 10 about 271,379 books.

6.2 Compared Models

⋆ GMF (Graph Regularized Weighted Nonnegative Matrix Factorization) [5]: A good single-domain model which constructs two graphs on user side and item side to exploit the internal and external information. In this model, the missing ratings are predicted by the graph regularized weighted nonnegative matrix tri-factorization.
⋆ SHI [20].
⋆ CBT (Codebook Based Transfer) [7]: a classical model for cross-domain CF recommendation which can only make use of the common rating pattern via the codebook information among different domains.
⋆ HPT:our proposed new model.

---

[1] www.grouplens.org/node/73/.
[2] www.epinions.com.
[3] http://www2.informatik.uni-freiburg.de/~cziegler/BX/.

6.3 Evaluation Metrics

Mean Absolute Error (MAE) and Root Mean Square Error (RMSE) are the two widely used evaluation metrics for evaluating CF algorithms. We adopt MAE to measure the prediction accuracy of multiple recommendation models. It is defined as follows:

$$MAE = \frac{\sum\limits_{i,j} |R_{ij} - R_{ij}^p|}{|T_E|}$$

where R_{ij} is the rating that user i gives to item j in test set, while R_{ij}^p is the predicted value of R_{ij}, $|T_E|$ is the number of ratings in test set. The smaller the value of MAE is, the better the recommendation model performs.

6.4 Experimental Results

To simulate the heterogeneous feedbacks, we reference to Weike Pan's method [14]. When the observed ratings range from 1 to 5, we set the ratings to 1 if they are 4 or 5 and set the ratings to 0 otherwise. When the observed ratings range from 1 to 10, we set the ratings to 1 if the ratings are 7, 8, 9 or 10 and set the ratings to 0 otherwise.

The data sets used in our experiments are constructed by the same strategy as [19]. Take the Book-Crossing data set as an example, we first randomly sample a $2N \times 2N(N \in N^+)$ dense rating matrix X, then take the submatrix $R = X_{1 \sim N, 1 \sim N}$ as the target rating matrix, the submatrix $R_1 = X_{1 \sim N, (N+1) \sim 2N}$ as the user-side auxiliary data matrix and the submatrix $R_2 = X_{(N+1) \sim 2N, 1 \sim N}$ as the item-side auxiliary data matrix. In this way, R and R_1 share common users, while R and R_2 share common items. And we apply the same construction strategy to Epinions data set and MovieLens10M data set.

The target ratings matrix R is randomly split into a training set and a test set, with the proportion of 60 % and 40 % respectively. Let l be the sparseness level of the data set. To evaluate the performance of each model in different sparse data, we sample the target training set randomly with various sparseness levels ranging from 0.01 % to 1 %. The utilized auxiliary data is always much denser than the target data. Different dimensions of latent factors such as {10, 20, 50, 100, 150, 200, 300, 500, 800} and different values of regularization parameters such as {0.01, 0.1, 0.5, 1, 5, 10, 50, 80} of each model are tried, the best of which are selected for comparisons. Given that the major algorithms in the models are iterative, we run each model 5 repeated times and report the average results of MAE. In different data sets with various sparseness levels l, varies of the values of MAE in different models with respect to N are as follows:

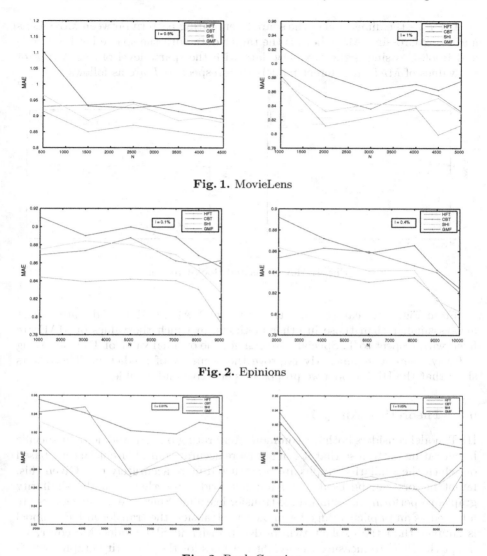

Fig. 1. MovieLens

Fig. 2. Epinions

Fig. 3. Book-Crossing

In each one of Figs. 1, 2 and 3, the data sparseness level l in the left subgraph is smaller than that in the right subgraph, which means the sampled data in the left subgraph are sparser than that in the right subgraph. The above experimental results show that the values of MAE in our proposed HFT model are almost always smaller than those in other models in all cases, which is more significant when the data is sparser in each data set. So it is demonstrated that HFT model is more effective than others, especially when the data is sparser. Also it performs well in the case of heterogeneous user feedbacks.

Let T be the dimension of the share latent rating pattern between MovieLens and Book-Crossing. MovieLens is the target data with the sparse level of 0.5 %, and Book-Crossing is the auxiliary data with the sparse level of 8 %. Varies of the values of MAE in different models with respect to T are as follows:

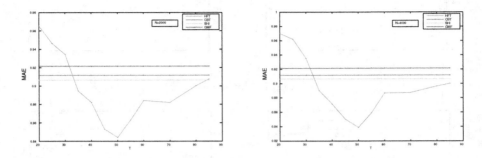

Fig. 4. MovieLens and Book-Crossing

From Fig. 4, we can see that the values of MAE in HFT model are almost always smaller than those in other methods, in which the values of MAE are fixed with respect to T. Specially we can get the optimal value of T by adjusting it freely, thereby significantly improve the accuracy of prediction. The results show that the HFT model we proposed is effective and flexible.

6.5   Theoretical Analysis

HFT model considers both the common latent rating pattern and domain-specific latent rating pattern so that it can get more accurate initial latent factors, based on which the similarity graphs are constructed more accurately too. Given this, latent factors can be used as an implicit bridge together with the similarity graphs to perform the twin bridge transfer learning. Since the latent factors are adopted as an implicit bridge for knowledge transfer, the transfer in HFT model is simpler than the explicit twin-bridge transfer in SHI model. Moreover, we adopt the PCC to measure the weight matrices for the similarity graphs, which is time-saving and more accurate than the way based on the p-nearest neighbors in SHI model. So HFT model owns a lower time complexity but higher recommendation accuracy than SHI model. By the twin bridge transfer learning, HFT model can transfer more useful knowledge, while alleviate negative transfer by regularizing the learning model with the similarity graphs, which can effectively filter out the negative information contained in the latent factors. So it is easy to understand that HFT model has higher recommendation accuracy than GMF model and CBT model, which not only are based on the single bridge transfer learning, but also don't consider the domain-specific latent rating patterns. In addition, HFT model applies to the transfer of heterogeneous user feedbacks across multiple domains by data normalization and the collective factorization, which makes the advantages of it more significant in the multi-source and heterogeneous environment of big data.

7   Conclusions and Future Work

We propose a novel HFT model of transfer learning for cross-domain CF recommendation, which not only performs well in the case of heterogeneous feedbacks, but also improves the recommendation by transferring more useful knowledge considering both the common latent rating pattern and domain-specific latent rating pattern and learning knowledge from the twin bridge of latent factors and similarity graphs. In addition, in HFT model the latent factors of users and items can be extracted by the collective factorization, so HFT model can be flexible to data quality across multiple domains. Extensive experiments show that our proposed HFT model is more efficient in sparse data and more flexible across domains than other three excellent models.

In the future, we try to exploit more auxiliary information for transfer. Social information, topic information and context information can be used for regularization fitting in the process of optimization. Moreover, given that recommendation tasks are often diverse, we intent to research multi-view and multi-task cross-domain transfer learning [21,22] for the cross-domain recommendation.

Acknowledgments. This work is supported by the National Natural Science Foundations of China (61272109 and 61502350), the Fundamental Research Funds for the Central Universities (2042014kf0057) and the National Natural Science Foundation of Hubei Province of China (2014CFB289).

References

1. Qi, Q., Chen, Z., Liu, J., Hui, C., Wu, Q.: Using inferred tag ratings to improve user-based collaborative filtering. In: Proceedings of the 27th Annual ACM Symposium on Applied Computing, pp. 2008–2013. ACM (2012)
2. Sarwar, B., Karypis, G., Konstan, J., Riedl, J.: Item-based collaborative filtering recommendation algorithms. In: Proceedings of the 10th International Conference on World Wide Web, pp. 285–295. ACM (2001)
3. Pan, S.J., Yang, Q.: A survey on transfer learning. IEEE Trans. Knowl. Data **22**(10), 1345–1359 (2010)
4. Gao, S., Luo, H., Chen, D., Li, S., Gallinari, P., Guo, J.: Cross-domain recommendation via cluster-level latent factor model. In: Blockeel, H., Kersting, K., Nijssen, S., Železný, F. (eds.) ECML PKDD 2013, Part II. LNCS, vol. 8189, pp. 161–176. Springer, Heidelberg (2013)
5. Gu, Q., Zhou, J., Ding, C.H.: Collaborative filtering: Weighted nonnegative matrix factorization incorporating user and item graphs. In: Proceedings of the 10th SIAM International Conference on Data, pp. 199–210 (2010)
6. Su, X., Khoshgoftaar, T.M.: A survey of collaborative filtering techniques. Adv. Artif. Intell. **2009**, 4 (2009)

7. Li, B., Yang, Q., Xue, X.: Can movies and books collaborate? cross-domain collaborative filtering for sparsity reduction. IJCAI **9**, 2052–2057 (2009)
8. Fernández-Tobías, I., Cantador, I., Kaminskas, M., Ricci, F.: Cross-domain recommender systems: a survey of the state of the art. In: Spanish Conference on Information Retrieval (2012)
9. Li, B.: Cross-domain collaborative filtering: a brief survey. In: 23rd IEEE International Conference on Tools with Artificial Intelligence (ICTAI), pp. 1085–1086. IEEE (2011)
10. Li, C.Y., Lin, S.D.: Matching users and items across domains to improve the recommendation quality. In: Proceedings of the 20th ACM SIGKDD international conference on Knowledge discovery and data mining, pp. 801–810. ACM (2014)
11. Wang, W., Chen, Z., Liu, J., Qi, Q., Zhao, Z.: User-based collaborative filtering on cross domain by tag transfer learning. In: Proceedings of the 1st International Workshop on Cross Domain Knowledge Discovery in Web and Social Network Mining, pp. 10–17. ACM (2012)
12. Lu, Z., Pan, W., Xiang, E.W., Yang, Q., Zhao, L., Zhong, E.: Selective transfer learning for cross domain recommendation. In: SDM, pp. 641–649. SIAM (2013)
13. Pan, W., Xiang, E.W., Yang, Q.: Transfer learning in collaborative filtering with uncertain ratings. In: AAAI (2012)
14. Pan, W., Yang, Q.: Transfer learning in heterogeneous collaborative filtering domains. Artif. Intell. **197**, 39–55 (2013)
15. Pan, W., Liu, N.N., Xiang, E.W., Yang, Q.: Transfer learning to predict missing ratings via heterogeneous user feedbacks. In: Proceedings of the 22th International Joint Conference on Artificial Intelligence, vol. 22, pp. 2318–2323 (2011)
16. Shi, Y., Larson, M., Hanjalic, A.: Generalized tag-induced cross-domain collaborative filtering. CoRR, abs/1302.4888 (2013)
17. Li, B., Yang, Q., Xue, X.: Transfer learning for collaborative filtering via a rating-matrix generative model. In: Proceedings of the 26th Annual International Conference on Machine Learning, pp. 617–624. ACM (2009)
18. Singh, A.P., Gordon, G.J.: Relational learning via collective matrix factorization. In: Proceedings of the 14th ACM SIGKDD international conference on Knowledge discovery and data mining, pp. 650–658. ACM (2008)
19. Pan, W., Xiang, E.W., Liu, N.N., Yang, Q.: Transfer learning in collaborative filtering for sparsity reduction. In: AAAI, vol. 10, pp. 230–235 (2010)
20. Shi, J., Long, M., Liu, Q., Ding, G., Wang, J.: Twin bridge transfer learning for sparse collaborative filtering. In: Pei, J., Tseng, V.S., Cao, L., Motoda, H., Xu, G. (eds.) PAKDD 2013, Part I. LNCS, vol. 7818, pp. 496–507. Springer, Heidelberg (2013)
21. Maurer, A., Pontil, M., Romera-Paredes, B.: Sparse coding for multitask and transfer learning. CoRR, abs/1209.0738 (2012)
22. Fang, Z., Zhang, Z.M.: Discriminative feature selection for multi-view cross-domain learning. In: Proceedings of the 22Nd ACM International Conference on Conference on Information Knowledge Management, pp. 1321–1330. ACM (2013)

Research on Automate Discovery of Deep Web Interfaces

Feiyue Ye and Hang Yu$^{(\boxtimes)}$

School of Computer Engineering and Science,
Shanghai University, Shanghai, China
{yefy,yuhang13}@shu.edu.cn

Abstract. The main means to obtain information from Deep Web is submitting query condition through the provided query interfaces, so it is the first problem that needs to be solved for Deep Web data integration system. At present, most researchers think of query interface is merely defined within the form html tag. This paper firstly proposes the concept of interface block, then designs the interface block location method based on page and vision information, and finally takes the judgment of whether interface block is a query interface or not as the special multi-class classification problems and by applying classification algorithm combining C4.5 decision tree and SVM. The experiment adopts TEL-8 data sets of UIUC, and the findings indicate that the method in this paper get an accuracy of 97.30%, and has good feasibility and practicability.

Keywords: Deep Web · Query interface · Interface block · Multi-class classification

1 Introduction

According to the distribution of information that the website carries, it can be divided into two parts: one is called the Surface Web, which the users can directly perceive from web pages; the other is called the Deep Web, which information is stored in back database [1]. The research of Michael K. Bergman [2] and others show that data volume stored in Deep Web is 500 times that of Surface Web, and these data cannot only serve as the source of good data, but also its structure and semantic features can help us to carry out more effective knowledge discovery. The main means to obtain data from the Deep Web is submitting a query condition through the provided query interfaces, therefore the first step to access Deep Web's resources is effectively discover query interface. However, due to the web page contains a lot of non-query interface and the structure of those is similar to the query interface, which makes it very difficult to automate discovery of query interface.

At present, many researchers have conducted study in this field [3–13]. Cope and others [3] adopt the method of C4.5 decision tree to realize the identification of query interface. The accuracy of this method is only 85%. Barbosa and

© Springer International Publishing Switzerland 2015
J. Wang et al. (Eds.): WISE 2015, Part II, LNCS 9419, pp. 191–198, 2015.
DOI: 10.1007/978-3-319-26187-4_14

others [4] improve the defects existing in the methods conducted by Cope and extract a total of 14 characteristics in the form. This improved method raises classification accuracy to 90.95%, but this method doesn't differentiate between query interface and search engine satisfactorily. Wang and others [6] built some rules to identify query interface, but this is not always true because a simple query interface having only one or two attributes. Marin-Castro and others [7] propose an algorithm of automatically discovering query interfaces based on machine learning. But this method assumes that query interface only designed by form html tag. Xu and others [13] put forward the idea that the page can be divided into several segments, and then build a mode for a specific domain, but the accuracy is low for forms with complex structure.

This paper proposes a new method of discovering interfaces. This method firstly transforms a web page to a DOM tree, and then transverses the nodes in the DOM tree to determine whether the content contained by this node in the web page needs judging or not by analyzing the CSS style information and the DOM tree structure. Secondly, it takes the judging process as a particular multi-type classification problem. This paper proposes an interface block classification algorithm based on the C4.5 decision tree and SVM. Among them, the essence of adopting the classification based on the C4.5 decision tree is to transform the query interface judgment issue into several issues of two-type classifications and finally adopting the SVM algorithm to further classify each of these two-type classification issues. Through the above methods, Deep Web search interface is enabled to get independence from the form html tag. Resultantly, the unique characteristics of various types in the non-query interface are fully utilized, improving the accuracy of classification.

2   The Interface Location Method

Some researches have been conducted on how to locate contents in Deep Web page, but these researches are about the location of the search result [14–16]. But through these researches, we find that DOM tree structure and CSS style information could reflect the page and visual information. The following content will have a detailed explanation of the method.

2.1   Related Definitions

Definition 1. *An interface block is a segment of HTML code in a Web page P delimited by HTML tags. This segment contains a set of control elements $C = \{c_i | 1 \leq i \leq n\}$ that can be described using the 3-tuple <name, label, domain>, where name corresponds to the name of the field associated to this element, label is a string that describes this element and the domain is the set of valid values that the field can take. Some examples of HTML control elements must be $C = \{input, button, select, option, option group and textarea\}$.*

Definition 2. *A Query Interface (QI) is defined as a HTML searchable form that is intended for users that want to query a Hidden-Web database. QIs have a* *heterogeneous schema of design, semantic and value; they may change frequently and without notice. Moreover, each QI may have a limited query capability.*

Definition 3. *Let T_i be a node in the DOM tree T_d, then a node density:*

$$D\left(T_i\right) = \frac{Count(Tag_{special})}{Count(Tag_{all})} \tag{1}$$

The numerator represents the number of leaf nodes which are label input, select, button and the textarea. The denominator represents the number of all leaf nodes.

Definition 4. *Let T_i and T_j for the DOM tree T_d two nodes, the similarity of T_i and T_j's visual information is defined as:*

$$Sim_{visual}\left(T_i, T_j\right) = \frac{sum_1 + sum_2 - p}{sum_1 + sum_2} \tag{2}$$

Sum_1 which represents the number of CSS style information contained T_i, sum_2 represent the number of CSS style information contained T_j, p represents the number of different CSS style information contained in both.

Definition 5. *Let T_i and T_j for the DOM tree T_d two nodes, the similarity of T_i and T_j's page information is defined as:*

$$Sim_{page}\left(T_i, T_j\right) = Sim_{Xpath}(T_i, T_j) * 2^{(1-Sim_{child}(T_i, T_j))} \tag{3}$$

Sim_{xpath} represents the similarity of T_i and T_j's Xpath and Sim_{child} represents the similarity of Ti and Tj's child node. Both use the calculation formula 4 in Yue K's paper [14]. We can extend Sim_{visual} and Sim_{page} formula to find the similarity between a plurality of nodes, The expansion formula is defined as:

$$multiSim\left(T_1 \ldots T_n\right) = \frac{\sum\limits_{i<j}^{n} Sim(T_i, T_j)}{C_N^2} \tag{4}$$

2.2   The Interface Location Algorithm

Next, we give the description of interface block location algorithm. As is shown in Algorithm I, this algorithm uses the root node of the DOM tree of the current page as input and the output is the corresponding tag containing the interface block content.

Algorithm I is mainly composed of a recursive function FindInterFace-Block( ). Among them, α, β, γ are the control parameters of the algorithm, and its value is determined by the experiment, which mainly completes the calculation of node density, similarity of page information and visual information. If all similarity greater than threshold, it means that the content contained by this node is the interface block; otherwise we can search the child node of this node.

Algorithm I. The interface location

Input: Node N_d, The number of the current node **Level**;

Output: the corresponding **Tag** containing the interface block content

1 FindInterfaceBlock(Level , Nd){

2 nds[n] <− getChildNodes(Nd)

3 **for** each node k from nds[n] **do**

4   density = Density(k)

5   Simcss = multiSimcss (getChildNodes(k))

6   Simpage = multiSimpage (getChildNodes(k))

7   **if**(density≥ α && Simcss ≥ β && Simpage ≥ γ)

8   **return** k

9   **else** FindInterfaceBlock(Level + 1, k)}

---

3   The Classification Method Based on C4.5 Decision Tree and SVM

We have found that it is much easier to judge whether an interface is of query interface or login interface than to directly judge whether an interface of query interface or non-query interface. Based on the considerations above, this paper transformed the problem into classifying between query interface and other non-query interfaces like register, mail or login. Moreover, as long as we confirm that an interface is of non-query interface, we do not need to further classify it within the range of non-query interface.

3.1   Feature Selection of Interface Block

In this period, we choose that seven common non-query interface, register, login, feedback, Newsletters, vote, mail and comment types, because they are often wrong as the query interface. Among them, query interface must hold the control of text, but comment interface and vote interface do not have such control, also mail interface must have the control of email, whereas query interface do not have such control, so it is very easy to distinguish query interface from mail, vote and comment types. In this paper, in order to choose some better features, we collect a total of 420 interface, of which there are 210 query interface, register(62), login(82), feedback(42), newsletters (24), and we can draw following conclusions from further analysis in the number of occurrences of HTML elements:

1. A total number of C_2 = {hidden, image, reset, submit, a, button} controls appears in the login interface must be less than or equal to 3
2. A total number of C_2 controls appears in the newsletters interface must be equal to 1
3. A total number of C2 controls appears in the feedback interface must be more than
4. A total number of C2 controls appears in the registration interface must be more than
5. Query interface must not contain C_1 = {textarea, password, Email} controls.

6. The WordList$_1$ = {email, password, phone, register, login, forgot, comment, message, feedback, newsletters, subscribe} must not appear in the query interface

7. The WordList$_2$ = {search, sort, keyword, category} must appear in the query interface. While the search word appears in the types of non-query interface, the page must include some words will appear in the non-query interface.

3.2   Classification Methods

C4.5 can handle well the differences among data sets brought by different attribute selection methods; and it has a low time complexity. SVM is a two-class classifier in essence. As for two-class classification, its time complexity is linear, and it can find out the globally optimal solution. Therefore, this paper combines the above two kinds of classification algorithms to solve the multi-class classification problem. The classification algorithm is described as follows.

Algorithm II. Automatic Identification of Deep Web interface

Input: InterfaceHtml, WordList$_1$, WordList$_2$, C_1, C_2, C_3: {checkbox, hidden, image, radio, reset, submit, text, a, button, select}, C_4: {clear, title, address, button, a, go, submit}

Output: The type of interface block

1 TextCount = SearchTextTag(InterfaceHtml)
2 **if**(TextCont > 0)
3   ControlCount = C4.5_DecisionTree (InterfaceHtml, C1, C_2)
4   **if**(ControlCount > 3)
5   type$_1$ = Registration-Query-SVM(InterfaceHtml, C_3)
6   type$_2$ = Feedback-Query-SVM(InterfaceHtml, C_3)
7   **if**(type$_1$ == query && type$_2$ == query)
8   type = SetQueryInterface(InterfaceHtml)
9   **else** type = SetNoQueryInterface (InterfaceHtml)
10   **else if**(ControlCount > 1)
11   type = Login-Query-SVM(InterfaceHtml, C_3)
12   **else** type = Newsletters-Query-SVM(InterfaceHtml, C_4)
13 **else** SetNoQueryInterface (InterfaceHtml)
14 **return** type

In algorithm II, First, we count the number of text tags. Then through the C4.5 decision tree, part of the query interface and non-query interface are discovered first, and the decision problem is then transformed into two two-class classification problems and one three-class classification problem. In C4.5 decision tree, we first judge whether or not the word is contained in WordList$_1$. Second, we judge whether or not the control elements are contained in C_1. Third, we judge whether or not the word is contained within WordList$_2$. Fourth, we count the number of controls that not are contained in C_2. Finally, with the SVM, further classification is carried out and the parameters of SVM being the number of controls in C_3 or C_4. As for three-class classification problem, only when the two classifiers both set the interface block into query interface at the same time, can we confirm it as a query interface.

4   Experimental Results

To prove the validity of the method, which is, automatically finding query interface, we use TEL-8 as the test set, and two experiments are conducted to assess the validity of the method: (1) the first experiment assesses the validity of interface block location algorithm that we proposed; (2) the second experiment assesses the validity of the interface block classification algorithm.

4.1   Experiment 1

We extracted 400 pages from TEL-8 data set, in which query interfaces have been classified artificially, and then recall rate of interfaces was calculated. To confirm the controlling parameters α, β and γ and their optimal combination in algorithm I, we use orthogonal experiment to determine the approximate optimal combination of the three parameters. Table 1 gives the experiment level and factor, and Table 2 gives the experiment results.

Table 1. Experiment level and factor

Level	Factor 1	Factor 2	Factor 3
	α	β	γ
1	0.32	0.42	0.52
2	0.35	0.38	0.54
3	0.38	0.35	0.57

Table 2. Experiment results

No	Factor 1	Factor 2	Factor 3	Results (Recall)	No	Factor 1	Factor 2	Factor 3	Results (Recall)
	α	β	γ			α	β	γ	
1	1	1	1	79%	6	2	3	1	82%
2	1	2	2	85%	7	3	1	2	78%
3	1	3	3	75%	8	3	2	1	80%
4	2	1	2	73%	9	3	3	3	65%
5	2	2	3	69%	Optimal combination	1	2	2	

Based on preliminary experiments and experience, we first determine the approximate value scope of the three parameters, among which the value of α is between 0.30–0.40; the value of β is between 0.35–0.45; and the value of γ is between 0.50–0.60. Orthogonal table of L_9 (3^4) is chosen, that is, with 3 levels and 4 factors, which means at most 9 separate experiments can be carried out, so as to confirm the optimal combination of parameters. As for this paper, we only choose 3 factors. Table 2 shows that the optimal parameters are $(\alpha, \beta, \gamma) = (0.32, 0.38, 0.54)$.

4.2   Experiment 2

In the experiment, we divide the 425 classified and marked query interfaces into a training set of 209 interfaces and a test set of 216 interfaces. Then we extracted 212 non-query interfaces as counterexamples of the training set, and 229 non-query interfaces as counterexamples of the test set. The experiment results are shown in Table 3.

Table 3. Experiment results

Interface	Predicted Query	Predicted Non-query
Query	209	7
Non-query	5	224

From Table 3, we can see our method attains high accuracy when distinguishing between query interface and non-query interface. Then we compare the experiment results in Table 3 with Barbosa's method [4] and Marin-Castro [7]'s method, and the results of comparison are shown in Fig. 1.

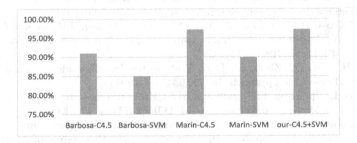

Fig. 1. Comparison of accuracy.

From Fig. 1, we can see that the rate of correct classification through our method is almost the same with the optimal result in Marin-Castro's method. However, in Marin-Castro's method, the discovery of query interfaces is dependent upon the tag of "<form></form>". Compared with that, our method therefore has witnessed a better result.

5   Conclusion and Future Work

The problem of discovering query interface automatically is the first problem that needs to be solved for Deep Web data integration system. This paper firstly proposes the concept of interface block, then designs the interface location method based on page and vision information, and finally applying classification algorithm combining C4.5 decision tree and SVM to obtain the query interface with

Deep Web. The results reported in this work give evidence of the effectiveness and usefulness of the proposed strategy. The simple query interface cannot be correctly distinguished only by extracting structural features, so we can incorporate semantic information into the words in interface to further enhance the efficiency and accuracy of determining query interfaces.

References

1. Madhavan, J., Ko, D., Kot, Ł., et al.: Google's Deep Web crawl. Proc. VLDB Endowment **1**(2), 1241–1252 (2008)
2. Bergman, M.K.: White paper: the Deep Web: surfacing hidden value. J. Electron. Publishing **7**(1), 1–17 (2001)
3. Cope, J., Craswell, N., Hawking, D.: Automated discovery of search interfaces on the web. In: Proceedings of the 14th Australasian database conference, ADC 2003, vol. 17, pp. 181–189 (2003)
4. Barbosa, L., Freire, J.: Combining classifiers to identify online databases. In: Proceedings of the 16th international conference on World Wide Web, WWW 2007, pp. 431–440. ACM, New York (2007a). ISBN 978-1-59593-654-7
5. Jiang, L., Wu, Z., Feng, Q., Liu, J., Zheng, Q.: Efficient Deep Web crawling using reinforcement learning. In: Zaki, M.J., Yu, J.X., Ravindran, B., Pudi, V. (eds.) PAKDD 2010, Part I. LNCS, vol. 6118, pp. 428–439. Springer, Heidelberg (2010)
6. Wang, Y., Li, H., Zuo, W., He, F., Wang, X., Chen, K.: Research on discovering Deep Web entries. Comput. Sci. Inf. Syst. **8**(3), 779–799 (2011)
7. Marin-Castro, H.M., Sosa-Sosa, V.J., Martinez-Trinidad, J.F., et al.: Automatic discovery of Web Query Interfaces using machine learning techniques. J. Intell. Inf. Syst. **40**(1), 85–108 (2013)
8. Wang, H., Xu, Q., Zhou, L.: Deep Web search interface identification: a semi-supervised ensemble approach. Inf. **5**, 634–651 (2014)
9. Gravano, L., Ipeirotis, P.G., Sahami, M.: QProber: a system for automatic classification of hidden-web databases. ACM TOIS **21**(1), 1–41 (2003)
10. He, B., Tao, T., Chang, K.C.C.: organizing structured web sources by query schemas: a clustering approach. In: Gravano, L. (ed.) Proceeding of ACM the 13th Conference on Information and Knowlege Management, pp. 22–31, ACM Press, Washington (2004)
11. Barbosa, L., Freire, J., Silva, A.: Organizing hidden-web databases by clustering visible web documents. In: Doqac, A. (ed.) Proceeding of IEEE the 23rd Internatiobnal Conference on Data Engineering, pp. 326–335, IEEE Computer Society, Istanbul (2007)
12. Shestakov, D.: On building a search interface discovery system. In: Lacroix, Z. (ed.) RED 2009. LNCS, vol. 6162, pp. 81–93. Springer, Heidelberg (2010)
13. Du, X., Zheng, Y.Q., Yan, Z.M.: Automate discovery of Deep Web interfaces. In: Information Science and Engineering (ICISE), pp. 3572–3575 (2010)
14. Yue, K., Dong, L., Derong, S., et al.: D-EEM: a DOM-Tree based entity extraction mechanism for Deep Web. J. Comput. Res. Dev. **5**, 014 (2010)
15. Lu, Y., He, H., Zhao, H., et al.: Annotating search results from web databases. IEEE Trans. Knowl. Data Eng. **25**(3), 514–527 (2013)
16. He, Y., Xin, D., Ganti, V., et al.: Crawling Deep Web entity pages. In: Proceedings of the sixth ACM International Conference on Web Search and Data Mining, pp. 355–364. ACM (2013)

Discovering Functional Dependencies
in Vertically Distributed Big Data

Weibang Li[✉], Zhanhuai Li, Qun Chen, Tao Jiang, and Hailong Liu

Northwestern Polytechnical University, Xi'an 710072, Shaanxi, China
{liweibang,jiangtao}@mail.nwpu.edu.cn,
{lizhh,chenbenben,liuhailong}@nwpu.edu.cn

Abstract. The issue of discovering FDs has received a great deal of attention in the database research community. However, as the problem is exponential in the number of attributes, existing approaches can only be applied on small centralized datasets. It is challenging to discover FDs from big data, especially if data is distributed. We present a new algorithm DFDD for discovering all functional dependencies in parallel in vertically distributed big data following a breadth-first traversal strategy of the attribute lattice that combines efficient pruning. We verify experimentally that our approach can process distributed big datasets and it is scalable with the number of cluster nodes and the size of datasets.

Keywords: Distributed data · Big data · Association rule mining

1 Introduction

Functional dependencies (FDs) discovery is a well-studied aspect of relational database theory. A functional dependency (FD) between two sets of attributes (X, Y) holds in a relation if values of the latter set are fully determined by the values of the former set [1]. Formally, for a relation schema R with $X \subseteq R$ and $Y \subseteq R$, functional dependency $X \rightarrow Y$ is satisfied if for all pairs of tuples t_i and t_j over R, we have: if $t_i[X] = t_j[X]$ then $t_i[Y] = t_j[Y]$. The FDs discovery problem is as follows: Given a relation schema R and an instance r of schema R, we wish to find all FDs which hold over r [2]. In this paper we address the problem of discovering functional dependencies in distributed big data.

FDs play an important role in relational theory and relational database design [3]. Discovering FDs from databases is an important issue and is investigated for many years. Given a database D, if D is a centralized database, there are some algorithms that can be used to discover FDs from D, e.g. the partition based methods such as TANE [4], the free-set based methods such as FUN [7], etc.

This work was supported in part by National Basic Research Program 973 of China (No. 2012CB316203), Natural Science Foundation of China (Nos. 61033007, 61272121, 61332006, 61472321), National High Technology Research and Development Program 863 of China (No. 2012AA011004), Basic Research Fund of Northwestern Polytechnical University (No. 3102014JSJ0005, 3102014JSJ0013).

J. Wang et al. (Eds.): WISE 2015, Part II, LNCS 9419, pp. 199–207, 2015.
DOI: 10.1007/978-3-319-26187-4_15

These methods are designed for centralized databases. In practice, however, a relation is often fragmented and distributed across different sites [6]. In these settings the FDs discovering problem is much harder and thus they could not be applied directly, unless all the data fragments are transported to one site.

Example 1. Figure 1 gives an relation R, R is vertically fragmented into three fragments (Fig. 1(b)), and each fragment R_i resides at site S_i. Here TID is a key of R, and A, B, C, D, E are attributes of R.

As is shown in Fig. 1, it is easy to discover FDs from R if data is centralized by existing FDs discovering algorithms. However, if R is fragmented and resides at different sites, we cannot apply the existing algorithms. For example, to determine whether or not candidate FD $A \rightarrow C$ is a FD, we cannot use existing methods and need to transport data from S_2 to S_1 or from S_1 to S_2 first of all.

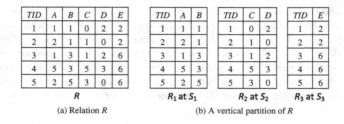

(a) Relation R (b) A vertical partition of R

Fig. 1. A relation R and its vertical partitions

The remainder of this paper is organized as follows. After the related work introduced in Sect. 2, some basic notations are introduced in Sect. 3. In Sect. 4, we lay out the principles of searching the space of FDs. The details of algorithm DFDD are given in Sect. 5. Experimental results are presented in Sect. 6, followed by conclusion in Sect. 7.

2　Related Work

Most of existing works on functional dependencies discovery were applied to centralized data [4,7,8]. These functional dependencies discovery approaches can be classified into two main groups.

The first group consists of approaches that traverse the search space in a breath-first manner and check the satisfaction of the candidate FDs level by level. The first group includes algorithms such as TANE [4], FUN [7], etc. The main characteristic of these algorithms is that it works level-wise when testing the satisfaction of the candidate FDs and uses the satisfied FDs to prune the candidate FDs at lower levels of the lattice to reduce the search space.

The second group of approaches are algorithms that traverse the search space in a depth-first manner. The key characteristic of the depth-first algorithm is that it starts with an individual attribute as LHS and appends more attributes until an FD for the selected RHS holds. The second group includes algorithms such as Miner [8]. Different sets are used in these algorithms and the candidate for LHSs are selected based on different sets.

All above works are related to FDs or other constraints discovering from centralized small data while in this paper we are discovering FDs from distributed big data that is vertically fragmented.

3   Preliminaries

Consider a relation schema R defined over a set of attributes: $A_1, \cdots, A_n$. The attributes set of R is denoted by $\text{attr}(R)$. Single attribute is denoted by the letters A, B, C, $\cdots$, while X, Y, Z stand for sets of attributes. The cardinality of an attribute set X in relation R is denoted by $|X|_R$, which gives the number of distinct values of X in R. $|X|$ stands for the number of attributes in X, while $|R|$ represents the cardinality of R (i.e. the number of tuples in R).

Definition 1 [3] (Functional Dependencies, FDs). Let X, $Y \subseteq \text{attr}(R)$ be two sets of attributes. The functional dependency between X and Y, denoted by $X \rightarrow Y$, holds in R if and only if: $\forall t_1, t_2 \in R$, if $t_1(X) = t_2(X)$, then $t_1(Y) = t_2(Y)$.

Definition 2 (Key). Let $X \subseteq \text{attr}(R)$ be an attribute set. X is a key for R if and only if: $X \rightarrow \text{attr}(R)$.

Remark. If X is a key, then $|X|_R = |R|$.

Definition 3 (Equivalence Classes). Let $X \subset \text{attr}(R)$ be an attribute set, denote the equivalence classes of a tuple $t \in R$ with respect to the given set X, i.e. $[t]_X = \{p \in R \mid p[A] = t[A] \text{ for all } A \in \text{attr}(R)\}$, where $t[A]$ is the attribute value of tuple t under attribute A.

Definition 4 (Partition). Let $X \subset \text{attr}(R)$ be an attribute set, denote $\prod_X = \{[t]_X \mid t \in R\}$ as a partition of R under X.

It is obvious that $\prod_X$ is composed of disjoint equivalence classes of tuples in R, and in each equivalence class the tuples under attribute set X have the same attribute value and the attribute value is unique. The union of the equivalence classes equals the relation R.

Definition 5 (Stripped Partition) [4]. Let $X \subset \text{attr}(R)$ be an attribute set, denote $\prod'_X = \{[t]_X \mid t \in R, |[t]_X| > 1\}$ as a stripped partition of R under X.

4   Search and Prune

This section we formulate the search strategies of candidate FDs and the pruning strategies of candidate FDs.

4.1 Search Strategy

To discover all non-trivial FDs, DFDD works as follows. It starts with candidate FDs generation, firstly generates and verifies candidate FDs with one attribute at the left hand side (LHS), then works its way to larger LHS attribute sets through the attribute lattice level by level from top to down. When processing a set X, this algorithm tests candidate dependencies of the form $X \setminus \{A\} \to A$, where $A \in X$. This guarantees that only non-trivial candidate dependencies are tested. This searching method can also be used to prune the search space efficiently.

Given schema $R = \{A_1, A_2, \cdots, A_m\}$, candidate dependencies are calculated using all possible attribute combinations of R as LHS. In this paper, we only pay attention to minimal FDs with single attribute on the left hand side (LHS), thus the number of attributes for the LHS of a candidate FD is at most $m-1$. Figure 2 shows a lattice of $attr(R) = \{A, B, C, D\}$, and the lattice is composed of all candidate FDs. The edge of the lattice between the first node from left at Level-1 and the first node from left at Level-2 represents the candidate FD $A \to B$, and the edge between the first node from left at Level-2 and the first node from left at Level-3 represents the candidate FD $AB \to C$, etc.

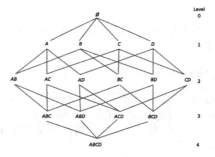

Fig. 2. Attribute lattice for $attr(R) = \{A, B, C, D\}$

Lemma 1. Given a relation schema R defined over $\{A_1, \cdots, A_m\}$, the total number of non-trivial candidate dependencies is $m \cdot 2^{m-1} - m$.

4.2 Pruning

To reduce unnecessary identifying of candidate FDs and improve the discovering efficiency, we can prune the implied FDs from the lattice. Candidate FDs pruning is to remove the candidate FDs (edges) in the lattice implied by the discovered FDs so that we do not need to verify them.

Theorem 1 [4]. An FD $X \to Y$ is satisfied if and only if $|\prod_X| = |\prod_{XY}|$.

For example, in Fig. 1, FD $A \to D$ is satisfied on relation R, since $|\prod_A| = |\prod_{AD}|$.

Theorem 2. Let X, $Y \subseteq \mathrm{attr}(R)$. If $|\prod_X| < |\prod_Y|$, then $X \nrightarrow Y$.

Theorem 3. Let $X \subseteq \mathrm{attr}(R)$, $A \in \mathrm{attr(R)} \setminus X$. If $X \nrightarrow A$, for each $Y \subset X$, we have $Y \nrightarrow A$.

Theorem 4. Let $X \subseteq \mathrm{attr}(R)$, $A \in \mathrm{attr}(R) \setminus X$. If $X \rightarrow A$, for each $X \subset Y$, we have $Y \rightarrow A$.

Theorem 5. The candidate FDs verifying tasks allocating problem is a NP-hard problem [5].

5   Algorithm

Algorithm Cet. The first algorithm, Cet, is a naïve approach: it reduces the FDs discovering problem for vertically fragmented distributed data to its counterpart for centralized databases. Cet first selects one site as the collaborate site, then transfers the data distributed at other sites to the collaborate site, and at which the candidate FDs are verified.

Algorithm DFDD. Algorithm 1 illustrates the main loop of DFDD. Its input is the set of columns and their partitions. Its output is the set of FDs discovered from distributed big data.

Algorithm 1. DFDD

Input: $R_1, R_1, \cdots, R_n$
Output: Ω
1 treeRoot $\leftarrow NULL$;
2 $\Omega \leftarrow \{\}, \Omega' \leftarrow \{\}$;
3 $\mathrm{attr}(R) \leftarrow \bigcup \mathrm{attr}(R_i), \Omega_i \leftarrow \{\}, i \in [1, n]$;
4 $\Omega_i' \leftarrow generate(\mathrm{attr}(R))$;
5 $allocate(\Omega_i')$;
6 **for** $i \in [1, n]$ **do**
7 $\quad$ **for** $\varphi_i' \in \Omega_i'$ **do**
8 $\quad\quad$ **if** $verify(\varphi_i') == TRUE$ **then**
9 $\quad\quad\quad$ $\Omega_i' \leftarrow \Omega_i' \setminus \{\varphi_i'\}$;
10 $\quad\quad\quad$ $\Omega_i \leftarrow \Omega_i \bigcup \{\varphi_i'\}$;
11 $\quad\quad\quad$ $prune(\Omega_i', \varphi_i', TRUE)$;
12 $\quad\quad\quad$ $broadcast(\varphi_i', TRUE)$;
13 $\quad\quad$ **else**
14 $\quad\quad\quad$ $\Omega_i' \leftarrow \Omega_i' \setminus \{\varphi_i'\}$;
15 $\quad\quad\quad$ $broadcast(\varphi_i', FALSE)$;
16 $\Omega \leftarrow \bigcup \Omega_i, i \in [1, n]$;
17 **return** Ω.

The FDs result set Ω and the candidate FDs set Ω' are initialized at first. DFDD then gets the union of distributed attributes, which is the preliminary of candidate FDs generating. Function $generate()$ generates candidate FDs Ω' set. After generating the candidate FDs, DFDD allocates verifying tasks to different sites by function $allocate()$. Then DFDD verifies allocated tasks and prunes the candidate FDs by the result of verifying and the message broadcasted by other sites at each site in parallel. For each candidate FD φ'_i in the candidate FDs set Ω'_i, DFDD verifies φ'_i at site S_i. If the verification result shows that φ'_i is a functional dependency, then DFDD removes φ'_i from candidate FDs set Ω'_i, and then adds φ'_i to FDs result set Ω_i. DFDD prunes the candidate FDs set Ω'_i by the pruning strategies proposed in Sect. 4.2, then broadcasts φ'_i and its status (true or false) to other sites. If the verification result shows that φ'_i is not a functional dependency, then DFDD removes φ'_i from candidate FDs set Ω'_i and broadcasts φ'_i and its status to other sites. Upon receiving the broadcasting messages from other sites, the current site will prune the candidate FDs set.

6　Experimental Results

In this section we present an experimental study of our algorithms for discovering FDs in vertically fragmented data.

Experimental Setting. To evaluate DFDD, we performed several experiments with different datasets. We used a cluster with a dedicated master node and eight workers with 1.7 GHz Intel Xeon 2 processor and 16 GB of RAM, and the operating system of each machine is Ubuntu 10.4. All the algorithms are implemented in Java. We used Apache Hadoop 1.1.2 as the distributed processing framework and Hama 0.6.4 as the Bulk Synchronous Parallel (BSP) computing model.

Data. We use two different types of data: (1) Real-life data taken from the United States Department of Transportation [9]. We generate instance $alos_4$ of 4 Millions tuples, $alos_{12}$ of 12 Millions tuples and $alos_{204}$ of 20 Millions tuples. (2) Synthetic data representing a companys employee records, referred to as EMP. We created two instances of EMP containing 4 Millions tuples, 12 Millions tuples and 20 Millions tuples each. We refer to these instances as emp_4, emp_{12} and emp_{20}, respectively.

Experimental Results. We conducted three sets of experiments, evaluating the centralized algorithm Cet and distributed algorithm DFDD. We varied the number of sites ($|S|$), size of the distributed data ($|D|$), and the number of attributes ($|attr(R)|$). All experiments report the average over three runs.

Exp-1: Varying the Number of Sites. To evaluate the scalability of our algorithms with the number of sites, we fixed the total data size and increased the number of sites $|S|$ from 2 to 8. We used datasets $alos_{12}$ and emp_{12}. Figures 3 and 4 show response times for algorithm Cet and algorithm DFDD. As expected, the response time of algorithm DFDD decreases as $|S|$ increases. However, the response time of algorithm Cet increases slightly as $|S|$ increases. Since all the verifying tasks of algorithm Cet are executed in one collaborate site, the increase of $|S|$ has almost no much impact on the response time of algorithm Cet. Figures 3 and 4 show that DFDD outperforms Cet significantly in response time.

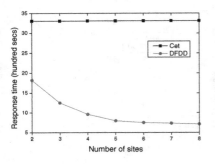

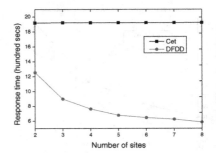

Fig. 3. Scalability with $|S|$ ($alos_{12}$)

Fig. 4. Scalability with $|S|$ (emp_{12})

Exp-2: Varying Data Size. To evaluate the scalability of our algorithms with data size $|D|$, we fixed the number of sites $|S|$ to 2 and 8 and increased the size of data $|D|$ from 20 % to 100 %. We used datasets $alos_{20}$ and emp_{20}. Figures 5 and 6 show response times for algorithm Cet and algorithm DFDD. $DFDD_2$ refers to the case that the number of sites $|S|$ is 2, while $DFDD_8$ refers to the case that the number of sites $|S|$ is 8. As expected, the response time of algorithm Cet and DFDD increases as $|D|$ increases. However, the increasing rate of Cet's response time is much higher than that of algorithm DFDD. Figures 5 and 6 show that DFDD outperforms Cet significantly in response time with data size $|D|$ increasing. And we can learn from Figs. 5 and 6 that with the increase of the number of sites $|S|$, the response time of algorithm DFDD decreases significantly.

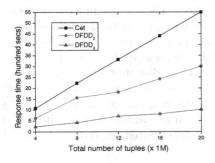

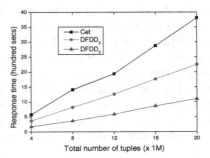

Fig. 5. Scalability with $|D|$ ($alos_{20}$)

Fig. 6. Scalability with $|D|$ (emp_{20})

Exp-3: Varying the Number of Attributes. To evaluate the scalability of our algorithms with the number of attributes $|attr(R)|$, we fixed the number of sites $|S|$ to 8 and fixed the size of data to 4 Millions and increased the number of attributes $|attr(R)|$ from 5 to 8. We used datasets $alos_4$ and emp_4. Figures 7

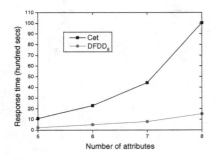
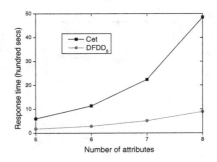

Fig. 7. Scalability with $|\mathrm{attr}(R)|(alos_4)$ **Fig. 8.** Scalability with $|\mathrm{attr}(R)|(emp_4)$

and 8 show response times for algorithm Cet and algorithm DFDD. As expected, the response time of algorithm Cet and DFDD increases as $|\mathrm{attr}(R)|$ increases. As is shown in Figs. 7 and 8, the response time of Cet grows exponentially with the increase of attributes $|\mathrm{attr}(R)|$. By Lemma 1, the number of candidate FDs is exponential to the number of attributes. With the increase of attributes $|\mathrm{attr}(R)|$, the verifying task increases exponentially, thus the response time grows exponentially. Figures 7 and 8 show that DFDD outperforms Cet significantly in response time with the number of attributes $|\mathrm{attr}(R)|$ increasing.

7   Conclusion

In this paper, we have proposed a distributed algorithm to discover FDs from vertically distributed data. The main feature of our algorithm emphasizes the verifying load balance among the sites of the distributed big data. To reduce the verifying tasks, we cluster and allocate the candidate FDs by the containment of the candidate FDs LHS. To prune the candidate FDs as early as possible, we broadcast the verifying results among the sites of the distributed big data. To improve the discovering efficiency, we discover the FDs at each site of the distributed big data in parallel. As the drawback of breath-first manner in discovering FDs, a future research is to combine the breath-first manner and depth-first manner when discovering the FDs from vertically distributed big data.

References

1. Codd, E.F.: Further normalization of the data base model. Technical report 909, IBM (1971)
2. Yao, H., Hamilton, H.J.: Mining functional dependencies from data. Data Min. Knowl. Disc. **16**(2), 197–219 (2008)
3. Maier, D.: The Theory of Relational Databases. Computer Science Press, Rockville (1983)

4. Huhtala, Y., Karkkainen, J., Porkka, P., Toivonen, H.: TANE: an efficient algorithm for discovering functional and approximate dependencies. Comput. J. **42**(2), 100–111 (1999)
5. Li, W., Li, Z., Chen, Q., Jiang, T., Liu, H., Pan, W.: Functional dependencies discovering in distributed big data. J. Comput. Res. Dev. **52**(2), 282–294 (2015)
6. Özsu, M.T., Valduriez, P.: Principles of Distributed Database Systems, 2nd edn. Prentice-Hall, Upper Saddle River (1999)
7. Novelli, N., Cicchetti, R.: FUN: an efficient algorithm for mining functional and embedded dependencies. In: Van den Bussche, J., Vianu, V. (eds.) ICDT 2001. LNCS, vol. 1973, pp. 189–203. Springer, Heidelberg (2000)
8. Lopes, S., Petit, J.-M., Lakhal, L.: Efficient discovery of functional dependencies and Armstrong relations. In: Zaniolo, C., Grust, T., Scholl, M.H., Lockemann, P.C. (eds.) EDBT 2000. LNCS, vol. 1777, pp. 350–364. Springer, Heidelberg (2000)
9. United States Department of Transportation. http://apps.bts.gov/xml/ontimesummarystatistics

Semi-supervised Document Clustering via Loci

Taufik Sutanto[1,2]([✉]) and Richi Nayak[1]

[1] Queensland University of Technology (QUT), Brisbane, Australia
[2] Syarif Hidayatullah State Islamic University, Jakarta, Indonesia
{taufik.sutanto,r.nayak}@qut.edu.au, taufik.sutanto@uinjkt.ac.id

Abstract. Document clustering is one of the prominent methods for mining important information from the vast amount of data available on the web. However, document clustering generally suffers from the *curse of dimensionality*. Providentially in high dimensional space, data points tend to be more concentrated in some areas of clusters. We take advantage of this phenomenon by introducing a novel concept of dynamic cluster representation named as *loci*. Clusters' loci are efficiently calculated using documents' ranking scores generated from a search engine. We propose a fast loci-based semi-supervised document clustering algorithm that uses clusters' loci instead of conventional centroids for assigning documents to clusters. Empirical analysis on real-world datasets shows that the proposed method produces cluster solutions with promising quality and is substantially faster than several benchmarked centroid-based semi-supervised document clustering methods.

Keywords: Loci · Ranking · Semi-supervised clustering

1 Introduction

In the large document corpora, it is common to find some documents with label or grouping information. Generalizing this valuable prior information to the larger part of the unlabeled documents is relevant to many real-world applications including web information system [7]. Semi-supervised clustering methods have been developed to solve this problem and have been reported to produce a better quality solution than the unsupervised methods [2,7]. Partitional centroid-based semi-supervised clustering algorithms such as seeded k-means or constrained k-means [2] are usually preferred due to their fast performance on large datasets. Nevertheless, these methods do not scale well for the problems with a large number of groupings [7].

Meanwhile, recent clustering studies on high dimensional data have shown the superiority of using the clusters' hub information instead of the conventional centroids for grouping the documents [3,8]. Hubness in clusters (Fig. 1a) is defined as the likelihood of data points in high dimensional data to occur more frequently in k-nearest neighborhood (k-NN) rather than occurring near the centroid point [8]. The behavior of these data points (i.e. hubs) is shown to be a property of high dimensional data and not merely a samples limitation [5]. Unfortunately, the

© Springer International Publishing Switzerland 2015
J. Wang et al. (Eds.): WISE 2015, Part II, LNCS 9419, pp. 208–215, 2015.
DOI: 10.1007/978-3-319-26187-4_16

hubness-based clustering algorithms have been limited to apply to a corpus with thousands of documents only [3,8]. The need of calculating the similarity between a document and at least the top-n hub points instead of a single centroid point, representing a cluster, makes the process cumbersome.

In this paper, we propose a novel concept of cluster *locus* and apply it in semi-supervised document clustering setting. The locus of a cluster is a hub-like neighborhood of most *relevant* documents to a target document (e.g. d_1 or d_2 in Fig. 1). In other words, the locus of a cluster is a dynamic cluster representation for a document. A locus can be viewed as a low-dimensional projection of a hub with regards to a document. The relevant documents can be efficiently chosen using the ranking scores generated by a scalable Information Retrieval (IR) system (e.g. search engine).

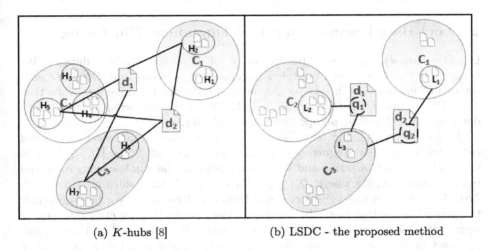

(a) K-hubs [8]   (b) LSDC - the proposed method

Fig. 1. The difference between hubs (a) and loci-based clustering (b).

Our previous clustering method, CICR [7], also uses the ranking scores produced by a search engine to improve the performance when the number of clusters in the dataset is large. However, there are substantial differences between CICR [7] and the proposed method LSDC. Not only does CICR use conventional centroids, it needs to update the cluster information in the search engine (*realtime index*) [7]. On the other hand, LSDC can work as a *plugin/add-on* to the existing IR system without any alteration to the IR system. Consequently, LSDC is more applicable at a production level.

The LSDC method has some important features that make it suitable for large datasets with a large number of clusters. It does not need centroid initialization or updates. LSDC does not need the input parameter k (i.e. the total number of clusters) like most other clustering algorithms. The total number of clusters that fit the dataset is automatically calculated by the algorithm. Moreover, as shown in Fig. 1b, in order to group a target document, the comparison is done with some clusters only and with some documents only within those clusters. This selective comparison reduces the need to scan all the documents (or clusters) in the corpus. As shown in our experiments in the latter section, a small size of relevant documents is enough to accurately cluster the documents. LSDC is able to generate new clustering labels that are not present in the initial labeled data (i.e.seed). Finally, a locus is not necessarily at the center of a cluster, hence LSDC does not have the tendency to form (hyper)sphere-like clusters.

Several real-world document datasets, exhibiting a large number of clusters, were used in the experiments. A comparison of LSDC with centroid-based semi-supervised clustering methods that utilize initial labels (seed) is reported. Empirical analysis reveals that LSDC is not only able to cluster data with a large number of clusters, but also produces fast and accurate clustering solution.

2   Loci Based Semi-supervised Document Clustering

Let $D = \{d_1, d_2, d_3, \ldots, d_N\}$ be the set of all N documents in a dataset. We denote $D^\ell \subset D$ as a set of labeled documents (i.e. seed) and $D^\mu \subset D$ as a set of documents that needs be clustered (i.e. target documents). Note that $D^\mu \cap D^\ell = \emptyset$ and $D^\mu \cup D^\ell = D$. Since LSDC is a *hard clustering* method, every document in D^μ will be mapped to a single cluster.

Using a search engine (e.g. *Sphinx*[1]), it is possible to get a set of ordered terms based on their frequency of occurrence in D. Let $F^t = \{(t_i, f_{t_i}) : t_i \in d, d \in D\}$ be a set of terms and their frequencies in the whole corpus imported from a search engine where $f_{t_i} \leq f_{t_j}$ for $i \leq j$. Given a document $d \in D^\mu$, *LSDC* extracts a set of s distinct terms from d to form a query q based on their frequency information in F^t. In this paper, the s terms in q are chosen from d as terms with s lowest frequencies and it occur more than once in the whole corpus.

A set of at most m relevant documents to the query q and its ranking score vector r is then generated by a search engine. Clusters' loci are calculated from the prior grouping information found in the relevant documents. Finally, similarities between d and the clusters' loci are then used to determine which cluster d should be grouped into.

2.1   Document Ranking Schemes

Let $q = \{t_1, t_2, \ldots, t_s\} \subseteq d$ be the query generated from d. Using a search engine, a set of m most relevant documents to q is identified and its ranking scores vector r is generated. A ranking function R_f employed in a search engine calculates the ranking scores r of m documents for query q generated from d as:

$$R_f : q \rightarrow D^q = \{(d_j^q, r_j) : j = 1, 2, \ldots, m'\}, \tag{1}$$

where $0 \leq m' \leq m$. If $m' = 0$, it indicates that there is no relevant document found in D for the given query q. In LSDC, this will trigger a new cluster

---

[1] http://sphinxsearch.com.

formation and enables LSDC to cluster the documents in D^μ to clusters beyond the existing initial prior information in D^ℓ.

There is a number of functions that can be used to calculate ranking score of a document with regards to a query. However, we have shown previously that the *Sphinx* search engine's specific ranking function called *SPH04* results in superior clustering quality compared to other ranking schemes such as weighted *tf-idf*, *BM25*, and *BM25* with proximity [7]. Therefore, we have used *SPH04* in all of our experiments. The ranking score *SPH04* of $d \in D$ given a query q of length s is defined in [1] as:

$$R_f^{SPH04}(q) = 1000 f_w(q, d) + \lceil 999 R_f^{BM^*}(q, d) \rceil, \tag{2}$$

with

$$f_w(q, d) = 4 \max\{LCS(q, d)\}(1 + U_w) + \begin{cases} 3, \text{ exact query match} \\ 2, \text{ first query term match} \end{cases} \tag{3}$$

and

$$R_f^{BM^*}(q, d) = \frac{1}{2} + \frac{\sum_{i=1}^{s}(\frac{t_f * idf}{t_f + 1.2})}{2s}, \text{ where } idf = \frac{log(\frac{N - n_t + 1}{N})}{1 + N}. \tag{4}$$

$LCS(q, d)$, the Longest Common Sub-sequence, is defined as the number of keywords that are present in d in the exact same order as in q. U_w is a user defined constant, t_f is a term frequency, the number of terms occurring inside document d, and n_t is the number of documents in D that have the term t.

2.2 Loci Based Document Clustering

Let D^q be the set of documents returned by a search engine in response to the query q. We obtain a set of relevant and labeled documents as $D_\ell^q = D^\ell \cap D^q$. Let C_ℓ^q be the set of distinct cluster label information within D_ℓ^q. For each cluster found in C_ℓ^q and assuming $D_\ell^q \neq \emptyset$, clusters' loci L_k of q are calculated as follows:

$$L_k = \{d : d \in D_k^q, k \in C_\ell^q\}. \tag{5}$$

Using the formulation given in (5), each relevant cluster in C_ℓ^q has only one cluster locus (shown in Fig. 1b). If $D_\ell^q = \emptyset$ then a new cluster is formed with d as its member. We would like to emphasize that in (5) the role of document ranking scores in calculating L_k is implicit. This would result in not only more efficient computation but also less communication cost.

Once the clusters' loci are formed, similarity values between d and each locus in L_k are calculated to decide the document's final clustering decision. Let ϕ denotes a similarity function, then the cluster decision is based on the solution of the following constrained optimization:

$$\begin{aligned} \max_k \quad & \{\phi(d, L_k) : k \in C_\ell^q\} \\ \text{subject to} \quad & \phi(d, L_k) \geq \rho \end{aligned}, \tag{6}$$

input : A set of indexed documents D, labeled documents D^ℓ, distinct initial
cluster labels C^ℓ extracted from D^ℓ, documents that are going to be
clustered D^μ, a similarity threshold ρ, and the maximum number of
relevant documents m.
output: Disjoint partitions of D^μ.

for *each $d \in D^\mu$* **do**
    Extract q from d;
    $D^q = R_f(q) = \{(d_j^q, r_j) : j = 1, 2, \ldots, m\};$ // Search engine's query
    $D_\ell^q = D^q \cap D^\ell;$ // Consider only labeled and relevant documents
    if $D_\ell^q = \emptyset$ **then**
       $C^\ell = C^\ell \cup \{d\};$ // Form a new cluster
    else
       $L_k = \{d : d \in D_k^q, k \in C_\ell^q\};$
       $\phi^* = \max_k \{\phi(d, L_k) : k \in C_\ell^q\};$
       if $\phi^* \geq \rho$ **then**
          $c_k = c_k \cup \{d\};$ // Assign d to c_k
       else
          $C^\ell = C^\ell \cup \{d\};$ // Form a new cluster
       end
    end
    $D^\ell = D^\ell \cup d;$
end

Algorithm 1. LSDC Algorithm.

where ρ is a user defined threshold to determine whether the similarity value
between a document and a locus is considered significant. The set of labeled
documents D^ℓ is then updated using the optimal decision (i.e. c_k). When the
optimization in (6) does not have a feasible solution, then d will form a new
unit cluster. The process is incrementally done for all of the documents in D^μ.
Algorithm 1 details the overall process.

3   Experiments

We used four openly available datasets: Reuters 21578[2], MediaEval Social Event
Detection (SED) 2013, SED 2014 development data[3], and Wikipedia (September
2014 dumps). The Reuters data has five different categories: topics, exchanges,
organizations, people, and places. Since LSDC is a *hard clustering* method, only
Reuter documents that have single topic category were used. Similarly, single
category documents from Wikipedia dumps were used. Short length documents
(usually pages referring to image files) were filtered. All of the text information
from SED 2013 and 2014 metadata were concatenated and used (except *URL*).
The datasets summary is given in Table 1.

---

[2] http://www.daviddlewis.com/resources/testcollections/reuters21578/.
[3] http://www.multimediaeval.org/.

Table 1. Summary of datasets in the experiments.

Dataset	#Docs	K	#terms	Raw size (MB)
Reuters	9446	66	28,614	5.4
SED'13	437,370	16,711	189,164	179
SED'14	362,578	17,834	158,272	105
Wikipedia	701,141	59,600	4,553,408	2,074

3.1 Pre-processing and Evaluation Criteria

Standard text pre-processing such as English stopwords and non-alpha-numeric characters filtering were applied. The document length normalized tf-idf (term frequency-inverse document frequency) weighting [6] was used to represent the document vectors. In all of the experiments the query length s was set to 30 and $\rho = 0$. The clustering evaluation metrics used were pairwise F1-score and Normalized Mutual Information (NMI) [4]. Running time was calculated for all of the clustering methods without the pre-processing steps. Included in the time measurements were initialization (in centroid based methods), communication cost (CICR and LSDC), and updating of the cluster information. Under the distributed computing environment, the time was measured using the real *CPU* time instead of wall time. Experiments were done using Matlab in a local area network connected machine (1Gbps) and no parallel processing has been utilized in all of the clustering algorithms.

3.2 Results and Discussion

Use of Label Information: LSDC as a semi-supervised clustering method depends on the availability of the prior information in the form of grouping information (labels) contained in D^{ℓ}. We evaluated *LSDC* with representing D^{ℓ} of different sizes from the data corpus D in order to analyze the impact of labeled data to the clustering quality. As shown in Fig. 2a, the *LSDC* performance is consistent with the increasing proportion of the supervision provided. Most importantly, *LSDC* does not require a large portion of labeled data to give a relatively acceptable clustering quality. With the exception of Reuters (i.e. small dataset), the NMI scores are all above 0.75 by using merely 10 % of D.

Loci Size: Across all four used datasets running time, NMI, and F1-score were recorded for all clustering solutions generated with varying values of m (i.e. the number of relevant documents returned from a search engine to be used in loci calculation). Figure 2b shows that as the size of m increases, the clustering quality is improves. However, the increment is getting smaller and reaches a plateau at around $m = 50$. This indicates that a relatively small value of m is generally enough for *LSDC* to produce *near-optimal* clustering quality.

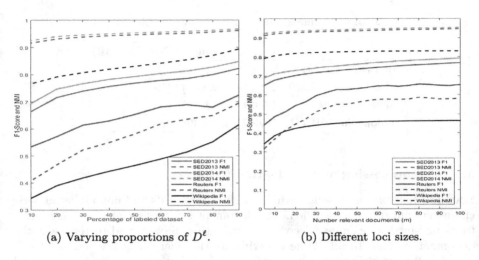

(a) Varying proportions of D^ℓ.   (b) Different loci sizes.

Fig. 2. The effect of different sizes of D^ℓ and m to LSDC clustering results.

Table 2. Performance comparison on different datasets, methods, and metrics.

	Time (hours:minutes)				NMI				F1-score			
	LS	CR	CK	SK	LS	CR	CK	SK	LS	CR	CK	SK
Reuters	00:01	00:02	00:01	00:04	.660	.714	.703	.676	.721	.747	.699	.657
SED 2013	00:18	29:41	182:56	>240:00	.946	.935	.965	.964	.768	.725	.838	.831
SED 2014	00:29	22:40	117:31	>240:00	.953	.946	.979	.978	.792	.767	.905	.899
Wikipedia	00:40	>240:00	>240:00	>240:00	.830	–	–	–	.464	–	–	–

Method: LSDC (LS), CICR (CR), constrained (CK) and seeded (SK) k-means.

Benchmarks: Comparisons were made with the centroid-based semi-supervised document clustering methods such as *CICR* [7], seeded k-means [2], and constrained k-means [2]. All documents in D^ℓ are used for centroid initialization in constrained k-means and seeded k-means. However, in constrained k-means only documents in D^μ are used to produce final cluster results while seeded k-means uses all of the documents in D. D is randomly split into two equal size of D^ℓ and D^μ. The evaluation on all methods are done using documents in D^μ against the available gold standard. We add an additional time stopper (240 h) as a time threshold.

Table 2 shows that overall LSDC is faster while producing comparable clustering quality. It appears that the discrepancy of speed between LSDC and the other methods is higher as the datasets get larger. This indicates that LSDC is more suitable for large size datasets. All other methods reach the time limit without producing any results for the large dataset (Wikipedia) while *LSDC* took only 40 min to finish. Seeded k-means even reached the time limit for medium size datasets (SED 2013 and SED 2014). The F-score and NMI values show that *LSDC* gives high-quality clustering solutions. In fact, *LSDC* gives marginally better results than *CICR* and almost similar results to constrained k-means and seeded k-means. We also compared the performance with the Euclidean distance measure, these clustering solutions consumed more time and yielded lower accuracy in comparison to the cosine similarity measure.

Further analysis on clustering quality: Although not shown in this paper (due to the space constraint), empirical analysis shows that longer queries do not necessarily produce better clustering quality. We also found that LSDC's clustering quality can be further improved by using the available structure information in the documents (e.g. *title* and *tags*) to build a customized ranking scheme.

4   Conclusions and Future Work

A novel concept of *loci* is introduced in this paper to dynamically represent document clusters. We proposed and evaluated a fast incremental loci based document clustering method (LSDC) on several real-world datasets. By extending a search engine capability to process a large set of documents and provide relevant documents to a query, LSDC is capable of clustering a large set of documents substantially faster than the benchmarked algorithms while retaining a comparable clustering quality. Analysis on the quality of prior information to the LSDC clustering quality and a possible extension of loci concept in classification problems with a large number of categories are subjects of our future investigations.

References

1. Aksyonoff, A.: Introduction to Search with Sphinx: From Installation to Relevance Tuning. O'Reilly, Sebastopol (2011)
2. Basu, S., Banerjee, A., Mooney, R.J.: Semi-supervised clustering by seeding. In: Proceedings of the Nineteenth International Conference on Machine Learning. ICML 2002, San Francisco, CA, USA, pp. 27–34 (2002)
3. Hou, J., Nayak, R.: The heterogeneous cluster ensemble method using hubness for clustering text documents. In: Lin, X., Manolopoulos, Y., Srivastava, D., Huang, G. (eds.) WISE 2013, Part I. LNCS, vol. 8180, pp. 102–110. Springer, Heidelberg (2013)
4. Manning, C.D., Raghavan, P., Schütze, H.: Introduction to Information Retrieval, vol. 1. Cambridge University Press, Cambridge (2008)
5. Radovanović, M., Nanopoulos, A., Ivanović, M.: Hubs in space: popular nearest neighbors in high-dimensional data. J. Mach. Learn. Res. 11, 2487–2531 (2010)
6. Singhal, A., Buckley, C., Mitra, M.: Pivoted document length normalization. In: Proceedings of the 19th Annual International ACM SIGIR Conference on Research and Development in Information Retrieval, SIGIR 1996, New York, NY, USA, pp. 21–29 (1996)
7. Sutanto, T., Nayak, R.: The ranking based constrained document clustering method and its application to social event detection. In: Bhowmick, S.S., Dyreson, C.E., Jensen, C.S., Lee, M.L., Muliantara, A., Thalheim, B. (eds.) DASFAA 2014, Part II. LNCS, vol. 8422, pp. 47–60. Springer, Heidelberg (2014)
8. Tomašev, N., Radovanović, M., Mladenić, D., Ivanović, M.: The role of hubness in clustering high-dimensional data. In: Huang, J.Z., Cao, L., Srivastava, J. (eds.) PAKDD 2011, Part I. LNCS, vol. 6634, pp. 183–195. Springer, Heidelberg (2011)

Differentiating Sub-groups of Online Depression-Related Communities Using Textual Cues

Thin Nguyen[1]([⊠]), Bridianne O'Dea[2], Mark Larsen[2], Dinh Phung[1],
Svetha Venkatesh[1], and Helen Christensen[2]

[1] Centre for Pattern Recognition and Data Analytics,
Deakin University, Geelong, Australia
{thin.nguyen,dinh.phung,svetha.venkatesh}@deakin.edu.au
[2] Black Dog Institute, University of New South Wales, Sydney, Australia
{b.odea,mark.larsen,h.christensen}@blackdog.org.au

Abstract. Depression is a highly prevalent mental illness and is a comorbidity of other mental and behavioural disorders. The Internet allows individuals who are depressed or caring for those who are depressed, to connect with others via online communities; however, the characteristics of these online conversations and the language styles of those interested in depression have not yet been fully explored. This work aims to explore the textual cues of online communities interested in depression. A random sample of 5,000 blog posts was crawled. Five groupings were identified: depression, bipolar, self-harm, grief, and suicide. Independent variables included psycholinguistic processes and content topics extracted from the posts. Machine learning techniques were used to discriminate messages posted in the depression sub-group from the others. Good predictive validity in depression classification using topics and psycholinguistic clues as features was found. Clear discrimination between writing styles and content, with good predictive power is an important step in understanding social media and its use in mental health.

Keywords: Web community · Feature extraction · Textual cues · Online depression

1 Introduction

Depression is a leading cause of disability worldwide and the prevalence of depression in developed countries constitutes a major health burden. Conservative estimates indicate that 3.2 % of individuals worldwide experience depression within any 12 month period. The Internet is increasingly used for the exchange of mental health information, support, and advice. Web communities are one such avenue for connecting with others over the Internet. These communities are used by individuals with depression and have been found to provide social support, reduce social isolation, and help people to cope more effectively whilst creating an online

© Springer International Publishing Switzerland 2015
J. Wang et al. (Eds.): WISE 2015, Part II, LNCS 9419, pp. 216–224, 2015.
DOI: 10.1007/978-3-319-26187-4_17

space to share ideas and experiences. As depression is a condition which significantly impacts social functioning, communities on the Web may have unique potential for reconnecting individuals and improving symptoms. However, to date, little is known about the topics discussed within these communities or the language features that characterise this discussion.

For healthcare research, applying machine learning techniques to online communities enables researchers to analyse the content of information exchanged as well as the sentiment and linguistic styles that users express in their online communication. For example, previous research has linked online emotional expression to global mental health data [4] and detected risk from potentially suicidal tweets [7]. For the analysis of textual content generated in social media, two feature sets have been widely used. Firstly, linguistic styles have been found to be indicative of depression in several studies in psychiatry [9,10]. Secondly, topic features have been used to characterise the content of text documents [5]. These two representations enable understanding of *what* people are writing about (topics) and *how* they write (language styles), potentially providing insights into the mental health status of individuals. Textual features have been found to be strong predictors of depression, and differentiate mental health communities from other online communities [6]. However, depressive symptomatology is transdiagnostic and central to a number of mental health disorders. Understanding the differences in the language styles may help to shed light on differences in diagnoses.

This study aims to examine the topics and linguistic features in a sample of online communities interested in depression. This study uses a large cohort of data from nearly 10,000 individual users in 24 online mental health communities. This work aims to explore the differences in language in the depression-specific communities with other depressive conditions such as bipolar disorder, self-harm, suicide, and grief. We present an analysis of these communities, focusing on the topics and psycholinguistic processes expressed in the content of users' posts, to engineer predictive feature sets. Both conventional statistical tests and machine learning algorithms are utilised in the analysis.

A key contribution of this work is to introduce a comprehensive view of bloggers in such mental health communities using topics of interest and language styles. Another contribution is to provide a set of predictors to differentiate the subgroups associated with the depressive communities. This work helps to improve our understanding of online mental health communities that are interested in depression and illustrates the potential of machine learning for improving psychiatric research and practice.

2   Methods

2.1   Datasets

This study focused exclusively on online communities within the Live Journal platform. Communities who listed 'depression' as an interest were identified

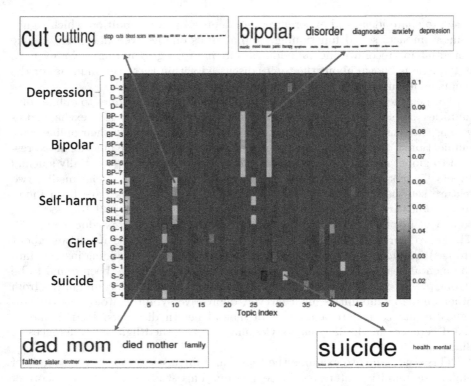

Fig. 1. The proportion of topics discussed among the communities.

through the 'search communities by interest'[1] function provided by the Live Journal website. This found a total of 352 communities. A sample of active communities was acquired, selecting those with at least 200 posts, and at least one update in the previous month. The final cohort consisted of 24 communities with a total of 38,401 posts. Based on their community names and descriptions, these 24 communities were then subgrouped into one of the following categories: *depression* (four communities), *bipolar disorder* (seven communities), *self-harm* (five communities), *grief/bereavement* (four communities) and *suicide* (four communities). To create a balanced dataset, 1,000 posts from each subgroup were randomly selected for machine learning. To ensure equal representation across each of the individual communities, the 1,000 post allowance was divided by the number of individual communities in that subgroup. For example, each of the four communities within the 'depression' subgroup contributed 250 posts.

2.2   Feature Extraction

To characterise the difference between online communities, two feature sets were extracted. (1) *Topics:* to extract topics, latent Dirichlet allocation (LDA) [1]

[1] http://www.livejournal.com/interests.bml.

Table 1. The prominent topics discussed within the subgroups.

No.	Word cloud	No.	Word cloud
8	**dad mom** died mother family father sister brother	29	**baby** pregnant weeks daughter child girl angel mommy birth husband born pregnancy children knew beautiful
10	**cut** cutting stop cuts blood scars	30	**anymore stop** worse cry crying boyfriend away afraid break wants start gets gotten handle and apart
13	**depression** fighting fight	31	**suicide** health mental illness prevention young person
18	**death** dead mark grave anniversary remember peace saturday passed	32	**cause** mom left wanted problem parents funny wrong tried thinks ask online
23	**meds** medication seroquel lithium lamictal psychiatrist effects doctor weeks	41	**school** college parents grade summer classes
28	**bipolar** disorder diagnosed anxiety depression manic mood issues panic therapy symptoms		

was used as a Bayesian probabilistic modelling framework. For the inference part, we implemented Gibbs inference detailed in [3] and the number of topics was set to 50; and (2) *Language styles:* the proportions of words associated with psycholinguistic categories as defined in the Linguistic Inquiry and Word Count (LIWC) package [8].

2.3 Classification and Feature Selection

We formulated four two-class classifiers to examine the usefulness of the extracted features in distinguishing posts made by the *depression* communities from those made by other subgroups. For each classifier, say *depression* versus *suicide*, we denote by $\mathcal{B}$ a corpus of posts made in *depression* and *suicide* community blogs. Given a blog post $d \in \mathcal{B}$, we are interested in predicting if the post belongs to *depression* or *suicide* subgroup, based on the textual features extracted from d.

We are interested in the features sets that are strongly predictive of each of the subgroups. For this purpose, Lasso, a regularised regression model [2], was chosen. Lasso performs logistic regression and selects features simultaneously, enabling an evaluation of both the classification performance and the importance of each feature in the classification process. Classification accuracy was used to evaluate the performance.

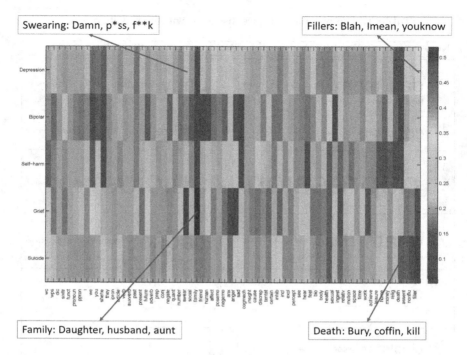

Fig. 2. LIWC styles among the five categories of communities (normalised for each feature).

3   Results

3.1   Analysis of Topics

Figure 1 illustrates the distribution of topics across the 24 communities. There were no topics which uniquely distinguished the depression subgroup from the other subgroups. The prominent topics discussed within the subgroups are shown in Table 1.[2] The bipolar disorder subgroup commonly discussed topic 23 (related to medication, and specifically Seroquel, an atypical antipsychotic drug used in the treatment of bipolar disorder), and topic 28 (which is predominately about bipolar disorder itself). The self-harm communities strongly expressed topic 10, which is related to cutting. The grief groups have strong expressions of topic 8 (parents and death) or topic 29 (babies and pregnancy).

Individual topics were expressed in each of the suicide-related communities. In community S-2, topic 31 (suicide and mental health) was commonly discussed, community S-3 had no dominant topic although the words "care" and "anymore" were expressed frequently. Similarly community S-4 did not have a dominant topic, although topic 8, associated with separation, was present and may reflect suicide bereavement.

---

[2] All 50 topics are placed at http://bit.ly/1JKY2vo.

Table 2. Models using topics to discriminate posts made by depression vs other subgroups.

Variable	Bipolar	Self-harm	Grief	Suicide	Variable	Bipolar	Self-harm	Grief	Suicide
(Intercept)	0.03	0.13	1.2	-0.34	T24	-0.97			
T01		-2.11		0.44	T25	0.28	-1.75		
T02	-1.6				T27	-0.19		-0.2	-3.03
T03	-0.15				T28	-5.62	1.95	3.64	3.54
T04				1.15	T29			-9.52	
T05		-0.3		1.8	T30	2.41	0.09	0.77	0.81
T06				0.7	T31	-1.72			-6.96
T07	-0.81				T32	1.78	1	1.54	2.68
T08	1.85	0.56	-9.32	-1.67	T33				0.78
T09			-6.66		T34	0.56	-0.25		0.52
T10	3.17	-15.99	4.26	0.29	T35	2.02	1.58	3.38	
T11	-0.03			-0.05	T37	-0.41	0.54	0.65	-0.07
T12		1.23		-0.95	T39	0.63	-3.13	0.52	0.38
T13	7.42	2.28	5.87	2.76	T40	-0.35		-5.5	-1.49
T14			0.62	0.86	T41	1.43	0.46	0.78	1.08
T15	2.07		0.18	0.45	T42	0.89	0.56		-0.64
T16	0.95				T44				0.44
T17			-1.98	-1.32	T45			-4.07	-0.06
T18	-0.51	-0.08	-4.95	-5.07	T46	-0.17			
T20	-1.54				T47	0.16			0.83
T21	-0.43	0.13			T48	-1.82	0.26		
T22				0.33	T49			0.05	0.48
T23	-5.5	1.64	2.06	3.76	T50	0.09			

3.2   Analysis of Linguistic Styles

Figure 2 shows the distribution of LIWC linguistic features across the five subgroups of communities. In comparison to the topic analysis, the depression subgroup had distinct linguistic features with frequent use of filler phrases (e.g., "blah", "I mean" or "you know") and swearing. The bipolar communities had a strong component of health-related features, which mirrors their discussion of medication-related topics. No individual features dominated in the self-harm subgroup, although expression of anger was relatively common. Within the grief communities, features related to the family were common, which matches the discussion of topics related to family members. Finally, the linguistic features associated with death are common in the suicide related subgroup.

3.3   Classification

The results above indicate that distinct topics and linguistic styles differentiate the subgroups, and there is some concordance between the discovered topics and the pre-defined topic-related linguistic features, such as health and family. We used Lasso to determine the pair-wise differentiating topic features between the

Table 3. Models using LIWC to discriminate posts made by depression vs other subgroups.

Variable	Bipolar	Self-harm	Grief	Suicide	Variable	Bipolar	Self-harm	Grief	Suicide
(Intercept)	−0.84	−1.85	1.24	0.06	Anxiety	−1.23		0.96	0.95
Words/post		1.36	1.03		Anger		−3.66	4.19	
Words/sentence	2.07	2.4	4.7		Sadness	6.25	6.28	−1.02	
Dictionary words	0.41	2.05			Cognitive			−1.1	
Words > 6 letters	−2.15	0.39	3.59	0.12	Insight			−0.7	
Function words			−2.94		Causation		−1.2	0.11	
Total pronouns		−0.24			Discrepancy	0.88		0.58	
1st pers singular	0.13		1.93	0.36	Tentative	−2.11			0.42
1st pers plural			−2.03		Certainty	1.1	1.69		−1.23
2nd person		−1.2	−1.53	−0.38	Inhibition		−1.77	−0.06	−0.51
3rd pers singular		0.91	−2.44	−1.23	Inclusive	0.12			
Impersonal pron.			−1.48		Exclusive	−0.14			
Articles	−0.39			−0.01	Perceptual	0.14			
Common verbs			−0.87		See			−2.38	−0.04
Auxiliary verbs	−0.33			0.09	Hear		1.36	−1.42	−0.34
Past tense		−0.81	−0.62		Feel	0.05	0.43	0.95	
Present tense		0.07	3.05	0.64	Body		−2.92	−0.98	
Future tense	0.25	−0.34	−0.69		Health	−5.44	0.03	−0.99	
Adverbs		−0.5	0.4		Sexual			−1.89	
Prepositions			1.15		Ingestion		1.36	2.19	0.18
Conjunctions	−0.11		1.83	0.52	Relativity		−0.23		−0.14
Negations	0.99		0.47	−0.41	Motion		0.92	−1.33	
Quantifiers		0.6	−0.12		Space	−0.17		0.21	
Numbers			−0.27		Time		−1.12	−3.58	
Swear words		2.46	4.88	0.77	Work		0.17	1.16	
Social process	1.56		0.45		Achievement			−1.24	
Family	0.73	1.71	−11.08	−0.05	Leisure	0.28	1.79	0.94	
Friends	1.83	0.39	0.74	1.01	Home		0.51	0.69	
Humans	1.55	−2.09	−1.88	0.2	Money		1.02		
Affective process	2.06		1.56		Religion			−4.69	−1.78
Positive emotion		0.04			Death	2.9	2.49	−7.28	−13.32
Negative emotion	0.2		0.74		Nonfluencies			0.25	2.11

depression and the other four subgroups. The coefficients are shown in Table 2. Individual coefficients that were not significant have been omitted, as have topics with no significant coefficients. Large positive coefficients indicate a weighting towards the depression class, and large negative coefficients indicate a weighting towards the alternative subgroup.

Table 4. The accuracy (%) in the two-class classifications of depression vs other categories.

Features	Bipolar	Self-harm	Grief	Suicide
LIWC	73.9	69.4	83.2	72
Topic	75	76.5	85.4	69.1
LIWC and Topic	77.6	78.4	88	73

Table 2 confirms that topics 23 and 28 are most indicative of the bipolar disorder subgroup, topic 10 is indicative of the self-harm subgroup, topics 8 and 29 are indicative of grief, and topic 31 is indicative of the suicide subgroup. Positive coefficients are present for each of the four alternative categories for topics 13 ("depression"), 30 ("anymore", "stop"), 32 ("cause") and 41 ("school"), indicating these topics are more indicative of the depression subgroup. Negative coefficients are consistently present for topic 18 ("death", "dead") indicating that these words are less indicative of the depression subgroup than the other subgroups.

Applying Lasso to the LIWC features, as shown in Table 3, confirms the dominance of features related to health, anger, family, and death for the bipolar, self-harm, bereavement, and suicide subgroups respectively. The feature associated with friends is consistently positively weighted, indicating that this feature is more associated with depression subgroup than the other subgroups.

Using the coefficients derived from the Lasso methods, we implemented four pair-wise classifiers between the depression subgroup and the other subgroups. The accuracy of these classifiers when using just the topics, just the LIWC features, and a combination of both are shown in Table 4. These results show that across each of the subgroups, classifying based on the derived topics outperforms the linguistic features, but that the accuracy is further improved when information from both domains are combined.

4   Conclusion

This study investigated the topics and linguistic features of the discussions in a sample of online communities that are interested in depression. Machine learning and statistical methods were used to discriminate the textual features among the communities. The results indicate that distinct topics and linguistic styles differentiate the subgroups, and that there is some concordance between the topics and linguistic features. Overall, latent topics were found to have greater predictive power than linguistic features for the prediction of the subgroups. The results of this study suggest that data mining of online communities has the potential to detect meaningful data for understanding the representations of depression and other mental health conditions online. The results highlight the potential applicability of machine learning to psychiatric practice and research.

References

1. Blei, D.M., Ng, A.Y., Jordan, M.I.: Latent dirichlet allocation. J. Mach. Learn. Res. **3**, 993–1022 (2003)
2. Friedman, J., Hastie, T., Tibshirani, R.: Regularization paths for generalized linear models via coordinate descent. J. Stat. Softw. **33**(1), 1–22 (2010)
3. Griffiths, T.L., Steyvers, M.: Finding scientific topics. PNAS **101**(90001), 5228–5235 (2004)
4. Larsen, M.E., Boonstra, T.W., Batterham, P.J., O'Dea, B., Paris, C., Christensen, H.: We Feel: Mapping emotion on Twitter. IEEE J. Biomed. Health **19**(4), 1246–1252 (2015)
5. Nguyen, T., Phung, D., Adams, B., Venkatesh, S.: Prediction of age, sentiment, and connectivity from social media text. In: Bouguettaya, A., Hauswirth, M., Liu, L. (eds.) WISE 2011. LNCS, vol. 6997, pp. 227–240. Springer, Heidelberg (2011)
6. Nguyen, T., Phung, D., Dao, B., Venkatesh, S., Berk, M.: Affective and content analysis of online depression communities. IEEE T. Affect. Comput. **5**(3), 217–226 (2014)
7. O'Dea, B., Wan, S., Batterham, P.J., Calear, A.L., Paris, C., Christensen, H.: Detecting suicidality on Twitter. Internet Interventions **2**(2), 183–188 (2015)
8. Pennebaker, J.W., Francis, M.E., Booth, R.J.: Linguistic inquiry and word count [Computer software] (2007)
9. Rude, S., Gortner, E.M., Pennebaker, J.W.: Language use of depressed and depression-vulnerable college students. Cogn. Emot. **18**(8), 1121–1133 (2004)
10. Stirman, S.W., Pennebaker, J.W.: Word use in the poetry of suicidal and nonsuicidal poets. Psychosom. Med. **63**(4), 517–522 (2001)

A New Webpage Classification Model Based on Visual Information Using Gestalt Laws of Grouping

Zhen Xu[✉] and James Miller

Department of Electrical and Computer Engineering, University of Alberta,
Edmonton, Alberta, Canada
{zxu3,jimm}@ualberta.ca

Abstract. Traditional text-based webpage classification fails to handle rich-information-embedded modern webpages. Current approaches regard webpages as either trees or images. However, the former only focuses on webpage structure, and the latter ignores internal connections among different webpage features. Therefore, they are not suitable for modern webpage classification. Hence, semantic-block trees are introduced as a new representation for webpages. They are constructed by extracting visual information from webpages, integrating the visual information into render-blocks, and merging render-blocks using the Gestalt laws of grouping. The block tree edit distance is then described to evaluate both structural and visual similarity of pages. Using this distance as a metric, a classification framework is proposed to classify webpages based upon their similarity.

Keywords: Webpage classification · Block tree · Gestalt laws of grouping · Normalized compression distance · Tree edit distance

1 Introduction

Webpage classification is becoming increasingly essential because it plays a substantial role in various information management and retrieval tasks, such as web data crawling and web document categorization [1]. Modern webpages, with much more abundant information, presents additional challenges to webpage classification [2]. Hence, traditional approaches that rely on text content cannot handle modern webpages. Nevertheless, people can get visual information directly. Specifically, people subconsciously follow the Gestalt laws of grouping for immediate content identification to perceive rich content [3]. Consequently, providing the machines with the visual features from webpages directly is a feasible way for them to "read and think" as people. Therefore, this paper proposes a methodology to evaluate webpage similarity by visual information using Gestalt laws of grouping, and classifies webpages in terms of their visual similarity.

© Springer International Publishing Switzerland 2015
J. Wang et al. (Eds.): WISE 2015, Part II, LNCS 9419, pp. 225–232, 2015.
DOI: 10.1007/978-3-319-26187-4_18

2   Related Work

To date, extensive work has been done on webpage classification [1]. In general, two major orientations are widely applied to explore webpage classification including treating webpages as images or trees.

In the first category, webpages are abstracted as images before computing their similarity. Recently, many scholars have focused their study on image similarity [4]. Liu et al. [5] proposed a feature-based image similarity measurement approach which uses image phase congruency measurements to compute the similarities between two images. Kwitt et al. [6] presented an image similarity model by using Kullback-Leibler divergences between complex wavelet sub band statistics for texture retrieval. Sampat et al. [7] put forward an image similarity method called the complex wavelet structural similarity. The theory behind it is that consistent phase changes in the local wavelet coefficients may arise owing to certain image distortions. Although image similarity techniques are very useful in searching for a similar image to the specified image, they are not suitable for webpage similarity assessment directly. This is because a specified webpage is an object embedded with a variety of elements and these elements can interact (such as overlap or partly overlap) with each other. It is, therefore, a different problem than pure image similarity assessment.

In the other category, a webpage is regarded as tree structured data. Thus, webpage similarity is studied through investigating tree similarity. With respect to tree structured data, a handful of tree distance functions are applied, such as tree edit distance [8], multisets distance [9], and entropy distance [10]. The tree edit distance is defined as the minimum cost of operations for transferring from one tree to another [11]. Tree edit distances can be further divided into different subcategories in terms of distinct mapping constraints. Mapping constraints include top-down, bottom-up, isolated subtree, etc. [12]. Müller-Molina et al. [9] propose a tree distance function with multisets, which are sets that allow repetitive elements. Based on multiset operations, they define a similarity measure for multisets. They did this by converting a tree into two multisets, with one multiset including complete subtrees and another consisting of all the nodes without children. Connor et al. [10] developed a bounded distance metric for comparing tree structures based on Shannon's entropy equations. Although the above achievements on tree similarity are significant, the theory cannot be used directly on webpage similarity research. The main reason is that the theme of tree similarity has always been structural similarity. However, our focus is on content similarity, in spite the obvious connection between structural and content similarity.

3   Render-Block Tree

Visual information of a webpage is retrieved and represented as the render-block tree by taking the webpage's DOM tree as a prototype instead of parsing sources code. This is because the DOM tree contains all information of a webpage, both textually and visually.

3.1   Render-Blocks

Each node of the render-block tree, i.e., a render-block, maps onto a DOM element. However, only visible DOM elements and their visible attributes are meaningful for analysis. Meanwhile, the semantic meaning of text in a webpage is not part of its visual features, so it is not considered. Hence, only text styles are of major concern. Properties of the render-block contain:

1. A render block always correlates to a DOM element;
2. A render block is always visible in the webpage;
3. A render block only contains visual features of corresponding DOM element.

The transformation from DOM elements to render-blocks only takes into account visible DOM elements, text content, and CSS attributes. The visibility of a DOM element is decided by its tag name, size, and styles. Elements with certain tag names, sizes, or styles are invisible, such as the elements with the tag name of SCRIPT or TITLE, width or height of 0, display style of none, etc. Most texts are displayed in a webpage, but some are not, such as texts of IMG elements. Visible CSS attributes refer to three sets of "front end" styles, namely, text styles (font, color, etc.), paragraph styles (direction, list-style, etc.), and background styles (background-color, border-width, etc.). On the contrary, the "back end" styles are not drawn by the browser, such as margin, cursor, etc. They are ignored during the transformation. Additionally, geometry information, such as top, left, width, and height, is kept during transformation. DOM elements keep their own offset positions inherited from their parent elements, but we convert them into absolute positions.

3.2   Tree Hierarchy

The render-block tree takes the DOM hierarchy as a prototype to illustrate the visual (render) layout. However, due to the flexibility of CSS, the DOM hierarchy sometimes is not consistent with the rendered layout. For example, a child DOM element by default overlaps its parent that is at the left top of the webpage, but a float command can move it onto a third element located at the right bottom. Therefore, in order to eliminate the inconsistency, the render-block tree hierarchy must be modified so that it always follows the rendered layout.

To construct a render-block tree, we manipulate nodes as follows:

1. Take the BODY render-block as the root node.
2. From the root node on, for every render-block, append all child render-blocks according to their corresponding DOM hierarchy.
3. If any render-block is completely located inside any of its sibling render-blocks, move it downward so that it becomes a child of that sibling. However, sibling nodes that geometrically overlap each other are acceptable in a render-block tree, and they are still considered as siblings.
4. If a parent DOM element is invisible or empty, then it has no corresponding render-block; however, its child DOM elements may have a block. In this condition, these child render-blocks shall become children of the render-block which is related to this parent DOM element's first visible parent element.

4   Semantic-Block Tree

The semantic-block tree shares the same hierarchy as the render-block tree. However, the nodes of this tree, i.e., the semantic-blocks, are achieved by merging semantically correlated render-blocks with Gestalt laws of grouping.

4.1   Interpreting Gestalt Law of Simplicity

Although the content of a DOM element can be further split, we do not split it in order to follow the Gestalt law of simplicity. As shown in Fig. 1a (the homepage of "google.ca"), the middle image above the search box contains multiple content (i.e., "GOOGLE" serves as the newspaper title and the three columns are utilized as texts, images, and animations, respectively), but when we read the whole webpage, we treat it as one large image instead of the aforementioned separated ones.

(a)   (b)

Fig. 1. Homepage of "Google.ca" and "Twitter.com"

4.2   Interpreting Gestalt Law of Closure

It is evident that an upper render-block will cover a lower one visually, leading to "incomplete" display of the latter. In this case, however, the lower render-blocks are still perceived as complete because of the Gestalt law of closure. As shown in Fig. 1b (the homepage of "twitter.com"), the upper right part of the background image is covered by two log-in boxes, but the image is still regarded as a complete rectangle (although we cannot see what is exactly covered). That is, the render-block remains as a complete rectangle.

4.3   Interpreting Gestalt Law of Proximity

In webpages, the size of the render-blocks cannot be ignored. They are grouped by distance in the Gestalt law of proximity. To measure distances between two non-zero-area render-blocks, a normalized Hausdorff distance (NHD) is employed. Consider two render-blocks R_1 and R_2:

1. For any point r_1 in R_1 and r_2 in R_2, the distance between them is the length of the corresponding line segment:

$$\|r_1 - r_2\| = \sqrt{(x_{r_1} - x_{r_2})^2 + (y_{r_1} - y_{r_2})^2};$$ (1)

2. For any point r_1 in R_1, the distance between itself and any point in R_2 is the infimum of distances between r_1 and all points in R_2:

$$d(r_1, R_2) = \inf_{r_1 \in R_2} \|r_1 - r_2\|;$$ (2)

3. The Hausdorff distance (HD) from R_1 to R_2 ($hd_{1,2}$) is the supremum of distances between all points in R_2 and all points in R_1:

$$hd_{1,2} = \sup_{r_1 \in R_1} d(r_1, R_2) = \sup_{r_1 \in R_1} \inf_{r_1 \in R_2} \|r_1 - r_2\|;$$ (3)

4. The Hausdorff distance [13] *between R_1 and R_2* is the maximum value between the HD from R_1 to R_2 ($hd_{1,2}$) and the HD from R_2 to R_1 ($hd_{2,1}$):

$$HD(R_1, R_2) = \max\{hd_{1,2}, hd_{2,1}\};$$ (4)

5. The normalized Hausdorff distance (NHD) is calculated by adding a normalizing factor f to HD:

$$NHD(R_1, R_2) = \max\left\{\frac{hd_{1,2}}{f_{R_1}}, \frac{hd_{2,1}}{f_{R_2}}\right\}.$$ (5)

The normalizing factor f can be the width, height, or diagonal distance of the render-block, depending on their relative position. As shown in Fig. 2, the surrounding region of R_0 is split by dashed lines. The normalizing factor f is calculated as: the height of R_2 (R_2 locates in the north/south region of R_0); the width of R_3 (R_3 locates in the west/east region of R_0); or the diagonal of R_4 (R_4 covers corner regions of R_0).

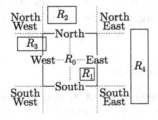

Fig. 2. NHD normalizing factor

4.4   Interpreting Gestalt Law of Similarity

The render-block similarity is divided into three parts: foreground similarity, background similarity, and size similarity. Due to most render-blocks being rectangles, shape similarity is not considered. Background similarity compares both the color and the image; foreground similarity includes textual and paragraph styles; and size similarity checks if the two render-bocks share the same width or height.

The CIE-Lab color space provides standard color difference. RGB colors are obtained directly from CSS and are translated into CIE-Lab colors [14].

Normalized compression distance (NCD) [15] is employed to calculate the similarity between two images x and y (in CIE-Lab color space) as shown in (6), where C calculates the compressed length of corresponding input.

$$NCD(x,y) = \frac{C(xy) - \min\{C(x), C(y)\}}{\max\{C(x), C(y)\}}. \tag{6}$$

4.5   Interpreting Reminder Terms of Gestalt Laws

The Gestalt law of common fate refers to the motion trend. In a group of render-blocks, if they are not placed on the same path, then the off-path render-blocks share no common fate with the other blocks. The Gestalt law of continuity is interpreted as alignment. If any of the four sides (left, right, top or bottom) of two render-blocks are aligned, then they are continuous.

The Gestalt law of symmetry tells that different but symmetrical render-blocks should be merged, however, there are very few webpages containing such instances. Hence, this law is not interpreted. Also, the Gestalt law of past experience is not considered because it refers to high level semantics and requires external knowledge.

5   Webpage Similarity Classification Model

Because a webpage can be ultimately represented by a semantic-block tree (or simply block tree), and each block contains all the visual information, visual similarity between two webpages can be reflected by block tree similarity. That is, visual similarity is evaluated by block tree edit distances.

5.1   Block Tree Edit Distance

Let T be a block tree, $|T|$ be the size of it, and t_i be its ith node. Two different block trees can then be denoted by T^p and T^q. The tree edit distance (TED) is then defined as the minimum cost of editing operations ("insert", "delete", and "relabel") when shifting from T^p to T^q [8]. This reflects the structural similarity between T^p and T^q by mapping node pairs.

Webpage similarity includes both structural and content similarity (visual similarity). To compare visual similarity between two webpages, a block tree edit distance (B-TED) is introduced. By encoding the content of each block into its label, the mapping procedure in TED calculation compares the blocks by their visual information. Same as comparing background images in Sect. 4.4, visual similarity between two blocks are evaluated by NCD. If they are not similar, then a "relabel" operation is needed.

People always see a subtree of a block rather than itself because it is overlaid by its descendants (if there is any). To simulate this, the content for encoding shall not be that of the block itself but of the complete subtree. In B-TED calculation, it can be achieved simply by encoding the screen capture image of a block.

5.2   Classification Using a Naive Bayes Classifier

The model adopts a naive Bayes classifier for classification. Through reading the feature vectors of two webpages whose categories are known, the classifier learns the connections between the features and categories. The feature vector contains three components: the block trees of the two webpages, and the B-TED value between them. The category variable is a Boolean; and its value is either T indicating the two webpages are similar or F indicating they are different. Details of this model are illustrated in Fig. 3.

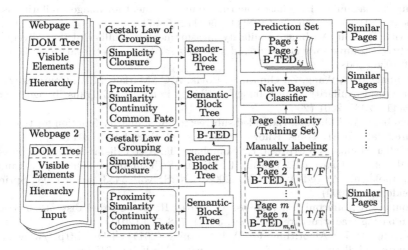

Fig. 3. Classification model

6   Conclusion

In this paper, a novel approach to evaluate webpage similarity is proposed. A render-block tree model is introduced to represent a webpage visually, and a semantic-block tree model is then retrieved by interpreting and applying the Gestalt laws of grouping. During interpretation, a normalized Hausdorff distance is introduced to evaluate proximities; the CIE-Lab color space and its color difference are used to find color similarities; and the normalized compression distance is employed to calculate image similarity. A classification model is finally proposed to evaluate webpage similarity. Block tree edit distance can be applied to recognize both structural and visual similar-ity of webpages.

Acknowledgment. The authors give thanks to China Scholarship Council (CSC) for their financial support.

References

1. Qi, X., Davison, B.D.: Web page classification: features and algorithms. J. ACM **41**(2), 12:1–12:31 (2009)
2. Wei, Y., Wang, B., Liu, Y., Lv, F.: Research on webpage similarity computing technology based on visual blocks. SMP **2014**, 187–197 (2014)
3. Wertheimer, M.: Laws of organization in perceptual forms (1938)
4. Rohlfing, T.: Image similarity and tissue overlaps as surrogates for image registration accuracy: widely used but unreliable. IEEE Trans. Med. Imaging **31**(2), 153–163 (2012)
5. Liu, Z., Laganière, R.: Phase congruence measurement for image similarity assessment. Pattern Recogn. Lett. **28**(1), 166–172 (2007)
6. Kwitt, R., Uhl, A.: Image similarity measurement by kullback-leibler divergences between complex wavelet subband statistics for texture retrieval. ICIP 2008, pp. 933–936 (2008)
7. Sampat, M.P., Wang, Z., Gupta, S., Bovik, A.C., Markey, M.K.: Complex wavelet structural similarity: a new image similarity index. IEEE Trans. Image Process. **18**(11), 2385–2401 (2009)
8. Shahbazi, A., Miller, J.: Extended subtree: a new similarity function for tree structured data. IEEE Trans. Knowl. Data Eng. **26**(4), 864–877 (2014)
9. M"uller-Molina, A.J., Hirata, K., Shinohara, T.: A tree distance function based on multisets. In: Chawla, S., Washio, T., Minato, S., Tsumoto, S., Onoda, T., Yamada, S., Inokuchi, A. (eds.) PAKDD 2008. LNCS, vol. 5433, pp. 87–98. Springer, Heidelberg (2009)
10. Connor, R., Simeoni, F., Iakovos, M., Moss, R.: A bounded distance metric for comparing tree structure. Inf. Syst. **36**(4), 748–764 (2011)
11. Cording, P. H.: Algorithms for Web Scraping (2011). [PDF] http://www2.imm.dtu.dk/pubdb/views/edoc_download.php/6183/pdf/imm6183.pdf
12. Zhai, Y., Liu, B.: Structured data extraction from the web based on partial tree alignment. IEEE Trans. Knowl. Data Eng. **18**(12), 1614–1628 (2006)
13. Chaudhuri, B.B., Rosenfeld, A.: A modified Hausdorff distance between fuzzy sets. Inf. Sci. **118**(1), 159–171 (1999)
14. Johnson, G.M., Fairchild, M.D.: A top down description of SCIELAB and CIEDE2000. Color Res. Appl. **28**(6), 425–435 (2003)
15. Li, M., Chen, X., Li, X., Ma, B., Vitányi, P.M.: The similarity metric. IEEE Trans. Inf. Theory **50**(12), 3250–3264 (2004)

Extracting Records and Posts from Forum Pages with Limited Supervision

Luciano Barbosa[✉] and Guilherme Ferreira

IBM Research – Brazil, Av. Pasteur, 138, Rio de Janeiro, Brazil
{lucianoa,guiferre}@br.ibm.com

Abstract. Internet forums are rich sources of human-generated content. Many applications, such as opinion mining and question answering, can greatly benefit from mining and exploring such useful content. An important step towards making user content from forums more easily accessible is to extract it from forum pages. We propose REPEX (REcord and Post EXtractor), a two-step solution that uses limited supervision to achieve this goal. Given a forum page, REPEX first extracts data records that contain human-generated content and then, from these records, extracts their user content. The record extraction assumes that (1) a record is composed of an automatic-generated part, which we call record template, and a human-generated part; and (2) the structure of record templates are usually consistent across records. Based on those, the record extractor initially locates the subtree that contains all records in the forum page, using an information-theoretic measure, and then identifies the template of the records in this subtree, modelling this as an outlier detection problem. Finally, starting from the templates, REPEX determines the boundaries of the records. For the post extraction, REPEX applies an information extraction approach that performs this task by identifying the posts' string boundaries.

Keywords: Forum · Record extraction · Post extraction · Data mining

1 Introduction

Internet forums contain user-generated content and address many different subjects and topics (e.g. games, movies, travel, computers, health etc.). To take advantage of such rich content, methods to collect and process forum data have been previously introduced [1,2,6,8]. In this paper, we focus on the particular problem of extracting human-generated content from conversational pages of forums, also known as thread pages.

Thread pages are composed of data records that contain the human-generated part (the user post), and the automatic-generated (or template) part that contains information such as date/time of the post, the user who posted it and the title of the posts. Similar to [3,6], we are interested in building a solution that is not specific for a particular layout template (template-independent). We also want to perform this task with limited supervision, i.e., without any training

© Springer International Publishing Switzerland 2015
J. Wang et al. (Eds.): WISE 2015, Part II, LNCS 9419, pp. 233–240, 2015.
DOI: 10.1007/978-3-319-26187-4_19

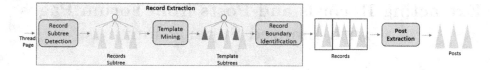

Fig. 1. Overview of REPEX's pipeline: given a thread page, Record Subtree Detection locates the subtree of records; Template Mining identifies the record template subtrees (in blue); and Record Boundary Identification determines the boundaries of each record; finally, from the records, Post Extraction extracts the posts (Colour figure online).

data, to avoid having to label a large amount of data, which is a laborious and time-consuming task. The main challenge of building such solution is that the structure of thread pages vary significantly across forum sites. To deal with all this variability, we propose a REPEX (REcord and Post EXtractor), see Fig. 1. Given a thread page, our method first extracts data records that contain the user posts (Record Extraction), and then the posts within these records (Post Extraction). In the remaining of this paper, we present REPEX in details and present results showing that REPEX is highly effective, obtaining high values of precision and recall for both tasks.

2 Record Extraction

As presented in Fig. 1, the first step of REPEX is to extract the records from the thread pages based on the record templates. The Record Extraction is composed of 3 sub-tasks: Record Subtree Detection, Template Mining and Record Boundary Identification. Record Subtree Detection locates the subtree in the DOM tree where all records are located. Within this subtree, Template Mining identifies the record templates and, based on the templates, Record Boundary Identification determines the boundaries of the records. In this section, we first describe the type detectors used to identify the record templates and, subsequently, explain each one of the components of the record extraction.

2.1 Type Detectors

Our first assumption regarding the problem of record extraction is that a record contains a template-generated part, composed of basic types: the date and time that the record was posted, its title and the user who posted it. Based on that, we implemented 3 type detectors to identify them in a thread page: date-time, user and record title. The date-time detector was built from regular expressions. For that, we started from date and time examples of regular expressions available on specialized websites[1]. Then, we improved the quality of these expressions using

---

[1] http://www.regxlib.com/.
http://www.regular-expressions.info/.

a validation set, described in Sect. 4. The user detector uses simple heuristics to detect user information. It checks for URLs with words such as "'member"', "profile" and "user". The title detector assumes the record title is similar to the title of the thread page. To measure that, we calculate the Jaccard similarity between the title of the thread page and a given text. We consider a similarity of 0.3 as a match. The text used as input to the detectors is segmented based on the DOM tree structure: all the text within a leaf text node is considered as a single sentence.

The great advantage of using a set of detectors, instead of a single one as [6] did, is that individual detectors can complement each other, and consequently produce better results. For instance, the user detector might work in sites in which the date-time detector might not. Another advantage is that building a strong detector is a laborious task. Thus, instead of having a single strong detector, one can build weak detectors, which need less effort to be implemented. Our experimental evaluation confirms all these observations.

2.2   Record Subtree Detection

The first step of Record Extraction is Record Subtree Detection. Given the thread page's DOM tree T, it identifies the subtree T' of T that contains all the records. To achieve this goal, we assume that the nodes that contain the template data types (date-time, user and title) are evenly distributed throughout the child subtrees of T'. Concretely, the algorithm works as follows. First, given T, the algorithm performs a complete scan of T, labelling nodes that match the three basic types. When a type detector matches a node, the types' counter in that node is incremented. Next, based on these counts, for each subtree T' in T, it measures how balanced the child subtrees CS of T' are with respect to the detected data type nodes. Only T's with a balance value higher than a threshold are considered candidate subtrees for the next steps. The tree balance is measured using an information-theoretic approach. More formally, consider p the probability of a child subtree c of T' having detected data type nodes. We calculate p of c by dividing the number of detected nodes in c over the total number of detected nodes in T'. If T' is balanced, the entropy of T' would be high, since p for all children would have a similar value. To have a value between 0 and 1, we define *Balance*, which is the normalized entropy of T':

$$Balance(T') = -\frac{\sum_{c \in CS} p_c log(p_c)}{log(|CS|)} \tag{1}$$

2.3   Template Mining

The goal of Template Mining is to identify the template part of the records. For that, we assume that the data types are more concentrated in nodes belonging to the template part of records than in other parts. Another assumption, similar to [6], is that the tree structure of the templates of the records is similar to each other. Based on those observations, we model this task as an outlier detection problem, in which detected nodes outside the template part of records are considered outliers. The algorithm works as follows. Initially, given the subtree T', identified in Record Subtree Detection, the algorithm obtains the candidate templates CT, i.e., the children of T' that contains detected nodes. For each child c of CT, it generates a signature composed of the HTML tags of the detected nodes of c in the depth-first search order. This signature represents a flat representation of the tree structure of c with respect to its detected nodes. Next, these signatures are provided as input to the Hierarchical Aglomerative Clustering (HAC) [7]. HAC starts with $|CT|$ clusters (a single cluster corresponds to a child signature), where $|CT|$ is the number of elements of CT. The two closest clusters are merged, resulting in $|CC| - 1$ clusters. Next, the two closest of the $|CC| - 1$ clusters are merged, and then the process continues until a stop condition. The output of this process is a set of clusters. The algorithm considers that the record templates are in the cluster with the highest number of elements, and the remaining clusters are discarded. The subtrees belonging to this cluster are returned, if the cluster has more than 2 elements. We adopted as stop condition a similarity threshold, defined experimentally. We use as similarity measure the levenshtein distance [4].

2.4   Record Boundary

Template Mining selects the template subtrees $\{t_1, ..., t_n\}$ in T'. These subtrees, however, do not necessarily contain the whole record. There might be cases, for instance, in which the human-generated content of a record is in a separated subtree of T'. In other cases, a record might be composed of multiple subtrees of T'. The task is, therefore, to define how we segment the children of T' in order to extract the records. First, the algorithm determines the record size, i.e., how many consecutive child subtrees of T' compose a record. It does so by calculating the distance (i.e., how many subtrees are) between each pair of consecutive template subtrees. It considers the distance with the highest frequency as the record size. Then, it defines in which position to the left of the template subtree the records start. For that, it goes backward from the first two template subtrees (t_1 and t_2) until it finds subtrees of T' with different child signatures. We define a child signature as the string composed of the subtree HTML tag concatenated with the tags of its children. Finally, the records are extracted from T' for each template subtree t_i.

3   Post Extraction

The Post Extraction is the final step of REPEX (see Fig. 1). It extracts the human-generated content from the data records. Instead of only relying on the DOM tree structure to perform this task, as we did for record extraction, we handle this task as an unstructured information extraction problem. For that, we look at regularities in the text resulting from the records. This illustrates the main assumption of this algorithm: posts are delimited between common types/strings across records. The goal of Post Extraction is then to identify these delimiters, and then extract all text between them. Concretely, the algorithm works as follows. First, it segments the record sentences, using their structure on the DOM tree. All the text in the same leaf node is considered a single sentence. Next, it runs a post-text detector over the sentences. Similar to the other detectors presented in this paper, the post-text detector uses simple rules to perform the detection. For instance, it looks for characters such as "." or "?" at the end of the phrase along with personal pronouns as "I" or "you" or "it". Here we assume that the text in posts have a good chance of having personal references. From all the records with detected texts, the algorithm selects the one that it has a high confidence of having in fact a post text: the record r with the largest detected text in phrase position p in r. From p, the algorithm goes backwards until it finds a type/string in r that matches in all others records. The positions $\{s_1, ..., s_n\}$ of these matches in the records represent where the posts should start. Conversely, the algorithm does the same procedure going forward from p. The positions $\{e_1, ..., e_n\}$ of these matches in the records represent where the posts should end. To perform this match, it verifies whether the strings are from the same data type (date-time, user and title) or if they share some prefix of size greater than 1. Finally, it extracts the posts using $\{s_1, ..., s_n\}$ and $\{e_1, ..., e_n\}$ as delimiters.

4 Experimental Evaluation

4.1 Experimental Setup

Data. For the evaluation, we collected thread pages from 118 forum sites. We tried to collect a set as diverse as possible. For instance, these websites are not restricted to any particular topic: they are discussions about games, cancer, psychology etc. In addition, similar to [5], we also tried to select as many forums as possible that use different softwares to publish their content. Out of the 118 sites, 72 were used in the validation set and 46 in the test set. For each one of the websites, we collected at most 5 thread pages, resulting in a set of 282 pages in the validation set and 200 in the test set. Then, we manually extracted the text in the records and posts from these pages, resulting in a total of 2,449 records and the same number of posts. Since there might be small differences between the way records and posts are extracted, we consider a match when the cosine similarity between the approach's record and the gold data's record is higher than 0.6 for records and 0.3 for posts.

Record Extraction Approaches. For comparison, we implemented another proposed solution to extract records from thread pages: MiBAT [6]. MiBAT uses a date-time detector to identify the template part of the records, which they call anchor trees. Then it aligns anchor trees using a tree matching algorithm [9]. The matched anchor trees compose the templates of the records. For this matching, the authors proposed two similarity measures. We used Pivot and Siblings (PS) similarity, since it showed the best results in their experiments. A similarity higher than a given threshold is considered a match. We used the validation set to tune this parameter. For further details, we refer the reader to [6]. We also used the validation set to tune two parameters of our approach: the minimum entropy of a parent node being considered relevant, and the similarity threshold in the HAC algorithm's stop condition.

Post Extraction Approaches. In addition to the post extraction approach proposed in this paper, which will we call String-based Extraction for the remaining of this section, we implemented two other strategies:

- Text Detection: this approach scans the records, and only considers as posts the text detected within the records by the Text Detector.
- Tree-based extraction: this algorithm works as follows. Given the subtree that contains all the records T', first it uses the Text Detector to identify text nodes in T'. For the child subtrees CS of T' that contain text nodes, it identifies the largest common subtree LCS of all CS. Since we assume the post part of the record subtree might not have much regularity, for each record, the algorithm considers the post part the tree structure of the record that does not belong to the LCS. This method has not been proposed previously in the literature. We implemented it to have a reasonable baseline for post extraction.

4.2　Record Extraction Results

For each approach, we measured precision, recall and F-Measure over the records in the test set. We also calculated the proportion of pages that had at least 1 record extracted by each approach. Table 1 presents the results. Our approach obtained high values of recall (0.94), precision (0.92), F-measure (0.93), and also

Table 1. Recall, precision, F-Measure and proportion of pages with at least 1 record extracted by each approach.

	Rec	Prec	F-Measure	Prop. of Pages
RecExt	0.94	0.92	0.93	0.97
MiBAT	0.51	0.94	0.66	0.53

Table 2. Results of our approach using different combinations of type detectors.

	Rec	Prec	F-Measure
User,Date-Time	0.86	0.92	0.89
Date-Time,Title	0.82	0.93	0.87
User,Title	0.76	0.96	0.85
User	0.67	0.96	0.79
Date-Time	0.68	0.93	0.79
Title	0.32	0.99	0.48

extracted records from the vast majority of the pages (0.94). The numbers also show that our approach outperforms the baseline in all measures.

We investigated possible causes for this difference in performance. For that, we calculated the performance of MiBAT over only the 53 % of the test set that it was able to extract records. As expected, its results are much better: recall = 0.92, precision = 0.97 and F-Measure = 0.95. For comparison, we also ran our approach over the same 53 % set. It obtained recall = 0.96, precision = 0.95 and F-Measure = 0.96. Our approach obtained higher recall (0.96 vs 0.92) but lower precision (0.95 vs 0.97). Overall, our approach obtained a slightly better F-Measure (0.96 vs 0.95).

We also evaluated the contribution of each detector for the final result. Table 2 presents the recall, precision and F-Measure for all the possible combinations of detectors. The combination of user and date-time detectors obtained the best results as well as these two detectors considered individually. Although the title detector individually obtained a poor result in terms of recall, combining it with the other detectors, it boosted the overall performance of our approach. These numbers clearly show that a combination of "weak" detectors that complement each other, i.e., covering different sets of pages, leads to an effective extractor.

Since MiBAT only uses a single date-time detector, we can compare its performance in Table 1 with our approach using only this detector (Table 2) over the entire test set. Our approach obtained a much higher recall than MiBAT (0.68 vs 0.51) and a slightly smaller precision (0.93 vs 0.94), as a result a higher F-Measure (0.79 vs 0.66). The main reason for this advantage in coverage is that the proportion of pages that our approach with only a date-time detector detected at least one record was much higher than MiBAT's: 0.68 vs 0.51. From this, we can conclude that MiBAT was not able to extract records even in pages that the date-time detector worked.

4.3 Post Extraction Results

The results of the post extraction approaches are presented in Table 3. The String-based approach obtained the highest values of recall (0.86), precision (0.93) and F-Measure (0.89), followed by the Tree-based approach. The numbers show that our approach of post extraction is in fact effective for this task. The lowest result was obtained by the approach that only uses the text detector to extract the posts. The main reason for this poor performance is that a reasonable portion of the text in posts are not detected by the Text Detector

Table 3. Results of post extraction.

	Rec	Prec	F-Measure
String-based Extraction	0.86	0.93	0.89
Tree-based Extraction	0.82	0.92	0.87
Text Detection	0.57	0.56	0.56

(low recall), and also much of the text detected by the Text Detector does not belong to the posts (low precision). We can conclude from this that using the text detection itself is not enough for this task, but it is very useful when used with our proposed strategy. Regarding the recall of all approaches, an important observation is that the post extraction is performed after the record extraction. As a result, the upper bound of recall is the one obtained by our record extraction technique: 0.94. The precision of the record extraction also has influence over the precision results for post extraction.

5 Conclusions

In this paper, we present REPEX, a solution for extracting data records and user posts from forum pages. To locate the data record subtree, it uses an information-theoretic approach. Next, within this subtree, it identifies the template part of the records using a clustering algorithm. Finally, it determines the boundaries of the records expanding from the templates. The extracted records are then passed to the post extraction, that uses an unstructured information extraction strategy to define the boundaries of posts, and extract them.

References

1. Cong, G., Wang, L., Lin, C.-Y., Song, Y.-I., Sun, Y.: Finding question-answer pairs from online forums. In: Proceedings of the 31st annual international ACM SIGIR conference on Research and Development in Information Retrieval, pp. 467–474. ACM (2008)
2. Jiang, J., Song, X., Yu, N., Lin, C.-Y.: Focus: learning to crawl web forums. IEEE Trans. Knowl. Data Eng. **25**(6), 1293–1306 (2013)
3. Liu, B., Grossman, R., Zhai, Y.: Mining data records in web pages. In: Proceedings of the ninth ACM SIGKDD international conference on Knowledge Discovery and Data Mining, pp. 601–606. ACM (2003)
4. Navarro, G., Baeza-Yates, R., Sutinen, E., Tarhio, J.: Indexing methods for approximate string matching. IEEE Data Eng. Bull. **24**(4), 19–27 (2001)
5. Seo, J., Croft, W.B., Smith, D.A.: Online community search using thread structure. In: Proceedings of the 18th ACM Conference on Information and Knowledge Management, pp. 1907–1910. ACM (2009)
6. Song, X., Liu, J., Cao, Y., Lin, C.-Y., Hon, H.-W.: Automatic extraction of web data records containing user-generated content. In: Proceedings of the 19th ACM international conference on Information and Knowledge Management, pp. 39–48. ACM (2010)
7. Tan, P.-N., Steinbach, M., Kumar, V., et al.: Introduction to data mining, vol. 1. Pearson Addison Wesley, Boston (2006)
8. Wang, H., Wang, C., Zhai, C., Han, J.: Learning online discussion structures by conditional random fields. In: Proceedings of the 34th international ACM SIGIR conference on Research and development in Information Retrieval, pp. 435–444. ACM (2011)
9. Yang, W.: Identifying syntactic differences between two programs. Soft. Pract. Experience **21**(7), 739–755 (1991)

Improving Relation Extraction by Using an Ontology Class Hierarchy Feature

Pedro H.R. Assis[1]([✉]), Marco A. Casanova[1],
Alberto H.F. Laender[2], and Ruy Milidiu[1]

[1] Department of Informatics, Pontifícia Universidade Católica do Rio de Janeiro,
Rio de Janeiro, RJ, Brazil
{passis,casanova,milidiu}@inf.puc-rio.br
[2] Department of Computer Science,
Universidade Federal de Minas Gerais, Belo Horizonte, MG, Brazil
laender@dcc.ufmg.br

Abstract. Relation extraction is a key step to address the problem of
structuring natural language text. This paper proposes a new ontology
class hierarchy feature to improve relation extraction when applying a
method based on the distant supervision approach. It argues in favour
of the expressiveness of the feature, in multi-class perceptrons, by experimentally showing its effectiveness when compared with combinations of
(regular) lexical features.

Keywords: Relation extraction · Distant supervision · Semantic Web ·
Machine learning · Natural language processing

1 Introduction

A considerable fraction of the information available on the Web is under the form
of natural language, unstructured text. While this format suits human consumption, it is not convenient for data analysis algorithms, which calls for methods
and tools to structure natural language text. Among the many key problems this
task poses, *relation extraction*, i.e., the problem of finding relationships among
entities present in a natural language sentence, stands out.

The most successful approaches to address the relation extraction problem apply supervised machine learning to construct classifiers using features
extracted from hand-labeled sentences of a training corpus [5,10]. However,
supervised methods suffer from several problems, such as the limited number
of examples in the training corpus, due to the expensive cost of manually annotating sentences. Such limitations hinder their use in the context of Web-scale
knowledge bases. Distant supervision, an alternative paradigm introduced by
Mintz et al. [9], addresses the problem of creating examples, in sufficient number,
by automatically generating training data with the help of a sample database.

© Springer International Publishing Switzerland 2015
J. Wang et al. (Eds.): WISE 2015, Part II, LNCS 9419, pp. 241–249, 2015.
DOI: 10.1007/978-3-319-26187-4_20

In this paper, we first discuss how to apply the distant supervision approach to develop a multi-class perceptron[1] for relation extraction. Then, we present new *semantic features*, defined based on a pair of entities e_1 and e_2 identified in the sentence. The semantic features associate classes C_1 and C_2 to the sentence, where C_1 and C_2 are derived from the class hierarchy of an ontology and the original classes of e_1 and e_2 in the hierarchy. The main contribution of the paper is the proposal of these semantic features.

Finally, we describe experiments to evaluate the effectiveness of our semantic based features, using a corpus extracted from the English Wikipedia and instances of the DBpedia Ontology. We conducted two types of experiments, adopting the automatic held-out evaluation strategy and human evaluation. In the held-out evaluation experiments, the multi-class perceptron identified, with an F-measure greater than 70 %, a total of 88 relations out of the 480 relations featured in the version of the DBpedia adopted. In the human evaluation experiments, it achieved an average accuracy greater than 70 % for 9 out of the top 10 relations, in the number of instances, selected for manual labeling. An early and short version of these results appeared in [2].

This paper is structured as follows. Section 2 discusses related work. Section 3 describes the approach adopted to construct multi-class perceptron for relation extraction and the definition of the ontology classes hierarchy feature. Section 4 contains the experimental results. Finally, Sect. 5 presents the conclusions and suggestions for future work.

2   Related Work

Soderland et al. [11] introduced supervised-learning methods as approaches for information extraction. They are the most precise methods for relation extraction [5,10], but they are not scalable to the Web due to the expensive cost of production and the dependency on an annotated corpus for the specific application domain. In order to address the scalability problem in relation extraction frameworks, weak supervision methods were introduced, based on the idea of using a database with structured data to heuristically label a text corpus [4,13,14].

Mintz et al. [9] coined the term distant supervision to replace the term weak supervision. They applied Freebase facts to create relation extractors from Wikipedia, achieving an average precision of approximately 67.6 % for the top 100 relations. The popularity of distant supervision methods increased rapidly since its introduction. Unfortunately, depending on the domain of the relation database and the text corpus, heuristics can lead to noisy data and poor extraction performance.

Finally, classifiers can be improved with the help of Semantic Web resources and, conversely, new Semantic Web resources can be generated by using relation extraction classifiers. For example, Gerber et al. [6] used DBpedia as background knowledge to generate several thousands of new facts in DBpedia from Wikipedia

---

[1] *Perceptron* is a linear classifier for supervised machine learning. It is an assembly of linear-discriminant representations in which learning is based on error-correction.

articles, using distant supervision methods. For relation extraction they used a pattern matching approach. In this work, instead of relying on the generation of relation patterns, we used DBpedia as background knowledge to generate an annotated dataset to construct a multi-class perceptron for relation extraction.

3   The Distant Supervision Approach

We transform the relation extraction problem into a classification problem by treating each relation r as a class $\mathbf{r}$ of a multi-class perceptron. To construct the perceptron, we feed a machine learning algorithm with sentences in a corpus C, together with their feature vectors, where the sentences are heuristically annotated with relations using the distant supervision approach. In this paper, we adopt a non-memory-based machine learning method, called Multinomial Logistic Regression [8], which computes a multi-class perceptron. This section covers the major points of the approach, referring the reader to [1] for the full details.

3.1   Distant Supervision

The approach we adopt to generate a dataset is based on distant supervision [9]. The main assumption is that a sentence might express a relation if it contains two entities that participate in that relation.

Formally, given an ontology O, we say that e_i is an entity *defined in* O iff there is a triple of the form $(e_i, \text{rdf:type}, K_i)$ in O such that K_i is a class in the vocabulary of O. The *relation database* of O is the set R_O such that a triple $(e_1, r_i, e_2) \in O$ is in R_O iff e_1 and e_2 are entities defined in O and r_i is an object property in the vocabulary of O. For example, if *"Barack Obama"* and *"United States"* are entities in O and there is a triple $t = ($ *"Barack Obama"*, *"president of"*, *"United States"*$)$, then $t \in R_O$.

Let C be a corpus of sentences each of which is annotated with two entities defined in O. Suppose that a sentence $s \in C$ is annotated with entities e_1 and e_2 and that there is a triple (e_1, r, e_2) in R_O. Then, we consider that s is *heuristically labeled* as an example of the relation r. For example, suppose that R_O contains the triple: *(Led Zeppelin, genre, Rock Music)*, where the rock band *Led Zeppelin* and the music genre *Rock Music* are defined in O. Then, every sentence annotated with *Led Zeppelin* and *Rock Music* is a prospective example of the relation *genre*, such as: *"**Led Zeppelin** is a british rock band that plays **rock music**."*

The approach is applicable for inverse relations if they are explicitly declared in the ontology O. They will be simply treated as new classes.

3.2   Features

We associate a feature vector with each sentence s in the corpus C. Feature vectors will have dimension 12, comprising 10 lexical features, as in [9], and two features based on the class structure of the ontology O.

For *lexical features*, let s be a sentence in a corpus C annotated with two entities e_1 and e_2. We break s into five components, $(w_l, e_1, w_m, e_2, w_r)$, where w_l comprehends the subsentence to the left of the entity e_1, w_m the subsentence between the entities e_1 and e_2 and w_r the subsentence to the right of e_2. For example, the sentence s_A *"Her most famous temple, the **Parthenon**, on the Acropolis in **Athens** takes its name from that title."* is represented as ("Her most famous temple, the", **Parthenon**, ", on the Acropolis in", **Athens**, " takes its name from that title."). Lexical features contemplate the sequence of words in w_l, w_m, and w_r and their part-of-speech; but not all the words in w_l and w_r are used. Indeed, let $w_l(1)$ and $w_l(2)$ denote the first and the first two rightmost words in w_l, respectively. Analogously, let $w_r(1)$ and $w_r(2)$ denote the first and the first two leftmost words in w_r, respectively. In the example, the corresponding sequences of length 1 and 2 are: $w_l(1) = $ "the", $w_l(2) = $ "temple, the", $w_r(1) = $ "takes" and $w_r(2) = $ "takes its". The part-of-speech tags cover 9 lexical categories: NOUN, VERB, ADVERB, PREPosition, ADJective, NUMbers, FOReign words, POSSessive ending and everything ELSE (including articles).

For *class-based features*, we propose to use as a feature of an entity e (and of the sentences where it occurs) the class that best represents e in the class structure of the ontology O. We claim that the chosen class must not be too general, since we want to avoid losing the specificities of the semantics of e that are not shared with the other entities of the superclasses. On the other hand, a class that is too specific is also not a good choice. Very specific classes restrict the accuracy of classifiers, since they probably contain fewer entities than more general classes. In other words, the number of entities in a class is likely to be inversely proportional to the class specificity.

Therefore, we propose to use as a feature of an entity e (and of the sentences where it occurs) the class associated with e that intuitively lies in the mid-level of the ontology class structure. For example, suppose we have the entity *Barack Obama*, with class hierarchy *President $\subset$ Politician $\subset$ Office_holder $\subset$ Person $\subset$ Agent $\subset$ owl:Thing*. We have to choose one class to represent the entity *Barack Obama*. If we choose the class *Agent*, for example, which is too general, all relations involving a president will be assign to every example of agents in our dataset, which therefore not a good choice. On the other hand, if we choose the class *President*, which is too specific, we will be missing several relations shared by politicians or office holders. Therefore, we choose the class at the middle level of the hierarchy, which in this example is *Office_holder*.

More precisely, given an ontology O, the *class structure* of O is the directed graph $G_O = (V_O, E_O)$ such that V_O is the set of classes defined in O and there is an edge $< C, D >$ in E_O iff there is a triple *(C, owl:SubClassOf, D)* in O. We assume that G_O is acyclic and that G_O has a single sink, the class *owl:Thing*. This assumption is consistent with the usual practice of constructing ontologies and the definition of *owl:Thing*. By analogy with trees, the *height* of G_O is the length of the longest path from a source of G_O to *owl:Thing* and the *level* of a class C in G_O is the length of the shortest path in G_O from C to *owl:Thing*.

We also assume that O is equipped with a service that, given an entity e, classifies e into a single class C_e. Assume that the shortest path in G_O from C_e to *owl:Thing* is $(C_k, \cdots, C_i, \cdots, C_0)$, where $C_k = C_e$ and $C_0 = $ *owl:Thing*. Then, we define the *class-based feature* of e as the class C_i, where $i = min(k, h/2)$, where h is the height of G_O. Note that we take the minimum of k and $h/2$ since the level of C_k may be smaller than half of the height of G_O.

Finally, let s be a sentence in the corpus C, annotated with two entities e_1 and e_2. We define the *class-based features* of s as the class-based features of e_1 and e_2.

4   Experiments

We adopted a version of DBpedia [3] as our ontology, which features 359 classes, organized into hierarchies, 2,350,000 instances and more than 480 different relations. We used all Wikipedia articles in English as a source of unstructured text. We annotated a Wikipedia article A with an entity e from DBpedia if there is a link in the text of A pointing to the article corresponding to e. For sentence boundary detection, we used the algorithm proposed by Gillick [7]. We also applied heuristics in order to increase the number of acceptable sentences. We annotated references to the main subject of an article by string matching between the article text and the article title. Also, for sentences with more than two instances annotated, we considered combinations of all pairs of instances.

Applying all strategies described above, we generated a corpus of 2,276,647 sentences with annotated entities, for which we obtained lexical and class-based features as described in Sects. 3.2 and 4. We used the Stanford Part of Speech Tagger [12] and the WSJ 0.18 Bidirectional model for POS features to extract the lexical features, but we simplified the POS tags into 9 categories, as already indicated in Sect. 3.2.

4.1   Held-Out Evaluation

We ran experiments to assess the impact of the class-based features by training the Multinomial Logistic Regression classifier [8] using only lexical features, only class-based features and both sets of features. Half of the sentences for each relation were randomly chosen not to be used in the training step. They are later used in the testing step.

For this kind of extraction task, final users usually consider an acceptable performance if it predicts classes with an F-measure greater than 70 %. Therefore, the comparison between the various options took into account the number of classes for which the perceptron achieved an F-measure greater than 70 %. Table 1 show the top 10 classes for each combination of features, with the classes identified by their suffixes, since they all share the same prefix in their URI: http://dbpedia.org/ontology. Also, Table 1 shows that class-based features were able to predict over 6 times more classes than our baseline (lexical features only)

Table 1. Top 10 classes for a perceptron trained with different feature set.

Features	No.	Class	Precision	Recall	F-measure
Lexical	1	/targetSpaceStation	1.00	1.00	1.00
	2	/department	0.98	0.86	0.92
	3	/discoverer	1.00	0.81	0.90
	4	/militaryBranch	0.94	0.83	0.88
	5	/notableWine	0.99	0.75	0.85
	6	/programmeFormat	0.87	0.77	0.82
	7	/type	0.69	0.83	0.75
	8	/license	0.98	0.58	0.73
	9	/sport	0.81	0.63	0.71
	10	/composer	0.95	0.54	0.69
		average:	**0.921**	**0.760**	**0.825**
		number of classes > 70% F-measure:		**6**	
Class-based	1	/areaOfSearch	1.00	0.98	0.99
	2	/ground	0.96	1.00	0.98
	3	/mission	0.97	1.00	0.98
	4	/politicalPartyInLegislature	1.00	0.95	0.97
	5	/precursor	0.99	0.96	0.97
	6	/sport	0.96	0.97	0.97
	7	/targetSpaceStation	0.94	1.00	0.97
	8	/discoverer	0.93	1.00	0.96
	9	/drainsTo	0.97	0.93	0.95
	10	/isPartOfAnatomicalStructure	0.91	1.00	0.95
		average:	**0.963**	**0.979**	**0.969**
		number of classes > 70% F-measure:		**60**	
Lexical and Class-based	1	/areaOfSearch	1.00	0.97	0.98
	2	/ground	0.97	1.00	0.98
	3	/mission	0.99	0.96	0.97
	4	/sport	0.97	0.97	0.97
	5	/targetSpaceStation	1.00	0.93	0.97
	6	/academicDiscipline	0.93	0.99	0.96
	7	/discoverer	0.99	0.93	0.96
	8	/locatedInArea	0.93	0.98	0.96
	9	/programmeFormat	0.93	0.99	0.96
	10	/politicalPartyInLegislature	1.00	0.91	0.95
		average:	**0.971**	**0.963**	**0.966**
		number of classes > 70% F-measure:		**88**	

and the inclusion of lexical features can improve the previous result in 32 %, predicting a total of 88 classes with more than 70 % of F-measure.

Although, in general, there is a considerable gain by using both sets of features, the perceptron trained using both sets of features had a worse performance than that trained using only class-based features for some classes. For example, /aircraftFighter is identified with a F-measure of 50 % using both sets of features, whereas it was identified with 77 % using only class-based features.

Table 2. Average accuracy for the top 10 relations in examples in our dataset for human evaluation of a sample of 100 predictions.

Relation	Number of instances	Average accuracy
http://dbpedia.org/ontology/country	607,380	73 %
http://dbpedia.org/ontology/family	159,717	75 %
http://dbpedia.org/ontology/isPartOf	139,694	90 %
http://dbpedia.org/ontology/birthPlace	138,797	76 %
http://dbpedia.org/ontology/genre	109,813	77 %
http://dbpedia.org/ontology/location	96,516	76 %
http://dbpedia.org/ontology/type	72,942	80 %
http://dbpedia.org/ontology/order	53,421	81 %
http://dbpedia.org/ontology/occupation	48,859	87 %
http://dbpedia.org/ontology/hometown	34,010	68 %

This shows that for some classes, our lexical features reduces the generalization of our model of classification, but overall they increase the robustness of predictions for the majority of classes.

4.2 Human Evaluation

For the human evaluation experiments, we also separated the sentences, annotated with pairs of entities, into training and testing data. We randomly chose half of the sentences not to be used in the training step, for each relation (in this section we again use the term "relation" instead of "class"). For each of the top 10 relations (in the number of instances in our dataset), we extracted random samples of 100 sentences from the remaining sentences and forwarded to two evaluators to manually label the sentences with relations. Finally, we compared the manually labeled sentences with the labeling obtained by a perceptron trained using both lexical and class-based features, as shown in Table 2, where the average accuracy is percentage of the sentences that the automatic labeling coincided with the manual labeling, for each relation. Note that the average accuracy ranged from 90 % for http://dbpedia.org/ontology/isPartOf to 68 % for http://dbpedia.org/ontology/hometown.

5   Conclusions

In this paper, we introduced a feature defined by ontology class hierarchies to improve relation extraction methods based on the distant supervision approach.

To demonstrate the effectiveness of class-based features, we presented experiments involving articles in the English Wikipedia and triples from DBpedia. We first heuristically labeled a corpus of sentences with relations, using the distant supervision method. We then used the class-based features, combined with common lexical features adopted for relation extraction, to train a multi-class perceptron. The held-out experiments demonstrated a substantial gain in how many relations could be identified (with an F-measure greater than 70 %), when the class-based features are adopted. We also conducted a human evaluation experiment to further assess the accuracy of the perceptron.

As future work, we plan to explore how sensitive the perceptrons are to the choice of the classes that annotate a sentence and define our semantic feature. Also, we intend to extend the feature vector extracted from sentences by adding more lexical features, such as dependencies path. Finally, we intend to improve the annotation of self-links (match between the article text and its title) by using co-reference resolution, synonyms, pronouns, etc.

Acknowledgments. This work was partly funded by CNPq, under grants 312138/2013-0 and 303332/2013-1, and by FAPERJ, under grant E-26/201.337 /2014.

References

1. Assis, P.H.R.: Distant supervision for relation extraction using ontology class hierarchy-based features. Master's thesis, Pontifícia Universidade Católica do Rio de Janeiro (2014)
2. Assis, P.H.R., Casanova, M.: Distant supervision for relation extraction using ontology class hierarchy-based features. In: Poster and Demo Track of the 11th Extended Semantic Web Conference (2014)
3. Auer, S., Bizer, C., Kobilarov, G., Lehmann, J., Cyganiak, R., Ives, Z.G.: DBpedia: a nucleus for a web of open data. In: Aberer, K., Choi, K.-S., Noy, N., Allemang, D., Lee, K.-I., Nixon, L.J.B., Golbeck, J., Mika, P., Maynard, D., Mizoguchi, R., Schreiber, G., Cudré-Mauroux, P. (eds.) ASWC 2007 and ISWC 2007. LNCS, vol. 4825, pp. 722–735. Springer, Heidelberg (2007)
4. Craven, M., Kumlien, J.: Constructing biological knowledge bases by extracting information from text sources. In: Proceedings of the 7th International Conference on Intelligent Systems for Molecular Biology (1999)
5. Finkel, J.R., Grenager, T., Manning, C.: Incorporating non-local information into information extraction systems by gibbs sampling. In: Proceedings of the 43rd Annual Meeting on Association for Computational Linguistics, pp. 363–370 (2005)
6. Gerber, D., Ngonga Ngomo, A.C.: Bootstrapping the linked data web. In: Proceedings of the 1st Workshop on Web Scale Knowledge Extraction, ISWC 2011 (2011)
7. Gillick, D.: Sentence boundary detection and the problem with the U.S. In: Proceedings of Human Language Technologies: The 2009 Annual Conference of the North American Chapter of the Association for Computational Linguistics (Short Papers), pp. 241–244 (2009)
8. McCullagh, P., Nelder, J.A.: Generalized Linear Models (1989)
9. Mintz, M., Bills, S., Snow, R., Jurafsky, D.: Distant supervision for relation extraction without labeled data. In: Proceedings of the Joint Conference of the 47th Annual Meeting of the ACL and the 4th International Joint Conference on Natural Language Processing of the AFNLP: vol. 2, pp. 1003–1011. ACL (2009)

10. Nguyen, T.D., yen Kan, M.: Keyphrase extraction in scientific publications. In: Proceedings of International Conference on Asian Digital Libraries, pp. 317–326 (2007)
11. Soderland, S.: Learning information extraction rules for semi-structured and free text. Mach. Learn. **34**, 233–272 (1999)
12. Toutanova, K., Manning, C.D.: Enriching the knowledge sources used in a maximum entropy part-of-speech tagger. In: Proceedings of the Joint SIGDAT Conference on Empirical Methods in Natural Language Processing and Very Large Corpora, pp. 63–70 (2000)
13. Wu, F., Weld, D.S.: Autonomously semantifying wikipedia. In: Proceedings of the 16th ACM Conference on Information and Knowledge Management, pp. 41–50 (2007)
14. Wu, F., Weld, D.S.: Automatically refining the wikipedia infobox ontology. In: Proceedings of the 17th International World Wide Web Conference (2008)

Crisis Mapping During Natural Disasters via Text Analysis of Social Media Messages

Stefano Cresci[1,2](✉), Andrea Cimino[3],
Felice Dell'Orletta[3], and Maurizio Tesconi[2]

[1] Bell Labs, Alcatel-Lucent, Paris, France
[2] Institute for Informatics and Telematics, IIT-CNR, Pisa, Italy
{stefano.cresci,maurizio.tesconi}@iit.cnr.it
[3] Institute for Computational Linguistics, ILC-CNR, Pisa, Italy
{andrea.cimino,felice.dellorletta}@ilc.cnr.it

Abstract. Recent disasters demonstrated the central role of social media during emergencies thus motivating the exploitation of such data for crisis mapping. We propose a crisis mapping system that addresses limitations of current state-of-the-art approaches by analyzing the textual content of disaster reports from a twofold perspective. A damage detection component employs a SVM classifier to detect mentions of damage among emergency reports. A novel geoparsing technique is proposed and used to perform message geolocation. We report on a case study to show how the information extracted through damage detection and message geolocation can be combined to produce accurate crisis maps. Our crisis maps clearly detect both highly and lightly damaged areas, thus opening up the possibility to prioritize rescue efforts where they are most needed.

Keywords: Twitter · Social media mining · Emergency management · Crisis mapping · Geoparsing

1 Introduction

Nowadays, a large number of people turns to social media in the aftermath of disasters to seek and publish critical and up to date information [13]. This emerging role of social media as a privileged channel for live information is favored by the pervasive diffusion of mobile devices, often equipped with advanced sensing and communication capabilities [20]. Recently, decision makers and emergency responders have envisioned innovative approaches to exploit the information shared on social media during disasters such as earthquakes and floods [16]. However, such information is often unstructured, heterogeneous and fragmented over a large number of messages in such a way that it cannot be directly used. It is therefore mandatory to turn that messy data into a number of clear and concise messages for emergency responders. Academia showed great interest for such an issue [14], setting up studies and developing experimental solutions along

© Springer International Publishing Switzerland 2015
J. Wang et al. (Eds.): WISE 2015, Part II, LNCS 9419, pp. 250–258, 2015.
DOI: 10.1007/978-3-319-26187-4_21

several research directions, such as emergency event detection [2,19], situational awareness [6] and crisis mapping [5,18]. Among these, the task of crisis mapping is of the utmost importance [12], as demonstrated during recent disasters such as the Tōhoku earthquake and tsunami (Japan – 2011), the Emilia earthquake (Italy – 2012) and the Hurricane Sandy (US – 2012).

In order to produce crisis maps, traditional systems only rely on geotag metadata of social media messages [18]. However, statistics report that only 1 % to 4 % of all social media messages natively carry geotag metadata [7]. This limitation drastically reduces the number of useful messages and results in very sparse maps. Recent work have instead demonstrated that emergency reports frequently carry textual references to locations and places [3,4], as shown in Fig. 1. Therefore, a fundamental challenge of novel crisis mapping systems is that of *geoparsing* the textual content of emergency reports to extract mentions to places/locations thus increasing the number of messages to exploit. Geoparsing involves binding a textual document to a likely geographic location which is mentioned in the document itself. State-of-the-art systems, such as [18], perform the geoparsing task by resorting to a number of preloaded geographic resources containing all the possible matches between a set of place names (toponyms) and their geographic coordinates. This approach requires an offline phase where the system is specifically set to work in a geographically-limited region. Indeed, it would be practically infeasible to load associations between toponyms and coordinates for a wide region or for a whole country. Moreover, all crisis mapping systems detect the most stricken areas by considering the number of messages shared and by following the assumption that more emergency reports equals to more damage [18]. Although this relation exists when considering densely and uniformly populated areas [15], it becomes gradually weaker when considering wider regions or rural areas.

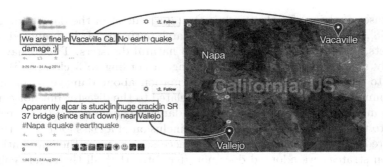

Fig. 1. Tweets shared in the aftermath of the 6.0 magnitude earthquake occurred in the South Napa region, California, US – August 24, 2014. These messages convey both situation assessments (green and red colored) and position information (blue colored) (Color figure online).

Contributions. Our proposed system exploits both situation assessments and position information contained in Twitter emergency reports. Figure 1 shows an example of Twitter emergency reports, highlighting the pieces of information we aim to extract. Overall, the main contributions of this work are summarized as in the following:

- We train and validate a machine learning classifier to detect messages conveying information about damage to infrastructures or communities. The classifier exploits a wide set of linguistic features qualifying the lexical and grammatical structure of a text. To our knowledge this is the first work employing a damage detection component in a crisis mapping task.
- We propose a novel geoparsing technique which exploits semantic annotation tools. By resorting to the Wikipedia and DBpedia collaborative knowledge-bases, it potentially allows to geocode messages from all over the world, thus overcoming the restriction of working on a specific geographic area. The semantic annotation process also alleviates the problem of toponymic polysemy (the word "Washington" may refer to the first US president, to the US capital, to the US state, etc.) by disambiguating the textual content of emergency reports. We propose and validate 2 implementations of this technique which respectively exploit TagMe [10] and DBpedia Spotlight [17].
- We leverage information visualization techniques and combine message geolocation and damage detection to produce crisis maps. We exploit D3.js[1] to build interactive, Web-based visualizations where geographic regions are colored according to the likelihood of damage. Our crisis maps can be easily embedded into Web emergency management systems, such as [3].
- We investigate a real case study to demonstrate the effectiveness of our system in detecting both highly and lightly damaged areas.

2   The Text Analysis System

The dataset exploited for this work is an improvement of the dataset originally used in [8], which is freely available for research purposes[2]. It is composed of Italian tweets, collected in the aftermath of 3 natural disasters. Tweets have been manually annotated for mentions of damage according to 3 classes: (i) tweets related to the disaster and carrying information about damage to infrastructures/communities (*damage*); (ii) tweets related to the disaster but not carrying relevant information for the assessment of damage (*no damage*); (iii) tweets not related to the disaster (*not relevant*). The inclusion of a class for tweets that are not related to a disaster (*not relevant*) is necessary because the automatic data collection strategy we adopted does not guarantee that all the tweets collected are actually related to the disaster under investigation.

Damage Detection. The goal of this component is that of automatically detecting mentions of damage in tweets. To perform this task we trained a machine learning classifier operating on morpho-syntactically tagged and dependency parsed

---

[1] http://d3js.org/.
[2] http://socialsensing.eu/datasets.

Table 1. Results of the damage detection task.

Dataset	Accuracy	damage			no damage			not relevant		
		Prec.	Rec.	F-M.	Prec.	Rec.	F-M.	Prec.	Rec.	F-M.
L'Aquila	0.83	0.92	0.87	0.89	0.81	0.87	0.84	0.77	0.71	0.74
Emilia	0.82	0.91	0.88	0.90	0.85	0.89	0.87	0.54	0.46	0.50
Sardegna	0.78	0.86	0.93	0.89	0.50	0.46	0.48	0.31	0.14	0.19

texts. Given a set of features and a learning corpus (i.e. the annotated dataset), the classifier trains a statistical model using the feature statistics extracted from the corpus. This trained model is then employed in the classification of unseen tweets and, for each tweet, it assigns the probability of belonging to a class: *damage, no damage, not relevant*. Our classifier exploits linear Support Vector Machines (SVM) using LIBSVM as the machine learning algorithm. Since our approach relies on multi-level linguistic analysis, both training and test data were automatically morpho-syntactically tagged by the POS tagger described in [9] and dependency-parsed by the DeSR parser using Multi-Layer Perceptron as the learning algorithm [1].

We focused on a wide set of features ranging across different levels of linguistic description [8]. The whole set of features is organized into 5 categories: *raw and lexical text features, morpho-syntactic features, syntactic features, lexical expansion features* and *sentiment analysis features*. This partition closely follows the different levels of linguistic analysis automatically carried out on the text being evaluated, (i.e. tokenization, lemmatization, morpho-syntactic tagging and dependency parsing) and the use of external lexical resources.

As shown in Table 1, we devised 3 experiments to test the performance of the damage detection component, one for each disaster covered by our dataset. Table 1 shows that the system achieved a good global accuracy for damage detection, ranging from 0.78 (Sardegna) to 0.83 (L'Aquila). Particularly interesting for this work are the scores obtained in the classification of the *damage* class. The F-Measure score for this class is always higher than 0.89 thus showing that the damage detection component is accurate enough to be integrated in a crisis mapping system.

Message Geolocation. Our proposed geoparsing technique builds on readily available semantic annotation tools and collaborative knowledge-bases. Semantic annotation is a process aimed at augmenting a plain-text with pertinent references to resources contained in knowledge-bases such as Wikipedia and DBpedia. The result of this process is an enriched (annotated) text where mentions of knowledge-bases entities have been linked to the corresponding Wikipedia/ DBpedia resource. Here, we aim to exploit semantic annotations for our geoparsing task by checking whether knowledge-bases entities, which have been linked to our tweet disaster reports, are actually places or locations.

We implemented our geoparsing technique with 2 state-of-the-art semantic annotation systems, namely TagMe [10] and DBpedia Spotlight [17]. However, it is worth noting that our proposed geoparsing technique does not depend on the annotators exploited in our prototypical implementations. Indeed it can be implemented with any annotator currently available, or with a combination of them. Thus, for each tweet we query the semantic annotators and we analyze the returned annotated texts. Such annotated texts come with the ID/name of the linked Wikipedia/DBpedia pages. Semantic annotation systems also provide a confidence score for every annotation. Higher confidence values mean annotations which are more likely to be correct. Thus, after annotating a tweet, we resort to Wikipedia/DBpedia crawlers in order to fetch information about all the entities associated to the annotated tweet. In our implementation we sort all the annotations on a tweet in descending order according to their confidence value, so that annotations which are more likely to be correct are processed first. We then fetch information from Wikipedia/DBpedia for every annotation and check whether it is a place or location. The check for places/locations can be simply achieved by checking for *coordinates* fields among entity metadata. We stop processing annotations when we find the first Wikipedia/DBpedia entity which is related to a place or location and we geolocate the tweet with the coordinates of that entity.

Table 2. Results of the message geolocation task.

	Precision	Recall	Accuracy	F-Measure	MCC
TagMe [10]	0.88	0.80	0.86	0.84	0.72
DBpedia Spotlight [17]	0.85	0.51	0.74	0.64	0.49

Then, following the approach used in [11,18], we manually annotated a random subsample of tweets to validate the geoparsing operation. Table 2 shows the results of the validation phase in terms of well-known metrics of information retrieval. Noticeably, the TagMe implementation achieves results comparable to those of the best-of-breed geoparsers with an F-Measure = 0.84, whether the systems described in [11,18] scored in the region of 0.80. Although exhibiting encouraging Precision, the DBpedia Spotlight implementation has a much lower Recall value (0.51 vs 0.80 of TagMe) which results in degraded performances in terms of Accuracy, F-Measure and Mathews Correlation Coefficient (MCC).

3   Case Study

In this section we validate our crisis mapping system on a real case study by combining information extracted by the damage detection and the message geolocation components. The 5.9 magnitude earthquake that struck Northern Italy on May the 20th 2012, is among the strongest in recent Italian history[3]. The shaking was clearly perceived in all Central and Northern Italy and caused 7 deaths and severe damage to the villages of the epicentral area[4]. The epicenter was located near the village of Finale Emilia in a rural and sparsely populated area. This represents an added challenge for the crisis mapping task since most of the tweets came from the big urban centers in Northern Italy, such as Milan, Venice and Genoa.

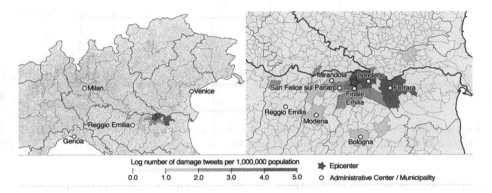

Fig. 2. Choropleth map for the Emilia 2012 earthquake showing the distribution of damage tweets among the municipalities of Northern Italy (Color figure online).

Figure 2 shows a choropleth map of Northern Italy where municipalities are colored so as to represent mentions of damage among tweet reports. Specifically, a color is assigned to a municipality according to the log number of tweets of the *damage* class geolocated in that municipality, normalized by its population. Data about the population of municipalities has been automatically fetched from Wikipedia and DBpedia during the crawling and query operations described in Sect. 2. The normalization allows to highlight the risk of damage also for rural and sparsely populated municipalities, where the number of available tweets is very low. Areas in which our system did not geolocate any *damage* tweet are grey colored in Fig. 2. As shown, despite geolocating tweets in all Northern Italy, our system only highlighted municipalities around the epicenter, clearly pointing to the damaged area. In figure, reddish colors are assigned to the municipalities of Bondeno, Ferrara, Finale Emilia and San Felice sul Panaro, thus accurately matching the most damaged locations.

Figure 3 shows 2 polar (radar) plots in which tweets of both the *damage* and *no damage* classes are considered. The plots are centered on the epicenter and present dotted concentric circles marking distances from the epicenter with a

---

[3] http://en.wikipedia.org/wiki/2012_Northern_Italy_earthquakes.

[4] http://www.reuters.com/article/2012/05/20/us-quake-italy-idUSBRE84J01K20120520.

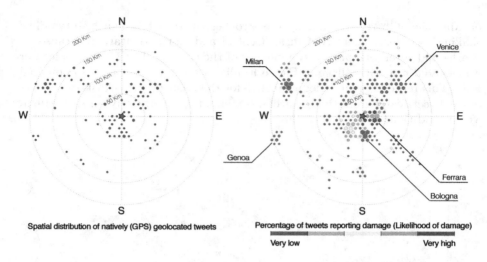

Fig. 3. Polar plots showing the hexbinned spatial distribution of tweets in Northern Italy for the Emilia 2012 earthquake (Color figure online).

step of 50 km. To avoid overplotting, we used an hexagonal binning (hexbinning) technique: the plot's surface is divided into hexagons whose area is proportional to the number of tweets geolocated in that space. Thus, areas with more tweets have bigger hexagons (e.g.: Milan, Bologna). On the left-hand side of Fig. 3 is a polar plot showing the spatial distribution of natively (GPS) geolocated tweets. The low number of natively geolocated tweets is represented by both the sparsity and the small size of the hexagons. This was the starting situation for our work, with a small number of geolocated messages and no damage label. Instead, on the right-hand side of Fig. 3 is the polar plot produced by our system in which hexagons are colored according the ratio of $\frac{damage}{damage+no\ damage}$ tweets. Big urban centers, such as Milan, Venice and Genoa, are represented by clusters of big sized hexagons reflecting the high number of tweets geolocated in those areas. However, despite the number of tweets, only the areas that actually suffered damage are yellow- and red-colored.

4   Conclusions and Future Work

In this work we proposed a novel crisis mapping system that overcomes the main limitations of current state-of-the-art solutions while still producing accurate maps. By introducing damage detection and by proposing a novel geoparsing technique, we are confident that our system represents a seminal work in the research field of crisis mapping. As such, possible directions for future work are manifold. The linguistic analysis carried out as part of the damage detection task could be brought to a deeper level. In fact, the damage detection classifier could be trained to detect the *object* that suffered the damage and such information could be used to enrich the maps. Results for the message geolocation task are already promising, but the validation process showed that there is still space for considerable improvements, especially regarding the Recall metric. Recent developments of semantic annotation tools open up the possibility to provide more implementations of our proposed geoparsing technique. Therefore we envision the possibility to simultaneously exploit multiple semantic annotators in a voting system. Finally, we are currently working to embed the hereby discussed system into the Web emergency management system described in [3].

Acknowledgements. The authors would like to thank Matteo Abrate and Salvatore Rinzivillo, for their insightful suggestions about data visualization. This research was supported by the .it domain registration authority (Registro .it) funded project SoS - Social Sensing (http://socialsensing.it/en).

References

1. Attardi, G., Dell'Orletta, F., Simi, M., Turian, J.: Accurate dependency parsing with a stacked multilayer perceptron. In: Evalita 2009 (2009)
2. Avvenuti, M., Cresci, S., La Polla, M., Marchetti, A., Tesconi, M.: Earthquake emergency management by social sensing. In: PERCOM 2014 Workshops. IEEE (2014)
3. Avvenuti, M., Cresci, S., Marchetti, A., Meletti, C., Tesconi, M.: EARS (earthquake alert and report system): a real time decision support system for earthquake crisis management. In: KDD 2014. ACM (2014)
4. Avvenuti, M., Del Vigna, F., Cresci, S., Marchetti, A., Tesconi, M.: Pulling information from social media in the aftermath of unpredictable disasters. In: ICT-DM 2015. IEEE (2015)
5. Birregah, B., Top, T., Perez, C., Châtelet, E., Matta, N., Lemercier, M., Snoussi, H.: Multi-layer crisis mapping: a social media-based approach. In: WETICE 2012. IEEE (2012)
6. Cameron, M.A., Power, R., Robinson, B., Yin, J.: Emergency situation awareness from twitter for crisis management. In: WWW 2012 Companion. ACM (2012)
7. Cheng, Z., Caverlee, J., Lee, K.: You are where you tweet: a content-based approach to geo-locating twitter users. In: CIKM 2010. ACM (2010)
8. Cresci, S., Tesconi, M., Cimino, A., Dell'Orletta, F.: A linguistically-driven approach to cross-event damage assessment of natural disasters from social media messages. In: WWW 2015 Companion. ACM (2015)
9. Dell'Orletta, F.: Ensemble system for part-of-speech tagging. In: Evalita 2009 (2009)
10. Ferragina, P., Scaiella, U.: Tagme: on-the-fly annotation of short text fragments (by wikipedia entities). In: CIKM 2010. ACM (2010)
11. Gelernter, J., Balaji, S.: An algorithm for local geoparsing of microtext. GeoInformatica **17**(4), 635–667 (2013)
12. Goolsby, R.: Social media as crisis platform: the future of community maps/crisis maps. ACM Trans. Intell. Syst. Technol. **1**(1), 1–11 (2010)
13. Hughes, A.L., Palen, L.: Twitter adoption and use in mass convergence and emergency events. Int. J. Emerg. Manag. **6**(3–4), 248–260 (2009)
14. Imran, M., Castillo, C., Diaz, F., Vieweg, S.: Processing social media messages in mass emergency: a survey. ACM Comput. Surv. **47**(4), 67 (2015)

15. Liang, Y., Caverlee, J., Mander, J.: Text vs. images: on the viability of social media to assess earthquake damage. In WWW 2013 Companion. ACM (2013)
16. Meier, P.: New information technologies and their impact on the humanitarian sector. Int. Rev. Red Cross **93**, 1239–1263 (2011)
17. Mendes, P.N., Jakob, M., García-Silva, A., Bizer, C.: Dbpedia spotlight: shedding-light on the web of documents. In: Semantics 2011. ACM (2011)
18. Middleton, S.E., Middleton, L., Modafferi, S.: Realtime crisis mapping of natural disasters using social media. IEEE Intell. Syst. **29**(2), 9–17 (2014)
19. Sakaki, T., Okazaki, M., Matsuo, Y.: Earthquake shakes twitter users: real-time event detection by social sensors. In: WWW 2010. ACM (2010)
20. Mora Segura, Á., de Lara, J., Sánchez Cuadrado, J.: Rapid development of interactive applications based on online social networks. In: Benatallah, B., Bestavros, A., Manolopoulos, Y., Vakali, A., Zhang, Y. (eds.) WISE 2014, Part II. LNCS, vol. 8787, pp. 505–520. Springer, Heidelberg (2014)

Tweet Location Inference
Based on Contents and Temporal Association

Saki Ueda[✉], Yuto Yamaguchi, Hiroyuki Kitagawa, and Toshiyuki Amagasa

University of Tsukuba, 1-1-1 Tennodai, Tsukuba, Ibaraki, Japan
{braose,yuto}@kde.cs.tsukuba.ac.jp, {kitagawa,amagasa}@cs.tsukuba.ac.jp

Abstract. How can we infer a tweet location? Are timestamps of tweets effective for the location inference? In this study, we propose a novel method for tweet location inference based on contents and timestamps of tweets. It is important to infer the locations of tweets for the services related to locations such as recommending restaurants, sending disaster-related information to users, and providing commercial messages to users. This study has two contributions: (1) we propose a novel method to infer tweet locations based on the contents and timestamps of tweets, andbreak (2) we experimentally demonstrate the effectiveness of the proposed method using Twitter data. The experimental results suggest that the proposed method can infer tweet locations more precisely than a baseline that does not take the temporal association into account.

Keywords: Location inference · Twitter

1   Introduction

How can we infer a tweet location? Are timestamps of tweets effective for the location inference? In this study, we propose a method for tweet location inference based on contents and timestamps of tweets. Due to the development of the social media, many people are publishing various information on the social media services. Twitter[1] is one of the most successful social media services, and mainly provides short message (tweet) service. Users can access Twitter easily from devises such as smart phones. Hence, many users frequently send tweets from places they are visiting and mention the topics about those places. Therefore, many tweets include timely information such as their locations and events around them.

Users can assign geotags, which are pairs of latitude and longitude, to tweets to explicitly show the places where they are. By using geotag information, we can provide services related to locations such as recommending restaurants, sending disaster-related information, and providing commercial messages to users. However, in reality, a small portion of tweets have geotag information for reasons such as privacy concerns. According to Cheng et al. [1], less than 0.42 % of all

---

[1] https://twitter.com/.

© Springer International Publishing Switzerland 2015
J. Wang et al. (Eds.): WISE 2015, Part II, LNCS 9419, pp. 259–266, 2015.
DOI: 10.1007/978-3-319-26187-4_22

tweets have their geotags. Therefore, to make the above services possible, it is important to infer tweet locations.

Current Work: In this study, we propose a novel method to infer tweet locations based on contents and *timestamps* of tweets. The proposed method infer the locations by *propagating* location information from tweets to tweets on the *similarity graph* that is constructed from the content similarity and the temporal similarity. We put the following two assumptions: (1) If contents of two tweets are similar in terms of their words, locations of those two tweets are likely to be close to each other. (2) When a user publishes two tweets in a short time, locations of these two tweets are likely to be close to each other. Based on these two assumptions, our method calculates two similarities, and then construct the similarity graph.

Contributions: The contributions of this work are summarized as follows:break (1) *Novel Method*: We propose a novel method for tweet location inference, which is based on the contents and timestamps of tweets. (2) *Experiments*: We experimentally demonstrate the effectiveness of the proposed method using Twitter data. Concretely, the experimental results suggest that the timestamps are important clues to infer tweet locations. The proposed method can infer tweet locations more precisely than a baseline that dose not take the temporal association into account.

2   Background

In this section, we briefly overview the related work in the areas of user location inference and the tweets location inference in Sect. 2.1. In Sect. 2.2, we explain label propagation.

2.1   Related Work

The problem of location inference in social media can be divided into two categories: (1) User location inference: inferring user's location of residence (2) Tweet location inference: inferring the location the tweet was posted. The problem we address in this paper is categorized in the latter.

User Location Inference: Many methods for user location inference has been proposed. These methods are divided into two types, where the content-based approaches (e.g., [1,2,7,8]) use the contents users posted, and the graph-based approaches (e.g., [3,9,10]) use the social graph of users. Cheng et al. [1] proposed a method of the content-based approach. They use local words to infer user location. Local words are the words with the deflection in the locations of the users who posted them. Chang et al. [2] proposed a method based on a model of words distributions using a Gaussian Mixture Model to infer user locations. Chang's method can extract local words without using training data. Yamaguchi et al. [7] proposed a method using local events detected in social streams to infer user locations. As for the graph-based approach, Backstrom et al. [3] and Jurgens

[10] proposed methods which focus on the friend relationship in social media. They assumed that a user's home location is close to his friends' home locations in social media. These studies mainly intended to infer users' home locations, while our method focuses on tweet location inference.

Tweet Location Inference: Kinsella et al. [4] proposed a method which create language models of locations from geotagged tweets and infer tweet locations by using these models. But this study differs from our proposed method because they only use the contents of tweets. Ikawa et al. [6] proposed a method which associates location information posted from location-based services with keywords by using contents of past users' tweets. This method uses the contents and timestamps of tweets like our method, but it is different from ours in that they only use the words as training data.

2.2   Label Propagation

Label propagation [5] (LP) is one of the semi-supervised learning methods. By learning from labeled and unlabeled data, the label propagation achieves high precision even when the number of labeled data items is small.

Given a set of n data points $X = \{x_1, x_2, ..., x_n\}$, LP calculates the label probability F_{ik} that data i has label k as follows: $F \leftarrow \lambda LF + (1 - \lambda)Y$. An $n \times c$ matrix F is the label probability matrix whose ik-th element is F_{ik}. An $n \times c$ matrix Y stores the explicit labels of nodes, where $Y_{ij} = 1$ if the node x_i has a label l_j and otherwise $Y_{ij} = 0$. $L = I - D^{-1/2}WD^{-1/2}$ is the normalized Laplacian matrix, where W is the similarity matrix whose ij-th element W_{ij} is the similarity between data i and j, D is a diagonal matrix whose diagonal component is the sum of the row of W, and I is an identity. λ is a parameter ($0 < \lambda < 1$). The above recursive calculation is guaranteed to converge. The predicted label of data i is given by $\text{argmax}_j F_{ij}$.

3   Problem Statement

This section defines some terminologies we use in this paper and state the problem of tweet location inference. Each tweet $p = (u, s, t, l)$ is composed of user u, timestamp s, text t and tweet location $l \in L$, where L is a set of predefined locations. Text t is represented by a bag of words w. If the tweet location is unknown, $l = NULL$. Let $P = P^L \cup P^U$ be a set of tweets, where P^L is a set of tweets whose locations are known, while P^U is a set of tweets whose locations are unknown. Using these terminologies, the tweet location inference problem is stated as follows:

Problem 1 (Tweet Location Inference). *Given a set of tweets P, inferbreak locations of tweets $p_i \in P^U$.*

We mention that we tackle the problem of inferring locations based solely on the contents and timestamps. However, the proposed method in this paper can be easily combined with location inference methods using other information such as user profiles and social graphs.

4   Proposed Method

Our proposed method infer locations of tweets based on their contents and timestamps. The intuition behind the proposed method is that locations of *similar* tweets are close to each other. To calculate the similarity, we put the following two assumptions: (1) If contents of two tweets are similar in terms of their words, locations of those two tweets are likely to be close to each other. (2) When a user publishes two tweets in a short time, locations of these two tweets are likely to be close to each other. The former is called *content similarity*, and the latter is called *temporal similarity*.

4.1   Constructing Feature Vectors

First, we extract feature words in geotaged tweets (i.e., training set) to generate feature vectors. Here we use spatially characteristic words as the features words. For example, names of places (e.g., New York and Seattle), or names of landmarks (e.g., Statue of Liberty) are regarded as feature words.

Let $P(X)$ be a probability distribution of all tweets over the predefined location set L and $Q_w(X)$ be a probability distribution of word w. We define feature words as follows:

Definition 1 (Feature words). *Feature words are the top r words that have the largest KL divergence $D_{KL}(Q_w \parallel P) = \sum_{l \in L} Q_w(l) log \frac{Q_w(l)}{P(l)}$.*

The intuition to use KL divergence to extract feature words is that spatially characteristic words have significantly different spatial distributions from the distribution of general words such as 'a' or 'the'. Note that we tested four types of feature words: extracting only place names by using dictionaries, mutual information, information gain, and KL divergence. By this preliminary investigation, we found that KL divergence is the most effective measure to extract feature words. We omit the results for brevity in this paper.

After determining r feature words, we make a feature vector for each tweet p. Feature vector $v(p)$ of tweet p is a r-dimensional vector where each element contains the tf-idf value of the corresponding word in tweet p.

4.2   Calculating Similarity

Content Similarity: We adopt the RBF kernel to calculate the content similarity. Letting γ be a parameter of RBF kernel, the content similarity S_{ij} between tweet p_i and p_j is calculated as follows: $S_{ij} = \exp(-\gamma \parallel v(p_i) - v(p_j) \parallel^2)$

Temporal Similarity: We calculate the temporal similarity assuming that when a user publishes two tweets in a short time, the locations of the two tweets are likely to be close to each other. In addition to the content similarity, we add the temporal similarity based on the above assumption. Letting θ be a parameter, the temporal similarity T_{ij} between tweets p_i and p_j *from the same user* is calculated as follows: $T_{ij} = h$ if $|s_i - s_j| < \theta$ and otherwise $T_{ij} = 0$ where

Fig. 1. *Calculating the temporal similarity*: $\theta = 60\,m$. The temporal similarity between nodes connected by an arrow is $T_{ij} = 1$ and otherwise $T_{ij} = 0$.

s_i is the timestamp of tweet p_i. Note that the temporal similarity between tweets from the different users is always 0.

Figure 1 shows four cases of calculating the temporal similarity. The temporal similarity between circles connected by an arrow is $T_{ij} = 1$ and otherwise $T_{ij} = 0$. In Case I, the temporal similarity is 1 because tweets are published by the same user and the timestamp interval is shorter than the parameter θ. In Cases II and IV, the temporal similarity is 0 because tweets are published by different users. In Case III, the temporal similarity is also 0 because the interval between two timestamps is larger than θ.

4.3   Location Inference

Based on two similarities calculated above, we construct a similarity graph as follows. First, we construct a k-nearest neighbor (kNN) graph using the content similarity, where k edges go from a node to the k most similar nodes. The intuition here is to make the graph *sparse* to efficiently propagate location information throughout the graph. Then we further add edges between nodes i and j if $T_{ij} = 1$. The weight matrix W of this similarity graph can be written as follows: $W = kNN(S) + \alpha T$, where α is a parameter to adjust the effect of S and T, and $kNN()$ converts the original S into the weight matrix of the kNN graph.

After constructing the merged similarity graph, our proposed method propagates the location information throughout the graph using the label propagation. Then, we get the label probability matrix F as explained in Sect. 2.2. The final estimate for the location $\hat{l}_i$ of tweet p_i is obtained as follows: $\hat{l}_i = \mathrm{argmax}_{l \in L} F_{il}$.

5   Experiments

We examined the effectiveness of the proposed method from the three perspectives:

1. Temporal Association: How much the temporal similarity is effective?
2. Content Association: How much the content similarity is effective?
3. Parameter: Is the parameter θ really effective?
4. Comparison: Do the proposed method outperform the baseline?

5.1   Setup

Dataset: We collected 3,526,922 geotagged Japanese tweets[2] through Twitter Streaming API[3] from September 2nd to November 20th, 2014. The number of users who posted these tweets is 259,420. From these users, we randomly selected 10,000 users who published more than 2 geotagged tweets, and then collected their latest 200 tweets. After that, users who published less than 10 geotagged tweets were selected from these 10,000 users. As a result, we obtained 296,865 tweets (including 7,752 geotagged tweets) from 1,652 users. Note that this procedure of dataset making is required because (1) the number of allowed API call is limited, and (2) we need multiple tweets published by the same user to validate the effectiveness of the temporal information. We divided these tweets into three sets: training set (0.4 %), validation set (0.1 %), and test set (99.5 %).

Settings: To extract nouns from the Japanese texts, we employ $MeCab^4$, a Japanese morphological analysis tool. We set $r = 500$ for feature extraction by KL divergence. We tuned parameters using the training set and the validation set. The parameters we tuned are γ in RBF kernel, k in kNN graph, α, and θ. Since the parameter space of combination is huge, we tuned these parameters as follows. We first tuned γ and k with values of $\alpha = 1$ and $\theta = 60$ min. And then we tuned α and θ with determined values of γ and k. In addition, we set parameter $\lambda = 0.99$ in label propagation according to the original paper [5]. We divided the map of target area (Japan) into 552 cells. The target area is the rectangle from 23 degree latitude and 122 degree longitude to 43 degree latitude and 146 degree longitude. Each cell is the one degree square rectangle.

5.2   Results

Temporal Association: How effective the temporal similarity is? In this experiment, we compared the proposed method and the one without the temporal similarity (i.e., using only the content similarity), which is equivalent to the special case where $\alpha = 0$. Figure 2(a) shows the results. The y-axis denotes the average accuracy rate of 5 repeated runnings, where the upper is the better, and the error bars are standard deviations. We can see that the proposed method achieves better result than the method which use only the content similarity, which demonstrates that the temporal similarity is effective for location inference.

Content Association: How effective the content similarity is? In this experiment, we compared the proposed method and the one without the content similarity (i.e., using only the temporal similarity), which is equivalent to the special case where $\alpha = \infty$. Parameters are $\theta = 60$ min for each method, and $\alpha = 1$ for the full proposed method. Figure 2(a) shows the results. The y-axis denotes the average accuracy rate of 5 repeated runnings, where the upper is

---

[2] Tweets posted from Japan.
[3] http://dev.twitter.com/streaming/overview.
[4] http://taku910.github.io/mecab/.

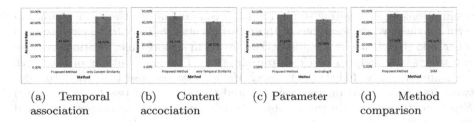

| (a)   Temporal | (b)   Content | (c) Parameter | (d)   Method |
| association | accociation | | comparison |

Fig. 2. *Results*: The x-axis denotes the each method and the y-axis denotes the average accuracy rate of 5 repeated runnings. The error bars show standard deviations.

the better, and the error bars are standard deviations. According this figure, the proposed method achieves better result than the method which use only the temporal similarity, which demonstrates the effectiveness of the content similarity. This result suggests that the contents of tweets are effective clue to infer tweet locations because the locations of two tweets including the same words are likely to be close each other.

Parameter: Is the parameter θ really effective? If users always publish tweets from the same location, we do not have to use the parameter θ. We resolved this question by this experiment. We compared the proposed method and the one which connected all nodes published by the same user in chronological order. The latter does not consider the content similarity and the parameter θ of the temporal similarity, assuming that tweets of the same user are published from the same location. Figure 2(c) shows the results. The y-axis denotes the average accuracy rate of 5 repeated runnings, where the upper is the better, and the error bars are standard deviations. According this figure, the proposed method achieves better result than the method which excludes the parameter θ, which demonstrates the effectiveness of the parameter θ. This result shows that the parameter θ is effective because tweets of the same user are published from different locations as time goes by.

Method Comparison: Do the proposed method outperform the baseline? We adopt a SVM with an RBF kernel as a baseline. Parameters γ of the RBF kernel and C, which is a cost parameter of misclassification, are tuned using the validation set. Note that the SVM uses only the content similarity. Figure 2(d) shows the results. The y-axis denotes the average accuracy rate of 5 repeated runnings, and the error bars show standard deviations. According to the results, the proposed method achieves higher accuracy rate than that of the baseline, suggesting that incorporating the temporal associations is effective. It is not easy for SVM to deal with the temporal information.

6   Conclusion

In this study, we proposed a novel method for tweet location inference, which is based on contents and timestamps of tweets. Our proposed method propagates the location information from tweets to tweets. We put the following two assumptions for calculating similarity: (a) if contents of two tweets are similar in terms of their words, locations of those two tweets are likely to be close to each other, and (b) when a user publishes two tweets in a short time, locations of these two tweets are likely to be close to each other. The contributions of this work are summarized as follows: (1) *Novel Method*: we propose a method to infer tweet locations based on the contents and timestamps of tweets (Sect. 4). (2) *Experiments*: we experimentally show the effectiveness of the proposed method using Twitter data (Sect. 5). Concretely, the experimental results suggest that the proposed method can infer tweet locations more precisely than the baseline that does not take the temporal association into account. Our future work includes, (1) improving the calculation method of the temporal similarity, and (2) examining the effectiveness of the proposed method in the case that the geographical range of labels change.

Acknowledgment. This research was partly supported by the program "Research and Development on Real World Big Data Integration and Analysis" of the Ministry of Education, Culture, Sports, Science and Technology, Japan.

References

1. Cheng, Z., Caverlee, J., Lee, K.: You are where you tweet: a content-based approach to geo-locating twitter users. In: Proceedings of the CIKM 2010, pp. 759–768 (2010)
2. Chang, H.-W., Lee, D., Eltaher, M., Lee, J.: @Phillies tweeting from philly? predicting twitter user locations with spatial word usage. In: Proceedings of the ASONAM 2012, pp. 111–118 (2012)
3. Backstrom, L., Sun, E., Marlow, C.: Find me if you can: improving geographical prediction with social and spatial proximity. In: Proceedings of the WWW 2010, pp. 61–70 (2010)
4. Kinsella, S., Murdock, V., O'Hare, N.: 'I'm eating a sandwich in glasgow': modeling locations with tweets. In: Proceedings of the SMUC 2011, pp. 61–68 (2011)
5. Zhou, D., Bousquet, O., Lal, T.N., Weston, J., Schölkopf, B.: Learning with local and global consistency. In: Proceeding of the NIPS 16, pp. 321–328 (2003)
6. Ikawa, Y., Enoki, M., Tatsubori, M.: Location inference using microblog messages. In: Proceedings of the WWW 2012, pp. 687–690 (2012)
7. Yamaguchi, Y., Amagasa, T., Kitagawa, H., Ikawa, Y.: Online user location inference exploiting spatiotemporal correlations in social streams. In: Proceedings of the CIKM 2014, pp. 1139–1148 (2014)
8. Eisenstein, J., O'Connor, B., Smith, N.A., Xing, E.P.: A latent variable model for geographic lexical variation. In: Proceedings of th EMNLP 2010, pp. 1277–1287 (2010)
9. Abrol, S., Khan, L.: Tweethood: agglomerative clustering on fuzzy k-closest friends with variable depth for location mining. In: Proceedings of the IEEE Second International Conference on Social Computing (SocialCom 2010), pp. 153–160 (2010)
10. Jurgens, D.: That's what friends are for: inferring location in online social media platforms based on social relationships. In: Proceedings of the Seventh International AAAI Conference on Weblogs and Social Media, pp. 273–282 (2013)

Finding Influential Users and Popular Contents on Twitter

Zhaoyun Ding[⊠], Hui Wang, Liang Guo, Fengcai Qiao, Jianping Cao, and Dayong Shen

College of Information System and Management, National University of Defense Technology, Changsha 410073, People's Republic of China
{zyding,huiwang,guoliang,fcqiao,jpcao,dyshen}@nudt.edu.cn

Abstract. On Twitter, People do not only find new friends by following others, but also propagation the information by retweeting. So, we can not measure the users' influence only by following relationships easily, also, it is not reasonable to measure tweets' popularity by the number of retweets. In this paper, a novel random walk model was proposed to measure the users' influence and tweets' popularity. In our model, the influence of users was measured not only by random walk of the following network, but also by the popularity of tweets. In fact, if a user often tweets popular contents firstly, we think this user is important and the influence of the user is higher. Moreover, if a content is retweeted by many high influencers, we think this content is important and popular. Experiments were conducted on a real dataset from Twitter containing about 0.26 million users and 10 million tweets, and results show that our method is consistently better than PageRank method with the network of following and the method of retweetNum which measures the popularity of contents according to the number of retweets.

Keywords: Influence · Popularity · Random walk model · Twitter

1 Introduction

Microblogs have rapidly become significant means for people to communicate with the world and each other. Unlike other social network services, the relationship of following between users can be unidirectional; a user is allowed to choose who she wants to follow without seeking any permission. Twitter employs a social network called following relationship; the user whose updates are being followed is called the friend, while the one who is following is called the follower. In case the author is not protecting his tweets, they appear in the so-called public timeline and his followers will receives all messages from him.

On Twitter, People do not only find new friends by following others, but also propagation the information by retweeting. Studies by Kwak et al. [1] have

D. Zhaoyun—This work was supported by National Natural Science Foundation of China (No. 71331008).

© Springer International Publishing Switzerland 2015
J. Wang et al. (Eds.): WISE 2015, Part II, LNCS 9419, pp. 267–275, 2015.
DOI: 10.1007/978-3-319-26187-4_23

shown that Twitter is more likely to be a news media. Users can not only find friends by microblogs, but also can publish or forward messages by themselves, and microblogs transform people from content consumers into content producers and proliferators. So, the influence of a user on Twitter should be measured by the relationship of following and the spreadability of contents comprehensively.

In this paper, a novel random walk model was proposed to measure the users' influence and tweets' popularity. In our model, the influence of users was measured not only by random walk of the following network, but also by the popularity of tweets. In fact, if a user often tweets popular contents firstly, we think this user is important and the influence of the user is higher. Moreover, if a content is retweeted by many high influencers, we think this content is important and popular. Figure 1 gives an example of the influence model.

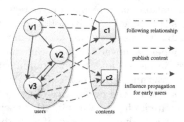

Fig. 1. An example of the influence model.

2   Related Work

With the popularity of Microblogs, there have been lots of studies about Microblogs, focusing on the influence of users and the popularity of contents. Most of the previous researches about the influence in Microblogs were based on the relationship of following or retweeting. Tunkelang [2] originally proposed a method analog to PageRank later named as TunkRank in order to measure the influence of users in microblogs. Weng et al. [3] proposed an algorithm called TwitterRank to measure the influence taking both the topical similarity between users and the link structure into account. Cha et al. [4] presented an in-depth comparison of three measures of influence: indegree, retweets, and mentions. Lee et al. [5] found influential individuals based on the temporal order of information adoption in Twitter. Pal and Counts [6] categorized tweets into three categories: Original tweet (OT), Conversational tweet (CT), Repeated tweet (RT) to iden-tify topical authorities in microblogs. Bakshy et al. [7] referred narrowly the influencer as the ability to consistently seed cascades that spread further than others, in which seed nodes had higher influence. M. Romero et al. [8] proposed an algorithm that determined the influence and passivity of users based on their information forwarding activity.

Yang and Counts [9] constructed a novel model to capture the three major properties of information diffusion — speed, scale, and range — by analyzing information diffusion on Twitter, via users' ongoing social interactions as denoted by mentions. The most similar work was proposed by Silva et al. [10]. They proposed ProfileRank, a new information diffusion model based on random walks over a user-content graph. However, the relationship of following was neglected in their work. So, the influence of users was not propagated from users to contents in the model of random walks.

3   Model

The following networks on Twitter were defined as $G = (V, E)$, where V was the set of users on Twitter, and E was the set of relationships of followings. Moreover, for the $G = (V, E)$, the adjacency matrix $G = (g_{ij})$ was defined.

$$g_{ij} = \begin{cases} 1 & ,if\ the\ user\ j\ follows\ the\ user\ i \\ 0 & others \end{cases} \tag{1}$$

A content can be propagated by many users. if a content is retweeted by many high influencers, we think this content is important and popular. So, the popularity of a content was decided by the influence of each user who tweets or retweets the content.

The bipartite graph was defined as $G_{fc} = (V, V_c, E_{fc})$, where V_c was the set of contents on Twitter, and E_{fc} was the set of edges where users publish contents. For the $G_{fc} = (V, V_c, E_{fc})$, the adjacency matrix $X = (x_{ij})$ was defined.

$$x_{ij} = \begin{cases} 1 & ,if\ the\ user\ j\ publishes\ the\ content\ i \\ 0 & others \end{cases} \tag{2}$$

Moreover, the popularity of contents can react to the influence of users. However, not all of users was reacted by the popularity of contents. In fact, we think this user who often tweets popular contents firstly is important and the influence of the user is higher. So, we think the influence only propagated from contents to starting users. Attentively, there is not only a starting user for a content usually. Because the same content was usually published by different users on different time sometimes.

The bipartite graph was defined as $G_{cf} = (V, V_c, E_{cf})$, where E_{cf} was the set of edges where contents react to starting users. Moreover, for the $G_{cf} = (V, V_c, E_{cf})$, the adjacency matrix $Y = (y_{ij})$ was defined as follows.

$$y_{ij} = \begin{cases} 1 & ,if\ the\ user\ i\ publishes\ the\ content\ j\ firstly \\ 0 & others \end{cases} \tag{3}$$

In this paper, the random walk model was constructed on multiple networks, including following graph and two bipartite graphs.

For the following graph $G = (V, E)$ and corresponding adjacency matrix $G = (g_{ij})$, we constructed transition probability matrix $P = \{p_{ij}\}$ as follows.

$$p_{ij} = \begin{cases} \frac{G(i,j)}{\sum_{v_k \in outlink[v_i]} G(i,k)} & outlink[v_i] \neq 0 \\ G(i,j) = 0 & otherwise \end{cases} \tag{4}$$

For the bipartite graph $G_{fc} = (V, V_c, E_{fc})$ and corresponding adjacency matrix $X = (x_{ij})$, we constructed transition probability matrix $M = \{M_{ij}\}$.

$$M_{ij} = \begin{cases} \frac{X(i,j)}{\sum_{v_k \in outlink[v_i]} X(i,k)} & outlink[v_i] \neq 0 \\ X(i,j) = 0 & otherwise \end{cases}. \tag{5}$$

At last, the popularity of contents will react to the influence of users. However, not all of users was reacted by the popularity of contents. In fact, we think this user who often tweets popular contents firstly is important and the influence of the user is higher. So, we think the influence only propagated from contents to starting users. Attentively, there is not only a starting user for a content usually. Because the same content was usually published by different users on different time sometimes. Figure 2 gives two common cascaded trees on Twitter. How to assign values of reacting to different starting users for each content? In fact, the earlier the content which was published by a starting user, the more important this user was. So, the importance of users became lower and lower as time goes on. In this paper, we think the importance of starting users decay with an exponential distribution with the parameter λ.

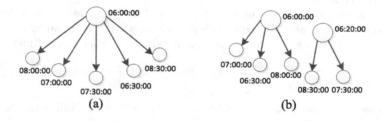

(a)   (b)

Fig. 2. Two common cascaded trees on Twitter.

For each content i, we consume the influence of the earliest user was equal 1. For others, the reduced value of influence was computed as follows according to the exponential distribution with the parameter λ.

$$\Delta I_i = (1 - e^{-\lambda t_{ij}}) - (1 - e^{-\lambda t_{i0}}). \tag{6}$$

The t_{i0} is the earliest time for a content i, and the t_{ij} is the time of a user j for the content i. So, exceptional for the earliest user for a content, the influence of others was as follows.

$$\begin{aligned} I_{ij} &= 1 - [(1 - e^{-\lambda t_{ij}}) - (1 - e^{-\lambda t_{i0}})] \\ &= 1 - (e^{-\lambda t_{i0}} - e^{-\lambda t_{ij}}) \end{aligned}. \tag{7}$$

For the bipartite graph $G_{cf} = (V, V_c, E_{cf})$ and corresponding adjacency matrix $Y = (y_{ij})$, we constructed transition probability matrix $L = \{L_{ij}\}$ as follows.

$$L_{ij} = \begin{cases} \frac{I_{ij}}{\sum_{v_k \in outlink[v_i]} I_{ik}} & outlink[v_i] \neq 0 \\ I_{ij} = 0 & otherwise \end{cases}. \tag{8}$$

Then, we will combine three transition probability matrixes to construct random walk models for multiple networks. For the transition probability matrix $P = \{p_{ij}\}$, the random walk model was constructed as follows.

$$\pi_{k+1} = \alpha P^T \pi_k + (1 - \alpha)\frac{1}{n}e, e = (1, 1, ..., 1)^T. \tag{9}$$

In the transition probability matrix $P = \{p_{ij}\}$, if a user was visited by more users, the influence of this user was higher. Also, if a user was visited by users whose influence were high, the influence of this user was higher.

For the transition probability matrix $M = \{M_{ij}\}$, the random walk model was constructed as follows.

$$\rho_{k+1} = \alpha M^T \pi_k + (1 - \alpha)\frac{1}{n}e, e = (1, 1, ..., 1)^T. \tag{10}$$

In the $M = \{M_{ij}\}$, if a content was published or retweeted by more users, the popularity of the content was higher. Also, if a content was published or retweeted by users whose influence were high, the popularity of this content was higher.

For the transition probability matrix $L = \{L_{ij}\}$, the random walk model was constructed as follows: $\pi_{k+1} = \alpha L^T \rho_k$.

In the transition probability matrix $L = \{L_{ij}\}$, if a starting user was reacted by more contents, the influence of the user was higher. Also, if a user was reacted by contents whose popularity were high, the influence of this user was higher.

We combine the above formulas to construct a new random walk model as follows.

$$\pi_{k+1} = \alpha P^T \pi_k + \beta L^T \rho_k + (1 - \alpha - \beta)\frac{1}{n}e, e = (1, 1, ..., 1)^T. \tag{11}$$

Random walk models were constructed by combining the above formulas.

$$\begin{cases} \pi_{k+1} = \alpha P^T \pi_k + \beta L^T \rho_k + (1 - \alpha - \beta)\frac{1}{n}e, e = (1, 1, ..., 1)^T \\ \rho_{k+1} = \alpha M^T \pi_k + (1 - \alpha)\frac{1}{n}e, e = (1, 1, ..., 1)^T \end{cases} \tag{12}$$

The formula 12 can be reconstructed as follows.

$$\pi = \alpha P^T \pi + \beta L^T M^T \pi + (1 - \alpha - \beta)\frac{1}{n}e, e = (1, 1, ..., 1)^T. \tag{13}$$

For n users in the following networks $G = (V, E)$ and m contents in the bipartite graph $G_{fc} = (V, V_c, E_{fc})$, P is a $n \times n$ matrix, L^T is a $n \times m$ matrix, and M^T is a $m \times n$ matrix. So, the $L^T M^T$ is a $n \times n$ matrix.

So, the formula 13 can be reconstructed as follows:$\pi = Q\pi$.

Where $Q = \alpha P^T + \beta L^T M^T + (1 - \alpha - \beta)\frac{1}{n}E$ and E is a $n \times n$ unit matrix. Clearly, Q is a stochastic $n \times n$ matrix that parameterizes the combined random walk. It is also easy to see that this Markov Chain is ergodic. Thus, the stationary probabilities can be found as, for any initial vector π_0.

4   Experimental Studies

4.1   Datasets

For the purpose of this study, a set of Twitter data about Chinese-based twitters who have published at least one Chinese tweet was prepared as follows. About 0.38 million users were collected through the API of Twitter. Moreover, we get tweets published from 2014-03-01 to 2014-04-30. About 1.5 million tweets published by only about 35 thousands were collected in these two months.

In order to test the effectiveness of our method, we extract 2000 contents and corresponding cascaded trees with timestamps which was similar to Fig. 2. Then, we constructed the bipartite graph $G_{fc} = (V, V_c, E_{fc})$ and the bipartite graph $G_{cf} = (V, V_c, E_{cf})$ respectively. Also, we collected about 12 million following relationships of these 0.38 million users and constructed following networks $G = (V, E)$.

4.2   Verifying the Convergence of Our Method

In order to verify the convergence of our method, we get the difference in value $\Delta\pi_i$ of succession each user's influence score. Figure 3 gives average difference in value $\Delta\bar{\pi}$ of all users for six experiments. Our method was named as *UserConRank*. Also, we compare with the convergence of PageRank.

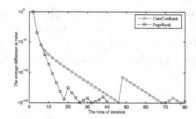

Fig. 3. Average difference in value $\Delta\bar{\pi}$ of all users for six experiments.

Experimental results showed our method was convergent after about 80 iterations, although there were some minor instabilities in the circulation process. On the other hand, the convergent rate of our method was slower than PageRank. Because random walk models were constructed by combining tree graphs in our method, the complexity of our method was higher than PageRank.

4.3   Verifying the Important of Contents

In this subsection, we get top-85 important contents respectively according to the number of retweets and our method. First, we verify how much difference for our method and the number of retweets according to the quantile-quantile plot. The quantile-quantile plot was described by the important ranking of contents for our method and the number of retweets. Figure 4 gave experimental results.

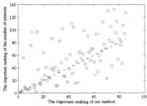
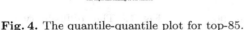

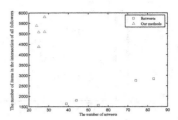

Fig. 4. The quantile-quantile plot for top-85.   **Fig. 5.** The number of retweets.

Experimental results showed most of important contents got by our method were similar to the method by the number of retweets. However, some contents which were retweeted by less users were more important. In fact, although some contents were retweeted by less users, some of these users had large number of followers who could read contents published by their friends. So, these contents were known by more users and they were more important. In order to verify the effectiveness of our method, we get 5 contents which were important by the number of retweets and were less important by our methods tagged as square in Fig. 5. Also, we get 5 contents which were important by our method and were less important by the number of retweets tagged as triangle in Fig. 5. We compared the number of retweets for each content and the number of items in the intersection of all followers for these users who retweeted the content.

Figure 5 gives experimental results. Although the number of retweets for our method was lower, the number of items in the intersection of all followers was higher and more users could read these contents. So, these contents were more important.

4.4 Verifying the Popularity of Users

In this subsection, we get top-100 influential users respectively according to PageRank and our method. First, we verify how much difference for our method and PageRank according to the quantile-quantile plot. The quantile-quantile plot was described by the influential ranking of users for our method and PageRank. Figure 6 gave experimental results.

Experimental results showed most of influential users got by our method were similar to the method by PageRank. However, some users were less influential than PageRank. In fact, although some users like the reality of celebrities had large number of followers, they published less contents which were not important. So, these users did not play an important role on information diffusion, and these users were less influential. In order to verify the effectiveness of our method, we get 10 users which were more influential by PageRank and were less influential by our methods. Also, we get 10 users which were less influential by PageRank and were more influential by our method. For each user, we count the diffusibility for 100 contents published by these users. The diffusibility was defined as how

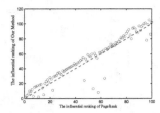
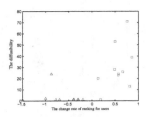

Fig. 6. The quantile-quantile plot for top-100.

Fig. 7. The diffusibility for 100 contents.

many users retweeting a content for one user. Moreover, we gave change rate of ranking for top-100 users as follows: $r_c = \frac{r_p - r_m}{r_p}$.

Where r_p was the ranking of PageRank, r_m was the ranking of our method, and r_c was the change rate of ranking for users.

Figure 7 gave experimental results. We found that the ranking of users could be enhanced since they played an important role on information diffusion and most of these users whose change rate of ranking were more than 0 had stronger diffusibility. In contrast, The ranking of users whose change rate of ranking were less than 0 could be reduced since these users were not almost involved in the diffusion of information and their diffusibility were lower.

5   Conclusions

A novel random walk model was proposed to measure the users' influence and tweets' popularity. In the future work, we will apply our algorithm to other dataset such as Facebook, DBLP, etc. Moreover, the influence of group is also an interesting work.

References

1. Kwak, H., Lee, C., Park, H., Moon, S.: What is twitter, a social network or a news media? In: Proceedings of the 19th International World Wide Web Conference (WWW 2010), pp. 591–600, Raleigh, USA, April 2010
2. Tunkelang, D.: A twitter analog to pagerank (2009)
3. Weng, J., Lim, E.P., Jiang, J., He, Q.: Twitterrank: finding topic-sensitive influential twitters. In: Proceedings of the 3th ACM International Conference on Web Search and Data Mining (WSDM 2010), pp. 261–270, New York, USA, February 2010
4. Cha, M., Haddadi, H., Benevenuto, F., Gummadi, K.P.: Measuring user influence in twitter: the million follower fallacy. In: Proceedings of the 4th International AAAI Conference on Weblogs and Social Media (ICWSM 2010), pp. 10–17, Washington, USA, May 2010

5. Lee, C., Kwak, H., Park, H., Moon, S.: Finding influentials based on the temporal order of information adoption in twitter. In: Proceedings of the 19th International Conference on World Wide Web (WWW-Poster 2010), pp. 1137–1138, Raleigh, USA, April 2010
6. Pal, A., Counts, S.: Identifying topical authorities in microblogs. In: Proceedings of the 4th ACM International Conference on Web Search and Data Mining (WSDM 2011), pp. 45–54, Hong Kong, February 2011
7. Bakshy, E., Hofman, J.M., Mason, W.A., Watts, D.J.: Everyone's an influencer: quantifying influence on twitter. In: Proceedings of the 4th ACM International Conference on Web Search and Data Mining (WSDM 2011), pp. 65–74, Hong Kong, February 2011
8. Romero, D.M., Galuba, W., Asur, S., Huberman, B.A.: Influence and passivity in social media. In: Proceedings of the 20th International Conference on World Wide Web (WWW-Poster 2011), pp. 113–114, Hyderabad, India, March 2011
9. Yang, J., Counts, S.: Predicting the speed, scale, and range of information diffusion in twitter. In: Proceedings of the 4th International AAAI Conference on Weblogs and Social Media (ICWSM 2010), pp. 355–358, Washington, USA, May 2010
10. Silva, A., Guimaraes, S., Zaki, M.: Profilerank: finding relevant content and influential users based on information diffusion. In: Proceedings of the 8th International Workshop on Social Network Mining and Analysis (SNAKDD 2013), pp. 11–20, Chicago, Illinois, USA, August 2013

Exploring Social Network Information for Solving Cold Start in Product Recommendation

Chaozhuo Li[(✉)], Fang Wang, Yang Yang, Zhoujun Li, and Xiaoming Zhang

State Key Laboratory of Software Development Environment,
Beihang University, Beijing, China
{lichaozhuo,fangwang,lizj}@buaa.edu.cn

Abstract. Cold start problem is a key challenge in recommendation system as new users are always present. Most of existing approaches address this problem by leveraging meta data to estimate the tastes of new user. Recently, social network has been becoming an integral part of daily life. Usually, social network information reflect users preferences to some extent, combining this kind of data would contribute to address the cold start problem. Existing approaches of this kind are either leverage relationships between users or utilize meta data such as demographic information. The huge textual information in social network has been neglected. In this paper, we propose a novel recommendation framework, in which the textual data in social network are used to improve the recommendation accuracy for new users. In particularly, both of new user's interests and items are modeled by mining the textual data in social network. Experimental results demonstrate that our approach is superior to other baseline methods in both precision and diversity.

Keywords: Recommendation system · Social network · Cold start

1 Introduction

In the recent decade, recommendation system has become an integral part of peoples lives, as a means to help users in information overload scenarios by proactively finding items or services on their behalf. Recommendation technology has been successfully applied in many e-commerce sites, such as Amazon.com, WalMart.com and Netflix.com. Existing work usually relies on user's historical item ratings, especially for the Collaborative Filtering (CF) based recommendation systems [12,15,16]. When it comes to new users without rating records, however, the performance of these strategies falls a great deal, which is known as the cold start problem and is one of the most challenging problems in recommendation systems.

Because of the prevalence of social network, many e-commerce systems allow users to provide their social network id (e.g., Twitter and Weibo). A large portion of costumers are pleasure to offer their social network URLs. For example, Douban website is a famous online bookstore in China, in which more than 23 percents of users has published social network URLs in their profiles. We argue

© Springer International Publishing Switzerland 2015
J. Wang et al. (Eds.): WISE 2015, Part II, LNCS 9419, pp. 276–283, 2015.
DOI: 10.1007/978-3-319-26187-4_24

that user's social network information can help to capture user's interests. For example, a microblog *"Captain Jack Sparrow in Pirates of the Caribbean"* implies the microbloger may be interested in fantasy-adventure films or books. Besides, many social network websites allow users to define their own tag of interests. This informations are also meaningful for product recommending. For example, social network users labeled with tag *"Comic"* are most likely to purchase comic related commodities.

In this paper, we design an efficient product recommendation system for handling the cold start problem, by leveraging users social network information. We run it on book domain as an example. The proposed techniques can also be adapted to recommending other kinds of products. Firstly, books with similar topics are aggregated into the same cluster which is used instead of specific books to represent user's reading interests. We propose an innovative approach to model the user's reading interests on book clusters by mining the textual data in social network. Specifically, a probability model used to mine user interest from user's tags and a classifier-based microblog mining method are proposed. We also construct item models based on book clusters using a matrix factorization model. Finally, the recommendations to new user can be made based on users interest model and the learned item models.

2 Recommendation System Using Social Network Textual Data

2.1 Problem Definition

In this paper, we aim to recommend bo oks to new users using their social network information. There are three major processes as below.

User interest modeling process can be represented as the constructing of a vector V: $V \in R^{1 \times p}$ for a single user, p is the count of book clusters. V_i indicates user's preference on book cluster i.

Item modeling process is represented as the constructing of a matrix I: $I \in R^{p \times n}$, n is the count of books. The item modeling process can be abstract into a mathematical expression: $S \approx UI$, where matrix $U : U \in R^{m \times p}$ stands for the interest models of m training users, S is the rating matrix. Given matrices U and S, we provide a matrix factorization model to build I by solving the following optimization problem:

$$\text{minimize} \ \frac{1}{2} \|S - UI\|_F^2 + \frac{\lambda}{2} \|I\|_F^{\,2} \tag{1}$$

After the construction of I, the final problem is how to make recommendations to new users with their interest models and I.

2.2 Overview of Our System

Figure 1 shows the detailed structure of our system. The major components are user interest modeling, item modeling and recommendation-making. In order to explain more clearly, framework is divided into two processes: training process and predicting process.

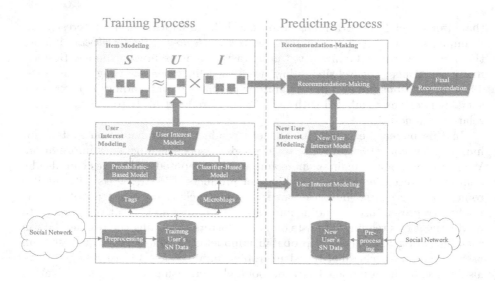

Fig. 1. Framework of our recommendation system contains three major parts: user interest modeling, item modeling and making recommendation. In order to explain more clearly, it is divided into two processes: training process and predicting process.

2.3   Crawling and Preprocessing

Data we used in this paper are crawled from two websites: Douban and Sina Weibo. Douban Website (http://book.douban.com/) is similar to Netflix, where users can label books with tags which describe book's contents by several terms, the users also can mark whether they have read a specific book. Thus in rating matrix S, if user u has read book i, $S_{u,i} = 1$, otherwise $S_{u,i} = 0$. Sina Weibo is the biggest social network platform in China, where users can label themselves with tags and publish microblogs. In addition, Douban users present their social network information in profiles. After matching Weibo users and Douban users by Weibo URLs provided in Douban users profiles, social network users reading list can be obtained.

2.4   User Interest Modeling

User interest modeling aims to extract user's interests from social network text. We propose "*book clusters*" as the intermediary between users and books. Each book cluster contains a group of similar books. The user interest modeling aims to mapping user's social network data to his tastes on book clusters.

User Interest Modeling Using Tags. In this section, we propose a probabilistic-based model using chi-square goodness score of fit test. Here gives some definitions of variables: T is the set of books read by users whose tags of interests contains tag t, C is the set of books in book cluster c. $C \cap T$ is the set of books which have been read by users who labeled with tag t and also belongs to book cluster c. $*$ is the set of all books. All sets but $*$ allow repeated records.

The global distribution of book clusters: $P(C|*) = \frac{Count(C)}{Count(*)}$. The conditional distribution of book clusters given readers with specific tag of interests: $P(C|T) = \frac{Count(C \cap T)}{Count(T)}$.

$S(T, C)$ is the degree of discrepancy between book cluster's distribution given all users and distribution given users labeled by specific tag.

$$S(T, C) = \frac{(P(C|T) - P(C|*))^2}{P(C|*)} \times Count(C \cap T) \tag{2}$$

To filter useless tags, we sort all scores in descending order and select top K records as representative tags and their correlation coefficients with book clusters. Considering the fact that users may have several different tags, we select the largest $S(T, C)$ for each tag t of the user as his interests score on book cluster c.

$$P_t{'}(U, C) = \underset{all\ T\ of\ U}{Max}(S(T, C)) \tag{3}$$

For each user, we can get a numeric vector $V : V \in R^{1 \times p}$ contains his preference on all book clusters. Then we normalize the vector using "$Min - Max\ Normalization$" strategy. Finally, we achieve user's interest scores on book clusters as user's interest model.

User Interest Modeling Using Microblogs. We design a classifier-based model to extract interest model from user's microblogs. For each book cluster, we train a binary classification model to decide whether the input microblog is related to the book cluster. The count of microblogs related to a book cluster are regarded as the user's interest score on this book cluster.

In training process, for each book cluster, using each book's tags, a SVM classifier is trained for each book cluster by one-vs-all multi-class classification. Books contained in a specific book cluster are represented in "Bag of Words". In predicting process, we take every single microblog as input and utilize the trained models to decide whether the microblog is related to the specific cluster. After processing all microblogs, we can get a vector $V \in R^{1*p}$ contains counts of microblogs related to every book cluster. Greater count values imply the user is more interested in this book cluster. After the above processes, we get user's interest scores on book clusters from microblogs as user's interests model.

Combination. Here we achieve two matrices representing user's interests models: U_{tag} and $U_{microblog}$. U_{tag} is predicted from user's tags, $U_{microblog}$ is predicted from user's microblogs. We choose a simple but effective method to combine these two matrices.

$$U = \lambda U_{tag} + (1 - \lambda)U_{microblog} \tag{4}$$

2.5   Item Modeling

We propose a novel way to combine book popularity and affiliation into item model. Item models are constructed from user's reading history and user's interest models. User's reading history can reflect book's popularity, combining with user's interests on book clusters, we can achieve book's relevance with book clusters.

Matrix S represents training users' rating scores, matrix U represents training users' interest models constructed according to Eq. (11). We propose to compute matrix $I : I \in R^{p \times n}$ given U and S by solving the optimization problem (1). Similar to solving optimization problem in [13], the optimal solution is given by

$$I = (U^T U + \lambda E)^{-1} U^T P \tag{5}$$

where E is a $k \times k$ identity matrix. λ is a small positive numeric avoiding the chance that determinant of $U^T U$ equals to 0. Each column in matrix I represents a model of book. Larger $I_{i,j}$ indicates the relevance between book cluster i and book j is tighter, users who interest in book cluster i are likely to enjoy book j.

2.6   Making Recommendation for New User

In this section, we show how our framework help solving cold start problem. Assume new users social network information (tags, microblogs) are available, we can construct his interest model $V_{cluster} : V_{cluster} \in R^{1 \times p}$, p is the count of book clusters. Combining new user's interest model and items' models, we can achieve his tastes on different books: $V_{book} = V_{cluster} \times I$. After sorting elements in V_{book} in descending order, we select top K books as the final recommendation list.

3   Experiments

3.1   Data Sets

We crawled user's social network information from Sina Weibo (http://weibo. com) and got users reading history from Douban Website (http://www.douban. com). We also crawled book's information from Douban Book Website (http:// book.douban.com). Finally we get 10242 active users and their related data. 2000 most popular books are picked as recommend targets. We select top K ranked books as final recommendation list. 10-fold cross validation method is used to assess the results.

3.2   Baseline Methods

Some frequently strategies used in cold start recommendation are shown below. **Random Strategy**: Recommended books are randomly selected from all books [12]. This is the simplest method to solve cold start problem. **Most Popular Strategy**: This is a naive method to select most popular books as recommendation list which is same to all new users [8,13]. **Tags Only Strategy**: User interest models are extracted only from user's social network tags. **Microblogs Only Strategy-trategy**: User interest models are extracted only from user's microblogs. **Nearest Neighborhood Strategy**:Ten nearest users are selected as new users neighbors according to their interest models and neighbor's most favorite books are picked as final recommendation list [6].

3.3   Evaluation Protocol

We adopt several binary relevance based information retrieval performance metrics. **Pre @ K**: In the K recommended books, this is the ration of books user has read. For all test users, we take the average of Pre @ K scores to evaluate different strategies. **MRR**: Mean Reciprocal Rank is a popular metric strategy used in information retrieval field, the reciprocal value of the mean reciprocal

Table 1. Experimental results

Strategy	K	PRE @ 10	MRR	MAP	DS
Random	3	0.01	0.018	0.018	7
Most popular	3	0.200	0.423	0.341	4
Tags Only	3	0.263	0.423	0.378	6.85
Microblogs Only	3	0.152	0.221	0.316	5.94
Nearest Neighborhood	3	0.156	0.196	0.211	4.33
Our Strategy	3	**0.266**	**0.434**	**0.399**	**8.53**
Random	5	0.015	0.036	0.036	12
Most popular	5	0.201	0.416	0.362	12
Tags Only	5	0.237	0.419	0.369	9.81
Microblogs Only	5	0.15	0.214	0.304	7.96
Nearest Neighborhood	5	0.163	0.204	0.214	6.8
Our Strategy	5	**0.249**	**0.447**	**0.426**	**12.61**
Random	10	0.022	0.056	0.025	**20.63**
Most popular	10	0.184	0.414	0.361	18
Tags Only	10	0.21	0.429	0.376	18.247
Microblogs Only	10	0.146	0.219	0.337	11.10
Nearest Neighborhood	10	0.163	0.217	0.242	17.12
Our Strategy	10	**0.229**	**0.459**	**0.405**	20.42
Random	20	0.019	0.071	0.015	**34.26**
Most popular	20	0.147	0.413	0.351	33
Tags Only	20	**0.197**	0.434	0.358	28.52
Microblogs Only	20	0.126	0.246	0.316	14.50
Nearest Neighborhood	20	0.148	0.208	0.227	22.75
Our Strategy	20	0.193	**0.456**	**0.397**	31.19

rank corresponds to the harmonic mean of the ranks. **MAP**: Mean Average Precision is a popular metric method used in IR domain. **Diversity**: Diversity is a very important quality in recommendation systems. A recommendation system with higher diversity can satisfy users' special tastes better. In fact there is no standard formula to evaluate the diversity score. Thus we simply choose the count of distinct book clusters in final recommendation list as diversity.

3.4 Experimental Results

The complete results are shown in Table 1. Given different size (K) of recommendation list, we compare performance of our framework with baseline methods. According to result, using users interest model extracted from his social network information, our strategy achieve a better performance.

4   Related Works

Cold start recommendation has received a lot of attentions in recent decades. In general, there are two main categories of research approach to solve the cold start problem. The first category is to make cold start recommendation without using external information. To improve precision of recommendation with few rating scores, Zhang [4] aimed to predict unrated items from like-minded user clusters and similar item clusters. The second category utilizes external information to recommend items for new users. Schein proposed an aspect model with latent variable method which combines both collaborative and content information in model fitting [5]. A predictive feature-based regression model that leverage information of users and items, was proposed by Park [8].

5   Conclusion

In this paper, we proposed a novel approach for solving cold start problem in production recommendation by leveraging social network textual information. Book clusters are viewed as the intermediaries between users and books. User interest models representing user's interests on book clusters are extracted from user's textual data (tags, microblogs). Item models representing correlations between book and book clusters are built by a well-designed matrix factorization method. Recommendations for new user without rating history are made based on his interest model and the trained item models. Experimental results show our framework is able to attain both high precision and diversity.

Acknowledgments. This work is supported in part by the National Natural Science Foundation of China (Grant Nos. 61170189, 61370126, 61202239), National High Technology Research and Development Program of China under grant (No. 2015AA016004), the Fund of the State Key Laboratory of Software Development Environment (No. SKLSDE-2015ZX-16), and Microsoft Research Asia Fund (No. FY14-RES-OPP-105).

References

1. Ahn, H.J.: A new similarity measure for collaborative filtering to alleviate the new user cold-starting problem. Inf. Sci. **178**(1), 37–51 (2008)
2. Chen, C.C., Wan, Y.H., Chung, M.C., Sun, Y.C.: An effective recommendation method for cold start new users using trust and distrust networks. Inf. Sci. **224**, 19–36 (2013)
3. Victor, P., Cornelis, C., De Cock, M., Teredesai, A.M.: Key figure impact in trust-enhanced recommender systems. AI Commun. **21**(2), 127–143 (2008)
4. Zhang, D.Q., Hsu, C.H., Chen, M., Chen, Q., Xiong, N., Lloret, J.: Cold-start recommendation using bi-clustering and fusion for large-scale social recommender systems. IEEE Trans. Emerg. Top.Comput. **2**(2), 239–250 (2014)
5. Schein, A.I., Popescul, A., Ungar, L., Ungar, H., Pennock, D.M.: Methods and metrics for cold-start recommendations. In: Proceedings of the 25th Annual International ACM SIGIR Conference on Research and Development in Information Retrieval, pp. 253–260, (2002)
6. Bobadilla, J., Ortega, F., Hernando, A., Bernal, J.: A collaborative filtering approach to mitigate the new user cold start problem. Knowl.-Based Syst. **26**, 225–238 (2012)
7. Phelan, O., McCarthy, K., Bennett, M., Smyth, B.: Terms of a feather: content-based news recommendation and discovery using twitter. In: Clough, P., Foley, C., Gurrin, C., Jones, G.J.F., Kraaij, W., Lee, H., Mudoch, V. (eds.) ECIR 2011. LNCS, vol. 6611, pp. 448–459. Springer, Heidelberg (2011)
8. Park, S., Chu, W.: Pairwise preference regression for cold-start recommendation. In: Proceedings of the third ACM Conference on Recommender Systems, pp. 21–28 (2009)
9. Zhou, K., Yang, S.H., Zha, H.Y.: Functional matrix factorizations for cold-start recommendation. In: Proceedings of the 34th international ACM SIGIR conference on Research and development in Information Retrieval, pp. 315–324 (2011)
10. Claypool, M., Gokhale, A., Miranda, T., Murnikov, P., Netes, D., Sartin, M.: Combining content-based and collaborative filters in an online newspaper. In: Proceedings of ACM SIGIR Workshop on Recommender Systems, vol. 60 (1999)
11. Lin, J., Sugiyama, K., Kan, M., Chua, T.: Addressing cold-start in app recommendation: Latent user models constructed from twitter followers. In: Proceedings of the 36th International ACM SIGIR Conference on Research and Development in Information Retrieval, pp. 283–292 (2013)
12. Su, X.Y., Khoshgoftaar, T.: A survey of collaborative filtering techniques. Adv. Artif. Intell. **4**, 2009 (2009)
13. Liu, N., Meng, X.R., Liu, C., Yang, Q.: Wisdom of the better few: cold start recommendation via representative based rating elicitation. In: Proceedings of the fifth ACM Conference on Recommender Systems, pp. 37–44 (2011)
14. Zhang, M., Tang, J., Zhang, X.C., Xue, X.Y.: Addressing cold start in recommender systems: a semi-supervised co-training algorithm. In: Proceedings of the 37th International ACM SIGIR Conference on Research and Development in Information Retrieval, pp. 73–82 (2014)
15. Kim, B.M., Li, Q.: Probabilistic model estimation for collaborative filtering based on items attributes. In: Proceedings of the 2004 IEEE/WIC/ACM International Conference on Web Intelligence, pp. 185–191 (2004)
16. Yu, K., Schwaighofer, A., Tresp, V.: Probabilistic memory-based collaborative filtering. IEEE Trans. Knowl. Data Eng. **16**(1), 56–69 (2004)

The Detection of Multiple Dim Small Targets Based on Iterative Density Clustering and SURF Descriptor

Haiying Zhang[1(✉)], Yang Gao[1], and Tao Li[2]

[1] Software School of Xiamen University, Fujian 361005, China
Zhang2002@xmu.edu.cn
[2] School of Computing and Information Sciences,
Florida International University, Miami, FL 33199, USA
taoli@cs.fiu.edu

Abstract. Following the classical TBD (Track before Detection) framework popular used in ATR (Automatic Target Recognition), a fast algorithm is proposed in this paper. Different from the classical data association methods, the extraction of target trajectory is converted into clustering process of searching density peaks. At first, the 3D time sequence is projected into 2D plane and then it is segmented into multiple zones either targets or clutter. Finally, the target trace is discriminated further by continuous and consistency constraints. During preprocessing, in order to guarantee the sparse of the background and the intensive of the targets SURF (Speeded up Robust Features) detector is introduced. The experiments result shows that the algorithm can detect small targets with lower SCR both in cloudy sky and sea background, compared with most recent algorithms it has a priority in time complexity and false alarm suppression ratio.

Keywords: Dim small target · SURF · Density clustering · ROI

1 Introduction

The dim small targets detection and tracking series problem are originated from early space warning system in military. As a main research aspect of ATR (Automatic Target Recognition), it had undergone a long research course. From the DBT (Detect before Tracking) to TBD (Track before detect), many classical methods and important schemes are presented, Such as dynamic programming [1], matching filter [2], top-hat filter [3], NN [4], correlation filter [5], and HMM filter [6] etc. For the inherent low SNR or SCR and small size, it is difficult to perform hard decision based on a single frame so that the framework whose emphasis on TBD is most popular. The main idea of the framework is to enhance the SNR of the image by constantly carrying out data association to create the candidate trajectory, and accordingly performing the soft decision. A series of methods are gathered in a handbook written by Bar shalom [7]. However, the methods were built on the premise that the probability distribution and dynamic of the targets are known. In many cases it is difficult to predict the distribution of the target and dynamic. Further, the decision relies on the candidate trajectories created during data association process and the time complexity will be exponential order.

© Springer International Publishing Switzerland 2015
J. Wang et al. (Eds.): WISE 2015,Part II,LNCS 9419, pp. 284–291, 2015.
DOI: 10.1007/978-3-319-26187-4_25

During recent five years, with the development of new technology and integrated methods, many researchers shift back to the scheme of DBT based on single frames. Tae designed two kinds of new filter to detect small target [8, 9], in Ref. [10], the original image is regarded as a sparse matrix plus a low-rank matrix, and the detection of the target is transformed as a matrix decomposition problem. Motivated by the robust properties of HVS, the detection of small target is solved by using scale-space theory and optimization method [11]. Assuming the target has a higher local contrast compared to the non-target, the derived kernel model [12] and local contrast map [13] methods are presented. Besides the improvements in technology, a new sensor is given in [14] for selecting new features to finish the detection progress. All the methods will be efficient in some situations; however, with the increase of the disturbance in real applications the performance will decline quickly and cause large quantity of false alarms which seriously affect the recognition process.

Aimed at suppressing false alarm and reducing time complexity a new method is proposed in this paper. Different from the classical data association methods, a framework of density clustering is introduced in which the target trajectories are segmented and extracted by constantly clustering without considering complex data association process, additional, SURF detector is adopted to suppress background to guarantee the result of clustering.

The paper is organized as follows. In Sect. 2, we give the description of the algorithm. Section 3 we analyze the algorithm in theory. Section 4 presents the experimental results to verify the algorithm compared with present algorithms. Finally we conclude this paper is Sect. 5.

2 Algorithm Description

The algorithm is composed of three steps: extract of ROI (Region of Interest), intensity clustering and discrimination of targets. We will explain it in the following chapters.

2.1 Extraction of ROI Based on SURF

The key to carry out density-based clustering is that the interested object areas have a higher density than the disturbance irritated by the noise and background clutter, so it is necessary to suppress large scale background and make the image matrix sparse. The SURF detector uses the multi-resolution pyramid technique to detect points of interest in images which can remove the highly structural and correlated background [15]. During preprocessing, depending on the good performance of SURF we suppress the background and make the image sparse.

2.2 Clustering by Fast Searching Density Peak

After preprocessing, we project the 3D image sequence into 2D image plane and in the accumulated image we can find the target trajectory present the state of intensity clustering but the isolated noise and clutter edge will have no the property. For the purpose, we carry out density clustering to extract the trajectories. In 2014 a fast clustering method to find density peaks is given by [16]. Unlike the mean-shift method [17], the procedure does not require embedding the data in a vector space and maximizing explicitly the density field for each data point. For different target traces will locate in a local density peak which corresponds to one cluster so we can use the method to extract target traces faster.

The algorithm assumes that cluster centers are surrounded by neighbors with lower local density and that they are at a relative large distance from any points with a higher local density. For any point x_i of the data set, we can define two qualities ρ_i and δ_i.

- Local density: we choose Gaussian kernel shown in Eq. (1)

$$\text{Gaussian kernel } \rho_i = \sum_{j \in I_s \setminus \{i\}} e^{-\left(\frac{d_{ij}}{d_c}\right)^2} \tag{1}$$

The dc is the cut-off distance.

- *Distance* δ_i: assuming $\{q_i\}_{i=1}^N$ is a index decedent sequence of $\{\rho_i\}_{i=1}^N$ which satisfied that

$$\delta_{qi} = \begin{cases} \min_{\substack{q_j \\ j < i_i}} \{d_{q_i q_j}\}, & i \geq 2 \\ \max_{j \geq 2} \{\delta_{q_j}\}, & i = 1 \end{cases} \tag{2}$$

When (ρ_i, σ_i) is the max then cluster center is defined.

In order to give a quantity of cluster center a new variable gamma is introduced that:

$$\gamma_i = \rho_i \delta_i, \ i \in I_S \tag{3}$$

The bigger gamma is, the more possibility it is assumed as cluster center, so that depending on gamma we can identify the center automatically.

2.3   Iterate Density Clustering

Ideally, one cluster will stand for one potential target trajectory, however, according to the algorithm principle in our application multiple adjacent target traces will be attributed into one cluster. To solve the problem we update the algorithm to an iterate version. Corresponding, we give the convergence criterion.

- Convergence criterion: after per iteration if the cluster number tends to one or in other words there is only one prominent gamma then the iterative process will stop.

3 Discrimination of the Target from the Clutter

After density-clustering the candidate target zone has been segmented multiple smaller areas which correspond to target or clutter. During this stage we will complete the discrimination process. For the slowly moving targets have smooth line or curve trajectory segment so it will meet with continuous and consistency in time spatial space.

- Definition *1: continuous*

  For a real trace every point belongs to the trace will present by times.

- Definition 2: *consistency*

  Because the target is moving slowly from far distance, in generally the movement offset between the adjacent frame is within several pixels so for every pair of adjacent points belongs to the trace, they will meet the following equation:

$$|point(x1, y1, t) - point(x2, y2, t - 1)| \leq velocity \qquad (4)$$

  Wherein velocity is decided by the movement range of the targets along x and y axis and $(x1, y1)$, $(x2, y2)$ are the point in frame t and $t - 1$ respectively.

4 Experiments and Analysis

4.1 Algorithm Performance Testing

In order to evaluate the performance of the algorithm we experimented on many simulated synthetic sequences with different background. In this section, we select two kinds of representative sequences for experimental evaluation.

Simulated sequence 1: the background is real cloudy sky and embedded 5 targets with different dynamics. The noise is Gaussian ($\mu = 0, \sigma^2 = 0.03$) (Fig. 1).

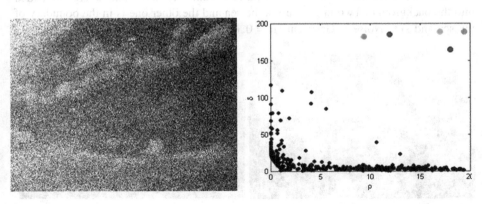

Fig. 1. One synthetic image embedded Gaussian noise and five targets

Fig. 2. Decision graph of accumulated image sequence based on density clustering

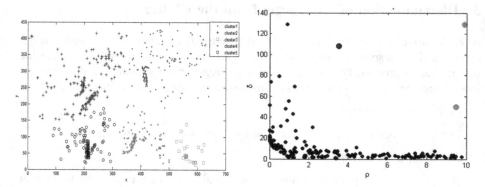

Fig. 3. The achieved five clusters after first density clustering

Fig. 4. Decision graph of cluster 2 after first clustering

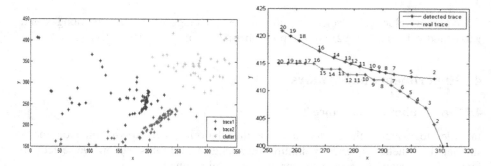

Fig. 5. The separated clusters for cluster 2 after second clustering

Fig. 6. One target trace compared with the real one

Simulated sequence 2: the background is sea and sky; three targets are embedded into the background. Two targets are in the sea and the other one is in the boundary of the sea and sky. Noise is Gaussian ($\mu = 0, \sigma^2 = 0.02$).

Fig. 7. One synthesis image

Fig. 8. The blobs after using SURF detector in one frame

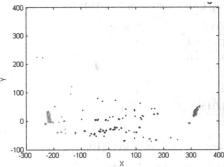

Fig. 9. The result of first clustering

Fig. 10. The result of overlapping the first clusters on the original image

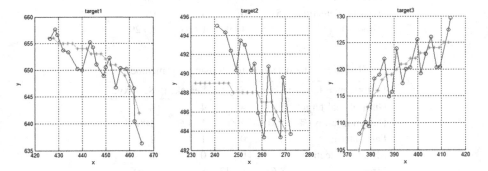

Fig. 11. The comparison of the real one and the detected target trajectory

From Figs. 2, 3, 4 and 5 we give the clustering process of *cluster* 2 in which two targets are clustered into one cluster and we can see that after iterative clustering it is segmented correctly.

The clustering result is given in Figs. 8, 9 and 10, we can find the good performance of the algorithm and final detected result is shown in Fig. 11. Due to target 2 moves towards the boundary of the sea and sky so we only obtain part of its trace and with the affection of sea clutter it has a little bigger offset from the real one (Fig. 11).

4.2 Comparison Experiments

In order to verify the performance presented in this paper many comparison experiments are carried out based on different background and SCR. Some results are shown in Tables 1, 2 and 3. For all of the four selected sequences, the noise embedded is Gaussian (0, 0.02) and the SCR is between $0.9 \sim 5.0$.

During the Tables 1 and 2 the '−' means the algorithm is invalid. From the experiments results we find that the methods proposed in [10] and [14] are invalid in

Table 1. The detection result using LMC method in Ref. [14]

Performance factor	Sky1	Sky2	Sea-sky1	Sea-sky2
Num of real target	5	4	4	4
Detection ratio (%)	100 %	0	0	0
False alarm	20	–	–	–
Missing detection	0	–	4	4
Time (s)	7.169	8.045	6.211	9.741

Table 2. The detection result using the method in Ref. [10]

Performance factor	Sky1	Sky2	Sea-sky1	Sea-sky2
Num of real target	5	4	4	4
Detection ratio (%)	0	100 %	0	100 %
False alarm	–	0	–	10
Missing detection	5	0	4	0
Time (s)	44.55	44.97	19.69	50.29

Table 3. The detection result using our method

Performance factor	Sky1	Sky2	Sea-sky1	Sea-sky2
Num of real target	5	4	4	4
Detection ratio (%)	100 %	100 %	100 %	100 %
False alarm	0	0	1	1
Missing detection	0	0	0	0
Time (s)	8.616	8.443	8.518	9.307

some situations and our method is more popular in dealing with different background. It is closer to the real time of LMC [14] and more efficient than the other two methods in detection ratio and false alarm suppression (Table 3).

5 Conclusions

In this paper a framework based on clustering analysis is given to solve the ATR problem. Based on iterative density clustering we can divide the candidate target trace from the background step by step. Compared with the traditional data association algorithm, the time complexity of our proposed algorithm is reduced heavily and the false alarm suppression is obvious. Next step we will solve the missing detection and trace smoothing problems.

Acknowledgements. The work is supported by National Natural Science Foundation of China (No: 61303080) and Natural Science Foundation of Fujian Province, China (No: 2013J01249).

References

1. Poret, B., Friedlander, B.: A frequency domain approach to multi-frame detection and estimation of dim targets. IEEE Trans. Pattern Recogn. **12**(4), 398–401 (1990)
2. Barniv, Y.: Dynamic programming solution for detecting dim moving targets. IEEE Trans. Aerosp. Electron. Syst. **21**(1), 144–156 (1985)
3. Henry, L., Neville, D.: Detection of small objects in clutter using GA-RBF neural network. IEEE Trans. Aerosp. Electron. Syst. **38**(1), 98–118 (2002)
4. Tom, V., Peili, T., Leung, M., Bondaryk, J.: Morphology-based algorithm for point target detection in infrared backgrounds. In: Proceedings of SPIE 1954, pp. 25–32, October 1993
5. Liou, R.J., Azimi-Sadjadi, M.R.: Dim target detection using high order correlation method. IEEE Trans. AES **29**(3), 841–855 (1993)
6. Lai, J., Ford, J.J., O'Shea, P., Walker, R.: Hidden Markov model filter banks for dim target detection from image sequences. In: IEEE Conference on Digital Image Computing (2008)
7. Bar-shalom, Y., Willett, P.K., Tian, X.: Tracking and Data Fusion. A Handbook of Algorithms. YBS Publishing (2011)
8. Bae, T.-W., Zhang, F., Kweon, I.-S.: Edge directional 2D LMS filter for infrared small target detection. Infrared Phys. Technol. **55**(1), 137–145 (2012)
9. Bae, T.-W.: Small target detection using bilateral filter and temporal cross product in infrared images. Infrared phys. Technol. **54**(5), 403–411 (2011)
10. Gao, C., Meng, D., Yang, Y.: Infrared patch-image model for small target detection in a single image. IEEE Trans. Image Process. **22**(12), 4996–5009 (2013)
11. Sui, X., Chen, Q., Bai, L.: Detection algorithm of targets for infrared search system based on area infrared focal plane array under complicated background. Optik **123**, 235–239 (2012)
12. Li, H., Wei, Y., Li, L., Yuan, Y.: Similarity learning for object recognition based on derived kernel. Neurocomputing **83**, 110–120 (2012)
13. Chen, C.P., Li, H., Wei, Y.: A local contrast method for small infrared target detection. IEEE Trans. Geosci. Remote Sens. **52**(1), 574–581 (2014)
14. Kim, S., Lee, J.: Scale invariance small target detection by optimizing signal-to-clutter ratio in heterogeneous background for infrared search and track. Pattern Recogn. **45**(1), 393–406 (2012)
15. Speeded up robust features. http://en.wikipedia.org/wiki/SURF
16. Rodriguez, A., Laio, A.: Clustering by fast search and find of density peaks. Science 1492–1496 (2014)
17. Han, J., Kamber, M.: Data Mining: Concepts and Techniques, 2nd edn. Morgan Kaufmann Publishers, San Francisco (2006)

Collaborative Content-Based Method
for Estimating User Reputation
in Online Forums

Amine Abdaoui[(✉)], Jérôme Azé, Sandra Bringay,
and Pascal Poncelet

LIRMM B5 UM CNRS, UMR 5506, 161 Rue Ada, 34095 Montpellier, France
{amin.abdaoui,jerome.aze,sandra.bringay,
pascal.poncelet}@lirmm.fr

Abstract. Collaborative ratings of forum posts have been successfully applied in order to infer the reputations of forum users. Famous websites such as *Slashdot* or *Stack Exchange* allow their users to score messages in order to evaluate their content. These scores can be aggregated for each user in order to compute a reputation value in the forum. However, explicit rating functionalities are rarely used in many online communities such as health forums. At the same time, the textual content of the messages can reveal a lot of information regarding the trust that users have in the posted information. In this work, we propose to use these hidden expressions of trust in order to estimate user reputation in online forums.

Keywords: Trust · Reputation · Online forums · Social networks

1 Introduction

Online forums are areas of exchange generated by their own users. Therefore, the veracity and the quality of the posted information vary wildly according to their author. With the massive and rapid growth of these conversational social spaces, it becomes very difficult for human moderators to separate good posts from bad ones. Consequently, more and more forums are implementing automated trust and reputation metrics to infer the trustworthiness of posts and the reputation of their authors. These metrics vary from ranks based on a simple post count to more elaborated reputation systems based on collaborative ratings. If the first category of metrics tries simply to reward users according to the number of their posts, the second category uses collaborative intelligence to rate a user's posts and then aggregate these ratings to give him a reputation value [1]. This idea has been successfully applied in many online forums such as news groups (*Slashdot*[1]), question-answering websites (*Stack Exchange*[2]), etc. However, collaborative rating is not so popular in other communities such as health forums, where users prefer to post a new message in order to thank each other rather than clicking the 'like' or 'vote up' button. The objective of this work is to use this

---

[1] http://www.slashdot.org/.
[2] http://www.stackexchange.com/.

© Springer International Publishing Switzerland 2015
J. Wang et al. (Eds.): WISE 2015, Part II, LNCS 9419, pp. 292–299, 2015.
DOI: 10.1007/978-3-319-26187-4_26

implicit collaborative intelligence hidden in the textual content of the replies in order to infer user reputations.

Many definitions of trust and computational trust exist in the literature [2, 3]. Here we define the trust that a user A has in another user B as: *"the belief of A in the veracity of the information posted by B"*, and the reputation of a user A as *"the aggregation of trust values given to user A"*. To infer such trust from textual replies and aggregate user reputations, we need to know both the recipient of each forum message and the trust expressed in it. However, the forum structure does not always provide explicit quoting or direct answering functionalities. Besides, when these functionalities are provided, many users prefer posting a message answering the whole thread rather than a one answering or quoting another specific message. In order to deal with this issue, we propose a rule based heuristic to extract an interaction network where the nodes are the users and the edges are the replying posts. Regarding the semantic evaluation of each post's content, the features that we are looking for are agreement and valorization for trust, and disagreement and depreciation for distrust. The rest of posts are considered as neutral. Finally, we propose a metric to aggregate trust and distrust replies that a user receives and infer his reputation in the forum. The proposed reputation metric considers propagation aspects by giving more weight to the replies posted by trusted users and less to the replies posted by untrusted ones.

The rest of the paper is organized as follows: Sect. 2 presents a summary of related work that match our methods. Section 3 gives the theoretical framework, presents the corpus of our study and describes the proposed approach. Section 4 presents and discusses the obtained results using manual annotations. Finally, Sect. 5 gives our main perspectives.

2 Related Work

Most of the methods found in the literature in order to extract interaction networks from online communities use the HTML structure of the web page [4–7]. They try to identify explicit message quoting. However, explicit quoting functionality is not always provided in online forums, and even when it exists many discussion participants do not use it. Moreover, a message may have many recipients. Consequently, posting it as an answer to another specific one may be insufficient. Gruzd [8] presented an automatic approach to discover and analysize social networks from threaded discussions in online courses. The authors proposed a Name Entity Recognition system to extract name mentions inside the textual content of posts. After a preprocessing step (removing quotations, stop words, etc.), their method used a dictionary of names combined with manually designed linguistic rules. Another textual based method has been proposed by Forestier et al. [9] to extract a network of user interactions. They suggested to infer three types of interactions: structural relations, name citations, and text quotations. While structural relations can be inferred directly from the structure of the forum, name citations and text quotations require analyzing the textual contents. First, name citation relations have been extracted by searching pseudonyms of authors inside the posts.

Then, text quotations are extracted by comparing sequences of words inside a message and the messages that have been posted before in the same thread.

On the other hand, existing trust metrics dealing with online forums can be organized in two main categories: structure-based trust metrics and content-based trust metrics. The first category focus on the structure of the website (including the number of postings, the distance between messages, quotes, citations, etc.) [10], while the second one use the textual content of messages to infer trust and reputation. For example Wanas et al. [11] automatically score posts based on their textual content. Their method is inspired from forums that use collaborative intelligence to rate posts. They tried to model how users would perceive a post as good or as bad. However, unlike Wanas, we believe that the textual content of the messages that reply to a user's post may reveal a lot of information regarding the trust or the distrust that the other users have in this post and therefore in its author. Consequently, instead of inferring a user's reputation from his own posts, we suggest to consider the messages replying to his posts. Moreover, we would like to give more importance to a reply made by a trusted user and less to a reply made by an untrusted user. A large effort has been done to include propagation aspects in order to rank webpages [12]. Similarly, we propose a reputation metric that include these propagation aspects.

3 Materials and Methods

3.1 Corpus of Study

CancerDuSein.org is a French health forum specialized in breast cancer. 1,050 threads have been collected which amounts 16,961 messages posted by 675 users. It represents all the data that have been posted between October 2011 and November 2013. This forum allows users to thank each other using a "like" button, but this functionally is rarely used. Less than 1.4 % of messages received at least one "like". On the other hand, *CancerDuSein.org* gives a rank to each user based on the number of posts since his registration. However, we believe that these ranks are not sufficient to infer reputations.

3.2 Theoretical Framework

Let $G = (V, E, t, r)$ be a multigraph where: V is the set of users, E is the multiset of 'reply-to' edges between these users, t is a function that returns the transmitter of a reply, and r is a function that returns the recipient of a reply:

$$t : E \to V \quad r : E \to V$$

$$e \to t(e) \quad e \to r(e)$$

Let $v \in V$ be a user. Then $E_v \subseteq E$ is the set of edges that reply to the user v:

$$E_v = \{e \in E : r(e) = v\}$$

Let E_v^+, E_v^- and $E_v^n \subseteq E_v$ be the subsets of trust, distrust and neutral edges that reply to the user v. Note that $E_v = E_v^+ \cup E_v^- \cup E_v^n$ and $E_v^+ \cap E_v^- \cap E_v^n = \emptyset$.

3.3 Extracting the Interaction Network

We suggest searching nine types of relations using manually designed heuristic rules, checked sequentially in the following order:

Explicit Quoting: *CancerDuSein.org* allows users to explicitly quote another user's post. However, only 349 posts on the Website are explicit quoting. They have been detected automatically using the HTML tag *<quote>*.

Second Posts: Messages posted at the second place in each thread have been considered as replying to the first one.

Names and Pseudonyms: If a message contains the pseudonym or the name of a user who previously posted a message in the same thread, then this user is considered as the recipient of the message. The following preprocessing steps were been applied to detect names and pseudonyms: (1) Remove all non-alphabetic characters except spaces; (2) Replace all accented characters by the corresponding non-accented ones; (3) Lowercasing.

Grouped Posts: If a message contains a group marker ("hello everyone", "Hi girls", "Thank you all", etc.) then all the users who previously posted in the same thread are considered as recipients for this post.

Second Person Pronouns: In French, singular second person pronouns and plural second person pronouns are different. If a singular second person pronoun is used then the recipient is considered to be the author of the previous post.

Activator Posts: If the activator[3] posts a new message in the same thread, we consider that his new message is addressed to all the users who posted after him.

Questions: If the message contains a question, then the message is addressed to all the users who previously posted in the same thread.

Answers: If there is a question posted before in the thread, the recipient is the user who posted this question.

Default: If none of the above rules before are satisfied, we consider that the recipient of the message is the activator.

3.4 Predicting Trust and Distrust

Once the interaction network is constructed, we need to classify each post with one of the following three classes: (1) Positive: the post expresses trust to its recipient; (2) Negative: the post expresses distrust to its recipient; (3) Neutral: otherwise.

---

[3] The user who opened the thread by posting the first message.

Building Lists of Trust and Distrust Expressions: We manually created two lists of expressions that should indicate if a message expresses trust (or distrust) to its recipient. These lists have been obtained by manual annotations of a set of threads using the brat tool[4]. The annotators were asked to choose trust, distrust or neutral for each thread post and to indicate the expressions that justify their choice. These expressions have been manually validated, and then corrected, lowercased and lemmatized.

Handling Negation: If a trust expression is under the scope of a negation term, it is considered as a distrust expression and vice versa.

Computing the Frequencies and Classifying the Posts: All posts have been automatically lowercased, lemmatized, and corrected using the Aspell[5] spell checker. Then, each post is assigned to the majority category carried by its words.

3.5 Proposed Metrics

For each user v, we define a reputation value $R(v)$ as follows:

$$R_{n+1}(v) = \begin{cases} \dfrac{\sum_{e \in E_v^+} R_n(t(e))}{\sum_{e \in E_v^+} R_n(t(e)) + \sum_{e \in E_v^-} R_n(t(e))}, & if E_v^n \neq E_v \\ 0.5, & Otherwise \end{cases}$$

This equation is recursive and can be computed by starting with reputations equal to 1 and iterating until it converges. The proposed reputation equation depends on both the number of trust and distrust replies a user receives and the reputations of the users who posted these replies.

We also define two complementary metrics: the neutral rate of the user $NR(v)$, and the reliability of the computed reputation value $Rel(R(v))$.

$$NR(v) = \begin{cases} \frac{|E_v^n|}{|E_v|}, & if\ |E_v| \neq \emptyset \\ 0, & Otherwise \end{cases}, \quad Rel(R(v)) = \begin{cases} \frac{|E_v|}{maxR}, & if\ |E_v| < maxR \\ 1, & Otherwise \end{cases}$$

Where $maxR$ is a constant that represents the maximum replies that a user should receive in order to have a reliability of one in his reputation.

4 Results

4.1 Evaluating the Network Extraction Step

Two datasets were used to test our rule based heuristic. The rules have been designed according to a development set (10 threads) and tested on other 10 unseen threads.

---

[4] www.brat.nlplab.org.

[5] www.aspell.net.

Prior-assessment: 15 non-expert annotators, unaware of the designed rules, annotated our two datasets. Each one annotated between 1 and 5 threads so that each thread had 3 different annotators. The goal was to find the recipient(s) of each post without knowing the results of our heuristic.

Post-assessment: Three expert annotators (the authors) annotated the links found by the heuristic in the two datasets. The goal was to validate or not the links found automatically with the possibility of adding a link which was not found by the heuristic.

Evaluation: Using these annotations, the quality of the developed heuristic was evaluated. The links obtained automatically were compared with those obtained from the annotations by considering only those that have been validated by two or more annotators (a majority vote). We compare the results of the prior-assesment and the post-assesment with two baselines. The first one considers the activator of the thread as the recipeint of all the messages posted in this thread (activator). The second baseline considers the author of the previous message as the recipient (previous) (Table 1).

Table 1. Precision (P), recall (R) and F1-score (F1) of baselines and our heuristic obtained on both dataset using prior and post assessments

		P	R	F1
Development set	Baseline 1 (activator)	0.39	0.24	0.30
	Baseline 2 (previous)	0.76	0.45	0.57
	Prior-assessment	0.70	0.68	0.69
	Post-assessment	0.80	0.84	0.82
Test set	Baseline 1 (activator)	0.55	0.35	0.45
	Baseline 2 (previous)	0.63	0.43	0.51
	Prior-assessment	0.81	0.83	0.82
	Post-assessment	0.83	1	0.91

Discussion: Our heuristic obtained higher F1-scores than both baselines. The results obtained using a post-assessment are better than those obtained using prior-assessment. This observation can be explained by the nature of the prior-assessment itself which gives much more freedom in choosing the links. Surprisingly, the results obtained on the test set have been better than those obtained on the development set.

4.2   Evaluating the Trust Prediction Step

Two new datasets have been used to evaluate the automatic trust inference. Unlike the first step where both datasets had prior-assessment and post-assessment, here prior-assessment has been done only for the first dataset and post-assessment has been done only for the second one.

Prior-assessment: Three annotators annotated the trust expressed in 97 messages without knowing the results of the automatic system. The agreement between them was less than the recipient assessment but still acceptable.

Table 2. Precision, recall and F1-score of the trust inference system using prior-assessment and post-assessment

Datasets	Class	P	R	F
Prior-assessment	Trust	0.67	0.93	0.78
	Distrust	0.50	0.25	0.33
	Neutral	0.96	0.83	0.89
	Global	**0.86**	**0.84**	**0.84**
Post-assessment	Trust	0.77	0.87	0.82
	Distrust	0.23	0.60	0.33
	Neutral	0.92	0.75	0.83
	Global	**0.84**	**0.78**	**0.80**

Post-assessment: The same three annotators annotated the trust expressed in 102 other messages. The results of the automatic system have been displayed, and annotators can chose the same value of trust or another one.

Evaluation: The results obtained by comparing the classification made by the system with the annotations (majority vote) are presented Table 2.

Discussion: The results obtained for the trust class are good but the recall is higher than the precision using both assessments. Therefore, our list of trust expressions seems to be sufficient to find the majority of trust posts. Moreover, the results obtained on the neutral class are also good, but the precision is higher than the recall. Finally, the results obtained on the distrust class have been the worst but it is difficult to make conclusions regarding the small number of distrust posts.

4.3   Evaluating the Proposed Metric

In our experiments, the constant *maxR* has been fixed to the average number of replies received by each user. The reputations of 157 users had reliabilities greater than 0.5.

Discussion: Figure 1 shows that all considered reputations are greater than 0.7. This observation can be explained by the fact that *CancerDuSein*.org is a forum where little

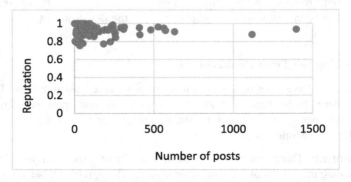

Fig. 1. User reputations that had more than 0.5 of reliability according to the number of posts

distrust is expressed, since the users aim at first to exchange emotional support. Moreover, the user reputations seem to be independent from the number of posted messages, which reinforces our opinion that the number of postings do not represent a good estimation of user reputations or ranks.

5   Conclusion

Many perspectives can be considered in order to improve the work and to better explore the idea. First, the user's reputation can be computed for each thread topic in addition to the global reputation in the whole forum. In fact, the user's expertise may change according to the discussed topic. Then, we are now scrolling other French forums in order to apply our method on a larger number of forums. Finally, we are planning to compare ourselves to PageRank or HITS based models built on the user interaction network.

References

1. Lampe, C., Resnick, P.: Slash(Dot) and burn: distributed moderation in a large online conversation space. In: Proceedings of the SIGCHI Conference on Human Factors in Computing Systems, pp. 543–550, New York, NY, USA (2004)
2. Sztompka, P.: Trust: a Sociological Theory. Cambridge University Press, Cambridge (1999)
3. Golbeck, J.: Trust and nuanced profile similarity in online social networks. ACM Trans. Web 3(4), 12:1–12:33 (2009)
4. Adamic, L.A., Zhang, J., Bakshy, E., Ackerman, M.S.: Knowledge sharing and yahoo answers: everyone knows something. In: Proceedings of the 17th International Conference on World Wide Web, pp. 665–674, New York, NY, USA (2008)
5. Stavrianou, A., Velcin, J., Chauchat, J.-H.: Definition and measures of an opinion model for mining forums. In: International Conference on Advances in Social Network Analysis and Mining, ASONAM 2009, pp. 188–193 (2009)
6. Welser, H., Gleave, E., Fisher, D., Smith, M.: Visualizing the signatures of social roles in online discussion groups. J. Soc. Struct. 8(2), 1–32 (2007)
7. Zhang, J., Ackerman, M.S., Adamic, L.: Expertise networks in online communities: structure and algorithms. In: Proceedings of the 16th International Conference on World Wide Web, pp. 221–230, New York, NY, USA (2007)
8. Gruzd, A.: Automated Discovery and Analysis of Social Networks from Threaded Discussions. Paper presented at, University of Illinois, Urbana-Champaign (2009)
9. Forestier, M., Velcin, J., Zighed, D.: Extracting social networks enriched by using text. In: Foundations of Intelligent Systems, pp. 140–145, Poland (2011)
10. Skopik, F., Truong, H.-L., Dustdar, S.: Trust and reputation mining in professional virtual communities. In: Gaedke, M., Grossniklaus, M., Díaz, O. (eds.) ICWE 2009. LNCS, vol. 5648, pp. 76–90. Springer, Heidelberg (2009)
11. Wanas, N., El-Saban, M., Ashour, H., Ammar, W.: Automatic scoring of online discussion posts. In: Proceedings of the 2nd ACM Workshop on Information Credibility on the Web, pp. 19–26, New York, NY, USA (2008)
12. Page, L., Brin, S., Motwani, R., Winograd, T.: The PageRank Citation Ranking: Bringing Order to the Web. Stanford InfoLab, Stanford (1999–66)

Fusion of Big RDF Data: A Semantic Entity Resolution and Query Rewriting-Based Inference Approach

Salima Benbernou, Xin Huang, and Mourad Ouziri[✉]

Université Sorbone Paris Cité, Paris Descartes, Paris, France
{salima.benbernou,xin.huang,mourad.ouziri}@parisdescartes.fr

Abstract. This paper presents an efficient approach to query big RDF datasources in order to get more relevant and complete results. The approach deals with two important heterogeneities in huge amount of data: semantic and URI-based entity identification heterogeneities. The paper proposes: (1) a semantic entity resolution approach based on inference mechanism to manage ambiguity of real world entities for linking data at the semantic and URI levels (2) a MapReduce-based query rewriting approach based on entity resolution results to include implicit data into query results (3) algorithms based on MapReduce paradigm to deal with huge amounts of data.

1 Introduction

Today, the need of organisations is to handle efficiently the volume of big data coming from not only proprietary data sources but also data from other heteregeneous souces developped by other organisations including government data named *open data*. The fusion of such data is needed to extract appropriate mutli sourced information and knowledge. Therefore, many challenges issue from that fusion including how to integrate data from multiple and heterogeneous data sources, how to identify the meaning between entities of different sources, how to handle the inconsistent naming styles in different data sources, and the conflicting data types for the same entity. Moreover, the Linked Data paradigm allows to describe a recommended best practice for exposing, sharing, and connecting data, information, and knowledge on the Semantic Web using URIs and RDF format, consequently create collections of interrelated datasets on the Web. The SPARQL language has been developped to get access data and draw inferences using vocabularies. Besides, some technologies that drive and enable open data exist, and is known as Linked Open Data such as DBpedia, FoaF, Geonames etc. We guess in our work that linked data contributes to solving the problem of structural heterogeneity and identifying entities. However, RDF and its URI mechanism are not sufficient to solve the semantic heterogeneity problem. For example, consider two real data sources Datasource1 (*ds1*) and Datasource2 (*ds2*) as depicted in Fig. 1 (both sources are represented in RDF data) related to insurance application of housing. The query over the data sources is to retreive

© Springer International Publishing Switzerland 2015
J. Wang et al. (Eds.): WISE 2015, Part II, LNCS 9419, pp. 300–307, 2015.
DOI: 10.1007/978-3-319-26187-4_27

all information about the house $h03$. By mean inference mechanism, $ds1$ will give an incomplete information - "Bob is living in the house $h03$" - because of the heterogeneity of the terminology used to describe the data from these two sources. Once a linking is processed between $ds1$ and $ds2$ through *entity resolution* method between $h03$ and $h25$ related to the same entity, the query result is complete {*bob, nboccupants*}. Entity Resolution (ER) is the task of disambiguating manifestations of real world entities in various resources by linking through inference across networks and semantic relationships in application.

Traditional evaluation techniques of a query (i.e., Jena[1] or Sesame[2]) are not suited to Big data since they require the loading of data previously established in memory before making its evaluation. It is then necessary to develop a SPARQL query execution engine adapted to big data with the help of MapReduce. For that purpose, several studies have been conducted such as HadoopRDF [1], Cliquesquare [2], H2RDF [3], but they are focusing on RDF storage.

To tackle with the aforementionned issues, the paper proposes an efficient querying approach over big RDF data including (1) inference mechanism to infer implicit data embedded in big data based on Mapreduce paradigm (2) entity resolution algorithm to cope with the ambiguity of real world entities in various sources that happens by linking (3) a rewriting SPARQL query mode evaluation over RDF big data to include implicit data into query result.

The rest of the paper is organized as follows. Section 2 presents an overview of the semantic big data fusion architecture. Inference-based semantic ER and linking approaches are discussed in Sect. 3. The query rewriting based inference over big RDF data is presented in Sect. 4. Section 5 concludes our work.

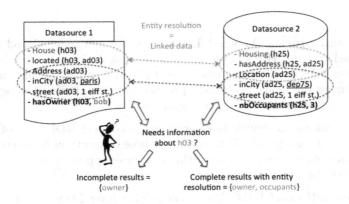

Fig. 1. Motivation example of semantic big data fusion

---

2   Architecture of Semantic Big Data Fusion Framework

The fusion architecture we propose in this paper supports four layers and is depicted in Fig. 2.

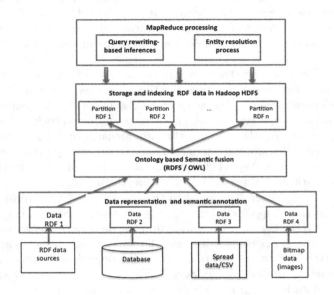

Fig. 2. Overall architecture for the semantic fusion of big RDF data

Indexing and Storing Merged Data: RDF data is indexed and stored in a distributed way on HDFS. The Indexing process deals with partitioning the merged large RDF data volume into several files to restrict the amount of data explored during the query evaluation process. Any exsiting technology is used for storing RDF such as the one developped in [2].

Semantic Linking: Semantic links are built to connect RDF data of different sources with the concepts of OWL ontology. Those connections are used by entity resolution and query rewriting algorithms we developped in order to generate the integrated data.

Entity Resolution of RDF Resources: This layer identifies and connects RDF data provided by multiple datasources that refers to the same real world entity. Such resolution is processed using semantic connections and domain-expert inference rules. Those rules will be evaluated on each pair of resources, and then MapReduce jobs are used due to the huge volume of data to be handled.

MapReduce-Based Query Rewriting: There are much implicit data embedded in the large volume of data obtained from the fusion, then this layer aims to infer the implicit data and include them in the query evaluation results. The inference is achieved by a query rewriting process. The rewriten query is a plan of MapReduce jobs.

3 Inference-Based Semantic Entity Resolution and Linking

Each datasource uses its own OWL ontology (as conceptual model) and identifies the resource using internal URIs (as entity identification). Therefore, same entities may be described using different or equivalent concepts (semantics) identified by different URIs among different data sources. As a real world example, *Paris* is identified in *INSEE* source (National Institute for Statistic and Economics Studies-France) by the URI http://id.insee.fr/geo/departement/75, whereas *Paris* is identified in *DBpedia* source by the URI http://fr.dbpedia. org/page/Paris. To reconcile such entities, we present in this section an inference mechanism to connect semantically all heterogeneous RDF fragments to the same entity. For illustration, Fig. 3 shows fragments of two RDF sources, $ds1 : h03$ and $ds2 : h25$, describing the same house provided by *Insurance company* and *INSEE* datasources, respectively. The RDF fragments are serialized by facts, some of them are as follows: (1) $ds1 : h03$ is a $ds1 : House$ and located at $ds1 : ad03$, (2) $ds2 : h25$ is a $ds2 : Housing$ and has address $ds2 : ad25$, (3) $ds1 : ad03$ is in Street *1 eiffel st*, inCity of $ds1 : paris$, (4) $ds2 : ad25$ is in *1 eiffel st.*, inCity $ds2 : dep75$, (5) $ds1 : paris$ is same as $ds2 : dep75$, (6) $ds1 : House$ is a $ds2 : Housing$. When propagating Fact (5) on facts (3) and (4), it is inferred that $ds1 : ad03$ and $ds2 : ad25$ represent the same address. This resolution will be propagated to facts (1) and (2) to infer that $ds1 : h03$ and $ds2 : h25$ represent the same house by considering the semantic linking given by axiom (6) and the given domain rule: *there can be only one house at a given address.*

We give in the following some useful definitions before presenting the inference approach for big RDF data.

Definition 1. *(Entity). An Entity is a real world object described by set of non disjoint concepts $E_c = \{c_1, c_2, ..., c_n\}$ and properties $E_p = \{p_1, p_2, ..., p_n\}$. The Entity may be maintained in multiple heterogeneous datasources and identified using different URIs-based identifiers, $E_{id} = \{uri_1, uri_2, ..., uri_n\}$.*

Definition 2. *(An entity fragment). An entity fragment EF is RDF resource that represents a part of entity E. An entity fragment is maintained in a single datasource and identified by a unique URI $uri_{fe} \in E_{id}$ and described using subset of concepts $EF_c \in E_c$ and subset properties $EF_p \in E_p$.*

Definition 3. *(Functional key of entity resolution). A functional key of entity resolution is a set of properties that identify a unique entity. If we consider a functional key $f_k = \{p_1, ..., p_n\}$, then we have:*

$$\forall(i,j), \forall k \in [1,n], (?x_i\ ?p_k\ ?v_{ki}\ \wedge\ ?y_j\ ?p_k\ ?v_{kj}\ \wedge\ ?v_{ki}\ SameAs\ ?v_{kj}) \Rightarrow (?x_i\ SameAs\ ?y_j)$$

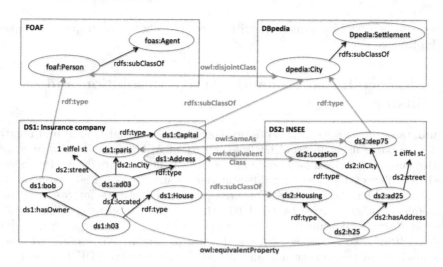

Fig. 3. Semantic connections between multiple datasources

where ($?x_i$ $?p_k$ $?v_{ki}$) and ($?y_j$ $?p_k$ $?v_{kj}$) are RDF triples stating that entity fragments $?x_i$ has property $?p_k$, which takes the value $?v_{ki}$. We use ? to specify variable.

In general way, f_k of entity resolution is given by business experts, and is a necessary but not sufficient condition for semantic entity resolution. That is, the fragments of same entity must be described using equivalent concepts.

Definition 4. (Semantic entity resolution). The semantic entity resolution is based on a functional key that includes the property $rdf : type$. That is, given a functional key $fk = \{p_1, ..., p_n\}$ and two entity fragments EF_i and EF_j then:

$$\forall k \in [1, n], (\exists r \in [1, n], \ p_r = rdf : type) \ then \ (?x_i \ ?p_k \ ?v_{ki} \ \wedge \ ?y_j \ ?p_k \\ ?v_{kJ} \ \wedge \ ?v_{ki} \ SameAs \ ?v_{kj}) \Rightarrow ?x_i \ SameAs \ ?y_j$$

Let us consider in Fig. 3 two domain rules, there is only one house at a given address and a street name is unique in a city. These two rules define respectively, the functional keys ($rdf : typeHousing, located$) and ($typeLocation, inCity$) for entity resolution. The underlying resolution rules is RDF N-Triples given as follows:

R1 ($?x_i$ rdf:type $ds2 : Housing \wedge ?y_j$ rdf:type $ds2 : Housing$) $\wedge$ ($?x_i$ ds1:located $?a_i \wedge ?y_j$ ds1:located $?a_j$) $\wedge$ ($?a_i$ owl:SameAs $?a_j$) $\Rightarrow$ ($?x_i$ owl:SameAs $?y_j$)

R2 ($?x_i$ rdf:type ds1:Location) $\wedge$ ($?y_j$ rdf:type ds1:Location) $\wedge$ ($?x_i$ ds2:street $?s_i$) $\wedge$ ($?y_j$ ds2:street $?s_j$) $\wedge$ ($?s_i$ owl:SameAs $?s_j$) $\Rightarrow$ ($?x_i$ owl:SameAs $?y_j$)

The entity resolution rules are applied on RDF data using a resolution algorithm. In traditional programming, the complexity of the algorithm is factorial with respect to the size of data, since it combines all data triples in order to trigger rules. This complexity makes the resolution algorithm not appropriate to deal with big data. Therefore, to address this limitation, we propose a MapReduce based algorithm that trigger entity resolution rules in parallel way on distributed small pieces of data. The algorithm reconciles pairs of entity fragments matching a functional key that appears in the antecedent of resolution rules. It is processed by connecting URIs using the ontological "owl:SameAs" relationship. The Map function groups the entity fragments related to the same entity by assigning them the same key. For that, the Map function transforms each serialized entity fragments into $< key, value >$. The key part is composed of properties of serialized entity fragment that appears in the antecedents of entity resolution rules and the $value$ part is the URI of the entity fragment to be reconcilied. The Reduce function gets as input a list of $< key, List < value >>$ where key is the key resolution defined by the Map function and $List < value >$ is the list of URIs of entity fragments sharing the key, thus representing fragments of the same entity. The Reduce function reconciles URIs of same key by connecting them using "owl:SameAs" relationship. Due to the space limitation the algorithm is presented through an example depicted in Fig. 4. Only the R2 rule is used in this example.

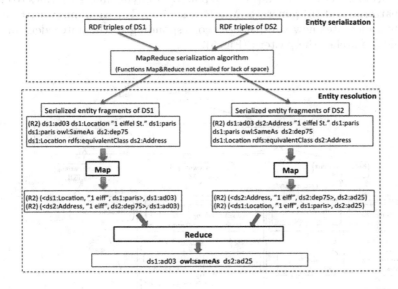

Fig. 4. MapReduce process of entity resolution

4   MapReduce-Based Query Rewriting on Big RDF Data

We present in this section a MapReduce query rewriting approach to compute a complete query results by including implicit data. Let us consider a SPARQL query Q2: *?x rdf:type ds2:Housing* $\wedge$ *?x ?p ?y)* with the aim is to get any information about housings.Traditional processing of this query returns only the address of housing *ds2 : h25*. However, when we consider: (1) entity resolution result of previous section, namely *ds1:h03 owl:SameAs ds2:h25*, result will be completed by the *ds1:h03* owner property and (2) the semantic connection *ds1:House rdfs:subClassOf ds2:Housing*, the result is completed by *ds1:h03*.

We propose a query rewriting algorithm based on MapReduce paradigm in order to enrich user query by adding more RDF patterns that explicitly refer to implicit data. It is processed into two steps. In the first step, a query plan composed of MapReduce jobs is generated for the query. In the second step, the generated query plan is evaluated in Hadoop framework to produce results.

The user query is rewritten using inference rules, including entity resolution as *SameAs* relationship rules. Inference rules are of the form: *antecedent* $\Rightarrow$ *goal*. The list of inference rules contains RDFS, OWL and the axioms rules defined by the user. To illustrate the proposed approach, we give only one typing and one entity resolution rule: *(R1): RDFS typing inference rule:* $(?x \ rdf : type \ ?y) \wedge (?y \ rdfs : subClassOf \ ?z) \Rightarrow (?x \ rdf : type \ ?z)$ and *(R2): Entity resolution inference rule*: $(?x \ ?p \ ?v) \wedge (?x \ owl : SameAs?y) \Rightarrow (?y \ ?p \ ?v)$. Inference rules are applied by backward reasoning algorithm. For a given query, the algorithm generates (i) MapReduce plan by applying inference rules to enrich query patterns and (ii) generates MapReduce jobs. For each query pattern, the algorithm generates new sub-patterns corresponding to the antecedent of rules whose goal matches the pattern (Fig. 5).

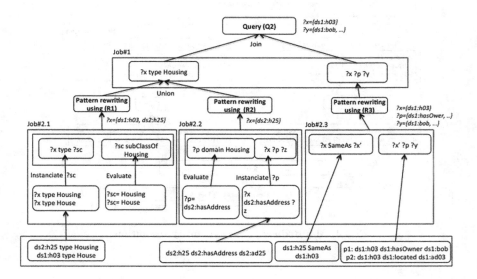

Fig. 5. Example of query rewriting plan in MapReduce jobs

Afterthat, MapReduce rewriting plan is evaluated on MapReduce framework in bottom up mode. MapReduce jobs on leaves are evaluted on data nodes. The output results are used as inputs for internal Jobs. Synchronization mechanism is developped to process job evaluations of MapReduce query rewriting plan. Experiments have been conducted on distributed clusters to show the efficiency of the algorithm by measuring the query time processing when applying inference mechanisms in Hadoop MapReduce framework with using real data from insurance company and open data *INSEE*[3].

5   Conclusion

We proposed a framework to process semantic fusion over big RDF data. We have developped an inference based semantic entity resolution and linking mechanism when the same entity is identified in different data sources and described using different semantics. We also have developed a rewriting SPARQL query evaluation approach over RDF data based on MapReduce paradigm to deal with the big RDF triplets and infer implicit data to include them in query results. We have verified experimentally, the effectiveness of the approaches using real data sets from insurance application.

Acknowledgment. The research leading to these results has received funding for Square Predict project from Fonds National pour la Société Numrique (FSN)- project investissement d'avenir (PIA 2013) Cloud Computing n 3 - Big Data program.

References

1. Du, J.-H., Wang, H.-F., Ni, Y., Yu, Y.: HadoopRDF: a scalable semantic data analytical engine. In: Huang, D.-S., Ma, J., Jo, K.-H., Gromiha, M.M. (eds.) ICIC 2012. LNCS, vol. 7390, pp. 633–641. Springer, Heidelberg (2012)
2. Goasdoué, F., Karanasos, K., Katsis, Y., Leblay, J., Manolescu, I., Zampetakis, S.: Growing triples on trees: an XML-RDF hybrid model for annotated documents. VLDB J. **22**(5), 589–613 (2013)
3. Papailiou, N., Tsoumakos, D., Konstantinou, L., Karras, P., Koziris, N: H₂rdf+: an efficient data management system for big RDF graphs. In: International Conference on Management of Data, SIGMOD 2014, Snowbird, UT, USA, 22–27 June 2014, pp. 909–912 (2014)

---

[3] http://www.insee.fr/en/bases-de-donnees/.

Enhancing Social Recommendation with Sentiment Communities

Davide Feltoni Gurini[✉], Fabio Gasparetti, Alessandro Micarelli,
and Giuseppe Sansonetti

Artificial Intelligence Laboratory, Department of Engineering,
Roma Tre University, Via Della Vasca Navale, 79, 00146 Rome, Italy
{feltoni,gaspare,micarel,gsansone}@dia.uniroma3.it
http://ai-lab-03.dia.uniroma3.it/people/

Abstract. Among the various recommender systems proposed in the literature, there is an increase in relevance and number of those that suggest users of possible interest to the target user. In this article, we propose a new algorithm for realizing user recommenders, named *SCORES (Sentiment COmmunities REcommender System)*. This algorithm relies on the identification of sentiment communities in which, for each topic cited by the user, we consider not only the relative sentiment, but also the volume and the objectivity of contents generated by him. The graph related to each topic is obtained by considering the Tanimoto similarity between users. The recommendation process occurs by clustering the obtained graph to detect latent communities, and suggesting to the target user the most similar K users based on tie strength measures. A comparative analysis between *SCORES* and some state-of-the-art approaches shows the benefits in term of performance.

Keywords: Sentiment analysis · Recommender system · Community detection

1 Introduction

With the proliferation of user-generated contents on social media such as reviews, discussion forums, blogs, and tweets, detecting sentiments and opinions from the Web and especially from Social Media is becoming an increasingly widespread form of data interpretation.

In this article, we exploit sentiment analysis for a new task: the identification of latent communities and their subsequent use in recommending similar users to the target user, that is, the user we want to suggest someone to follow. The research questions underlying our work are, therefore, the following:

1. Can the consideration of sentiment bring benefits for recommending users to follow?
2. If so, what is the best approach to do this?
3. Are there differences depending on the category of topics dealt with by the user?

© Springer International Publishing Switzerland 2015
J. Wang et al. (Eds.): WISE 2015, Part II, LNCS 9419, pp. 308–315, 2015.
DOI: 10.1007/978-3-319-26187-4_28

In particular, we show how this can be done by means of the identification of latent communities of users that, instead of considering social relationships, takes into account of the user's sentiment and interest towards specific topics. Based on such contributions, we define a *sentiment-volume-objectivity (SVO)* function. Hence, our method relies on (i) the construction of graphs, one for each topic cited by the user, (ii) the detection of the SVO-based latent communities through clustering techniques, (iii) the use of different measures of tie strength to adopt in the computation of the global similarity between users.

The experimental tests were performed on different real-world datasets, obtained by monitoring the traffic produced by users on Twitter[1]. Such data enabled us to realize a comparative analysis of our system, called *SCORES (Sentiment COmmunities REcommender System)*, with different approaches proposed in the literature.

The paper is structured as follows. Section 2 reviews some related works. Community detection in Sect. 3, whilst the user recommendation process is presented in Sect. 4. Section 5 reports the results of the experimental efforts. Section 6 concludes and outlines some possible future works.

2   Related Work

With the exponential advancement of social media and networking sites, sentiment analysis is increasingly being applied for the task of social network analysis for several purposes such as prediction in political elections [9], and event identification [8]. As far as we are aware, however, there have been few attempts to consider user attitudes in micro-posts for community detection or user recommendation. In [10] the authors view the problem of community sentiment discovery as a semidefinite programming (SDP) problem and as an optimization problem, and solve both of them through a SDP-based rounding method. Yuan *et al.* [11] provide an interesting study on how to make use of sentiment towards topics of common interest for link prediction between users. They put forward different techniques to assess how the *sentiment homophily* (i.e., the tendency of people to express similar levels of sentiment to that expressed by their friends) can improve the prediction of the likelihood of two users to follow each other. User recommendation approaches that ignore sentiment opinions have been proposed by Chen *et al.* [3], and Arru *et al.* [2] exploring different recommendations strategies.

3   Community Detection

The idea behind this work is that taking into account user attitudes towards his own interests can yield benefits in recommending friends to follow. Specifically, we consider (i) which is the sentiment expressed by the user for a given concept, (ii) how much he is interested in that concept, and (iii) how much he expresses objective comments on it. For *concept* we mean any entity hashtag extracted from a tweet that can somehow characterize it.

---

[1] twitter.com.

In our model the first contribution is the *sentiment* $S(u, c)$, which represents a feeling or opinion about a concept c expressed by the user u, with $u \in U$ (set of all users) and $c \in C_u$ (set of all concepts expressed by the user u). The goal of our sentiment analysis system is to obtain an output value that represents how much positive, negative or neutral is the sentiment expressed in a tweet. For this reason, we implemented a supervised machine learning algorithm based on a Naïve Bayes classifier. The second contribution in our model is the *volume* $V(u, c)$, that is, how much a user u wrote about a specific concept c. The third and last contribution is the *objectivity* $O(u, c)$, which expresses how many tweets about a concept c do not contain sentiments.

All of this contribution are entirely explained in [4]. Based on such contributions, we define a *sentiment-volume-objectivity (SVO)* vector, which takes into account all of them. If we consider a user $u \in U$ and a concept $c \in C_u$, it is defined as follows:

$$\boldsymbol{SVO(u,c)} = [\alpha S(u, c), \beta V(u, c), \gamma O(u, c)] \tag{1}$$

where α, β, and γ are three constants in the $[0, 1]$ interval, such that $\alpha + \beta + \gamma = 1$. In order to determine the optimal values of those parameters, we implemented a *mini-batch gradient descent* algorithm. In Sect. 5 we will see how such values depend on the category of topics mentioned by the user.

For each concept c we compute the Tanimoto similarity [7] between users u and $v \in U$ as follows:

$$sim(u, v, c) = \frac{\boldsymbol{SVO(u,c)} \cdot \boldsymbol{SVO(v,c)}}{\| \boldsymbol{SVO(u,c)} \|^2 + \| \boldsymbol{SVO(v,c)} \|^2 - \boldsymbol{SVO(u,c)} \cdot \boldsymbol{SVO(v,c)}} \tag{2}$$

The similarity value lies in between $[0, 1]$.

Once the similarities between users are computed, for each concept c we build a graph $G_c(V, E)$, where V represents the set of users, E the set of edges between them. We consider the similarity value as an edge between them, only if the similarity value between two users exceeds a threshold value Θ. Also the optimal value for Θ was determined through a gradient descent algorithm that maximizes the recommender precision. Afterwards we implemented a clustering algorithm based on modularity optimization that allows us to detect the latent communities for the considered concept c. This algorithm tends to optimize the modularity value Q that has the following expression:

$$Q = \frac{1}{2m} \sum_{u,v} [A_{uv} - \frac{k_u k_v}{2m}] \cdot \delta(g_u, g_v) \tag{3}$$

where A_{uv} represents the weight of the edge between u and v, $k_u = \sum_v A_{uv}$ is the sum of the weights of the edges linked to the user u, g_u is the community to which user u is assigned, $m = \frac{1}{2} \sum_{uv} A_{uv}$, and δ-function $\delta(s, t)$ is 1 if $s = t$ and 0 otherwise.

4 User Recommendation

Once identified the communities for all concepts mentioned by the target user u, the user recommender system works as follows. For every user v in the dataset, for each mentioned concept c we verify if it was also mentioned by the user u. In the positive case, we consider the related graph and calculate the measure of tie strength between u and v to obtain the recommendation score.

The notion of *tie strength* in social networks was introduced in [5]. Since a lot of *tie strength* measures have been proposed in the literature, we introduce in Table 1 which of those measures we employed in our recommender system.

Given a graph $G_c(V, E)$, we define *neighbor* of its node u a node with a direct link to u (i.e., a node with a path distance equal to one), and denote by $\Gamma(u)$ the set of its neighbors.

The first *tie strength* measure we employed is the GRAPH DISTANCE, that is, the number of hops of the shortest path between node u and node v. COMMON NEIGHBORS represents the shared neighbors between u and v, meanwhile JACCARD INDEX normalizes the common neighbors with the total neighbors of u and v. In the ADAMIC-ADAR tie strength, $\Gamma(u)$ and $\Gamma(v)$ are the neighborhoods of u and v respectively, and N is the number of nodes belonging to both of them. The WEIGHTED ADAMIC-ADAR is a modified version of the previous Adamic-Adar measure, where $sim(\Gamma(u) \cap \Gamma(v))$ is given by SVO Similarity in Eq. 2, that is, the average edge weight of the common neighbors between u and v. The PREFERENTIAL ATTACHMENT represents the number of neighbors of u multiplied by the number of neighbors of v. In this case we do not take the community

Table 1. Tie strength measures

GRAPH DISTANCE	$TS(u,v) = length(path(u,v))$
COMMON NEIGHBORS	$TS(u,v) = |\Gamma(u) \cap \Gamma(v)|$
JACCARD INDEX	$TS(u,v) = \dfrac{|\Gamma(u) \cap \Gamma(v)|}{|\Gamma(u) \cup \Gamma(v)|}$
ADAMIC-ADAR	$TS(u,v) = \displaystyle\sum_{N \in \Gamma(u) \cap \Gamma(v)} \dfrac{1}{log|N|}$
WEIGHTED ADAMIC-ADAR	$TS(u,v) = \displaystyle\sum_{N \in \Gamma(u) \cap \Gamma(v)} \dfrac{1}{log|N|} \cdot Avg(\dfrac{1}{sim(\Gamma(u) \cap \Gamma(v))})$
PREFERENTIAL ATTACHMENT	$TS(u,v) = |\Gamma(u)| \cdot |\Gamma(v)|$
KATZ	$TS(u,v) = \displaystyle\sum_{l=1}^{\infty} \tau^l \cdot |paths_{u,v}^{<l>}|$
SIMRANK	$TS^\phi(u,v) = \begin{cases} 1 & \text{if } u = v \\ \phi \cdot \dfrac{\sum_{a \in \Gamma(u)} \sum_{b \in \Gamma(v)} TS(a,b))}{|\Gamma(u) \cdot \Gamma(v)|} & \text{otherwise} \end{cases}$
RANDOM WALK	Starting from node u, two choices: − move to a random neighbor with probability $1 - \mu$ − jump back to u with probability μ

structure into account, but emphasize well-connected nodes over less connected ones. The KATZ measure sums over the entire collection of paths, each one exponentially damped by its length to emphasize short paths. In the equation $|paths_{u,v}^{<l>}|$ is the set of all length-l paths from u to v. A very small value of τ yields a tie strength much like common neighbors, since paths of length three or more yield a slight contribute to the summation. SIMRANK represents the similarity between two nodes u and v by recursively computing the similarity of their neighbors with $0 \leq \phi \leq 1$. In the RANDOM WALK the tie strength between u and v is the probability that the last node of the process is v.

To calculate the total score between two users u and v, we consider the sum of tie strength contributions for concepts mentioned by both of them:

$$Score(u,v) = \omega \cdot \sum_{c \in C_u \cap C_v} TS_c(u,v) \tag{4}$$

$$\omega = \frac{|C_u \cap C_v|}{|C_u \cup C_v|} \tag{5}$$

where ω is the ratio between the number of concepts shared by u and v and all the concepts cited by u and v. In this way the contribution of users sharing more concepts with the target user is greater than others.

We evaluate the total score between the target user u and all the users v in the dataset, and suggest to him a ranked list of the most K relevant users based on such value.

5   Experimental Evaluation

5.1   Datasets

In order to comprehensively evaluate SCORES, we considered three datasets obtained from Twitter. Those datasets were gathered using the Twitter APIs searching for specific hashtags.

- **Dataset 1** was obtained in 2013 during the Italian political elections. We retrieved the Twitter streams about politician leaders and Italian parties from Jan 25th to Feb 27th. The final dataset counts 1,085,121 tweets written in Italian language and 70,977 unique users.
- **Dataset 2** contains Italian tweets representing the most important mobile tech companies such as Samsung, Apple, Nokia, Huawei, LG, Motorola, and Blackberry. The dataset was gathered from Sep 2014 to Feb 2015, and counts 3,511,455 tweets from 181,000 unique users.
- **Dataset 3** was gathered searching automotive brands such as Audi, BMW, Ferrari, Jaguar, Mercedes, Toyota, and Porsche. The collection set was retrieved in English to facilitate the reproducibility of our approach from Dec 2014 to Feb 2015, and counts 2,915,131 tweets from 110,350 unique users.

5.2   Results

The goal of our recommender is to suggest to a target user someone to follow. To compare different profiling approaches and recommendation strategies, we need to determine when a user u is indeed relevant to another user v. We suppose that u is relevant to v if a *following relationship* exists between them. This assumption has already been proposed in literature [1,2,6] and is supported by the phenomenon of *homophily*, that is, the tendency of individuals with similar characteristics to associate with each other. In order to evaluate our system we selected, for each dataset, 1000 users that (i) posted at least 50 tweets in the observed period, and (ii) had more than 30 friends and followers. We used the *Success at Rank K (S@K)* metric, which provides the mean probability that a relevant user is located in the top K positions of the list of suggested users. Table 2 shows the performance of our recommendation algorithm for different tie strength measures. Interestingly, our experimental evaluation enabled us to notice strong correlations among communities related to a specific dataset. The first analysis indicates that the best measures among all datasets are Weighted Adamic-Adar and Katz, but these results change while varying the dataset and, therefore, the concepts. Katz measure works best for topics about politics ($Dataset_1$) where the strong ties are very important. Indeed, if we use Katz tie strength, we suggest the most similar SVO users and, therefore, the nearest users within the same SVO community. On the contrary, in the other two datasets Weighted Adamic-Adar resulted the best tie strength measure, which is inversely proportional to the number of common neighbors and the SVO similarity. In this case, we are indeed suggesting the weak ties, that is, users that belong to different SVO communities with low similarity. These findings highlight that in less opinion-oriented topic such as technology ($Dataset_2$) and automotive ($Dataset_3$), recommending users belonging to a different SVO communities might be more useful. We plan to further investigate this issue in order to fully understand the real nature of those interactions and exploit such knowledge in the recommendation process.

In Table 3 we report the results of a comparative analysis of our system with some state-of-the-art functions. More precisely, we considered the following functions: (i) a content-based function, called *S1-Twittomender* [6], where users are profiled through the content of their tweets, (ii) a collaborative filtering function, *S7-Twittomender* [6], where users are represented through a combination of followers and followees, (iii) a *VSM (Hashtag)* function representing cosine similarity in a vector space model, where vectors are weighted hashtags and (iv) a *Friend-of-Friend (FOF)* function proposed in [3], which only leverages the social network information (followers and followees).

As can be seen, our approach outperforms the other ones. These results confirm the potential of sentiment as a valuable feature for improving user recommender systems.

Finally, we also analyzed the user recommender performance in terms of variations of the three parameters α, β, and γ (see Eq. 1). In order to determine the

Table 2. Performance in terms of S@10 metric for different tie strength measures and different datasets (*$\tau = 0.2$; **$\phi = 0.8$).

Tie strength	$Dataset_1$	$Dataset_2$	$Dataset_3$	Average
GRAPH DISTANCE	0.155	0.151	0.159	0.155
COMMON NEIGHBORS	0.202	0.172	0.169	0.181
JACCARD INDEX	0.178	0.162	0.167	0.169
ADAMIC-ADAR	0.177	0.195	0.185	0.186
WEIGHTED ADAMIC-ADAR	0.179	**0.215**	**0.205**	**0.200**
PREFERENTIAL ATTACHMENT	0.152	0.158	0.161	0.157
KATZ*	**0.218**	0.191	0.189	**0.199**
SIMRANK**	0.165	0.156	0.159	0.160
RANDOM WALK	0.175	0.161	0.165	0.167

Table 3. A comparison among different state-of-the-art techniques. The values of Θ similarity threshold are 0.821 for $Dataset_1$, 0.630 for $Dataset_2$, and 0.711 for $Dataset_3$.

Recommender system	$Dataset_1$	$Dataset_2$	$Dataset_3$
SCORES	**0.218**	**0.215**	**0.205**
S1-TWITTOMENDER	0.130	0.118	0.115
S7-TWITTOMENDER	0.172	0.163	0.161
VSM (HASHTAG)	0.127	0.099	0.105
FOF	0.165	0.155	0.159

best values of those parameters, we implemented a *mini-batch gradient descent* algorithm and found the following values:

- $Dataset_1$: $\alpha_1 = 0.45$, $\beta_1 = 0.45$, and $\gamma_1 = 0.10$
- $Dataset_2$: $\alpha_2 = 0.25$, $\beta_2 = 0.50$, and $\gamma_2 = 0.25$
- $Dataset_3$: $\alpha_3 = 0.28$, $\beta_3 = 0.52$, and $\gamma_3 = 0.20$

Based on the proposed model and the used datasets, these weights appear to highlight the contribution of *volume* and *sentiment* in $Dataset_1$, and *objectivity* in $Dataset_2$ and $Dataset_3$. This can be explained because $Dataset_2$ (technology) and $Dataset_3$ (automotive) are likely to contain more news and articles with few opinions and sentiments than $Dataset_1$.

6   Conclusion

In this paper, we have described an approach to leveraging community detection for people recommendation. Our work emphasizes the use of implicit sentiment analysis in improving recommendation performance. The experimental results reveal the benefits of our approach compared with some state-of-the-art techniques. Such findings enable us to answer the research questions as follows:

1. Exploiting sentiments within user recommendation may indeed improve the system performance;
2. The best approach observed is a specific combination of SVO integrated into Weighed Adamic-Adar and Katz tie strength measures;
3. The aforementioned combination are topic dependent. Particularly, the contributions of sentiments are higher for politic-oriented topics instead of automotive and technology.

As future work, we plan to exploit temporal information for understanding the evolution of relationships between users over time. We also plan to further investigate how parameters α, β, and γ shape the formation of the communities and deploy SCORES in a wide domain such as movie or news recommendation.

References

1. Abel, F., Gao, Q., Houben, G.-J., Tao, K.: Analyzing user modeling on twitter for personalized news recommendations. In: Konstan, J.A., Conejo, R., Marzo, J.L., Oliver, N. (eds.) UMAP 2011. LNCS, vol. 6787, pp. 1–12. Springer, Heidelberg (2011)
2. Arru, G., Feltoni Gurini, D., Gasparetti, F., Micarelli, A., Sansonetti, G.: Signal-based user recommendation on twitter. In: Proceedings of the 22nd International Conference on World Wide Web Companion, pp. 941–944 (2013)
3. Chen, J., Geyer, W., Dugan, C., Muller, M., Guy, I.: Make new friends, but keep the old: recommending people on social networking sites. In: Proceedings of the 27th International Conference on Human Factors in Computing Systems, CHI 2009, pp. 201–210. ACM, New York (2009)
4. Feltoni Gurini, D., Gasparetti, F., Micarelli, A., Sansonetti, G.: A sentiment-based approach to twitter user recommendation. In: Proceedings of the 5th ACM RecSys Workshop on Recommender Systems and the Social Web (2013)
5. Granovetter, M.: The strength of weak ties. Am. J. Sociol. **78**(6), 1360–1380 (1973)
6. John, H., Mike, B., Barry, S.: Recommending twitter users to follow using content and collaborative filtering approaches. In: Proceedings of the 4th ACM Conference on Recommender Systems, RecSys 2010, 26–30 September 2010
7. Tanimoto, T.T.: An elementary mathematical theory of classification and prediction. IBM Internal Report (1957)
8. Thelwall, M., Buckley, K., Paltoglou, G.: Sentiment in twitter events. J. Am. Soc. Inf. Sci. Technol. **62**(2), 406–418 (2011). http://dx.doi.org/10.1002/asi.21462
9. Tumasjan, A., Sprenger, T., Sandner, P., Welpe, I.: Predicting elections with twitter: what 140 characters reveal about political sentiment. In: Proceedings of the Fourth International AAAI Conference on Weblogs and Social Media, pp. 178–185 (2010)
10. Xu, K., Li, J., Liao, S.S.: Sentiment community detection in social networks. In: Proceedings of the 2011 iConference, pp. 804–805. ACM, New York (2011)
11. Yuan, G., Murukannaiah, P.K., Zhang, Z., Singh, M.P.: Exploiting sentiment homophily for link prediction. In: Proceedings of the 8th ACM Conference on Recommender Systems, RecSys 2014, pp. 17–24. ACM, New York (2014)

Trending Topics Rank Prediction

Soyeon Caren Han, Hyunsuk Chung, and Byeong Ho Kang[✉]

School of Engineering and ICT, University of Tasmania,
Sandy Bay, Hobart, TAS 7005, Australia
{Soyeon.Han,David.Chung,Byeong.Kang}@utas.edu.au

Abstract. Many web services, such as Twitter and Google, provide a list of their most popular terms, called a trending topics list, in descending order of popularity ranking. The changes in people's interest in a specific trending topic are reflected in the changes of its popularity rank (up, down, and unchanged). This paper analyses the nature of trending topics and proposes a temporal modelling framework for predicting rank change of trending topics using historical rank data. Historical rank data show that almost 70 % of trending topics tend to disappear and reappear later. Therefore it is important to reflect this phenomenon in the prediction model, which is related to handling missing value and window size. Missing value handling approach was selected by using expectation maximization. An optimal window size is selected based on the minimum length of topic disappearance in the same topic but with a different context. We examined our approach with four machine-learning techniques using the U.S. twitter trending topics collected from 30th June 2012 to 30th June 2014. Our model achieved the highest prediction accuracy (94.01 %) with C4.5 decision tree algorithm.

Keywords: Trending topic · Temporal prediction · Trends prediction

1 Introduction

Many web services, including Google and Twitter, analyze their user data and provide a Trending Topics service, which displays the most popular terms that are discussed and searched within their community. One of these services, Twitter, monitors their social data, detects the terms currently most often mentioned by their users, and publishes these on their site. Kwak et al. [3] demonstrated that over 85 % of trending topics in Twitter are related to breaking news headlines, and the related tweets of each trending topic provides the people's opinion in real-world issue. Hence, being able to know which topics people are currently most interested in on Twitter, and their point of view, may lead to opportunities for analyzing the market share in almost every industry or research field, including marketing, politics, and economics. The 'Trending Topics' list shows the top 10 trending topics in descending order of popularity ranking. Based on the rank of a trending topic, it is possible to recognize the degree of current popularity of that topic in a specific geographic location, from individual cities

© Springer International Publishing Switzerland 2015
J. Wang et al. (Eds.): WISE 2015, Part II, LNCS 9419, pp. 316–323, 2015.
DOI: 10.1007/978-3-319-26187-4_29

to worldwide[1]. The lower the rank the higher the popularity, the higher the rank the lower the popularity. Based on people's interest change, the trending topic has different hourly ranking changes: up, down, and unchanged. Predicting the trending topics hourly ranking change can be helpful to identify the influence of the topic in the near future.

However, the 'Trending Topics' list displays only limited information, including the trending topic term, its rank, location, and updated date and time. We used only this available information to predict the future rank changes of trending topics. Therefore, the research aim of our study is to answer the following question: "how can we predict the change of trending topics' popularity (up, down, and unchanged) by using historical rank data?" In order to use the historical rank data for rank change prediction, there is an issue to investigate. Several trending topics tend to disappear and reappear from the trending topics list so it is impossible to know the exact rank when it disappears. Historical rank data show that almost 70 % of trending topics tend to disappear and reappear later. It is important to reflect this 'disappearance and reappearance' phenomenon in the prediction model, which is related to handling missing value and window size. First, we applied and compared four missing value-handling approaches in Sects. 2.1 and 4.2. Secondly, for selecting window size, we proposed a method to select the appropriate window size for predicting rank change of trending topics. It was found that the context can change while the trending topic has disappeared. We need to find the minimum length of topic disappearance hours in the same topic with different contexts, and apply it to the window size.

2 Temporal Modeling of Trending Topic Ranking Changes

The goal of this research is to predict the trend of trending topics rank change in the next hour. We propose a temporal modeling framework for predicting trending topics rank change using historical rank pattern and machine learning techniques. The proposed model can be described using the following equation:

$$FRC(T_x) = ML(PRP(T_x)) \tag{1}$$

$$PRP(T_x) = [r_{t-n}, ..., r_{t-1}, r_t] \tag{2}$$

$$FRC(T_x) = \begin{cases} up, & \text{if } r_t - r_{t+1} > 0 \\ down, & \text{if } r_t - r_{t+1} < 0 \\ unchanged, & \text{if } r_t - r_{t+1} = 0 \end{cases} \tag{3}$$

In order to predict the next rank change FRC of a specific trending topic T_x, we used past rank pattern data (PRP) of the topic T_x. Then, machine learning

---

techniques ML are applied for learning our model. Equation 3 describes the example of historical rank pattern PRP of a specific trending topic T_x at time t. It shows all historical rank patterns of a topic T_x in the specific period n. FRC represents the trends of the topic's ranking in the next hour. By comparing the current rank and the next-hour rank, the predicted rank change in the next hour will be one of three classes: up, down, and unchanged. For example, if the next-hour rank r_{t+1} is higher than the current rank r_t, the FRC will be 'down'. There are two main issues when we use the historical ranking data for our model: missing ranking handling and window size selection.

2.1 Missing Ranking Handling

As the 'Trending Topics' list displays the top 10 trending topics of the moment, it displays the topics from rank1 to rank10. In other words, if the topic is suddenly out of the 'Trending Topic' list, it is impossible to recognize the exact ranking, whether the topic is ranked 11th or 50th.

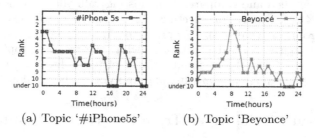

(a) Topic '#iPhone5s'   (b) Topic 'Beyonce'

Fig. 1. Topic disappearance and reappearance from 'Trending Topics' list

Figure 1 shows the example of the nature of trending topics disappearance and reappearance. The figure contains two sub-figures that display the hourly rank change of two different trending topics in 24 h; x-axis represents the 24 h from the point the trending topic initially appeared on the list, and y-axis shows the ranking, from rank1 to rank under 10th, of the trending topic. 'Under 10' in y-axis describes when the topic disappeared from the list. The first figure shows the hourly rank change of the topic '#iPhone5s', which appeared when Apple introduced the iPhone5. The topic was out of the list for three hours from the point it appeared in the 16th hour. The second figure shows the rank change of the topic 'Beyonce' that is referring to news that Beyonce had a fight with her husband. Manual inspection of the trending topics revealed that topic disappearance and reappearance is not limited to the type of topic. Various types of trending topics, including breaking news, persistent news (e.g. TV show or sport match) and hash tags seem to disappear and reappear randomly. We then analyzed how many trending topics actually disappear and reappears. Table 1 shows the percentage of trending topics that reappear and failed to reappear

Table 1. The percentage of trending topics that reappeared or non-reappeared

	Reappearance	No-reappearance
Percentage	66.28 %	34.82 %

after the topic disappeared. The proportion of reappearing trending topics is almost 70 %.

Based on this analysis, we claim that it is crucial to deal with missing rank data for our prediction model. We applied the following four missing value-handling approaches: (1) Deletion: we applied pairwise deletion approach, (2) Dummy variable control: we handle the missing value as zero (0), (3) Mean substitution: we replaced all missing rank in a variable by the mean of that variable, and (4) Expectation maximization (EM): This is a maximum likelihood approach that can be used to create a new data set in which all missing values are imputed with a maximum likelihood value. Based on the EM calculation, we found that the replaceable value for the missing value of our data should be the rank, lowest+1.

2.2   Window Size Selection

The proposed temporal model uses historical trending topics rank pattern, time-series data, so it is important that sequences of the same window size should be used in training and testing. However, the primary difficulty is selecting an optimal window size for prediction using a good learning technique instead of trial and error. We analyze the actual trending topic ranking data on USA Twitter. According to the data analysis result, we found that the same topic terms are sometimes referring to different events, and this normally occurs when the time length of the topic disappearance exceeds a certain time. For example, Table 2 shows the example of analyzing the same trending topic '#MalaysiaAirlines' that is about two different events. The table displays the collected date and representative content of each topic. In 2014, there were two sad events that are related to Malaysia Airline: Firstly, the Malaysia airline flight MH370 disappeared on 8 March carrying 227 passengers and 12 crews. The second referred to MH17 which is believed to have been downed by a surface-to-air missile in the eastern Ukraine on 17 July with 259 passengers on board. The table shows that the same topic '#MalaysiaAirlines' are about two different events based on the collected date. On the first row, the extracted content shows the words missing, disappear and loss, which are related to the missing airline but those on the second row has the words shot, missile, kill, crash, and attack which relates more to the airplane that was allegedly attacked by missile. Hence, before and after almost 4 months disappearance, the topic term '#MalaysiaAirlines' is separated into two different events.

We proposed the approach to identify the minimum length of topic disappearance that has different contexts by comparing the context similarity in two time-points (before-and-after the topic disappearance). As mentioned earlier, the

Table 2. The trending topic '#MalaysiaAirlines' with different events

Collected date	Extracted contents
2014/03/08	missing, flight, Malaysian, MH370, passenger, disappear, crash, pray, crew, lost, ocean, fail, safety, loss, airplane
2014/07/17	shot, down, missile, incident, kill, crash, attack, another, flight, victims, Malaysian, report, 259, explode

trending topic terms consist of words, hash-tags or short phrases but it does not provide any description. It is almost impossible to recognize the exact meaning of a trending topic without extracting its detailed information. Hence, we proposed an approach to extract the representative contents for each trending topic and compare the context similarity in two time points if the topic disappeared at one point. The proposed approach is conducted as follows: (1) collect the trending topic and related tweets of the topic published less than 1 h ago, (2) preprocess the related tweets by removing stop words and (3) extract the representative 15 (fifteen) terms using term frequency (TF), and (4) calculate the cosine similarity of context of a specific trending topic at two different time-points.

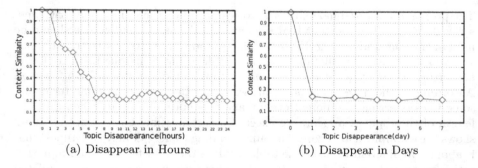

(a) Disappear in Hours   (b) Disappear in Days

Fig. 2. The average of content similarity based on the topic disappearance time

Figure 2 presents two sub-figures that show the result of context similarity based on the length of continuous disappearance; x-axis represents the length of topic continuous disappearance, and y-axis shows the cosine similarity rate (1 means exactly same and 0 is completely different). As you can see from the graphs, you can find that the context similarity is very low (0.2) if the topic continuously disappeared for over 7 h. Moreover, the similarity does not go down after 7 h, which is around 0.2. In other words, if a specific trending topic 'A' does not appear in the list for over 7 h and then reappears again, we can tell the first appeared topic 'A' and reappeared topic 'A' are talking about different contexts. In other words, if the topic disappears for less than 7 h and reappears, the topic can be considered as the same topic. If the topic disappears for more than 7 h, the topic is considered a different topic. Based on this result, the optimal window size for trending topic rank data can be the minimum length of continuous disappearance without different contexts in same topic term. The optimal size for U.S. twitter trending topic rank data should be 7 (seven). The evaluation of prediction with different window size will be conducted in the Sect. 4, evaluation result.

3   Experimental Setup

3.1   Evaluation Data

For evaluating the trending topics' rank prediction, we used Twitter API and collected trending topic terms, related tweets and ranking patterns for those topics for two years (from 30th June, 2012 to 30th June, 2014) in different countries (USA, UK, and Australia). For trending topic terms, we crawled the top 10 trending topics every hour. The API returns the trending topic term, the rank, the location and time of the API request. In total, we collected 57359 unique trending topics from U.S., 33400 topics from U.K., and 10039 topics from Australia. Since the trending topics list displays only the topics terms, with no detailed information, it is impossible to identify the meaning of trending topic until you examine related tweets for that topic. To solve this issue, we searched the related tweets of each trending topic. While collecting the related tweets, we aimed to avoid crawling the tweets that contain irrelevant contents. We used the published date-time to extract the appropriate related tweets. As we collect the top 10 trending topics on an hourly basis, we search the related tweets that users upload in the last one hour. For example, when 'Malaysia Airline' is on the trending topics list at 8pm, we search and collect the related tweets that users uploaded between 7pm and 8pm. This collecting approach minimizes irrelevant tweets. In order to achieve the Eq. 2 in Sect. 2, the training data contains historical rank pattern as features, and the predefined future rank as class. The number of features are changing based on the optimal window size.

4   Evaluation Result

4.1   Window Size Selection Examination

As we discussed in the Sect. 2.2, we proposed that the approach to selecting the optimal window size for trending topics' ranking change predictions. We found that optimal window size can be same as the minimum length of topic disappearance time that has same topic term with different meaning (see Sect. 2.2). We discovered the optimal window size for U.S. twitter data can be 7 (seven). In order to examine the proposed window size selection approach, we applied our approach to the trending topic rank data from U.K. and Australia Twitter. Based on this examination, we found that the optimal window sizes for U.K. and Australia was 6 (six) and 8 (eight) respectively, as shown in Table 3. We evaluate the prediction performance with those window sizes to examine whether the proposed approach selects the optimal window size of different data.

Table 3. Optimal window size for three countries (U.S., U.K., AU.)

	USA	UK	AU
Optimal Window Size	7	6	8

Table 4. U.S trending topics ranking change prediction accuracies with different missing ranking handling approaches and window sizes

	Window Size	Missing Value	NB	NN	SVM	C4.5
(1)	5	Zero(0)	79.71%	88.20%	79.91%	88.74%
(2)	5	Lowest+1	80.11%	88.92%	80.82%	89.85%
(3)	5	Mean	75.10%	86.56%	77.29%	87.49%
(4)	5	Deletion	75.91%	85.42%	77.52%	85.74%
(5)	7	Zero(0)	83.91%	93.56%	85.36%	93.08%
(6)	**7**	**Lowest+1**	83.03%	**93.68%**	86.04%	**94.01%**
(7)	7	Mean	80.23%	91.06%	83.22%	92.91%
(8)	7	Deletion	82.93%	92.76%	83.93%	90.10%
(9)	9	Zero(0)	83.88%	92.53%	85.31%	93.00%
(10)	9	Lowest+1	83.00%	92.54%	85.61%	93.88%
(11)	9	Mean	80.34%	91.40%	83.29%	92.14%
(12)	9	Deletion	82.91%	90.92%	83.91%	90.11%

4.2 Prediction Evaluation

The experiments were designed to test the proposed model. We use the prediction performance as an indication of the suitability, which is obtained from four machine-learning techniques we discussed in the previous section. We used 10-fold cross validation on two years of training data. Each experiment result has different window sizes and different missing ranking handling techniques. Table 4 shows the prediction result of U.S. trending topics ranking changes with different window sizes (5,7,9) and four different missing handling techniques (Zero, Lowest+1, Mean, and Deletion). As mentioned in Sect. 2.2, the optimal window size for U.S. data can be size 7. The result shows that the prediction with size 7 has the highest performance among 5, 7 and 9, which proves that our approach performs successfully. Since there is little difference in prediction accuracy of size 7 and 9, it is difficult to define whether 7 is better than 9. However, we can infer that if there is no difference, using size 7 is effective in performance, including data size and speed. For missing ranking handling, missing value imputation with lowest+1 achieve the best prediction performance. This is because the other three approaches, mean, zero, and deletion, are not considered the nature of trending topics ranking but the imputation with EM. Therefore, it shows the best prediction accuracy in all three instances. We applied four machine learning techniques, including Naive Bayes, Neural Network, Support Vector Machine, and C4.5 Decision Tree algorithm, and C4.5 achieved the highest performance (94.01 %). Finally, we analyzed the performance of other countries (U.K. and AU.) to make sure that our model performs well. Previously, we found that the optimal window sizes for two countries were size 6 and size 8 respectively. As a result, U.K. Trending Topic Ranking change prediction achieves the best prediction performance (92.54 %), and Australian results the highest accuracy (80.13 %) in 6 and 8 instances, and lowest+1 imputation.

5 Related Works

The Trending Topics list has received much attention [3]. 'Twitter Trending Topics', real-time event detection service provided by Twitter, shows the most often mentioned terms. They show only the topic term and its rank with no detailed explanation so various summarisation and extraction approaches are proposed to reveal the exact meaning of trending topics. Han and Chung [2] applied simple Term Frequency approach for extracting the representative keywords to disambiguate the approach. Han et al. [1] proved that the most successful approach to reveal the exact meaning of trending topics is simple Term Frequency. Lee et al. [4] classifies trending topics into 18 general categories by labeling and applying machine-learning techniques.

6 Conclusion

In this paper, we proposed a temporal modeling framework that predicts trending topics' hourly ranking change. The only available data for this problem is rank history data of each trending keywords. People may have question about whether any predication models using this data can suggest any promising prediction results. Surprisingly, our method achieved significant performance. It is possible to obtain additional feature but it would be very difficult to predict rank perfectly, which is not because of algorithmic factors but because of trending topics' irregularly changing nature.

References

1. Han, S.C., Chung, H., Kim, D.H., Lee, S., Kang, B.H.: Twitter trending topics meaning disambiguation. In: Kim, Y.S., Kang, B.H., Richards, D. (eds.) PKAW 2014. LNCS, vol. 8863, pp. 126–137. Springer, Heidelberg (2014)
2. Han, S.C., Chung, H.: Social issue gives you an opportunity: discovering the personalised relevance of social issues. In: Richards, D., Kang, B.H. (eds.) PKAW 2012. LNCS, vol. 7457, pp. 272–284. Springer, Heidelberg (2012)
3. Kwak, H., Lee, C., Park, H., Moon, S.: What is twitter, a social network or a news media? In: Proceedings of the 19th International Conference on World Wide Web, pp. 591–600. ACM, April 2010
4. Lee, K., Palsetia, D., Narayanan, R., Patwary, M.M.A., Agrawal, A., Choudhary, A.: Twitter trending topic classification. In: 2011 IEEE 11th International Conference on Data Mining Workshops (ICDMW), pp. 251–258. IEEE, December 2011

Moving Video Mapper and City Recorder with Geo-Referenced Videos

Guangqiang Zhao[✉], Mingjin Zhang, Tao Li, Shu-Ching Chen,
Ouri Wolfson, and Naphtali Rishe

School of Computing and Information Sciences,
Florida International University, Miami, FL 33199, USA
{gzhao002,zhangm,taoli,chens,wolfson,rishen}@cs.fiu.edu
http://terrafly.com/

Abstract. There has been growing interest in correlating co-visualizing a video stream to the dynamic geospatial attributes of the moving camera. Moving videos comes from various GPS-enabled video recording devices and can be uploaded to video-sharing websites. Such public website do not presently display dynamic spatial features correlated to a video player. Although some systems include map based playback products, there has been no unified platform for users to share geo-referenced videos that takes spatial characteristics into account. We present here Moving Video Mapper, which integrates both historical and live geo-referenced videos to give users an immersive experience in multidimensional perspectives. The platform has been evaluated using real data in an urban environment through several use cases.

Keywords: Geo-referenced video · Moving objects · Online map · Dashboard camera · Digital city

1 Introduction

A photograph is usually taken from a location. An action video, which can be seen as a series of photographs from different locations, usually reflects an object moving along a trajectory. In recent years, thanks to the price drop and miniaturization of Global Positioning System (GPS) chips, GPS-enabled video recording devices have proliferated. One remarkable example is the now widely used dashboard cameras (dashcam), which continuously record videos and GPS tracking while the users are driving [7]. Other examples include smartphones, unmanned

---

This material is based in part upon work supported by the National Science Foundation under Grant Nos. III-Large-MOD IIS-1213013, I/UCRC IIP-1338922, AIR IIP-1237818, SBIR IIP-1330943, III-Large IIS-1213026, MRI CNS-1429345, MRI CNS-0821345, MRI CNS-1126619, CREST HRD-0833093, I/UCRC IIP-0829576, MRI CNS-0959985, RAPID CNS-1507611 and U.S. DOT Grant ARI73. Includes material licensed by TerraFly (http://terrafly.com) and the NSF CAKE Center (http://cake.fiu.edu).

J. Wang et al. (Eds.): WISE 2015, Part II, LNCS 9419, pp. 324–331, 2015.
DOI: 10.1007/978-3-319-26187-4_30

aerial vehicles (UAVs), and sports (a.k.a. action) camcorders. Some devices with networking capabilities can even stream real-time video remotely. Videos taken by such devices comprise Big Data. However, the playback software modules are generally from different providers and they are often incompatible with each other. Integration is need to facilitate route preview, virtual sightseeing, crime monitoring, smart city applications [7] and data mining [13].

With the rapid development of online geographic information system (GIS) services, Web-based map applications have entered the everyday life. Many existing online map services include geo-tagged photos and videos, either uploaded by users or collected by the provider [6].

This paper proposes Moving Video Mapper, a framework to integrate geo-referenced videos from moving objects for immersive city observation through multidimensional perspective. Compared with the state of the art, our main contributions include:

- A smart data importing system able to recognize a wide variety of video and GPS track forms from devices of wide range of brands
- Supports real-time video streaming along with historical video archiving
- Geo-tagging service for video recording devices without GPS capabilities
- When possible, retrieve and display data from other sensors (e.g. accelerometer, gyroscope, and barometer)
- Automatically and intelligently classify related videos of specific locations by time of day, weather conditions, etc.
- Synchronous playback of both the video and the route, with cross-interactive capability

The rest of the paper is organized as follows. In Sect. 2, we introduce the related work. We then describe the architecture of our framework in Sect. 3. A prototype has been implemented and used to validate our system using volunteer collected data, which is discussed in Sect. 6. Finally, we conclude in Sect. 7.

2   Related Work

Displaying multimedia containing geographic information on a map is not a new idea. Many products on the market can display geo-tagged multimedia on a map [6,12], most of which are photo-based. For videos, they only reference them as points (markers on the map). It is difficult for users to understand how the videos were recorded by these scattered markers. Google developed a product named StreetView, based on GoogleMaps, which provides panoramic view-points along streets worldwide [1]. Although StreetView covers almost all sceneries along the road, it is still discrete photos. Some systems have been proposed to solve this problem by generating smooth videos or photos from panoramas along streets [2,5,8]. These methods depend on the intensity of the panoramas collected and therefore the quality of the videos or images generated from these panoramas can be compromised.

Recently, several platforms and frameworks have been proposed to utilize geo-referenced videos along with the map. These kinds of videos come with spatial and temporal information bound to frames of the video. PLOCAN [10] focuses on combining a Web-based video player and a map together to play dedicated videos with positions shown on the map. Citywatcher [7] lets users annotate dashcam videos and upload them to the City Manager. Chiang et al. [3] proposed a framework to share and search user-uploaded dashcam videos. However, it requires a specific application installed on the smartphone to record the video. Furthermore, there is no existing platform that integrates all kinds of geo-referenced videos.

The limitations of the aforementioned methods/systems have motivated us to find new ways of showing geo-referenced videos that provide the best user experience. Our proposed platform is based on our previous work named City Recorder [11]. We will introduce the architecture of Moving Object Mapper in the next section.

3   System Design

The Moving Object Mapper is an online GIS platform based on classic three-tier server-client. As shown in Fig. 1, the presentation tier resides on the client side while the logic tier and data tier run at the server side.

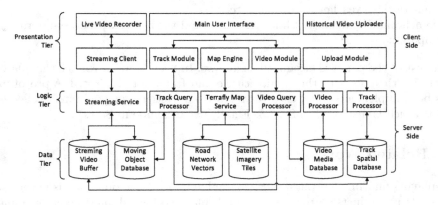

Fig. 1. System architecture

3.1   The Presentation Tier

The presentation tier collects data from users and respond with query results. The event-driven main user interface relies on three major modules. The track module and video module are core parts of the website to show tracks and videos to users, respectively. We use TerraFly [9] as our map engine to support the track visualization, and the video module utilizes the open-source jPlayer[1]. The historical uploader redirects users to the geo-referenced video uploading page and guides users through-out each step. Real-time videos will be transferred to the server by the streaming client.

3.2   The Logic Tier

The logic tier processes data between presentation tier and data tier. The historical data will be transferred to desired format and stored into the database, which covered in Sect. 4. In Sect. 5 we will show how to deal with data streaming. The track and video query processor mainly handles general queries from the main user interface. Section 6 will explain their visualization in depth.

3.3   The Data Tier

We use various types of databases to adapt different data in the data tier. The videos are stored in multi-media database. The tracks, road network vectors and satellite imagery tiles are stored in separate spatial databases. The recording locations of live videos are constantly updated into the MOD (Moving Object Database).

4   Historical Data Processing

Most GPS-enabled devices will store track files along with recorded videos in some kinds of storage media. Here the historical data specifically refers to the offline data that users uploaded manually.

4.1   Track Data Processor

The track data processor extracts tracks from the user-uploaded file and stores them into the track database. Since there are a variety of devices from different manufacturers on the market, the first step is converting these tracks to a device-independent format. The most common track file formats include: NMEA 0183[2], GPX, and KML. A track parse module inside this processor handles all the track extraction tasks.

For videos that recorded along streets (e.g. dashcam videos), we can align the track data based on the road network using a map matching algorithm (MMA). Regarding to the devices without GPS capabilities, there is also a geo-tagging tool for users to manually mark the track on the map.

---

[1] http://jplayer.org/.
[2] http://en.wikipedia.org/wiki/NMEA_0183.

4.2   Video Data Processor

Similar to track processings, we want a unified format for the videos which is suitable for Web-based players. The mp4 (MPEG-4 Part 14) is chosen considering its broad network compatibility. Another task is compression in order to reduce the video size. We utilize a free video processing tool FFmpeg[3] to perform video re-coding and compression tasks.

Then we utilize Support Vector Machines (SVM) to learn and classify the videos by different characteristics (e.g. weather, road condition, and time of day).

5   Live Video Streaming

We provide a mobile application for live video streaming purpose. After users correctly registered on the server, the application will notify the server whenever it is online. The video will be transferred to the server wirelessly along with current location from the GPS sensor. s The location of the moving object is updated in the MOD. Then the buffered video file on server is ready for responding to streaming requests from the users. The videos and tracks will be archived into the historical database in a certain time interval.

6   Use Case Study

In this section we demonstrate the effectiveness of our system and show how it is used in real life through a series of use cases.

(a)   (b)

Fig. 2. (a) Street tracks. (b) Route query.

6.1   Street Network Based Videos

Most street network based videos come from dashboard cameras. Figure 2a shows that a car is turning right. The line graph below the video indicates the speed drop. It also can be switched to the altitude graph as needed. Figure 2b is a map with tracks rendered as polylines in different colors. We did a route query between two points. After clicking the simulator button, it will open the driving simulator shown in Sect. 6.3.

---

[3] https://www.ffmpeg.org/.

6.2 Free Moving Objects

Free moving objects (e.g. pedestrians, boats, and aircrafts) usually have unpredictable tracks. Figure 3 shows a live streaming video from a balloon. The polygon represents the field of vision projected on the ground.

Fig. 3. Live video from UAV.

(a)   (b)

Fig. 4. (a) Car simulator. (b) Aircraft simulator.

6.3 Moving Video Simulator

The simulator is designed for providing the user an immersive experience. We have designed several different themes to simulate various life situations. Figure 4a simulate a car using the dashcam video. If other videos are found at the same location and have same directions, the application will notify users (See the blue "Night Mode" button). This allows users to preview a place under different time of day, season of year, weather conditions, etc. It will greatly improve the user experience. Figure 4b shows an aircraft simulator using the video from a hexacopter drone.

6.4   Data Management

Users can upload their geo-referenced videos to our system using a guided interface as shown in Fig. 5a. Then the data will be processed at the server side and the estimated time is displayed to users as shown in Fig. 5b. Users can choose between waiting online or getting a notification email. Users have the option to change privacy settings of videos as seen in Fig. 5c. They can also share a video with others using the system generated URLs, and the video will not available to public without this URL.

(a)   (b)   (c)

Fig. 5. (a) Data uploading. (b) Data processing. (c) Data management.

7   Discussion and Future Work

This study presented the Moving Object Mapper for geo-referenced video sharing. City Recorder is a public service that relies on contribution from and cooperation of voluntary users. This mode is proven to be successful in numerous geolocation-related domains in recent years (e.g. OpenStreetMap). However, it is difficult to ensure the quality of the videos or detect inappropriate contents. Another concern is privacy protection. We are working on utilizing some technologies like automatically blurring faces and license plates [4] to prevent publication of sensitive information. As we are entering an era of Virtual Reality (VR), we are also planning to deliver 3D contents via the Web.

References

1. Anguelov, D., Dulong, C., Filip, D., Frueh, C., Lafon, S., Lyon, R., Ogale, A., Vincent, L., Weaver, J.: Google street view: capturing the world at street level. Computer **6**, 32–38 (2010)
2. Chen, B., Neubert, B., Ofek, E., Deussen, O., Cohen, M.F.: Integrated videos and maps for driving directions. In: Proceedings of the 22nd Annual ACM Symposium on User Interface Software and Technology, pp. 223–232. ACM (2009)
3. Chiang, C.-Y., Yuan, S.-M., Yang, S.-B., Luo, G.-H., Chen, Y.-L.: Vehicle driving video sharing and search framework based on GPS data. In: Pan, J.-S., Krömer, P., Snášel, V. (eds.) Genetic and Evolutionary Computing. AISC, vol. 238, pp. 389–397. Springer, Heidelberg (2014)
4. Frome, A., Cheung, G., Abdulkader, A., Zennaro, M., Wu, B., Bissacco, A., Adam, H., Neven, H., Vincent, L.: Large-scale privacy protection in google street view. In: IEEE 12th International Conference on Computer Vision, 2009, pp. 2373–2380. IEEE (2009)
5. Kopf, J., Chen, B., Szeliski, R., Cohen, M.: Street slide: browsing street level imagery. ACM Trans. Graph. (TOG) **29**, 96 (2010)
6. Luo, J., Joshi, D., Yu, J., Gallagher, A.: Geotagging in multimedia and computer vision-a survey. Multimedia Tools Appl. **51**(1), 187–211 (2011)
7. Medvedev, A., Zaslavsky, A., Grudinin, V., Khoruzhnikov, S.: Citywatcher: annotating and searching video data streams for smart cities applications. In: Balandin, S., Andreev, S., Koucheryavy, Y. (eds.) NEW2AN/ruSMART 2014. LNCS, vol. 8638, pp. 144–155. Springer, Heidelberg (2014)
8. Peng, C., Chen, B.Y., Tsai, C.H.: Integrated google maps and smooth street view videos for route planning. In: International Computer Symposium (ICS 2010), pp. 319–324. IEEE (2010)
9. Rishe, N., Chen, S.C., Prabakar, N., Weiss, M.A., Sun, W., Selivonenko, A., Davis-Chu, D.: Terrafly: a high-performance web-based digital library system for spatial data access. In: ICDE Demo Sessions, pp. 17–19 (2001)
10. Rodríguez, J., Quesada-Arencibia, A., Horat, D., Quevedo, E.: Web georeferenced video player with super-resolution screenshot feature. In: Moreno-Díaz, R., Pichler, F., Quesada-Arencibia, A. (eds.) EUROCAST. LNCS, vol. 8112, pp. 87–92. Springer, Heidelberg (2013)
11. Zhao, G., Zhang, M., Li, T., Chen, S.C., Wolfson, O., Rishe, N.: City recorder: virtual city tour using geo-referenced videos. In: 2015 IEEE International Conference on Information Reuse and Integration (IRI) (2015)
12. Zheng, Y.T., Zha, Z.J., Chua, T.S.: Research and applications on georeferenced multimedia: a survey. Multimedia Tools Appl. **51**(1), 77–98 (2011)
13. Zheng, Y., Wang, L., Zhang, R., Xie, X., Ma, W.Y.: Geolife: managing and understanding your past life over maps. In: 9th International Conference on Mobile Data Management, MDM 2008, pp. 211–212. IEEE (2008)

WebGC Gossiping on Browsers Without a Server

[Live Demo/Poster]

Raziel Carvajal-Gómez[✉], Davide Frey, Matthieu Simonin,
and Anne-Marie Kermarrec

INRIA Rennes, Rennes, France
{raziel.carvajal-gomez,davide.frey,matthieu.simonin,
anne-marie.kermarrec}@inria.fr

Abstract. Decentralized social networks have attracted the attention of a large number of researchers with their promises of scalability, privacy, and ease of adoption. Yet, current implementations require users to install specific software to handle the protocols they rely on. The WebRTC framework holds the promise of removing this requirement by making it possible to run peer-to-peer applications directly within web browsers without the need of any external software or plugins. In this demo, we present WebGC, a WebRTC-based library that supports gossip-based communication between web browsers and enables them to operate with Node-JS applications. Due to their inherent scalability, gossip-based protocols constitute a key component of a large number of decentralized applications including social networks. We therefore hope that WebGC can represent a useful tool for developers and researchers. (A previous version of this demo appeared in [8]. Since then, we have integrated the library with a new decentralized signaling service and introduced support for web workers.)

1   Introduction

A number of authors have proposed the use of gossip-based protocols for the implementation of decentralized social networks [9,13,14], these protocols form an unstructured distributed system to disseminate information in a periodic way, they are easy to deploy and resilient to failures. Yet, like for other peer-to-peer (P2P) solutions, their implementations have always required users to install specific software to support decentralized protocols to enable users to access the social network from their web browsers. This constitutes a major show stopper for the adoption of decentralized social-network solutions.

In this demo, we present WebGC, a library for gossip-based communication between web browsers. Based on the WebRTC framework, WebGC has the potential to improve the applicability of decentralized user-centric applications like social networks by enabling them to run directly within web browsers. WebGC is a Javascript library to provide a simplified framework for building

© Springer International Publishing Switzerland 2015
J. Wang et al. (Eds.): WISE 2015, Part II, LNCS 9419, pp. 332–336, 2015.
DOI: 10.1007/978-3-319-26187-4_31

gossip-based applications. This includes the implementations of standard components such as random-peer-sampling [11] and clustering [19] protocols. Moreover, it augments the WebRTC connection-initiation protocol by means of a decentralized signaling mechanism.

WebRTC's primary goal consists in enabling direct communication between two browsers for media and real-time communication. For this reason, WebRTC relies on a signaling server that makes it possible to establish a connection between two browsers that do not know each other. While this feature is very convenient for applications that need to establish a small number of connections, the signaling server quickly becomes a bottleneck in the context of P2P applications. WebGC's decentralized signaling service establishes connections by exploiting the very operation of peer-sampling protocols.

The demo will present a running example of a WebGC-enabled application: a semantic overlay grouping users by interests. Attendees will be able to join the application using their own laptops and follow its evolution in real time on a web page.

2 WebGC Architecture

WebGC is a Javascript library and relies on two underlying frameworks: WebRTC, and SimplePeer. The former is directly provided by compatible web browsers and implements the low-level interaction primitives that make it possible to establish communication between browsers. The latter also takes the form of a Javascript library and provides a wrapper around WebRTC that simplifies the establishment of data connections between peers. The left diagram in Fig. 1 places WebGC in the context of these two libraries.

The figure also shows that WebGC comprises a decentralized signaling service that replaces the centralized signaling server used in WebRTC applications. This makes it possible to eliminate bottlenecks when operating in environments that do not involve NAT and firewalls. When these are present, WebGC relies on ICE, STUN, and TURN like standard WebRTC applications. However, we are currently considering augmenting the library with NAT traversal solutions such as Nylon [12] or Croupier [10]. In the following, we describe the gossip-based framework provided by our library and detail its decentralized signaling service.

WebGC Internals. The right diagram in Fig. 1 depicts WebGC's architecture. Its core consists of a COORDINATOR object that instantiates the gossip protocols and acts as a communication broker dispatching incoming messages to the various protocols. The library currently includes the implementation of two peer sampling protocols, CYCLON [18] and the generic protocol suite from [11], as well as a clustering protocol [7,19]. All protocols implement a GOSSIPPROTO-COL "interface"—since Javascript does not natively support interfaces, we adopt the interface pattern described in [15]. The COORDINATOR makes it possible to stack these protocols on top of each other [7] to implement applications.

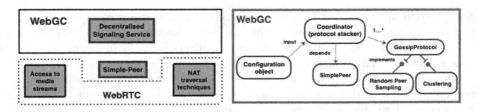

Fig. 1. General architecture

The GOSSIPPROTOCOL interface follows the scheme proposed in the literature [11,16,17] and defines the high-level operations that constitute a gossipprotocol. Developers can use the protocols provided by the library, but they
can also implement the operations in GOSSIPPROTOCOL to define new protocols. Finally, the latest version of WebGC also includes support for web workers.
This makes it possible to run complex CPU-intensive tasks without blocking the
browser's user interface.

Decentralized Signaling. A Signaling Service acts a mediator between two peers,
say P_1 and P_2, by assisting them during the establishment of a mutual connection. Let us assume P_1 wants to initiate a connection to P_2. WebRTC does
not allow nodes to contact each other using their IP addresses. So P_1 sends
its connection request to P_2 via the signaling service. P_2 responds by sending a
reply also through the signaling service, and finally P_1 initiates a direct link with
P_2 thanks to the data (IP address, communication protocol, etc.) contained in
the request and reply messages. WebRTC does not require a specific signaling
server but most existing solutions rely on a centralized architecture. However, a
centralized signaling server would easily become a bottleneck in a peer-to-peer
setting, particularly when using gossip-based protocols, which frequently need
to establish new connections.

While other decentralized signaling libraries exist [1], WebGC integrates signaling within the operation of peer-sampling and clustering protocols. In both
these types of protocols, which we refer to as overlay protocols, nodes maintain data structures called *views* that contain references to other nodes, and
periodically exchange messages that contain subsets of their views.

Our decentralized signaling solution maintains an additional routing table
that contains an entry for each of the node references that appear in any of
the views of running overlay protocols. Each such entry contains the node reference of a mediator node, i.e. a node that has an active connection with the node
in the entry. When a node needs to establish a connection to another node in its
table, it simply contacts the node's mediator and uses it as a signaling server.

WebGC maintains its routing tables by augmenting the messages sent by
the overlay protocols it hosts. Specifically, for each message sent by one such
protocol, WebGC also sends a routing table update that tells the receiving node
how to establish a connection which each of the nodes referenced in the overlay
message. Consider a node P_j sends to node P_i a message containing a reference

to node P_k. If P_j has an open connection with P_k, then P_j itself can act as a mediator between P_i and P_k, so it sends a routing table entry $< P_k, P_j >$. Otherwise, P_j simply forwards the information about its current mediator for P_k and thus sends an entry $< P_k, R_j[k] >$, where $R_j[k]$ is the mediator for P_k stored in P_js routing table.

3   Demo Timeline

Our demo will present a simple WebGC-based application consisting of two stacked overlay protocols and a chat service. We will provide one or two demo machines, but users will also be able to join the demo using their laptops. The application's web interface will require each user to specify his/her own interests in the form of keywords. This will organize users into a semantic overlay thereby identifying groups of users with similar interests.

Each of them will be able to follow the application's evolution on the web interface in the form of graphs that will display their RPS [11] and a clustering [19] views. In addition, users will be able to broadcast messages to each of the users in either view. Thanks to this feature, the demo might also provide workshop attendees with a way to exchange messages during the workshop.

4   Related Work

The introduction of WebRTC [2] has motivated a number of developers to design in-browser peer-to-peer solutions. PeerJS [3], SimplePeer [4], and P [1] all seek to offer a simplified API to program peer-to-peer applications on top of WebRTC. The previous version of WebGC [8] was built on top of PeerJS and exploited PeerServer, its associated signaling server. Since then, we have performed a complete refactoring to build our decentralized signaling solution, and to support multi-threading by means of web workers. In the process, we also migrated from PeerJS to SimplePeer, which simplified the library's requirements and made it possible to run WebGC applications without a browser on NodeJS [5].

While we designed our decentralized signaling solution to work in conjunction with gossip-based overlay maintenance, others have considered the idea of having a decentralized signaling server. P [1], for example, uses servers only for bootstrapping and then each peer can take up the role of a mediator. Unlike WebGC, however, P does not integrate signaling with gossip-based overlay maintenance. Finally, our work on WebGC is closely related to past efforts dedicated to the modular implementation of gossip protocols [6,11,16,17].

5   Conclusions

We presented WebGC, a library that simplifies the development of gossip-based applications based on the WebRTC framework. Built on top of SimplePeer [4], WebGC includes a decentralized signaling service as well as the implementation of several standard gossip protocols. Our demo demonstrates the effectiveness of our library in a real context, with a gossip-based chat application.

References

1. https://github.com/unsetbit/p/
2. http://www.webrtc.org/
3. http://peerjs.com/
4. https://github.com/feross/simple-peer/
5. https://nodejs.org/
6. http://gossiplib.gforge.inria.fr/
7. Bertier, M., Frey, D., Guerraoui, R., Kermarrec, A.-M., Leroy, V.: The gossple anonymous social network. In: Gupta, I., Mascolo, C. (eds.) Middleware 2010. LNCS, vol. 6452, pp. 191–211. Springer, Heidelberg (2010)
8. Carvajal-Gomez, R., Frey, D., Simonin, M., Kermarrec, A.: WebGC: browser-based gossiping. In: Proceedings of the Middleware 2014 Posters & Demos Session, 8–12 December 2014, Bordeaux, France, pp. 13–14 (2014). http://doi.acm.org/10.1145/2678508.2678515
9. Datta, A., Sharma, R.: GoDisco: selective gossip based dissemination of information in social community based overlays. In: Aguilera, M.K., Yu, H., Vaidya, N.H., Srinivasan, V., Choudhury, R.R. (eds.) ICDCN 2011. LNCS, vol. 6522, pp. 227–238. Springer, Heidelberg (2011)
10. Dowling, J., Payberah, A.H.: Shuffling with a croupier: nat-aware peer-sampling. In: 2012 IEEE 32nd International Conference on Distributed Computing Systems, 18–21 June 2012, Macau, China, pp. 102–111 (2012). http://dx.doi.org/10.1109/ICDCS.2012.19
11. Jelasity, M., Voulgaris, S., Guerraoui, R., Kermarrec, A.M., van Steen, M.: Gossip-based peer sampling. ACM TOCS 25(3), 8 (2007)
12. Kermarrec, A., Pace, A., Quéma, V., Schiavoni, V.: Nat-resilient gossip peer sampling. In: 29th IEEE International Conference on Distributed Computing Systems (ICDCS 2009), 22–26 June 2009, Montreal, Québec, Canada, pp. 360–367 (2009)
13. Mega, G., Montresor, A., Picco, G.: Efficient dissemination in decentralized social networks. In: 2011 IEEE International Conference on Peer-to-Peer Computing (P2P), pp. 338–347, August 2011
14. Nilizadeh, S., Jahid, S., Mittal, P., Borisov, N., Kapadia, A.: Cachet: a decentralized architecture for privacy preserving social networking with caching. In: Proceedings of the 8th International Conference on Emerging Networking Experiments and Technologies, CoNEXT 2012, pp. 337–348. ACM, New York (2012). http://doi.acm.org/10.1145/2413176.2413215
15. Osmani, A.: Learning JavaScript Design Patterns: JavaScript and jQuery Developer's Guide. O'Reilly Media Inc., Sebastopol (2012). http://books.google.fr/books?id=JYPEgK-1bZoC
16. Rivière, E., Baldoni, R., Li, H., Pereira, J.: Compositional gossip: a conceptual architecture for designing gossip-based applications. SIGOPS Oper. Syst. Rev. 41, 43–50 (2007)
17. Taïani, F., Lin, S., Blair, G.S.: Gossipkit: a unified componentframework for gossip. IEEE Trans. Softw. Eng. 40(2), 123–136 (2014). http://doi.ieeecomputersociety.org/10.1109/TSE.2013.50
18. Voulgaris, S., Gavidia, D., van Steen, M.: CYCLON: inexpensive membership management for unstructured P2P overlays. J. Netw. Syst. Manage. 13(2), 197–217 (2005). http://dx.doi.org/10.1007/s10922-005-4441-x
19. Voulgaris, S., van Steen, M.: Epidemic-style management of semantic overlays for content-based searching. In: Cunha, J.C., Medeiros, P.D. (eds.) Euro-Par 2005. LNCS, vol. 3648, pp. 1143–1152. Springer, Heidelberg (2005)

Service Discovery for Spontaneous Communities in Pervasive Environments

Ghada Ben Nejma[1(✉)], Philippe Roose[1], Marc Dalmau[1],
and Jérôme Gensel[2]

[1] LIUPPA Laboratory, T2I Research Team,
2 Allée du Parc de Montaury, 64600 Anglet, France
{gbennej,roose,dalmau}@iutbayonne.univ-pau.fr
[2] LIG Laboratory/STEAMER Research Team,
681 rue de la Passerelle, BP72, 38402 Saint Martin d'Hères Cedex, France
Jerome.Gensel@imag.fr

Abstract. In this paper, we propose a community application that helps users to access communities and organize social exchange between members. The key design of this application is to consider community as distinct social entity that should be supported with the services as a single user is. We propose a service discovery strategy based on semantic modeling of service, semantic distance computing and inference rules to offer the right service to communities members. Then, selected services will be deployed through the software platform for deploying reconfigurable distributed applications.

1 Introduction

The democratization of mobile devices has made information accessible to anyone at any time and from anywhere while facilitating the capture of physical contextual data, thereby justifying the growing interest for pervasive computing. Fostering the social dimension has given rise to an emerging field of research called Pervasive Social Computing [1]. The notion of "community" is one among many elements of this field. The contextual information associated with a community can be exploited for adaptability and dynamic deployment of services, which are important factors for Pervasive Computing. A community is considered in our approach as a set of distinct social entity that should be supported with the services as a single-user is.

Our work particularly focuses on spontaneous and short-lived communities. Intuitively, it is the type of community that best matches with circumstantial, accidental, incidental or fortuitous situations. We define this new type of communities as 'a spontaneous group of individuals having a common interest or purpose related to a circumstantial situation and relative to a geographical territory'. This kind of community has to meet specific needs (*i.e.* adapt to a circumstantial situations), which are not taken into account by perennial communities. In previous work [2], we described our community application, called Taldea, that helps users to access to communities and exchanges multimedia documents with other members. Services are associated with communities so that when a user creates or joins a community, the corresponding

© Springer International Publishing Switzerland 2015
J. Wang et al. (Eds.): WISE 2015, Part II, LNCS 9419, pp. 337–347, 2015.
DOI: 10.1007/978-3-319-26187-4_32

services can be deployed on his device. This raises the following issue: how to help users to discover services adapted to their social context (*i.e.* communities)?

Motivated by this and the fact that no studies, to our knowledge, have been done to discover services that match with community perceptions in term of interests and needs, we propose an approach that guides communities to discover services.

The paper is organized as follows. In Sect. 2, we propose a scenario of the community application Taldea showing two parts of our contribution. From this use case, seven requirements are identified in Sect. 3 to foster spontaneous communities' applications. The architecture and the service discovery strategy of Taldea are described in Sect. 4. Related work is presented in Sect. 5. Finally, a conclusion summarizes the presented approach and outlines some future work.

2   Motivating Scenario

In this paper, we propose an application called Taldea for fostering spontaneous communities.

As shown in Fig. 1, our contribution mainly consists in helping people to access communities and to facilitate service discovery for them. To motivate our work and to identify the main requirements for such a new collaborative environment, we illustrate the following scenario.

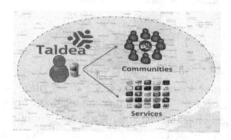

Fig. 1. Main contributions of Taldea

A visitor equipped with her Smartphone enters into a botanical park. Thanks to Geofencing[1] techniques, the Smartphone is automatically connected to Taldea. Once connected to Taldea, the user can join existing communities and access proposed services in three ways: through recommendation, by searching or creating of a new community. Communities that have semantically close interests and that are geographically close will be recommended to the visitor.

The user can also look for a specific community by formulating a natural language query. The results are those that semantically match the user query.

---

[1] The geo-fencing approach is based on the observation that users move in a virtual perimeter for a real-world geographic area.

On the occasion of the World Environment Day, the user decides to create a new community "Shrubs planting" and invites people to join him. The new community is recommended to relevant users of Taldea (they can be members of other communities) currently present and visiting the park. Services used by communities semantically close are recommended to the new community.

If the user is not satisfied with the recommendation results or his circumstances change, he can search for a specified situation with corresponding services by formulating a query. In this case, the user seeks to learn how to prune shrubs. The situation that meets to his query proposes a video showing how to prune shrubs. Due to low bandwidth, we deploy a new service reconfiguration like displaying subtitled image sequences of watering tree to ensure the continuity of service. This configuration is ensured by the platform Kalimucho for adaptation [2].

In the further alternative, the user can define a new situation with correspondent services to facilitate a further search. For example, he designs a set of service that will be deployed in case of accident situation.

3 Taldea Requirements

To address the issues stemming from the use case presented in the previous section, Taldea must fulfil the following requirements:

a. Community Management: creating a community, joining a community, exchanging and sharing data, managing the community space, *etc.*;
b. Community knowledge Management: using a knowledge base to store all social exchanges between users and capitalize knowledge produced by the communities;
c. Processing multimedia data and services: ensuring the management of multimedia data and provide the dedicated services for their exploitation in mobile computing environments;
d. User Context awareness: Acquiring and managing user context in order to dynamically deploy services that match with the social context of the user (*i.e.* communities);
e. Managing user profiles: managing information such as interests, preferences, privacy constraints;
f. Privacy issues: in the simplest form, reasoning with policies-based access control allows users to control their data and protect their privacy.

Nothing prevents from considering other requirements in order to adapt to the user preferences, or other dimensions of the context.

4 Contribution

4.1 Overall System Architecture

The architecture contains two types of structural entities: Community Manager and Member Device. Figure 2 show an overview of the Taldea Architecture:

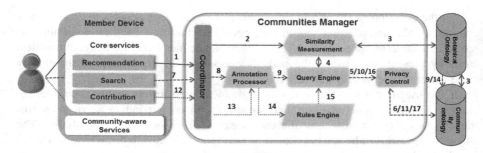

Fig. 2. Overview of Taldea's Architecture

- The **Community Manager** is the coordinator, which interacts with the knowledge base. The core of this manager is a set of services used to supply and to extract knowledge from ontologies and update them. It interacts with two types of ontologies: the communities' ontology and the domain ontology. The communities' ontology is used to formally describe the communities' resources. The domain ontology describes a specific domain with concepts and relationships that hold between those concepts; in our case, is the botanical domain.
- **The Member Device** allows the user to discover and interact with the communities using a set of services. We have classified services in two groups according to how they will be deployed: the core services and the communities' aware services. Core services are required for user connection and access to the communities. Regardless of the user or the type of community, these services must be automatically (at the runtime) deployed. On the other side, communities' aware services are deployed depending on the social context of the user. More details can be found in our previous work [2].

4.2   Community Access

As previously mentioned, our contribution consists mainly in helping people to access communities and to facilitate service discovery for them. The first part of our contribution facilitates the user access to communities. There are three ways to access a community in Taldea: recommendation, search and creation of a community. For more details, the reader can refer to our previous paper [2].

4.3   Service Discovery

In this paper, we highlight the role of communities as a source of contextual information. Services are deployed according to the social context of the user (*i.e.* specifically her communities). Services used to support exchanges with user 'communities are deployed on the user device. Once connected to communities, the user selects services to be deployed from the proposed list of communities' related-services. For a new community or new requirements, a large amount of services is available. Find the right service is a heavy task. In our work, we attempt to overcome service overload (*i.e.* growing number of proposed services compared to the resource-limited mobile devices) by helping the user to discover relevant services. For this purpose, we define three ways of service discovery: service recommendation, situation search and situation construction. User selected services will be deployed through Kalimucho [3], a service-based reconfiguration platform that provides a contextual-deployment of applications in order to meet context requirements. This platform is able to trigger the context changes and to modify the structure and the deployment of the application using five basic actions: add, remove, connect, disconnect and migrate component/connector. Our application is composed of interconnected services supervised by the Kalimucho platform.

4.3.1 Service Recommendation

In our work, a particular emphasis has been given to the dynamic deployment of service. Based on semantic service recommendation, we address the problem of service overload by presenting services that can be relevant to a given community. Taldea exploits the semantic expressed and structured in the ontologies in order to recommend to a new community services that are used by semantically closed communities (*i.e.* communities with close interests). A service is automatically annotated by the interests of communities that have used it. Each time a community use a service, its interest is added to the service description. Then, we explore semantic measures to discover services for the new community.

Community and service are annotated each one with one or more concepts. Thus, the problem consists in measuring the similarity between two sets of annotations. Different semantic measures have been proposed in the literature [4] to assess the semantic proximity between concepts. In this context of community applications, we evaluate the semantic similarity between the service's profile and community's profile, by the mapping of community interests and service annotations in the domain ontology (*i.e.* the botanic ontology). In this context, we adopted Wu and Palmer measure. It is simple to implement, and has good performance compared to other similarity measures [5]. We have adopted this approach based on arcs between concepts because other approaches used the frequency of information. This is not significant in the case of communities because a community is created around infrequent concepts in the corpus. The question of measuring the similarity between community interests and service annotations would come down to measuring the similarity between two sets of concepts. However to our knowledge, no existing research address the matching problem between two sets of concepts. We use the Hausdorff distance as metric that measures how far two subsets of a metric space are from each other. The problem is formulated as follows:

For a new community $C^k \in C$ and a given service $S^h \in S$. We consider a set of the community interests I^k and A^h the annotations' set of the service S^h where $I_i^k \in I^k$ and $A_j^h \in A^h$. We take $(1- wup (I_i^k, A_j^h))$ as the semantic distance between I_i^k and A_j^h. Wup is similarity measure between two concepts, where distance $= (1 -$ normalized similarity measure). The following formula presents Relevance (Community, Service) with Hausdorff distance.

$$\text{Relevance} \left(I^k, A^h\right) =$$

$$\max\left\{ \max_{I_i^k \in I^k}\left(\min_{A_j^h \in A^h}(1 - \text{wup}(I_i^k, A_j^h)) \right), \max_{A_j^h \in A^h}\left(\min_{I_i^k \in I^k}(1 - \text{wup}(I_i^k, A_j^h)) \right) \right\} \tag{1}$$

Calculate the relevance of a service compared to a community is calculating the semantic distance between service annotations and community interests. The Wu and palmer measure (wup) computes the distance between two concepts. Then, Hausdorff distance is used to compute the overall distance between the two sets. For example, a community that has the set A = {a1, a2} as interests and a service that have the set B = {b1, b2, b3} as annotations. We compare the minimum of {(1-wup (a1, b1)), (1-wup (a1, b2)), (1-wup (a1, b3))} with the minimum of {(1-wup (a2, b1)), (1-wup (a2, b2)), (1-wup (a2, b3))} and we keep the maximum of the two values. Using Hausdorff distance, we can say that any concept $I_i^k \in I^k$ is at most at distance Relevance $\left(I^k, A^h\right)$ to some concept of $A_j^h \in A^h$.

4.3.2   Situation Search

In the literature, service search is the process of finding appropriate services for a given request. The basic approach was proposed by Paolucci's [6]. Then, it was extended by others studies. The main idea of this approach is to annotate the service advertisement and the service request explicitly using semantic annotations. The service requester provides some functional properties (*i.e.* input, output, precondition, result) of the requested service. The similarity among requested properties and offered properties are based on subsumption reasoning between ontology concepts.

In a previous work, we adopted this approach, but ultimately, we have seen a difficulty for end-user to make a service request. The approach of service search is not intended for end-user but for advanced users. We propose a more user-friendly method to search services. The basic premise was to define user situation and correspondent services. Matheus [7] introduced the notion of situation as "a situation corresponds to the limited parts of reality we perceive, reason about, and live in. In situation semantics, basic properties, relations, events and even situations are reified as objects to be reasoned about." For this reason, we add the notion of situation to our community ontology as you see below (Fig. 3).

A situation is triggered by a query (contextual information captured by human) or reasoning on contextual information captured by sensor. In this paper, we consider only the case of natural language query. Situation deduction using context reasoning and situation awareness are presented in our work [8]. As mentioned in [7], "the primary basis for situation awareness is knowledge of the objects within the region of interest, typically provided by "sensors" (both mechanical and human) that perform object identification and characterization". In this context, we use the most sophisticated sensor, the human. The user describes his situation in natural language query. The query will be annotated through Textannot[2]. Using the results of annotation, a Sparql

---

[2] http://themat2i.univ-pau.fr:8080/TextAnnot-WWW/.

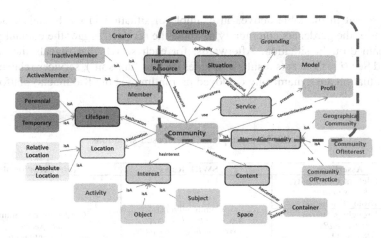

Fig. 3. Extract from the community ontology.

query is formulated to search situation and to deduce later the corresponding service. One or more services are attached for each situation.

A set of rules is defined for the concept situation. These rules propose a set of services (atomic or composite) to a community in a given situation. Example of an atomic service: returns the address when given a longitude and latitude. Example of a composite service: after pinpointing the user, the service pushes notification when user's friends passe nearby and gives them the ability to meet at the same bar (Fig. 4).

Fig. 4. Example of rules for situation.

4.3.3 Situation Construction

The end-user can define a situation and correspondent services to facilitate next search. For example, the user defines a situation of accident where a set of services are deployed once the situation is triggered. To define a situation, the concept situation and related concept and relationship are instantiated (Fig. 5).

Fig. 5. Situation and related concepts.

We present in the table above, an example of situation in the botanical domain. A member of the gardeners community attribute the service "trigger the coolant pump" to the loam drought situation. Afterwards, for each search that reveals the concept Loam and Sec (*i.e.* Concepts results of queries such as 'dry soil'), services related to the drought situation are dynamically deployed on runtime via the Kalimucho platform [3] (Table 1).

Table 1. Example of rules-based situations.

Assertions	SWRL Rules	Inferences' result
Gardeners_community *has_situation* Drought_situation	Community(?x), *has_situation* (?x, ?y), *corresponding_service* (?x, ?y) $\rightarrow$ *use_service*(?x, ?y)	Gardeners_community *use_service* Coolant pump
Drought_situation *definedBy* (Sec and Loam)		
Drought_situation *correspond_service* Coolant_pump		

5 Proof of Concept for Taldea

Taldea has been designed to gather the innovative aspects as data ephemerality, location based services, context awareness, semantic technologies, etc. for helping nomad user to join communities, to interact with people around and to have adapted services. Several scenarios highlight the importance of spontaneous community. Several use cases can be defined for Taldea like Conference, Exposition, Festival, Sport Event, Natural Disaster, etc. Screenshots in Fig. 6 shows some implemented features in the context of botanical Park.

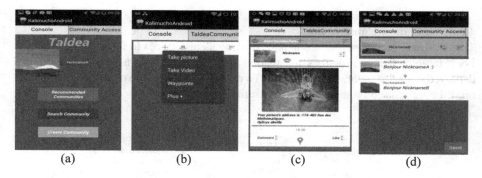

(a)	(b)	(c)	(d)

Fig. 6. Screenshots of the Taldea application running on a Samsung phone - showing some implemented features such as the communities access, the service recommendation, the community Space, the chat.

These screenshots show the community access module (a). Once connected to a community, the user can view the community space (c). He can access to the list of related services (b) as take picture or P2P chat (d).

6   Related Work

Several studies focused in community's applications but few of them are interested to ephemeral and spontaneous communities. In this paper, we propose an application for fostering spontaneous communities. Five features shape our system: community, the semantics, location awareness, ephemerality, context awareness. Taldea is not the first to incorporate one of these dimensions but it can be seen as one of the few attempts to combine them. Our contributions crosses several domains including semantic web, community application, location based service, semantic service discovery, ubiquitous computing, community application, temporary social network. In this section, we identify the main related works to the proposed contribution.

6.1   Community Oriented Application

Recently the notion of communities has gained worldwide popularity in different domains. Researches are more interested in discovering communities as in [9]. The challenge is to cluster users based on the semantics of data exchange and on the structure of communications to retrieve implicit communities and user interests. In pervasive social computing, applications are increasingly interested in communities. In some works (e.g. [10]), community is considered as some contextual information that can be harnessed for adaptability and personalization. On the other hand, some approaches like [11] are community-centered. They manage factors that affect directly communities and provide service for this granularity of users. The existing applications offer services and information related to a specific theme, which has to be defined in advance by developers. Therefore, community application represents new challenge in terms of the ability to have a spontaneous social experience.

6.2   Temporary Social Network

Mostly for privacy implication, the perception and orientation of users is directed to Temporary Social Network (TSN). Users, particularly the youngest ones, do not want their data to be online forever. The success of these platforms is explained by data ephemerality and user-friendly interaction. Among the researches that have been done targeting TSN: CALBA [12] is a framework designed for TSNs to select vendors as advertising sources for mobile users in a specific place. The selection is based on geographical proximity and user's preferences. Bfriend [13] is a location-aware ad-hoc social networking platform based on the Facebook social graph. Users receive push notification when a friend is in the same place at the same time. Snapchat [14] is a mobile messaging app for sharing pictures that disappear: it sends photos and then deletes them from the receiver's phone a few seconds after that are viewed. The user's content on Snapchat does not to end users online forever. Existent TSN offer the ability to have a private and spontaneous social experience, but does not and cannot entirely live up to this claim, giving users a false sense of security because there are some loopholes. For example recipients can take a screenshot or to plug the phone into a computer to recover the content from the local memory. We are also interested in ephemerality and one of the main characteristics of our community is short lifespan.

6.3 Semantic Service Discovery

Semantic service discovery is an important paradigm in distributed application development. Multiple researches studied this process and are interested to their architecture, algorithms, and tools. Klusch [15] presents a critical survey of different approaches in literature. In this field, several proposals have emerged. For instance to compute the semantic similarity between the service and the query [16, 17], to use alternative language to annotate services [18], to develop P2P architecture for service discovery [19], to consider other part of profile for matching [19], and to propose matchmakers (*i.e.* standalone tools) available to the end user [20].

7 Conclusion

The main contribution of this paper is to propose relevant services for spontaneous communities. To achieve this, we incorporate the notion of services in our proposed community ontology. This is particularly important for spontaneous communities to find the right service in circumstantial situation among the large amount of available services. We proposed three ways of service discovery: service recommendation, situation search and situation construction. User selected services will be deployed through Kalimucho. We used the semantic distance to recommend services for communities. For service search, we used a web service to annotate user query. We combine the obtained results with inference rules to find the desired services. To facilitate further search, we enabled the user to attribute some services to a particular situation. Our immediate plan is to evaluate Taldea usability by assessing the user satisfaction for the proposed services. In the future work, several issues will be investigated, we plan to integrate spatial and temporal dimensions for services recommendation and enrich the community description to include a spatio-temporal contextualization of social exchanges between users.

References

1. Schuster, D., et al.: Pervasive social context: taxonomy and survey. ACM Trans. Intell. Syst. Technol. (TIST) **4**(3), 46 (2013)
2. Nejma, G.B., et al.: Design and development of semantic application for communities. In: The International Workshop on Semantic and Social Media Adaptation and Personalization, 2014. Corfu, Greece (2014)

3. Dalmau, M., Roose, P.: Kalimucho: Plateforme logicielle distribuée de supervision d'applications (2013)
4. Slimani, T., B. BenYaghlane, and K. Mellouli. Une extension de mesure de similarité entre les concepts d'une ontologie. In International Conference on Sciences of Electronic, Technologies of Information and Telecommunications (2007)
5. Slimani, T., Yagahlane, B.B., Mellouli, K.: A new similarity measure based on edge counting. In: Proceedings of the World Academy of Science, Engineering and Technology, p. 17 (2006)
6. Paolucci, M., Kawamura, T., Payne, T.R., Sycara, K.: Semantic matching of web services capabilities. In: Horrocks, I., Hendler, J. (eds.) ISWC 2002. LNCS, vol. 2342, pp. 333–347. Springer, Heidelberg (2002)
7. Kokar, M.M., Matheus, C.J., Baclawski, K.: Ontology-based situation awareness. Inf. Fusion 10(1), 83–98 (2009)
8. Keling, D., Dalmau, M., Roose, P.: Kalimucho: middleware for mobile applications. In: The 29th Symposium on Applied Computing, 2014. Gyeongju, Korea (2014)
9. Leprovost, D., Abrouk, L., Gross-Amblard, D.: Discovering implicit communities in web forums through ontologies. Web Intell. Agent Syst. 10(1), 93–103 (2012)
10. Lima, C., Gomes, D., Aguiar, R.: Pervasive CSCW for smart spaces communities. In: IEEE International Conference on Pervasive Computing and Communications Workshops (PERCOM Workshops) (2012)
11. Maret, P., Laforest, F., Lanquetin, D.: A semantic web model for ad hoc context-aware communities: application to the smart place scenario. In: International Conference on Enterprise Information Systems, 2014, pp. 591–598, Portugal (2014)
12. Xu, W., Chow, C.-Y., Zhang, J.-D.: CALBA: capacity-aware location-based advertising in temporary social networks. In: Proceedings of the 21st ACM SIGSPATIAL International Conference on Advances in Geographic Information Systems. ACM (2013)
13. Smailovic, V., Podobnik, V.: Bfriend: context-aware ad-hoc social networking for mobile users. In: MIPRO, 2012. Proceedings of the 35th International Convention (2012)
14. Poltash, N.A.: Snapchat and sexting: a snapshot of bearing your bare essentials. Rich. JL & Tech. 19, 14 (2013)
15. Klusch, M.: Semantic web service coordination. In: Schumacher, M., Schuldt, H., Helin, H. (eds.) CASCOM: Intelligent Service Coordination in the Semantic Web, pp. 59–104. Springer, Berlin (2008)
16. Fu, P., et al.: Matching algorithm of web services based on semantic distance. In: Proceedings of the International Workshop on Information Security and Application (IWISA), Qingdao, China (2009)
17. Ge, J., Qiu, Y.: Concept similarity matching based on semantic distance. In: Fourth International Conference on Semantics, Knowledge and Grid, SKG 2008. IEEE (2008)
18. Cardoso, J., Miller, J.A., Emani, S.: Web services discovery utilizing semantically annotated WSDL. In: Baroglio, C., Bonatti, P.A., Małuszyński, J., Marchiori, M., Polleres, A., Schaffert, S. (eds.) Reasoning Web. LNCS, vol. 5224, pp. 240–268. Springer, Heidelberg (2008)
19. Skoutas, D.N., Sacharidis, D., Kantere, V., Sellis, T.K.: Efficient semantic web service discovery in centralized and P2P Environments. In: Sheth, A.P., Staab, S., Dean, M., Paolucci, M., Maynard, D., Finin, T., Thirunarayan, K. (eds.) ISWC 2008. LNCS, vol. 5318, pp. 583–598. Springer, Heidelberg (2008)
20. Klusch, M., Fries, B., Sycara, K.: OWLS-MX: a hybrid semantic web service matchmaker for OWL-S services. Web Seman. Sci. Serv. Agents World Wide Web 7(2), 121–133 (2009)

C3PO: A Network and Application Framework for Spontaneous and Ephemeral Social Networks

Antoine Boutet[2]([✉]), Stephane Frenot[3], Frederique Laforest[2], Pascale Launay[1], Nicolas Le Sommer[1], Yves Maheo[1], and Damien Reimert[3]

[1] IRISA, Universite de Bretagne-Sud, Lorient, France
{pascale.launay,nicolas.sommer,yves.maheo}@univ-ubs.fr
[2] Laboratoire Hubert Curien, Universite de Saint Etienne, Saint Ètienne, France
{antoine.boutet,frederique.laforest}@univ-st-etienne.fr
[3] DICE-INRIA, CITI, Universite de Lyon, Lyon, France
{stephane.frenot,damien.reimert}@inria.fr

Abstract. The C3PO project promotes the development of new kind of social networks called Spontaneous and Ephemeral Social Networks (SESNs) dedicated to happenings such as cultural or sport events. SESNs rely on both opportunistic networks formed dynamically by the mobile devices of event attendees, and on an event-based communication model. Therefore, user can exchange digital contents with the other members of their SESNs, even without Internet access. This paper presents the framework developed in the C3PO project to provide network and application supports in such challenged networks. This framework exploits the different wireless interfaces of the mobile devices to interconnect them and to disseminate content through the resulting opportunistic network. At the application layer, this framework is composed of plugins that process locally the data stream and interact with the network level to offer generic features, or to easily build applications dedicated to specific happenings.

1 Introduction

The massive adoption of Online Social Networks (OSNs) has changed the way the information is generated and exchanged between people. Besides the two giants Facebook and Twitter, a multitude of more or less specialized social platforms allow people to interact and share information via the Internet. Although users are faced with a diversity of purposes and types of platforms (business-oriented, dedicated to music, books or sports, designed to share photos, videos or news...), the vast majority of these social networks are build on the same architectural model: a centralized service provider acts as a broker to filter and disseminate the data produced by users. This architecture follows the traditional Web 2.0 model but is not without drawbacks from the user's point of view. First, it induces a dependence on a centralized authority that controls the data management and exchange (more than often, data ownership is even transferred to the service provider). Second, this architecture requires that the user has a permanent access to the Internet to exchange with others. Third, social acquaintances between

© Springer International Publishing Switzerland 2015
J. Wang et al. (Eds.): WISE 2015, Part II, LNCS 9419, pp. 348–358, 2015.
DOI: 10.1007/978-3-319-26187-4_33

users are mostly permanent and are used as pipes of communication: information transits among users following these social links.

To overcome these limitations, the C3PO project (Collaborative Creation of Contents and Publishing using Opportunistic networks) [9] investigates multimedia content production and exchange in a new type of social networks that we call Spontaneous and Ephemeral Social Networks (SESNs). SESNs rely on a peer-to-peer distributed architecture formed spontaneously by mobile devices carried by people and, optionally, by fixed devices that can be deployed to support such networks. Devices are interconnected using their wireless interfaces (e.g., Wi-Fi or Bluetooth). Opportunistic communication techniques are devised in order to support the connectivity disruptions resulting from the mobility of users and the short communication range of the radio interfaces. Thus, these techniques allow devices to exchange data even if they are out of the radio range of each others and no end-to-end path exist between them.

Due to their spontaneous and ephemeral nature, SESNs are suited to produce multimedia reports on conferences and cultural or sport events, such as a marathon as illustrated in Fig. 1. The spectators, competitors, organizers and other participants of a marathon can take photos, can write comments or generate information, and can publish these contents in the SESN dedicated to this event. During such large events, the geographic vicinity of users is a more appropriate paradigm than spreading content through social links in conventional online social networks. Moreover, relationships set up during the event can disappear at the end of the event.

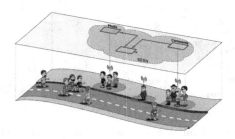

Fig. 1. Cultural and sport events are typical use cases of SESNs

In this paper, we present the framework designed in project C3PO to support both opportunistic networking and applications in SESNs. At the network level, this framework allows the building of a communication middleware that offers a message-oriented API (based on the concepts of topic and named channel) to application programmers. This communication middleware is done by implementing a small number of functions that constitute a set of reactions to standard communication events (reception of a message, appearance of a neighbor...). The programmer either develops original implementations of these functions, or uses the implementations provided in the C3PO toolkit. At the application level, the application programmers develop C3PO plugins adapted to specific needs. For instance, a plugin may be dedicated to the emission, reception and presentation of the official results of a race. To date, the C3PO framework proposes several plugins, from filtering schemes to present specific contents to more complex recommendation operations which highlight valuable contents over time.

2 Network Layer of C3PO

This section presents the network layer of the C3PO project. It considers both the network architecture and the framework that sustains this architecture.

2.1 Network Architecture

A SESN is formed by users carrying off-the-shelf mobile devices (smartphones, tablets) communicating with each others thanks to short range wireless interfaces implementing communication standard such as Bluetooth or IEEE 802.11 (Wi-Fi). In most cases, the users that compose the SESN can move, and are not necessarily all located in the same place at the same time. The topology of the network formed by their devices changes continuously and can be fragmented into communication islands. Thus, as depicted in Fig. 2, the network is structured as a collection of what we call *micronets*, grouped together in independent *macronets*.

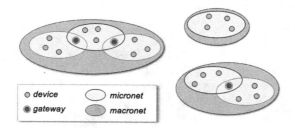

Fig. 2. SESNs rely on opportunistic networks formed dynamically by the mobile devices of happening attendees.

Micronets: We define a micronet as a subset of devices connected using a common communication technology, that are able to communicate directly with each others. A micronet is an abstraction of a piconet for Bluetooth, a Basic Service Set (BSS) for Wi-Fi legacy, or a group for Wi-Fi direct. We consider that devices in a micronet communicate *directly* as there exists a network layer that allows any device to address and send messages to any other device in its micronet, even if it is not the case at the MAC layer. Wireless communication technologies available for current off-the-shelf mobile devices allow only a limited number of devices to be interconnected and exchange data directly between them (e.g. at most 8 devices for a Bluetooth piconet). Consequently, micronets have to be interconnected in order to allow SESN users to communicate over the whole network.

Macronets: We define a macronet as a group of micronets interconnected through devices that are members of at least two micronets. The resulting communication network is a collection of independent macronets. Macronets can be analogized to Wi-Fi Direct multi-groups as defined in [3] or to Bluetooth scatternets. But the micronets forming a macronet can use different communication technologies.

It cannot be assumed that there is a common addressing scheme in a macronet nor that messages can be transmitted directly between the devices of a macronet, even in the case of a Bluetooth scatternet. The C3PO framework relies on the store-carry-and-forward principle for both intra-macronet and inter-macronet communication. Indeed, temporaneous end-to-end paths in the whole SESN (i.e., between macronets) cannot be established; therefore, a store-carry-and-forward layer must be developed for communication between macronets, and this layer can also be used for exchanging messages inside a macronet. Besides the simplification of the framework, this option is likely to provide a better tolerance against network disruptions.

2.2 Network Framework

The C3PO network level framework is a Java-based framework dedicated to opportunistic networking. It provides application developers with an API offering two distinct application-level communication paradigms, namely a topic-based publish/subscribe paradigm, and a channel-based send/receive paradigm. Its architecture, illustrated in Fig. 3, is organized in five main modules deployed on each device of the network.

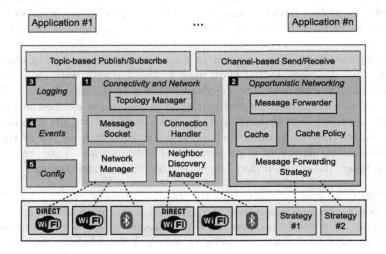

Fig. 3. General architecture of the C3PO network level framework

Modules 1 and 2 are related to the management of micronets and macronets. Module 1 manages wireless interfaces and sets up communications in micronets. It defines abstract neighbor discovery managers and network managers, that are respectively responsible for discovering neighbor devices (i.e., devices in radio range) to build micronets and for managing communications to support data exchanges between devices inside a micronet. Module 2 defines opportunistic networking mechanisms to perform data forwarding inside and between macronets. It includes a cache of messages, a message forwarder and abstract forwarding strategies in order to implement the store-carry-and-forward principle. The three other modules, namely modules 3, 4 and 5, composing the framework define functionalities that are respectively dedicated to the configuration of the framework, to the management of the events produced by the framework, and to log the traces that are generated by the framework (e.g., contact and message exchanges traces). Several concrete implementations of the abstract functionalities defined in the C3PO framework are provided in a toolkit. These implementations can be used as they are, or can be extended by developers in order to adapt them to their own needs.

Two distinct application-level communication paradigms are provided by the network level framework: a topic-based publish/subscribe model, and a channel-based send/receive model.

- The *topic-based publish/subscribe communication model* makes it possible to develop applications that can publish multimedia contents on specific topics, and subscribe in order to receive the contents related to given topics. The model implemented in the C3PO framework relies on a purely peer-to-peer decentralized approach. The subscription and the publication are local to each device. Thanks to the store-carry-and-forward principle, contents published in a topic by publishers are disseminated opportunistically in the communication network by mobile devices, being either devices hosting subscribers for this topic or ordinary intermediate devices, and are thus delivered to the topic subscribers.
- The *point-to-point communication paradigm* using the concept of channel is intended for applications that allow users to communicate with each others by sending messages addressed to specific recipients. In the framework, a channel between two devices is identified by the addresses of the devices and a channel ID. Messages sent through a channel are opportunistically forwarded by intermediate devices towards their destination according to one of the message forwarding strategies implemented in the framework. Thus, two devices can exchange data even if they are not within mutual radio range.

The C3PO framework implements the store-carry-and-forward principle. Different types of messages are currently considered: application-level messages and several types of control messages such as beacon messages that are used to exchange the neighbor lists, gossiping messages, acknowledgements, and drop messages that make it possible to implement network healing protocols in order to reduce the number of messages that are disseminated in the network.

The messages that are published or sent by the applications are exchanged by mobile devices according to a given forwarding strategy. In order to accomodate expected field conditions, a specialized instance of the framework can be built by programming (mainly by inheritance) a message forwarding strategy as a set of reactions to four main events: the emission of a message by a local application, the reception of a message from the network, the discovery and connection of a new neighbor device, and the disappearance and disconnection of a neighbor device. To date, we provide message forwarding strategies that replicate the messages on all the nodes forming the opportunistic network (i.e., epidemic strategies), provided these nodes have enough space in their local cache to store these messages. Compared to a plain flooding, our implementation reduces drastically the number of messages transmitted, by maintaining on each device information about the caches of its neighbors, through the exchange of catalogs of message IDs. More information on the C3PO network level framework can be found in [11].

3   Application Layer of C3PO

The application layer of the C3PO framework is built on top of the network layer and uses its API for receiving and sending messages throughout the SESN. This application layer is designed as an in-browser architecture, so as to comply to any device as soon as it provides a browser compatible with the Javascript technologies.

The main objective of the application layer is to offer a mean to define, adapt and extend the application to the target SESN. As a matter of fact, SESNs should provide generic functionalities that are common to any SESN, but also specific functionalities according to the happening it covers. An example of generic functionality is presenting the flow of contents exchanged by end-users, while a specific functionality can be the identification of numbers in runners pictures during a race. Another challenge in such systems is to help users read first the most valuable contents in the data stream. In C3PO, each functionality is developed as a single plugin while the whole application is done through plugins composition. In this section, we present the general architecture of the application layer of C3PO, as well as some examples of application plugins.

3.1   Application Layer Architecture

At the application level, the C3PO framework gets messages as a flow of events, processes them and presents the results to the end-users. This level is also responsible for the creation of new events reflecting some end-user actions. For example, end-users can create new contents to be exchanged with other end-users; they can also promote a received content to stress their special interests.

Depending on the SESN happening nature, the events may carry very different contents, from a short text message to information about a promote, and thus event processing requires extensibility. Devices in the same SESN have a

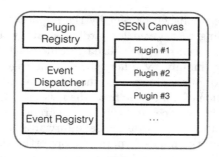

Fig. 4. C3PO application layer architecture.

set of identical processing units, but some more powerful devices may also have additional specific ones.

The global architecture of the application level framework part is illustrated on Fig. 4. Plugins are components responsible for the processing of incoming events and for the interaction with the end-user [7]. The event dispatcher filters incoming events with regular expressions (RE) stored in the plugin registry. For each matching RE, the corresponding plugin processing unit is invoked with the corresponding event. Plugins can store their processing results in their own context. When a device connects to a SESN, the SESN Manager registers the corresponding plugins in the registry. A SESN Canvas organizes plugins in the graphical user interface.

A plugin is defined with three elements: (1) the RE of the events contents it is interested in, (2) the processing modalities of the corresponding events, (3) the management of end-user interactions with the results of its process. For example, an image plugin can declare in its RE that it requires events containing references to images, process the events by uploading the images, interacting with the end-user by showing the list of images and proposing a "promote" button for each image.

Plugins must conform to the C3PO plugins API, that contains one standardized function invoked by the event dispatcher. This method has one parameter: a reference to an event in the event registry. An event can be processed by many plugins.

The plugin registry gathers all RE in a way that optimizes plugins selection, the event dispatcher uses this registry to trigger processing units in plugins, and the event registry stores all event that have to be handled. The architecture is built as a message bus. Plugins may be orchestrated with their RE specification. A first plugin generates messages whose RE is exclusively requested by another plugin.

Plugins are developed using the AngularJS framework. User interactions in a plugin are developed as an AngularJS directive. This Directive may also include an event creation. This new event is thrown into the events queue of the network level input. The graphical user interface is implemented using HTML5, CSS3, Javascript, AngularJS and Design Material. The binding between the user interface and the network level part is done using Apache Cordova. Several Cordova plugins have been developed in order to make accessible the functionalities provided by the framework in Javascript.

3.2   Application Plugins

In this section, we describe a subset of plugins that have already been developed in the C3PO project. Programmers can easily develop their own plugins to provide applications that match with the targeted SESNs.

- The "**Flow**" plugin takes all the received and emitted events and displays them as a list ordered with the last received first. Users can browse the list, and can click a "promote" icon that indicates that they are interesting in this event. Each promote done by the end-user generates and throws a new event including a #promote hashtag and the source event id. A screenshot is given on Fig. 5a.
- The "**Tags Cloud**" plugin takes all the incoming events and builds a tags cloud that reflects the frequency of each hashtag in the content of the events set. Its user interface displays the tags cloud. A screenshot of the Tags Cloud plugin interface is given on Fig. 5b.
- The "**Image**" plugin selects events containing a reference to an image (url) and displays them as a list ordered with the last received first.
- The "**My Journal**" plugin displays all events promoted by the user. Users can manipulate the displayed list to raise, diminish or remove contents. At the SESN termination it provides a summary of the happening. A scrennshot of the MyJournal plugin interface is given on Fig. 5c.
- The "**Popular**" plugin allows users to show the most popular events, based on the "promote" actions of the SESN members that have been received by the device. The processing of the most popular events is done independently by each device. So two devices in two isolated macronets will not have the same list of contents in this plugin. Two devices subscribing to the same topic and being in the same micronet have a bigger chance to get the same list of contents in this plugin.
- The "**Vote**" plugin allows users to submit a content to a collaborative vote, and to vote about a content. At the user interaction level, vertical swiping is used to vote: down to ignore, up to promote. The Vote plugin is associated to a "VoteResult" plugin that displays the results of the collaborative votes on each submitted content.
- The "**Live Recommendation**" plugin quickly identify the most valuable contents over time. To achieve that, the underlying algorithm of this plugin attributes a score to each content based on both criteria, its popularity and its freshness. While the former exploits the number of promotes, the latter captures the delay between the publication of the content in the system and their promotes. This plugin uses a sliding window, and refresh the score to each content after a certain time window of size w. The following equation computes the score attached to content c at the current date d:

$$score(c, d) = \sum_{P(c,d) \in [d0, d0+w]} \frac{P(c, d) \times w}{\Delta T}$$

where $P(c, d)$ is the number of promote that c has collected so far, $d0$ the starting date of the sliding window, and $\Delta T = d - d_t$ which reflects the freshness of c where d_c is the date when c has been created.

Each element of the sum follows the pattern of $\frac{1}{x}$. As a consequence, each element of the sum (i.e. each promote in the considered sliding window) increases if $\Delta T < w$. It means when the promote has happened before the considered time window w. In contrast, this score is minored if the promote happens after the time window w. In addition, this score is weighted by the number of promote c has obtained so far. Lastly, the score associated to c is cumulative and sums the score computed to each promote received in the considered time window w. The starting date of the time window is different for each content and is initialized at the reception of its first promote. The size of the time window defines the temporal granularity to follow the event, more the time window is small, more the event is largely detailed (by default, w is set at 10 min).

– The "**Config**" plugin helps user to activate or to deactivate each other plugin. When deactivated, the plugin is removed from the user interface but keeps its local context. If the user reactivates a plugin, it starts processing events with the previous saved context.

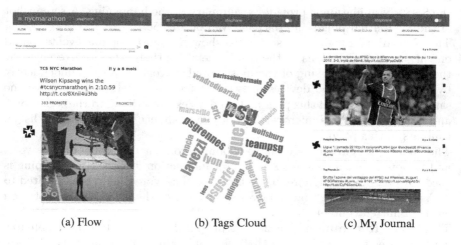

(a) Flow   (b) Tags Cloud   (c) My Journal

Fig. 5. C3PO application level is organized through plugins composition.

4   Related Work

Nowadays, it exists a plethora of social networks. Most of them rely on a centralized architecture and require connectivity to the Internet. Over the last years, decentralized social networks have been studied and developed by researchers and non-academic projects, especially to address privacy issues related to online social networks [5, 8]. However, none of them have considered social networks formed spontaneously by a set of mobile devices.

Data sharing in opportunistic networks have been studied in different projects. Haggle [10] takes advantage of contact opportunities and of device mobility in order to follow the "store, carry and forward" principle in the communication between devices. PodNet [6] implements a publish/subscribe model, where users can express their interests via keywords and receive content items accordingly. In Crowd [1], multimedia contents are exchanged via an online Web portal through Wi-Fi hotspots. SocialNet[1] mainly aims at detecting the social links between people to improve the opportunistic forwarding of contents. Social links rely on community memberships, history of contacts, recurrent mobility patterns and/or user interests.

In C3PO, people implicitly express their common interests by physically attending happenings, and by being members of the same SESNs. Due to the ephemeral nature of the SESNs, the exploitation of information such as the history of contacts or the recurrent mobility patterns is not relevant. Projects Sarah and Scampi [4] both have defined a middleware platform to support communication and service provision in opportunistic networks. They have also implemented Android applications based on these middleware platforms. Sarah allows to discover neighbors, manage a list of contacts, exchange contents in a secure manner. Nevertheless, none of these projects provides incentive tools to support a collaborative production of rich multimedia contents during ephemeral events.

5   Conclusion

In this paper, we have introduced a new type of social networks dedicated to happenings, namely Spontaneous and Ephemeral Social Networks (SESNs). A SESN has the characteristic of relying on the opportunistic network formed by the devices of the happening attendees. We have presented the framework developed in the C3PO project to support opportunistic network communication and application in SESNs. This framework is able to manage the different wireless interfaces of mobile devices so as to form an opportunistic network dynamically. At the application level, this framework exploits plugin composition to provide both generic functionalities and specific functionalities according to the targeted happenings. It is designed to be highly configurable and extensible, namely making it possible to define various message forwarding strategies and various application functionalities. Preliminary evaluation results we have obtained for this framework in real conditions confirmed the validity of the approach [2].

Acknowledgments. The C3PO project is supported by the French ANR (Agence Nationale de la Recherche) under contract ANR-13-CORD-0005. http://www.c3po-anr.fr/.

---

[1] http://www.social-nets.eu.

References

1. Belblidia, N., de Amorim, M.D., Costa, L.H.M., Leguay, J., Conan, V.: Part-whole dissemination of large multimedia contents in opportunistic networks. Comput. Commun. **35**(15), 1786–1797 (2012)
2. Boutet, A., Laforest, F., Frenot, S., Reimert, D.: MyStream: an in browser stream processing personalization service to follow events from Twitter (2015)
3. Casetti, C.E., Chiasserini, C.-F., Pelle, L.C., Valle, C.D., Duan, Y., Giaccone, P.: Content-centric routing in Wi-Fi direct multi-group networks. In: WoWMoM (2015)
4. Conti, M., Giordano, S., May, M., Passarella, A.: From opportunistic networks to opportunistic computing. IEEE Commun. Mag. **48**(9), 126–139 (2010)
5. Cutillo, L.A., Molva, R., Strufe, T.: Safebook: a privacy-preserving online social network leveraging on real-life trust. IEEE Commun. Mag. **47**(12), 94–101 (2009)
6. Distl, B., Csucs, G., Trifunovic, S., Legendre, F., Anastasiades, C.: Extending the reach of online social networks to opportunistic networks with PodNet. In: MobiOpp, pp. 179–181 (2010)
7. Frénot, S., Grumbach, S.: An in-browser microblog ranking engine. In: Castano, S., Vassiliadis, P., Lakshmanan, L.V.S., Lee, M.L. (eds.) ER 2012 Workshops 2012. LNCS, vol. 7518, pp. 78–88. Springer, Heidelberg (2012)
8. Kalofonos, D., Antoniou, Z., Reynolds, F., Van-Kleek, M., Strauss, J., Wisner, P.: Mynet: a platform for secure P2P personal and social networking services. In: PerCom, pp. 135–146 (2008)
9. Laforest, F., Le Sommer, N., Frenot, S., de Corbiere, F., Maheo, Y., Launay, P., Gravier, C., Subercaze, J., Reimert, D., Brodu, E., Daikh, I., Phelippeau, N., Adam, X., Guidec, F., Grumbach, S.: C3PO: a spontaneous and ephemeral social networking framework for a collaborative creation and publishing of multimedia contents. In: MoWNeT, pp. 214–219 (2014)
10. Scott, J., Hui, P., Crowcroft, J., Diot, C.: Haggle: a networking architecture designed around mobile users. In: WONS (2006)
11. Sommer, N.L., Launay, P., Maheo, Y.: A framework for opportunistic networking in spontaneous and ephemeral social networks. In: CHANTS (2015)

Towards a Unified Framework
for Data Cleaning and Data Privacy

Yu Huang[✉] and Fei Chiang

Computing and Software Department, McMaster University, Hamilton, Canada
{huang223,fchiang}@mcmaster.ca

Abstract. Data quality has become a pervasive challenge for organizations as they wrangle with large, heterogeneous datasets to extract value. Existing data cleaning solutions have focused on scalable techniques to resolve inconsistencies quickly. However, given the proliferation of sensitive, confidential user information, data privacy concerns have largely remained unexplored in data cleaning techniques. In this work, we present a new privacy-aware, data cleaning framework that aims to resolve data inconsistencies while minimizing the amount of information disclosed. We present a set of data disclosure operations that facilitate the data cleaning process, and propose two information-theoretic measures for privacy loss and data utility that are used to correct inconsistencies in the data.

Keywords: Data cleaning · Data quality · Information disclosure

1 Introduction

Organizations have found it increasingly difficult to extract value from their data due to increasing data volume and data heterogeneity. Many data processing tasks assume the data conforms to standard formats and data types, which is rare in real data. Most real datasets contain missing, duplicate, and incomplete values. Data quality is a pervasive problem that spans across all industries. Recent studies report that the annual cost of poor data quality for an organization ranges between $5M$–$20M$, with over half of these companies indicating that managing data quality remains their number one issue [6].

Given increasing amounts of personal and sensitive information that are collected online by social media sites and by organizations, there is growing concern of how to perform automated data cleaning tasks while ensuring sensitive and confidential information remains protected. Data privacy and data cleaning have becoming increasingly important, and new techniques for improving data quality while ensuring minimal data disclosure are strongly needed.

Developing automated techniques for data cleaning while providing data privacy guarantees is challenging, as these two tasks normally have competing goals. In data cleaning, the objective is to gain as much knowledge about the data as possible to correctly resolve data errors. In data privacy, the goals is to conceal

© Springer International Publishing Switzerland 2015
J. Wang et al. (Eds.): WISE 2015, Part II, LNCS 9419, pp. 359–365, 2015.
DOI: 10.1007/978-3-319-26187-4_34

Table 1. Master table M

ID	Name	Age	Nationality	Disease
m_1	Gorou	26	Japan	H1N1
m_2	Mary	26	Canada	H1N1
m_3	Joe	30	U.S	H1N1
m_4	Wang	24	China	H2N2
m_5	Zhang	24	China	H2N2
m_6	Zhao	35	China	Hepatitis B
m_7	Jirou	34	Japan	Hepatitis A
m_8	Bush	40	Canada	Skin cancer
m_9	Kate	40	U.S	Skin cancer
m_{10}	Smith	46	Canada	Leukemia

Table 2. Target table T

ID	Name	Age	Nationality	Disease
t_1	Gorou	26	Japan	**H1N1**
t_2	Mary	26	Canada	**colitis**
t_3	Bush	40	Canada	Skin cancer
t_4	Jirou	34	Japan	Hepatitis A
t_5	White	35	Canada	Hepatitis B

as much of the data as possible. For example, consider Tables 1 and 2, represent, respectively, a trusted (clean) master data source M (such as hospital medical records), and a potentially dirty target data source T (e.g., patient records from a local pharmacy).

Suppose there is a data quality (business) rule that states for each unique value in the attribute *age*, we expect to see a unique value in the attribute *disease*, as age influences the occurrence of particular diseases. This rule can be represented by $F_1 : age \rightarrow disease$[1]. We observe that Table 2 contains two records, t_1 and t_2 which violate rule F_1, as the unique value $age = 26$ corresponds to two non-unique values *H1N1* and *colitis*. For the pharmacy to correct these data inconsistencies, it requires consultation with a curated and clean master data source such as M in Table 1 to reveal its data values to T. Naturally, T would like M to disclose as much information as possible to maximize the data utility in T. However, given the sensitive nature of the data, M would like to disclose a minimal amount of information to maximize its data privacy (enough to help T resolve its data inconsistencies).

To solve this problem, we need to consider three challenges: (a) the design of a framework that combines data cleaning and data privacy; (b) define metrics to quantify privacy loss and data utility; and (c) define a set of operations that can be applied to M to control the amount of data disclosure. We present preliminary results of our work, and make the following contributions: (1) We propose a privacy-aware, data cleaning framework that aims to balance the tradeoff between data privacy and data utility; (2) We define information-theoretic metrics for privacy loss and for data utility; (3) We propose a set of data disclosure operations that control the amount of privacy loss incurred.

2   Related Work

Our work finds relation to two lines of work: (1) privacy preserving data publishing; and (2) data cleaning.

---

[1] For illustration purposes of this example, we assume persons of a unique age determine a unique disease.

Privacy preserving data publishing (PPDP) techniques aim to publish data in a manner that ensures sensitive information is not inadvertently released. Techniques such as *k-anonymity, l-diversity, (α, k)-anonymity* focus on protecting sensitive and *quasi-identifier* attribute values using various partitioning strategies [5,7]. However, these techniques only consider how to break the linkage between sensitive and quasi-identifier values, and do not consider how to manage and resolve inconsistent data values.

Data cleaning techniques have mainly focused on identifying a minimal number of updates to the data to correct the underlying inconsistencies [1–3]. However, these solutions aim to identify sufficiently clean portions of the (same) data instance to resolve the inconsistencies without considering data privacy nor sensitive attribute value requirements. We aim to bridge the gap between data cleaning and data privacy by proposing a framework that recommends updates to the data while being cognizant to reveal a minimal amount of information.

3 Preliminaries

We introduce definitions and concepts from information theory and dependency theory that will be used in our framework.

3.1 Self-information and Conditional Entropy

Self-information. It quantifies the information content associated with the occurrence of an event in a probability space [8]. $I(x) = \log(\frac{1}{p(x)}) = -\log(p(x))$ where $p(x)$ is the probability of event x. The smaller the probability, the larger the self-information. Let R be a relation, $X \in R$ is an attribute that generates a probability space. Hence, each value $x \in X$ corresponds to an event, and we use self-information to capture the information contained in each value x.

Conditional Entropy. Given a particular value of a random variable Y, the conditional entropy of X given $Y = y$ is defined as $H(X|y) = -\sum_{x \in X} p(x|y) \log p(x|y)$ where $p(x|y) = \frac{p(x,y)}{p(y)}$ is the conditional probability of x given y. The value $H(X|y)$ measures the uncertainty in X when y is given.

3.2 Functional Dependencies

For a relation R, a functional dependency $F: X \rightarrow Y$ is a constraint between two attribute sets X and Y. A data instance I of R satisfies F, denoted $I \vDash F$, if for every pair of tuples t_1 and t_2 in I, if $t_1[X] = t_2[X]$ then $t_1[Y] = t_2[Y]$. Consequently, if $t_1[X] = t_2[X]$ and $t_1[Y] \neq t_2[Y]$, then we say that t_1 and t_2 are inconsistent tuples w.r.t. F. For example, Table 1 satisfies F_1, whereas Table 2 does not. A data *repair* is an update to a data value that resolves an inconsistency w.r.t. F. Specifically, if $t_1[X] = t_2[X]$ then we can update $t_1[Y]$ to be equal to $t_2[Y]$ to resolve the inconsistency. We consider repairs only to the right hand side attributes of F.

4   Privacy Preserving Data Cleaning Framework

Figure 1 shows our privacy preserving data cleaning framework. We assume that T contains inconsistent tuples, and to clean these errors, requests that M disclose its data values to T. Both M and T each reveal a minimal amount of information to each other, that is, T only discloses its inconsistent tuples to M. In the following sections, we describe each module of our framework.

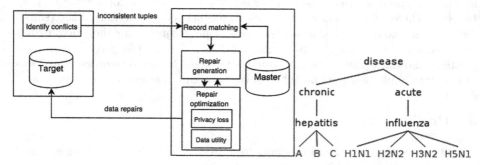

Fig. 1. Framework overview   **Fig. 2.** Ontology tree

4.1   Identify Conflicts and Record Matching

The *identify conflicts* module identifies all inconsistent tuples in T w.r.t. each defined FD. This can be done by partitioning the tuples in T according to the values of X, and identifying those tuples with different values in Y for the same value in X. For each group of values (that have the same X values) but different values in Y, T will send these records to M for record matching.

In the *record matching* module, tuples from M are matched with the inconsistent tuples from T according to the values in X. We use the string similarity measure cosine similarity. Records from M that satisfy a given minimum similarity threshold are returned as matching records. For example, in Table 2, the values $t_1[disease]$ and $t_2[disease]$ (the values in bold) are identified as inconsistent data values w.r.t. F_1. From the record matching, records $m_1, m_2 \in M$ are matched against inconsistent records $t_1, t_2 \in T$.

4.2   Repair Generation

In the *repair generation* module, we generate a set of repairs to the values in T based on the matched records from M. We propose three data disclosure operations for the data value $m[Y]$:

1. M discloses $m[Y]$ to T, e.g., M reveals $m_1[disease] = $ "H1N1" to T.
2. M discloses no values to T. In this case, M maintains its data privacy, but does not help to clean T's data.

3. M discloses a *generalization* of value $m[Y]$. Figure 2 shows an ontology tree containing the disease values in M, depicting relationships between various diseases and higher level categorizations of diseases. Instead of disclosing the specific disease value 'H1N1' to T, we disclose its parent value, which is the less specific (more generalized) value 'influenza'. This reveals less specific, sensitive information while still resolving the inconsistency in T.

M passes the disclosed data value(s) to T, who then applies these value(s) as repairs. To ensure that the tuples in T are consistent following a generalization repair operation, we slightly modify the definition of FD satisfaction to include parent (and ancestor) values from the ontology tree. Specifically, given $F : X \rightarrow Y$, let $y \in Y$ and $ancestor(y)$ be the ancestral node of y. A data instance I satisfies F if for every pair of tuples $t_1[X] = t_2[X]$ then $t_1[Y] = t_2[Y]$ or $t_2[Y] = ancestor(t_1[Y])$. In our example, if $t_2[disease]$ is updated to 'influenza' then t_1 and t_2 now satisfy F_1.

4.3 Repair Optimization

The *repair optimization* module quantifies the amount of privacy loss and data utility gain from a repair operation. We first describe our privacy loss and data utility metrics, and then show how we model the tradeoff between privacy loss and data utility via a multi-objective optimization function.

Privacy Loss Metric. We propose an information theoretic based metric to measure the amount of information disclosed by M. We consider two types of information loss: (1) Self-information loss: denoted as $I(y)$, to quantify the amount of information in an attribute value y. (2) Association information loss: denoted as $aI(y) = H(X) - H(X|y)$. Given an FD $F : X \rightarrow Y$, there is an association between the attributes X and Y. If M discloses a value $y \in Y$, this will disclose information about the associated X. The entropy value $H(X) - H(X|y)$ measures the reduction in uncertainty of X after y is known. A large value indicates a large reduction in the uncertainty of X, which consequently indicates that there has been a large information gain (i.e., more information about X has been revealed). Let y be a value from M, we define our privacy loss metric, $priv(y)$, as the sum of the information loss for a single attribute value and its associated values, $priv(y) = I(y) + (H(X) - H(X|y))$.

Data Utility Metric. Let y be the value disclosed by M to T. We want to measure the utility of y, $util(y)$. If M discloses y directly, we define $util(y) = \frac{1}{H(y)+1}$, where $H(y)$ is information entropy, and measures the amount of uncertainty in y. We take the inverse since the lower the uncertainty value, the greater the utility gain, and we add 1 to prevent a zero denominator. If M discloses a generalized value of y (an ancestor of y), we define $H(ancestor(y)) = -\sum p(u) \log p(u)$ for every descendant u of $ancestor(y)$ in the ontology tree.

Example: We continue our running example, and compute the privacy loss and data utility for each of the three disclosure operations discussed in Sect. 4.2. First, if M discloses $m_1[disease] = H1N1$, since $p(H1N1) = \frac{3}{10}$, we get $I(H1N1) = 0.523$ and $aI(H1N1) = H(Age) - H(Age|H1N1) = 0.543$, then $priv(H1N1) = 1.066$; and $util(H1N1) = \frac{1}{0+1}$, since there is no uncertainty. Second, if M discloses no attribute values then $priv = 0$, and there is also no data utility gain, hence $util = 0$. Lastly, instead of revealing $H1N1$, M discloses a generalization value $influenza$. We consider all descendant nodes of $influenza$ that occur in M to be an instance of influenza. Hence, $p(influenza) = p(H1N1) + p(H2N2) = 0.5$. We get $I(influenza) = 0.301$, and $H(Age|influenza) = 0.458$, and $aI(influenza) = 0.316$, and finally $priv(influenza) = 0.617$; and $H(influenza) = -(4 * (0.25 \log 0.25)) = 0.602$ (assuming equal likelihood of each influenza type), then $util = \frac{1}{0.602+1} = 0.624$. Disclosing a generalization repair significantly reduces the amount of information disclosed ($priv$: $1.066 \rightarrow 0.617$), and data utility metrics ($util$: $1 \rightarrow 0.624$) as expected.

Finally, to ensure a balanced tradeoff between privacy loss and data utility, we consider modelling our problem as a multi-objective function. Since we want to find a y which can minimize $priv(y)$ and maxmize $util(y)$, the objective function is $obj(y) = min(priv(y)) + max(util(y))$. We can convert it to weighted sum optimization function as $wsum(y) = (\alpha \, priv(y)) + (\beta \frac{1}{util(y)})$ [4]. The weights α and β represent the relative importance of privacy loss and data utility, respectively, in the optimization, and should sum to 1. We use the simulated annealing algorithm to navigate the search space for an optimal solution.

5 Conclusion

We presented our privacy-preserving data cleaning framework that recommends data repairs to correct inconsistencies while minimizing information disclosure. We proposed: (1) a set of disclosure operations that control the amount of information revealed; and (2) information-theoretic privacy loss and data utility metrics. In future work, we intend to extend the set of disclosure operations, and focus on evaluating the efficiency and effectiveness of our framework.

References

1. Beskales, G., Ilyas, I., Golab, L.: Sampling the repairs of functional dependency violations under hard constraints. In: VLDB, pp. 197–207 (2010)
2. Beskales, G., Ilyas, I., Golab, L., Galiullin, A.: On the relative trust between inconsistent data and inaccurate constraints. In: ICDE, pp. 541–552 (2013)
3. Chiang, F., Miller, R.J.: Active repair of data quality rules. In: ICIQ, pp. 174–188 (2011)
4. Deb, K.: Multi-objective optimization. In: Burke, E.K., Kendall, G. (eds.) Search Methodologies, pp. 403–449. Springer, New York (2014)

5. Fung, B., Wang, K., Chen, R., Yu, P.S.: Privacy-preserving data publishing: a survey of recent developments. ACM Comp. Surv. **42**(4), 14 (2010)
6. Howard, P.: The business case for data quality. Bloor Research, White Paper (2012)
7. Sweeney, L.: k-anonymity: a model for protecting privacy. J. Uncertainty Fuzziness Knowl. Based Syst. **10**(5), 557–570 (2002)
8. Thomas, J., Cover, T.: Elements of Information Theory, vol. 2. Wiley, New York (2006)

A Data Quality Framework for Customer Relationship Analytics

Fei Chiang[(✉)] and Siddharth Sitaramachandran

McMaster University, Hamilton, Canada
{fchiang,sitaras}@mcmaster.ca

Abstract. Poor data quality has become an increasingly pervasive problem for organizations leading to operational inefficiency, increased costs, and missed opportunities. As high quality data is a prerequisite to trusted data analysis, we propose a framework that focuses on improving the data model to improve data quality. In particular, we show how changes to the underlying data design can achieve key data quality properties. We conduct a case study that demonstrates the application of the framework to a customer relationship management (CRM) problem. Our evaluation shows that a set of CRM queries can be efficiently run over data sizes of up to 10 million records, and organizations can glean new insights about customer preferences and activity.

Keywords: Data quality · Data design · Data analytics

1 Introduction

Poor data quality is a serious problem for organizations as it has become increasingly difficult to reap value from their data, leading to operational inefficiency, increased costs, and missed sales opportunities. Studies have indicated that by 2017, 33 % of the largest global organizations will experience an information crisis due to poor data quality [9]. While recent legislations such as the US Sarbanes-Oxley Act, the European Basel Accords, and the Privacy Act in Australia, require organizations to maintain accurate and compliant records, there are no specific guidelines provided as to how to achieve high data quality.

Most real data often contain missing, incomplete, duplicate, and inconsistent values. To manage this problem, organizations focus on correcting data errors by implementing specific, often manual, cleansing routines [2,4]. For example, many existing solutions focus on identifying and correcting misspellings, syntactic inconsistencies, or data that does not conform to an expected format and structure. These solutions, however, do not address semantic notions of correctness that are often subjective, and specific to an individual user or organization. For example, consider an organization with a financial database that has an error in the *income* field. If the error is corrected by representatives from the Marketing department, the correction involves updating the income value to include all gross product sales. Alternatively, if the error is corrected by representatives

© Springer International Publishing Switzerland 2015
J. Wang et al. (Eds.): WISE 2015, Part II, LNCS 9419, pp. 366–378, 2015.
DOI: 10.1007/978-3-319-26187-4_35

from the Accounting department, the income value is updated to include gross sales and monetary donations, which are also considered sources of income. This semantic interpretation of data quality is often modelled via a set of *data quality rules*. These rules are often referred to as integrity rules as they are defined over a data set to preserve data integrity. They represent the domain specific relationships that should hold over the data and that the data values are expected to satisfy.

Having the correct data design is a critical component in the data quality management process. A data architect gathers the requirements for the system and its applications. A satisfying design ensures that the defined tables, attributes, and the relationships among the data values, model the application requirements. A data design that fails to capture the intended user and application requirements produces unreliable and irrelevant data.

As increasing amounts of data are generated, the importance of having high quality data for trusted data analysis has become critical for organizational success. Data driven decision making has become a requirement for many organizations. Examples of data analytic queries include mining data repositories to glean insights about customer preferences, identifying popular products, and forecasting customer behaviour and preferences to predict future sales. All these queries require high quality data to ensure the results are accurate, trusted and timely.

Unfortunately, many organizations today are faced with data deluge, with an overwhelming abundance of data and little insights on how to reap value from this data. Legacy systems with outdated data models, stale data quality rules, and adhoc data designs that are pieced together on an as-needed basis, all further exacerbate the problem. In this paper, we propose a framework that focuses on improving data quality by improving the data design with the objective of providing more accurate, less redundant, and correct information for customer relationship analytics. We make the following contributions:

1. A framework that improves the data design to improve data quality.
2. We conduct a case study showing the application of the framework to a customer relationship management (CRM) problem. We discuss how changes to the data model lead to knowledge about customer preferences and behaviour.
3. We conduct experiments to evaluate CRM query performance. In addition to having a more complete and accurate view of customers, we show that the queries run efficiently and produce higher quality results.

2   Related Work

Recent data cleaning systems provide a holistic view to data cleaning by using a variety of data quality rules to capture different application and user semantics [5,7]. Some of these systems have focused on building scalable, distributed platforms that provide ease-of-use and increased efficiency [10].

A recent research thrust in data cleaning has focused on statistical based techniques that consider different types of error, and aim to minimize the distance between a clean data distribution and an ideal data distribution [3,6]. Wang et al., take a diagnostic approach to data cleaning by proposing techniques that identify errors during the data generation process [14]. Our work follows a similar diagnostic thread where we aim to provide a more automated data cleaning process via: (1) focusing on data design errors that are a common source of data quality problems; and (2) leveraging integrity constraints to ensure intended data relationships are enforced in the data.

3 Preliminaries

3.1 Data Quality Properties

The notion of high quality data is often subjective based on user and application requirements. However, the following are data quality properties that organizations aim to maximize [12]:

- Existence: whether the organization has (or is able to collect) the data.
- Validity: whether the data values are within the acceptable domain.
- Consistency: the same piece of data contains the same value across different locations.
- Integrity: the completeness of relationships between data elements.
- Accuracy: the data describes the properties of the object it is meant to model.
- Relevance: whether the data is appropriate to support the desired objectives.

A data quality model must first address the existence and relevance properties prior to implementation. The remaining four properties are defined according to an organization's specific requirements and data usage.

3.2 Integrity Constraints

Integrity constraints (also known as data dependencies or data quality rules) are the primary means for preserving data integrity. Constraints represent domain specific rules and relationships that hold over any database instance that accurately reflects the domain. Several types of integrity constraints exist in databases. *Keys* represent a common constraint that consist of a set of one or more attributes that uniquely identify a record in a database. Normally, one key is designated a *primary key*, which is used to uniquely identify a record. To support references between different datasets (and tables), a *foreign key* consists of a set of attributes from table B that refers to the primary key in another table A. The use of primary and foreign keys support *referential integrity*, that is, to ensure that referenced values between tables are NOT NULL and are complete.

In relational data, a *functional dependency (FD) F* over a relation R is an integrity constraint represented as $F : X \rightarrow Y$, where X, Y are attributes in R. A data instance of R satisfies F if for every pair of records t_1, t_2 in the data instance, if $t_1[X] = t_2[X]$ then $t_1[Y] = t_2[Y]$. A functional dependency defines a relationship between two specific attribute sets, where all records in the data instance are expected to satisfy this relationship. By having the correct integrity constraints defined and enforced over the data, prevents inadvertent and anomalous data updates that can alter the underlying attribute relationships.

4   Framework Overview

Figure 1 presents an overview of our framework that improves data quality and CRM data analytics. We assume the input data consists of a relational database or data files that can be transformed into relational format. In the *Design Validation* component, we first validate the schema of the input data by: (1) checking whether appropriate integrity constraints[1] are defined to model the intended attribute relationships; (2) validating that the table is not 'overloaded' (i.e., modelling several entities in a single relation); (3) whether the data model (and data) support the intended application semantics. At the end of the Design Validation phase, we identify the gaps between the current and intended design, that is, the missing constraints, tables and data that are needed to capture the intended CRM application semantics.

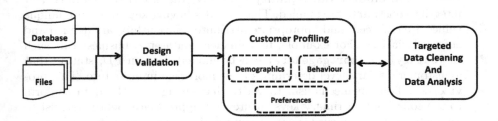

Fig. 1. Framework overview

In *Customer Profiling*, we fill the gaps by implementing the necessary tables and data quality rules to capture customer preferences and behaviour. This includes capturing customer demographic data, and information about a customer's preferences for particular products, areas of interest, and customer activity (e.g., the type of products recently purchased, the amount spent). Given the new design, a set of analytical queries that aggregate data across the tables can be run to glean insights about customers, their preferences, their recent activity, and anticipated future behaviour.

5   A Data Quality Framework for CRM Analytics

We describe how each component in our framework helps to improve overall data quality and CRM analytics.

---

[1] We refer to data quality rules and (integrity) constraints interchangeably.

5.1   Design Validation

In the *Design Validation* component, our objective is to develop a data model (schema) that satisfies as many of the data quality properties as possible (discussed in Sect. 3.1). We focus on addressing the following design issues commonly seen in practice leading to poor data quality [12]:

D1 **Minimize duplicate data.** We discourage the use of disparate database instances where information is duplicated. Unfortunately, this is common practice in many organizations due to siloed organizational structures, and poor data modelling. By minimizing the data duplication, we improve data consistency (by minimizing the number of occurrences that represent the same value), and data accuracy (we avoid having to reconcile duplicates).

D2 **Minimize overloading of tables.** We validate that information stored within each table represents a single entity and its properties, or represent the relationship between entities. Overloading a table to model multiple entities leads to "wide" tables containing many attributes. While this centralized design avoids the need to aggregate data from multiple tables, records that do not have values for all the attributes will contain missing or NULL values. Validation must be done to ensure that the attributes represent the desired properties and are of the correct data types.

D3 **Define and enforce data quality rules.** We validate that a set of correct integrity constraints are defined. This includes having keys to ensure unique values are enforced, and appropriate attribute relationships are defined and enforced via FDs. For example, in a financial dataset, a business rule such as [salary] → [mortgageRate] can be modelled as an FD, stating that a customer's salary level determines their mortgage interest rate. This provides a validation mechanism to verify that changes to the data conform to the constraints. Primary keys ensure that appropriate identifiers exist to uniquely identify a record. Foreign key relationships ensure that references between tables involve values that exist in the participating tables.

D4 **Minimize incomplete information.** Organizations often populate attributes with default or dummy values, or worse, leave them empty. This creates inconsistencies in the data, especially if there is duplication of data. In some cases, for a table containing many incomplete values across all records, it is preferable to partition the table into multiple, smaller, tables to minimize data incompleteness.

If the data is loaded in a database system, the above issues can be identified by running queries (with unique, non-distinct keywords) over tables suspected of containing redundant data to identify duplicate data records. We validated the data quality rules via: (1) validating that correct key and check constraints were defined; and (2) in consultation with domain experts, we defined the necessary FDs to ensure they aligned with the intended application semantics. An alternative would be to apply FD discovery algorithms [8,11], to identify the relationships that hold in the data. However, if the given data is potentially dirty, then caution must be taken to identify those rules that are most relevant and aligned with the intended application semantics.

5.2   Customer Profiling

To support CRM analytics, organizations require a holistic view of customers that provides complete information on their demographics, behaviour, and preferences. In Customer Profiling, we fill the gaps between the current data design and the desired CRM functionality. This includes capturing new data on customer preferences and behaviour, defining new integrity constraints, and redesigning existing tables to enable more efficient CRM queries. We discuss the type of data we collect in each of these categories and the resulting queries that are enabled.

Demographics. We first define tables that include customer identification and contact information. Supplementary data such as age, gender, marital status, and salary are additional characteristics that we use to segment customers for targeted marketing. We collect this information by aggregating data from existing customer tables, customer response surveys, and customer profile accounts. Depending on the source reliability of this data, some data values are susceptible to errors and omissions. For example, sensitive data such as ethnicity, and salary may be less trustworthy due to customer privacy concerns. To handle such cases, a reliability weight $w \in [0, 1]$ can be assigned to each attribute to represent the trustworthiness of the data source. Attribute values with $w = 1$ denote a lineage from a trustworthy data source, where a weight of $w = 0$ indicates an unreliable source.

Preferences. Targeted marketing towards individuals and customer segmentation groups requires a detailed understanding of customer preferences. This information is obtained by mining customer purchase data. We apply data mining algorithms such as the Apriori algorithm [1], and frequent item set mining [13] to extract correlated attribute value relationships from the data. This data includes customer purchase transactions showing the type of product, the purchase price, the quantity, date and time of the sale, correlated with demographic data, this data is used to derive a more comprehensive view of the customer.

Behaviour. This includes information such as customer purchase patterns, time between purchases, and mining for correlations among recent purchases. To capture this information, we adopt a similar model as data warehouse environments. We define tables that aggregate customer data and recent activity along a set of time dimensions (per hour, per day, per week, etc.). These tables support online analytical processing (OLAP) queries that aggregate information across a set of dimension and fact tables. In contrast, to support frequent customer activity such as purchase transactions, we store this data (e.g., transaction timestamps, product characteristics) in a transaction table. We use IBM DB2 v.10.5.3 to implement these two workload models and SQL scripts to populate the tables. We use data from both types of tables (OLAP and transactional OLTP) to gain a holistic view of recent customer activity and how this compares against past behaviour. This provides us with a temporal view of customer activity and preferences that previously were not available.

5.3   Data Cleaning and Data Analysis

Following the Design Validation and Customer Profiling stages (conducted during data design) we enter the Data Cleaning and Data Analysis stage. This stage is executed during runtime once the data has been consumed, processed and initial analysis has been done. Normally, data quality errors arise and are identified as new data is generated, and integrated with existing data. We use the defined integrity constraints to identify violating records and flag these inconsistencies for review by a user. Data cleaning algorithms that utilize integrity constraints can be applied to propose updates (also known as repairs) to the data to correct the inconsistencies [4]. These algorithms use constraints (such as FDs) as a benchmark of the conditions and relationships that the data should satisfy. Data that do not conform to these conditions and relationships are identified as inconsistencies. Repair algorithms propose data updates according to a cost function that aims to minimize or maximize an objective. For example, one such objective is to resolve all inconsistencies using at the fewest number of data updates. As the data is corrected and inconsistencies are removed, a set of OLAP and transactional OLTP queries are run over the data stores to identify customer purchase patterns and trends. As new data is generated and added to the data stores, the Customer Profile tables can be updated to include new (and revised) table schemas and integrity constraints, as needed to ensure relevant data is returned and efficient query performance is maintained.

6   Case Study

We conduct a case study to evaluate our framework by working with a telecommunications company named *AirWave*[2]. AirWave provides communication based services via chip technology embedded in electronic devices. Their clients (also referred to as vendors) apply this technology to market and sell products to customers (also referred to as users). Customer activity such as purchases and interactions with client products are recorded via chip enabled devices. To increase sales for their clients (and for AirWave), a re-design of their data model was needed. The objectives were to enable management to perform targeted marketing towards different customer segments, and identify data quality issues.

6.1   Redesign

Figure 2(a) shows a portion of AirWave's schema revealing several issues (as discussed in Sect. 5.1):

(a) Customer information is stored on a per client basis, each with a different schema, duplicating data and causing redundancy. In the `client 2` instance, customers are referred to as users. To reconcile whether two customers (across different client instances) are the same person will involve

---

[2] The name AirWave is used to protect the organization's identity.

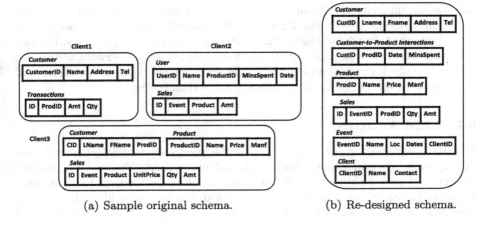

(a) Sample original schema. (b) Re-designed schema.

Fig. 2. Case study schema.

checking the appropriate attributes for equality (e.g., `Client2.UserID` and `client1.CustomerID`). The current design has limited scalability, as additional clients are added, customer data is further duplicated. Queries executed over the original design provide only a local (client) view of customer activity. Similarly, there are schema inconsistencies for capturing sales information. The design in Fig. 2(a) does not permit a global view of sales, and computing total sales for an Event is possible only for Clients 2 and 3.

(b) The `Client2.User` table contains information about users and the products that he/she has expressed interest (via `MinsSpent` interacting with the product). For each product that a user has expressed an interest, the user name is duplicated. This is an example of an overloaded table containing information that should be contained in two (smaller) tables.

(c) No integrity constraints are defined. For example, attribute `Client3.Customer.ProdID` is not defined as a foreign key referencing the primary key `Client3.Product.ProductID`. Changes to the primary key do not trigger an update to all foreign keys, thereby, causing inconsistencies.

Design Changes. Figure 2(b) shows a sample of the changes we implemented:

(i) Reconcile inconsistencies and standardize naming conventions across all tables such that entities have a consistent name (user vs. customer), and attributes represent the same information (e.g., LName and FName should be used to represent a customer's name).

(ii) Implement a global schema and remove the localized (per-client) schema model. Our global schema contains a table for each entity and its properties. For example, a single `Customer` table exists for all clients, a `Product` table to capture product information, and a `Customer-to-Product Interactions` table to capture the products for which customers have shown interest.

(iii) Define integrity constraints between foreign keys and primary keys. For example, the attributes `CustomerToProductInteractions.CustID` and `CustomerToProductInteractions.ProdID` reference, respectively, the primary keys `CustID` and `ProdID` in tables `Customer` and `Product`. We also define check constraints to ensure that specific data types are enforced, and out of range values for attributes are flagged as errors.

6.2 CRM Analytics

In CRM, a *customer lifecycle* describes the progression of steps a customer undergoes when evaluating, purchasing, and maintaining loyalty to a product or service. Figure 3 shows how our customer profiling data is used in a customer lifecycle. In Stage 1, a *lead* is the entry point to the customer lifecycle, and any user who registers with AirWave is considered a lead. As part of our redesign and profiling, AirWave can query basic user information from survey response and user registration data, to move a lead towards the next stage of the lifecycle.

In Stage 2, a lead is converted to a *prospect* when the data shows persons who have interacted with products, and engaged with client representatives. This is a critical stage in the lifecycle, as key information on user preferences, time spent with the product, and demographic data, can be used for personalized marketing to convert this prospect into a customer. Lastly, in Stage 3, our objective is to retain customer loyalty and increase the return on investment (ROI) for each customer. We mine the historical transaction data to gain a deeper understanding of a customer's buying habits, lifestyle, and product preferences. These factors are used to identify related products and services that can be leveraged in cross-selling and up-selling for increased sales.

6.3 Evaluation

We evaluated a set of CRM queries to test the efficiency of our framework. The final design consisted of 29 tables, and was implemented on an Intel Xeon E5-2670, 2.6 GHz, with 6 cores, and 32 GB RAM, using IBM relational database DB2 v10.5.3. Our evaluation focused on three objectives:

1. Evaluate the query performance and scalability of commonly run CRM queries to ensure the running times are within acceptable time limits for AirWave.
2. Compare the query performance using the original data design versus using the new, revised design.
3. A qualitative evaluation showing improved data quality via a reduction in the number of duplicate, and incomplete data records.

Query Performance and Scalability. In consultation with data architects and analysts at AirWave, we evaluated the performance of 13 commonly run

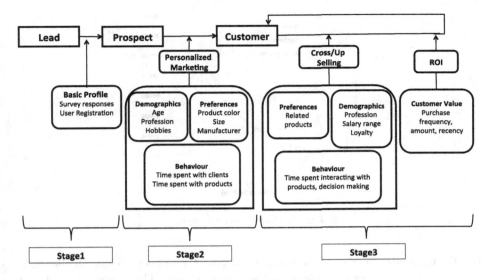

Fig. 3. Customer lifecycle

CRM queries. Due to space limitations, we provide full details of the queries on our website.[3] The queries are classified into three broad categories:

- Sales: Compute product sales across clients, events, and customer segments.
- Customer: Determine customers that belong to each stage of the lifecycle, and customer characteristics that lead to increased sales.
- Aggregation: Compute statistical summaries to extract information about historical sales trends, customer buying activity, and product popularity.

To evaluate scalability, we ran the queries over increasing data sizes ranging from 0.5 M to 10 M records. Due to space limitations, we report the main results (running times can be found on our website):

(a) For 6 M records, 95 % of the queries ran in less than 2 s. This set of queries included commonly run sales queries to determine total revenue, the customer ROI, and characteristics of loyal customers. The remaining 5 % of the queries ran in under 3m. These longer running queries included aggregation queries that extracted and summarized historical information.

(b) Figure 4 shows the running times in milliseconds (ms) for each query category. The aggregation queries running times are shown using the right y-axis scale. Due to larger table sizes and complex statistical summaries, we observe that the aggregation queries have higher running times than the sales and customer queries. Customer queries show improved performance over the sales queries due to a number of records pruned based on selective customer characteristics (e.g., identify popular products for only male customers).

---

[3] The CRM queries can be found at: www.cas.mcmaster.ca/~sitaras/casestudy/.

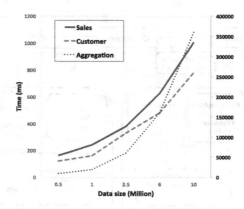

Fig. 4. Query performance.

(c) We measured the time to compute total sales for an event across all clients; a query that was not possible using the original design. The query ran in 425 ms over 10 M records, demonstrating the efficiency and value of our framework.

Comparative Performance. We evaluate the running time of eight queries (Q_1-Q_8) using the original data design versus using the new, revised design. Due to space limitations, the definitions and descriptions of the queries Q_1-Q_8 are given on our website. Figure 5 shows the comparative query performance. For all queries (except Q_5), we see a performance improvement of at least 75 % or more, demonstrating the effectiveness of our design changes. For query Q_5, we observe no difference in performance as the data records retrieved across multiple clients in the old design were equal to those retrieved in the new design.

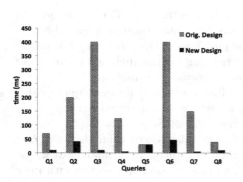

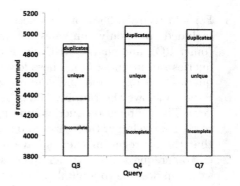

Fig. 5. Comparative performance   **Fig. 6.** Qualitative evaluation

Qualitative Evaluation. To study the data quality benefits of our design changes, we ran the queries Q_1–Q_8 (using the original data design versus the new design), and computed the number of records returned that were unique, and those that contained incomplete or duplicates data values. In the interest of clarity, Fig. 6 shows the returned records for the top-3 queries with the greatest benefit (Q_3, Q_4, Q_7), and the proportion of these records that were duplicates or had incomplete attribute values, using the old data design. Between 87–90 % of the records returned were either duplicates or contained missing values. When we ran the same queries using the new design (that included revised table schemas, and new integrity constraints), the duplicate and incomplete records were removed from the result (only the unique records were returned), demonstrating the quality improvement of the query results.

7   Conclusion

We presented a framework that focuses on improving data design to improve data quality. As high quality data is a prerequisite to trusted data analysis, we conduct a case study to show the value of our framework towards improving CRM data analytics. Our evaluation shows that common sales, customer, and aggregation queries can be efficiently run to glean insights that were not previously possible. As next steps, we intend to investigate the use of ontologies for improving data quality, to model and infer different interpretations of correctness.

References

1. Agrawal, R., Srikant, R.: Fast algorithms for mining association rules in large databases. In: VLDB, pp. 487–499 (1994)
2. Batini, C., Scannapieco, M.: Data Quality: Concepts, Methods and Techniques. Springer, Heidelberg (2006)
3. Berti-Equille, L., Dasu, T., Srivastava, D.: Discovery of complex glitch patterns: a novel approach to quantitative data cleaning. In: ICDE, pp. 733–744 (2011)
4. Chiang, F., Miller, R.J.: Active repair of data quality rules. In: IJIQ, pp. 174–188 (2011)
5. Dallachiesa, M., Ebaid, A., Eldawy, A., Elmagarmid, A., Ilyas, I.F., Ouzzani, M., Tang, N.: NADEEF: a commodity data cleaning system. In: SIGMOD, pp. 541–552 (2013)
6. Dasu, T., Loh, J.M.: Statistical distortion: consequences of data cleaning. PVLDB 5(11), 1674–1683 (2012)
7. Geerts, F., Mecca, G., Papotti, P., Santoro, D.: The LLUNATIC data-cleaning framework. PVLDB 6(9), 625–636 (2013)
8. Huhtala, Y., Kärkkäinen, J., Porkka, P., Toivonen, H.: Efficient discovery of functional and approximate dependencies using partitions. In: ICDE, pp. 392–401 (1998)
9. Judah, S., Friedman, T.: Twelve ways to improve your data quality. Gartner Research Report (2014)

10. Khayyat, Z., Ilyas, I., Jindal, A., Madden, S., Ouzzani, M., Papotti, P., Quiané-Ruiz, J., Tang, N., Yin, S.: Bigdansing: a system for big data cleansing. In: SIGMOD, pp. 1215–1230 (2015)
11. Lopes, S., Petit, J.-M., Lakhal, L.: Efficient discovery of functional dependencies and armstrong relations. In: Zaniolo, C., Grust, T., Scholl, M.H., Lockemann, P.C. (eds.) EDBT 2000. LNCS, vol. 1777, pp. 350–364. Springer, Heidelberg (2000)
12. Moore, M.: Dirty data is a business problem, not an it problem. Gartner (2007)
13. Pei, J., Han, J.: Constrained frequent pattern mining: a pattern-growth view. SIGKDD Explor. 4(1), 31–39 (2002)
14. Wang, X., Dong, X., Meliou, A.: Data x-ray: a diagnostic tool for data errors. In: SIGMOD, pp. 1231–1245 (2015)

Inter-rater Reliability in Determining the Types of Vegetation on Railway Trackbeds

Roger G. Nyberg[1,2]([✉]), Siril Yella[1], Narendra K. Gupta[2],
and Mark Dougherty[1]

[1] Department of Informatics and Computer Science,
Dalarna University, 78170 Borlange, Sweden
{rny,sye,mdo}@du.se
[2] School of Engineering and the Built Environment,
Edinburgh Napier University, Edinburgh EH10 5DT, UK
N.Gupta@napier.ac.uk

Abstract. Vegetation growing on railway trackbeds and embankments can present several potential problems. Consequently, such vegetation is controlled through various maintenance procedures. In order to investigate the extent of maintenance needed, one of the first steps in any maintenance procedure is to monitor or inspect the railway section in question. Monitoring is often carried out manually by sending out inspectors or by watching recorded video clips of the section in question. To facilitate maintenance planning, the ability to assess the extent of vegetation becomes important. This paper investigates the reliability of human assessments of vegetation on railway trackbeds.

In this study, five maintenance engineers made independent visual estimates of vegetation cover and counted the number of plant clusters from images.

The test results showed an inconsistency between the raters when it came to visually estimating plant cover and counting plant clusters. The results showed that caution should be exercised when interpreting individual raters' assessments of vegetation.

1 Introduction

Vegetation that grows on railway trackbeds and embankments can present potential problems. The presence of vegetation threatens the safety of personnel inspecting the railway infrastructure. In addition, vegetation growth clogs the ballast and results in inadequate track drainage, which in turn could lead to the collapse of the railway embankment. In the main, assessing vegetation within the realm of railway maintenance is carried out manually by making visual inspections along the track. If on-site inspections are not carried out, monitoring is carried out by watching videos recorded by maintenance vehicles operated by the national railway administrative body.

The true extent of vegetation is contained within the vegetation biomass. To collect this, one has to perform destructive tests, i.e. excavating plants (roots

© Springer International Publishing Switzerland 2015
J. Wang et al. (Eds.): WISE 2015, Part II, LNCS 9419, pp. 379–390, 2015.
DOI: 10.1007/978-3-319-26187-4_36

and shoots) and then weighing the dried or fresh roots and shoots on a scale. This is very time consuming and, thus, financially expensive. Instead, a visual estimate (VE) is often used in which humans visually make estimates of the extent of vegetation. Common measurable vegetation attributes include density of individual plants, cover, frequency and production [1,2]. In this investigation, VEs of cover were used, together with the counting of plant clusters. Carrying out VEs to measure vegetation can introduce problems, particularly with regard to the reliability of the measurements; such concerns are the focus of this paper and [3].

Several methods have been used to quantify whether the participating raters (variously known as judges, assessors, observers and coders) are *in agreement with each other*, and to assess the reliability of their estimates when making independent ratings about a set of *subjects* (here, plants). A concise summary of the many reliability measures is: *"All reliability measures are intended to express the degree to which several assessors, several measuring instruments, or several interrogations of the same units of analysis yield the same descriptive accounts, category assignments, quantitative measures or data for short"* [4].

Other investigations into raters' reliability in the railway domain have shown that human raters tend to have a low level of agreement [5,6]. The purpose of this investigation was to confirm and thereby strengthen previous findings or to question and reject them.

For more detailed information on methods for measuring raters' reliability and agreements refer to [7,8]. Details of how to use the ICC can be found in [9,10].

2   Method

Five maintenance engineer administrators representing four national regions of the Swedish national railway administration, the Swedish Transport Administration (STA), and Borlange municipality were asked to make VEs from images that showed the railway trackbed. Common to these maintenance engineer administrators was that each was ultimately responsible for determining whether or not vegetation management should be carried out within their region, or municipality.

The image data were acquired during a one-day field experiment at a railway section along the railway between Falun and Grycksbo, Sweden (WGS 84 decimal (lat, lon) coordinates 60.6657, 15.5437). Weather condition: Sunny, almost clear, 1/8th of the sky was covered with clouds, 25–27°C, Dry conditions. All images were acquired vertically from a nadir (bird's-eye) perspective with a Nikon D90 DSLR camera, which sensed the visual spectrum (approx. 390 to 750 nm). A tripod was used.

The investigation was carried out as an online Internet survey. First, each of the respondents was contacted by telephone. Then, each respondent was instructed through slideshow presentations and web conferencing software as to what to do and how to make the VE. Each respondent was able to browse

Fig. 1. (a) Sampling area, and (b) AFC vs ACC method

the images on a website and they had the opportunity to ask questions orally in real time while taking the survey. The respondents had no contact with each other.

All the participating raters were instructed on where to make their assessment, i.e., by defining the sampling area (see the non-yellow sampling area in Fig. 1a), and how to make their assessments using the aerial foliage cover (AFC) and aerial canopy cover (ACC) observation methods (see Fig. 1b). A review of these two methods of observation can be found in [2]. The order of the images was randomly re-arranged. Each rater then proceeded to make a series of VEs using AFC or ACC.

By using these methods, the raters were asked to make a VE of vegetation cover on a scale of 0 to 100 %. Vegetation was categorized as either: (1) *woody plants*, (2) *herbs*, (3) *grass*, and (4) *the remainder*. The latter category included gravel, soil, wood, and rocks, as well as litter, such as dead plants. After finishing making VEs of the attributes shown in the list above, the raters were also asked to count the number of vegetation clusters in each image.

Plant Cluster Definition: In this investigation, a plant cluster was defined as being an individual plant or a tightly clumped group of individual plants, as shown in Fig. 2. The number of clusters is dependent on the position of the centre of gravity of each observed sub-cluster, or plant.

It is relatively easy to assess whether a plot is devoid of vegetation, or is completely covered by the plant type in question; thus, images with 0 % and 100 % of the attribute in question were removed before the analysis. The purpose of this investigation was to see whether or not the raters would give the same estimate or whether there would be differences between them. It was not in the interests of this study to try to find out which raters differed in their estimates; therefore, no post-hoc tests were performed.

2.1 Inter-rater Agreements of Visual Estimates of Plant Cover and Counting Plant Clusters

Inter-rater agreement, (or inter-rater reliability) is the degree of agreement between the five raters who estimated the number of plant clusters from the

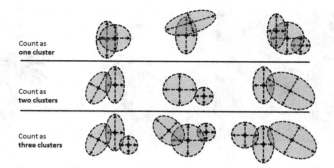

Fig. 2. Plant cluster definition

same set of images. The term "inter-rater" implies "between raters". The raters participating in this investigation were assumed to be representative of a larger number of similar raters in the population. Hence, the ICC(2,1) class was chosen (see Definition in Eq. 1).

$$ICC(2,1) = \frac{var(\beta)}{var(\alpha) + var(\beta) + var(\varepsilon)} \tag{1}$$

where var denotes the variance, $var(\alpha)$ denotes the variability due to differences in the rating scale used by the raters. For example, when considering a sample plot containing a "true" value of 5 % plant cover, rater A estimates this plot to contain 10 % cover; however, rater B estimates the same plot to contain 15 %. Here, $var(\beta)$ denotes the variability caused by differences in the observed phenomenon/subjects (e.g., the sample plots containing plants), and $var(\varepsilon)$ denotes the variability caused by differences in the evaluation of the observed phenomenon or subjects by the raters. For example, rater A finds that a sample plot contains 45 plants, but rater D finds the same sample plot contains 5 plants, because of different personal opinions on what and how to count. The ICC(2,1) class is generalisable, whereas ICC values from ICC(3,1) class are not. The ICC coefficient can theoretically vary between 0 and 1.0, where an ICC value of 0 indicates no agreement (i.e., no reliability), while an ICC value of 1.0 indicates perfect agreement (i.e., the raters were unanimous in their decisions). Qualitative ratings of ICC agreement based on the ICC values were suggested by [11] as follows: *Poor ICC-value < 0.40; Fair ICC-value 0.40 to 0.59; Good ICC-value 0.60 to 0.74; Excellent 0.75 to 1.0.* In addition to the ICC(2,1) method the Krippendorff's α was calculated for the ratio data using the general Eq. 2;

$$\alpha_{Kripp} = 1 - \frac{D_o}{D_e} \tag{2}$$

where D_o is the observed disagreement and D_e is the expected random disagreement.

Six sub-investigations were carried out (see Sects. 3.1 to 3.6). Each rater visually assessed the cover extent from images using two different methods: AFC and ACC, respectively. The type of target plants were woody plants, herbs and grass, respectively. After making VEs to assess the cover of woody plants, herbs, and grass, the raters were asked to count the number of plant clusters in each image (see Sect. 3.7). This counting procedure was carried out twice, once after they made their VE using AFC, and once after using ACC.

3   Results

The total number of images presented in each sub investigation was 51. In cases where all raters unanimously estimated an image to contain exactly 0 % cover of the target plant (i.e., woody plants, herbs or grass), these images were removed. This approach was chosen in favour of allowing the investigator to decide if the target plant was present or not.

Parametric methods were used for the analysis of the VEs. ANOVA tests at 95 % confidence level were performed on the raters' VEs of each target plant using ACC and AFC. All of the density plots indicated irregular, positively skewed distributions. Therefore, in order to perform a parametric analysis the data was $\log_{10}$-transformed, x_i', and then normalised by subtracting the mean $\log_{10}$-value, $\overline{log10(x)}_i$, of all the human raters individual estimate of the same image i. This type of normalisation, which is essentially a rescale operation, is commonly named *per-example mean subtraction, mean-centring,* or *normalisation by subtracting the mean.* In essence, it centres the distribution to give a zero mean value. This makes the distribution of the raters' estimates of each image more comparable with estimates of other images used in the experiment.

3.1   Visual Estimates of Woody Plants Using the ACC Method

The number of observed images in the analysis was 42. A *non-significant* difference in estimates was reported. $F = 1.499$ at $df = 4$ and 205, $p = 0.2037$. (H_0 could not be rejected at 0.05 significance level; see box-plot in Fig. 3a).

Plot-wise variation and differences are presented in Fig. 3b. The maximum difference when the raters made a VE of the same plot was 79 %, i.e., the highest estimate minus the lowest estimate in the same plot. The minimum difference was 0 %, i.e., total agreement. The median difference (over all plots) between the highest and lowest plot cover estimate: $Md = 14\%$ and the mean: $\overline{x} = 19.6\%$.

3.2   Visual Estimates of Woody Plants Using the AFC Method

The number of observed images in the analysis was 38. A *significant* difference in estimates was reported. $F = 5.219$ at $df = 4$ and 185, $p = 0.0005262$. (H_0 was rejected at 0.05 significance level; see box-plot in Fig. 4a).

Plot-wise variation and differences are presented in Fig. 4b. The maximum difference when the raters made an VE of the same plot was 35 %, i.e. the highest estimate minus the lowest estimate in the same plot. The minimum difference was 0 %, i.e. total agreement. The median difference (over all plots) between highest and lowest plot cover estimate: $Md = 6\%$ and the mean: $\overline{x} = 9.0\%$.

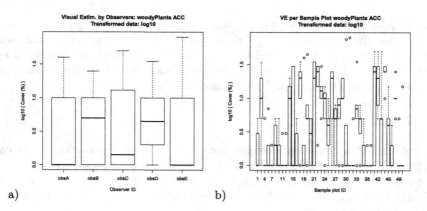

a)   b)

Fig. 3. Log$_{10}$ transformed data for: (a) VE by each rater, and (b) VE per sample plot.

3.3   Visual Estimates of Herbs Using the ACC Method

The number of observed images in the analysis was 42. A *non-significant* difference in estimates was reported. F = 1.227 at df = 4 and 245, p = 0.2998. (H_0 could not be rejected at 0.05 significance level: see box-plot in Fig. 5a).

Plot-wise variation and differences are presented in Fig. 5b. The maximum difference when the raters made a VE of the same plot was 70 %, i.e., the highest estimate minus the lowest estimate in the same plot. The minimum difference was 0 %, i.e., total agreement. The median difference (over all plots) between the highest and lowest plot cover estimate: Md = 26.5 % and the mean: $\overline{x}$ = 30.0 %.

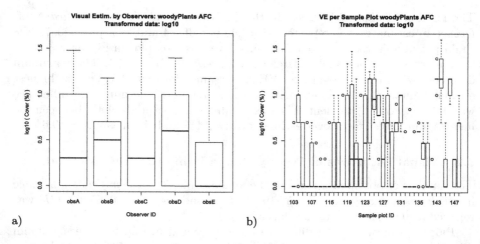

a)   b)

Fig. 4. Log$_{10}$ transformed data for: (a) VE by each rater, and (b) VE per sample plot.

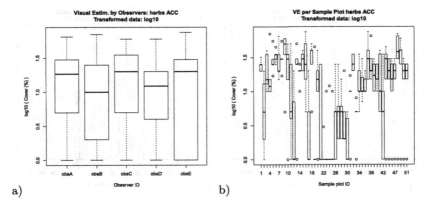

Fig. 5. Log$_{10}$ transformed data for: (a) VE by each rater, and (b) VE per sample plot.

3.4   Visual Estimates of Herbs Using the AFC Method

The number of observed images in the analysis was 47. A *significant* difference in estimates was reported. F = 4.674 at df = 4 and 230, p = 0.001206. (H_0 was rejected at 0.05 significance level; see box-plot in Fig. 6a).

Plot-wise variation and differences are presented in Fig. 6b. The maximum difference when the raters made a VE of the same plot was 44 %, i.e., the highest estimate minus the lowest estimate in the same plot. The minimum difference was 0 %, i.e., total agreement. The median difference (over all plots) between the highest and lowest plot cover estimate: Md = 9 % and the mean: $\overline{x} = 13.9$ %.

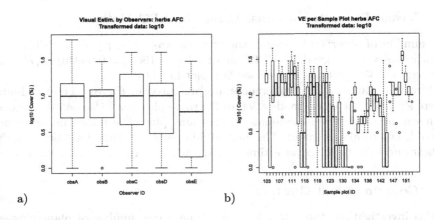

Fig. 6. Log$_{10}$ transformed data for: (a) VE by each rater, and (b) VE per sample plot.

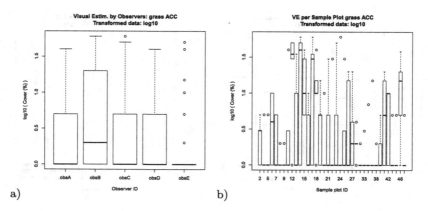

Fig. 7. Log$_{10}$ transformed data for: (a) VE by each rater, and (b) VE per sample plot.

3.5 Visual Estimates of Grass Using the ACC Method

The number of observed images in the analysis was 37. A *significant* difference in estimates was reported. F = 2.666 at df = 4 and 180, p = 0.03395. (H_0 was rejected at 0.05 significance level; see box-plot in Fig. 7a).

Plot-wise variation and differences are presented in Fig. 7b. The maximum difference when the raters made a VE of the same plot was 59 %, i.e., the highest estimate minus the lowest estimate in the same plot. The minimum difference was 0 %, i.e., total agreement. The median difference (over all plots) between the highest and lowest plot cover estimate: Md = 9 % and the mean: $\bar{x}$ = 19.3 %.

3.6 Visual Estimates of Grass Using the AFC Method

The number of observed images in the analysis was 33. A *significant* difference in estimates was reported. F = 5.427 at df = 4 and 160, p = 0.0004012. (H_0 was rejected at 0.05 significance level; see box-plot in Fig. 8a).

Plot-wise variation and differences are presented in Fig. 8b. The maximum difference when the raters made a VE of the same plot was 84 %, i.e., the highest estimate minus the lowest estimate in the same plot. The minimum difference was 0 %, i.e., total agreement. The median difference (over all plots) between the highest and lowest plot cover estimate: Md = 9 % and the mean: $\bar{x}$ = 22.2 %.

3.7 Counting Plant Clusters

In this investigation, the raters had to estimate the number of plant clusters by counting them in each image, as defined in Fig. 2. ANOVA tests at 95 % confidence level were performed on the raters' assessment of the number of plant clusters after the ACC and AFC sessions, respectively.

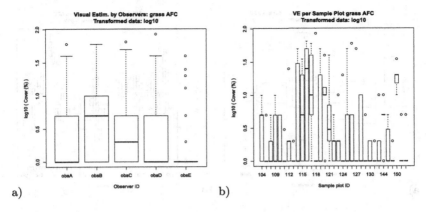

Fig. 8. Log$_{10}$ transformed data for: (a) VE by each rater, and (b) VE per sample plot.

Inter-rater Agreements in Counting Plant Clusters. Parametric methods were used for the analysis of the log$_{10}$-transformed data set. The original data was not normally distributed, the original data points x_i were transformed into $\hat{x} = log10(x_i)$, $\sum_{i=0}^{n=51} x_i$. A *significant difference* in count estimates between the raters was reported. $F = 5.579$ at $df = 4\,and\,250$ $df = 4\,and\,250$, $p = 0.0002566$. (H_0 was rejected at 0.05 significance level; see box-plot in Fig. 9a).

Plot-wise variation and differences are presented in Fig. 9b. The maximum difference between the raters in counting the same plot was 76 plant clusters, i.e., the highest count minus the lowest count in the same plot (see Fig. 9b).

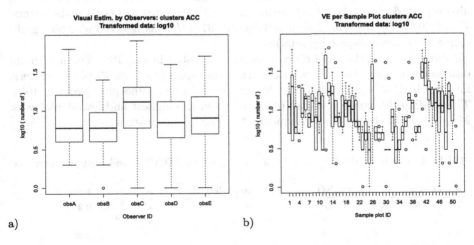

Fig. 9. Each rater's counting of: (a) plant clusters after the ACC session, and (b) plant cluster counts per sample plot

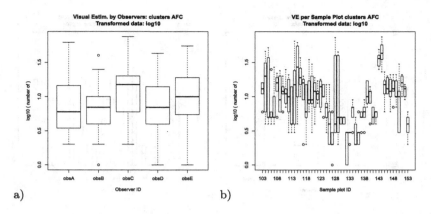

Fig. 10. Each rater's counting of: (a) plant clusters after the AFC session, and (b) plant cluster counts per sample plot

Table 1. The raters' levels of counting agreement as of the ICC(2,1) and Krippensdorff's α.

Count during session	n	Reliability ICC(2,1)	p-value	Krippendorff's α
ACC	50	0.25	$3*10^{-7}$	0.22
AFC	50	0.40	$2.7*10^{-10}$	0.36

The median difference (over all plots) between the highest and lowest plot was: Md = 16 plant clusters and the mean: $\bar{x} = 20.65$ plant clusters.

The rater's assessment of the number of plant clusters after the AFC session also reported a *significant difference* in count estimates in between the raters. $F = 8.093$ at $df = 4$ *and* 250, $p = 0.000003761$. (H_0 was rejected at 0.05 significance level; see box-plot Fig. 10a)

Plot-wise variation and differences are presented in Fig. 10b. The maximum difference between the raters in counting the same plot was 70 plant clusters, i.e., the highest count minus the lowest count in the same plot (see Fig. 10b). The median difference (over all plots) between the highest and lowest plot was: Md = 13 plant clusters and the mean: $\bar{x} = 17.8$ plant clusters.

Table 2. The five raters' reliability according to the ICC(2,1) and Krippendorrf's α

Target plants	VE method	NO. of images n	Reliability ICC(2,1)	ICC p-value	Krippendorff's α
Woody plants	ACC	42	0.27	$1*10^{-6}$	0.25
	AFC	38	0.38	$5*10^{-10}$	0.30
Herbs	ACC	50	0.36	$1*10^{-11}$	0.30
	AFC	47	0.41	$4*10^{-14}$	0.35
Grass	ACC	37	0.16	0.003	0.12
	AFC	33	0.13	0.010	0.11

To assess the reliability of the five raters, ICC(2,1) and Krippensdorff's α were calculated for the two plant cluster counting sessions (see Table 1, where n denotes the number of images).

4   Conclusion and Discussion

Visual Estimates of Plant Cover. The ANOVA test results showed a degree of inconsistency when it came to estimating plant cover. Four out of six tests were found to be significant concerning differences in the mean estimates of cover. Note that for each sub investigation a density plot of the residuals from the $\log_{10}$-transformed data was produced to qualitatively determine that the residuals were approximately normally distributed. The residuals in each sub investigation appeared to be approximately normally distributed, thereby justifying our choice of the ANOVA test. It should be noted that some of the raters sometimes found it difficult to differentiate between a pine and a tuft of grass. This could explain some of the fluctuations in cover estimates.

The raters' level of agreement was assessed by computing the intracorrelation coefficient ICC(2,1) and the Krippendorrf's α. The degrees of freedom used for the calculations in Table 2 are $df1 = (n-1)$ and $df2 = (o-1)(n-1)$, where o is the number of raters, and n is the number of estimated images.

Results for the ICC(2,1) and Krippendorrf's α presented in Table 2 shows values between $ICC(2,1) = 0.13$ and $\alpha_{Kripp} = 0.11$ (when estimating the cover of grass using AFC) up to $ICC(2,1) = 0.41$ and $\alpha_{Kripp} = 0.35$ (when estimating the cover of herbs using AFC). If values around the arithmetic mean (in between 0.4 to 0.6) are characterised as being moderate agreement, then the results obtained for raters using ICC(2,1) could be described as ranging from *poor agreement* to just about *fairly moderate agreement*. Similarly, the α_{Kripp}-values could be described as *poor agreement*.

Counting of Plant Clusters: The results of our investigation into the inter-rater reliability of raters' counting of plant clusters showed instability between individual raters. ANOVA tests showed that there were significant differences between the raters' counting of plant clusters. This applied both to the counting after the ACC session, as well as that carried out after the AFC session (see Sect. 3.7). In addition, ICC(2,1) and Krippensdorff's α values were calculated (see Table 1). Again, raters' levels of agreement ranged from *poor agreement* up to the lower boundary of *moderate* when counting plant clusters.

Thus, if individual raters were to estimate the amount of cover or count plant clusters, the inter-rater reliability results would show *poor* to *moderate* agreement. Hence, caution should be exercised when interpreting individual raters' results.

References

1. Coulloudon, B., Eshelman, K., Gianola, J., Habich, N., et al.: Sampling vegetation attributes. Interagency Technical Reference BLM/RS/ST-96/002+1730, Bureau of Land Management's National Applied Resource Sciences Center, BC-650B. P.O. Box 25047. Denver, Colorado 80225–0047 (1999)
2. Elzinga, C.L., Salzer, D.W., Willoughbyh, J.W.: Measuring and monitoring plant populations. Technical Reference 1730–1 BLM/RS/ST-98/005+1730 (5.1 MB), Bureau of Land Management. Denver, Colorado. USDI, BLM (1998)
3. Yella, S., Nyberg, R.G., Gupta, N., Dougherty, M.: Reliability of manual assessments in determining the types of vegetation on railway trackbeds. In: The 3rd International Workshop on Data Quality and Trust in Big Data (QUAT 2015) with the 16th International Conference on Web Information Systems Engineering (WISE 2015), Miami, Florida (2015)
4. Krippendorff, K.: Recent developments in reliability analysis. In: Annual Meeting of the International Communication Association, Miami, FL, 21–25 May 1992
5. Nyberg, R.G., Gupta, N.K., Yella, S., Dougherty, M.S.: Machine vision for condition monitoring vegetation on railway embankments. In: 6th IET Conference on Railway Condition Monitoring (RCM 2014), pp. 1–7, September 2014
6. Nyberg, R.G., Gupta, N.K., Yella, S., Dougherty, M.S.: Monitoring vegetation on railway embankments: supporting maintenance decisions. In: Proceedings of the International Conference on Ecology and Transportation, July 2013
7. Shoukri, M.M.: Measures of Interobserver Agreement and Reliability. CRC Press, Boca Raton (2003)
8. Martin, P., Bateson, P.: Measuring Behaviour: An Introductory Guide. Cambridge University Press, Cambridge (2007)
9. Rankin, G., Stokes, M.: Reliability of assessment tools in rehabilitation: an illustration of appropriate statistical analyses. Clin. Rehabil. **12**(3), 187–199 (1998)
10. Hallgren, K.A.: Computing inter-rater reliability for observational data: an overview and tutorial. Tutorials Quant. Methods Psychol. **8**(1), 23–34 (2012)
11. Cicchetti, D.V.: Guidelines, criteria, and rules of thumb for evaluating normed and standardized assessment instruments in psychology. Psychol. Assess. **6**(4), 284–290 (1994)

Reliability of Manual Assessments in Determining the Types of Vegetation on Railway Tracks

Siril Yella[1(✉)], Roger G. Nyberg[1,2], Narendra K. Gupta[2], and Mark Dougherty[1]

[1] Department of Computer Engineering,
Dalarna University, 78170 Borlänge, Sweden
{sye,rny,mdo}@du.se
[2] School of Engineering and Built Environment,
Edinburgh Napier University, Edinburgh EH10 5DT, UK
N.Gupta@napier.ac.uk

Abstract. Current day vegetation assessments within railway maintenance are (to a large extent) carried out manually. This study has investigated the reliability of such manual assessments by taking three non-domain experts into account. Thirty-five track images under different conditions were acquired for the purpose. For each image, the raters' were asked to estimate the cover of woody plants, herbs and grass separately (in %) using methods such as aerial canopy cover, aerial foliar cover and sub-plot frequency. Visual estimates of raters' were recorded and analysis-of-variance tests on the mean cover estimates were investigated to see whether if there were disagreements between the raters'. Intra-correlation coefficient was used to study the differences between the estimates. Results achieved in this work revealed that seven out of the nine analysis-of-variance tests conducted in this study have demonstrated significant difference in the mean estimates of cover ($p < 0.05$).

1 Introduction

Vegetation on and alongside railway tracks is a serious problem. The presence of vegetation on the tracks reduces the elasticity of the ballast. Vegetation alongside railway tracks (especially in curves and level crossings) severely challenges visibility, as a result of which, trains have to be slowed down. Proper control and maintenance is therefore necessary to ensure smooth operational routines [1–4]. Current day vegetation assessments within railway maintenance are largely carried out manually by visually inspecting the track on-site, or by looking at video clips collected by maintenance trains, or trailers as they run along the track. The aim of this study is to investigate and assess the reliability of such manual assessments; more specifically to compare the visual estimates reported by the different raters' to be able to evaluate disagreements (if any). Two different studies have been conducted in parallel to investigate the above. First study has mainly investigated issues relevant to quality and reliability of the visual estimates reported by domain experts [5]. Second study has investigated similar issues based on the visual estimates reported by non-domain experts and is the topic of this article.

J. Wang et al. (Eds.): WISE 2015, Part II, LNCS 9419, pp. 391–399, 2015.
DOI: 10.1007/978-3-319-26187-4_37

Work reported in this article is part of a major research project aimed at automating the process of detecting vegetation on railway embankments. A complete description of the research project is out of the scope of this article but could be found elsewhere [6–8]. The rest of the paper is organised as follows. Section 2 presents data acquisition and methodology. A brief introduction to the methods is also provided for the benefit of the readers unfamiliar with the methods. Section 3 presents results of the visual estimates reported by the different raters'. The paper finally presents concluding remarks.

2   Image Acquisition and Methodology

Images of railway tracks were acquired and presented to the raters' for the sake of simplicity. A DSLR Nikon D90 camera mounted on a tripod was used to acquire the relevant images in visible spectrum (400 to 700 nm). The camera was set up on a tripod at a height of 1.6 meters vertically above the ground to capture a nadir view of the track bed. Note that no additional lighting or flash was used and all the images were acquired under normal weathering conditions. All the images were of high (4288 * 2448 pixels) resolution and were saved as RAW files in the RGB colour space. A total of 35 images have been acquired to compensate for the ocular inspection onsite. Note that the limited number of images in the current work is mainly due to the operational constraints in the rail transportation domain. Collecting images of railway tracks, embankments, and other relevant track components demands rerouting or even cancellation of traffic operations and are expensive procedures. Three raters' were picked at random from among the academic staff at Dalarna University, Sweden. The raters' neither had experience in estimating plant cover nor did they have any experience of the railway domain. For each image, the raters' were asked to estimate the cover of woody plants, herbs and grass separately (in %) using three different methods as follows:

1. Aerial canopy cover (ACC)
2. Aerial foliar cover (AFC)
3. Sub-plot frequency (SF)

   In the context of assessing vegetation aerial canopy cover (ACC) is the area of ground covered by the vertical projection of the outermost perimeter of the natural spread of foliage of plants, also known as the convex hull. Small openings within the canopy are included. If more than one species is to be included in the total cover, the canopy cover may exceed 100 % because of overlapping. Aerial foliar cover (AFC) is the area of ground covered by the vertical projection of the aerial portions of the plants. Small openings in the canopy and intra-specific overlap are excluded. In contrast, sub-plot frequency is a measure of the number of sub-plots that contain the target species. A good discussion concerning the methods could be found elsewhere [9]. Note that the raters' were briefed about the aforementioned methods in prior and were asked to visually assess the extent and sub-plot frequency in a maximum of 35 images. At this stage it is worth mentioning that the images used for SF contained a 10×10 sub-plot sampling frame (Fig. 1). As mentioned earlier the three raters' were asked to estimate the presence of woody plants, herbs and grass in images and a total of nine investigations were recorded for further analysis. In this article the mean estimates reported by the raters' were computed and analysis-of-variance (ANOVA) tests were tried and tested to investigate whether if there were differences between the estimates and intra-correlation coefficient (ICC) was employed for the purpose. i.e. it tested the null hypothesis (H_0) that the means of estimates are equal between the raters' in order to assess reliability in terms of the consistency of measurements made by several raters' measuring the same quantity.

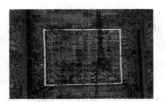

Fig. 1. Sub-plot sampling frame

ICC is deemed particularly useful while assessing the reliability of ratings by comparing the variability of different ratings of the same subject with the total variation across all ratings and all subjects. The ICC coefficient can theoretically vary between 0 and 1.0, where an ICC value of 0 indicates no agreement whereas an ICC value of 1.0 indicates perfect agreement/reliability. A complete discussion of the classed is out of the scope of this article but could be found elsewhere [10]. In this particular article, ICC (2,1) class was chosen (Eq. 1).

$$ICC(2,1) = \frac{var(\beta)}{var(\alpha) + var(\beta) + var(\varepsilon)} \tag{1}$$

Before proceeding any further it is worth mentioning that preliminary visual analysis of the (raters') mean and median histogram plots have indicated an irregular, positively skewed distribution. Therefore the data was $\log_{10}$ transformed for all further parametric analysis. The fact that the density plots of the residuals obtained from the $\log_{10}$ transformed data were approximately normally distributed further justifies the choice of the ANOVA test.

3   Results and Analysis

What follows next is a brief discussion of the visual estimates reported using the different methods.

3.1   Visual Estimates on Woody Plants

Estimates using ACC. An ANOVA test at 95 % confidence level was performed on the raters' visual estimates of woody plants using ACC. A significant difference in

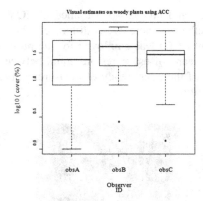

Fig. 2. Log10-transformed visual estimates the raters' on woody plants using ACC

estimates was reported. F = 4.943 at df = 2 and 60, p = 0.0103 (H_0 was rejected at 0.05 significance level, Fig. 2).

Estimates using the AFC. An ANOVA test at 95 % confidence level was performed on the raters' visual estimates of woody plants using AFC. A significant difference in estimates was reported. F = 10.5 at df = 2 and 63, p = 0.0001158. (H_0 was rejected at 0.05 significance level, Fig. 3).

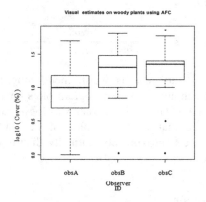

Fig. 3. Log10-transformed visual estimates the raters' on woody plants using AFC

Estimates using SF. An ANOVA test at 95 % confidence level was performed on the raters' visual estimates of woody plants using SF. A significant difference in estimates was reported. F = 9.897 at df = 2 and 66, p = 0.0001741. (H_0 was rejected at 0.05 significance level, Fig. 4).

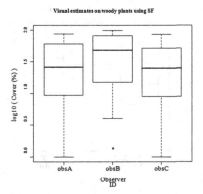

Fig. 4. Log10-transformed visual estimates the raters' on woody plants using SF

3.2   Visual Estimates on Herbs

Estimates using ACC. An ANOVA test at 95 % confidence level was performed on the raters' visual estimates of herbs using ACC. A non-significant difference in estimates was reported. F = 1.391 at df = 2 and 66, p = 0.256. (H_0 was not rejected at 0.05 significance level, Fig. 5).

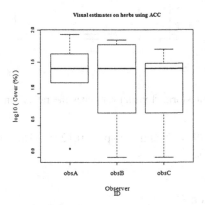

Fig. 5. Log10-transformed visual estimates the raters' on herbs using ACC

Estimates using AFC. An ANOVA test at 95 % confidence level was performed on the raters' visual estimates of herbs using AFC. A significant difference in estimates was reported. F = 3.955 at df = 2 and 60, p = 0.02435. (H_0 was rejected at 0.05 significance level, Fig. 6).

Estimates using SF. An ANOVA test at 95 % confidence level was performed on the raters' visual estimates of herbs using SF. A non-significant difference in estimates was

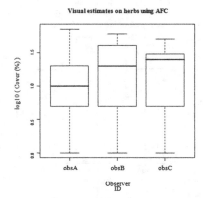

Fig. 6. Log10-transformed visual estimates the raters' on herbs using AFC

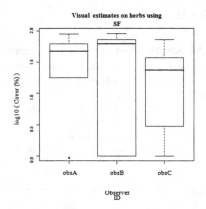

Fig. 7. Log10-transformed visual estimates the raters' on herbs using SF

reported. F = 2.129 at df = 2 and 60, p = 0.1278. (H_0 was not rejected at 0.05 significance level, Fig. 7).

3.3   Visual Estimates on Grass

Estimates using ACC. An ANOVA test at 95 % confidence level was performed on the raters' visual estimates of grass using ACC. A significant difference in estimates was reported. F = 55.62 at df = 2 and 54, p = 7.676 * 10^{-14}. (H_0 was rejected at 0.05 significance level, Fig. 8).

Estimates using AFC. An ANOVA test at 95 % confidence level was performed on the raters' visual estimates of grass using AFC. A significant difference in estimates was reported. F = 48.94 at df = 2 and 54, p = 7.472 * 10 − 13. (H_0 was rejected at 0.05 significance level, Fig. 9).

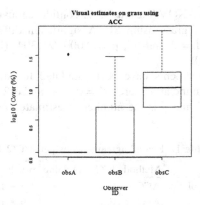

Fig. 8. Log10-transformed visual estimates the raters' on grass using ACC

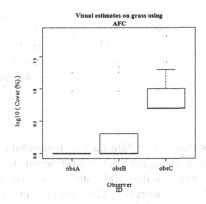

Fig. 9. Log10-transformed visual estimates the raters' on grass using AFC

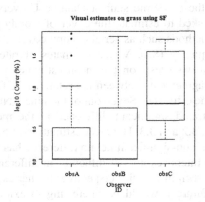

Fig. 10. Log10-transformed visual estimates the raters' on grass using SF

Estimates using SF. An ANOVA test at 95 % confidence level was performed on the raters' visual estimates of grass using SF. A significant difference in estimates was reported. F = 21.21 at df = 2 and 57, p = 0.0000001304. (H_0 was rejected at 0.05 significance level, Fig. 10).

Finally the raters' agreements have been tabulated for simplicity (Table 1). The degrees of freedom used in the article are df1 = (n − 1) and df2 = (o − 1) (n − 1), where o is the number of raters' and n is the number of estimated images.

Table 1. Raters' agreements as per ICC(2,1)

Type	Method	ICC(2,1) value	p-value
Woody plants	ACC	0.68	$1.5 * 10^{-8}$
	AFC	0.62	$3.9 * 10^{-8}$
	SF	0.76	$1.1 * 10^{-12}$
Herbs	ACC	0.58	$2.4 * 10^{-6}$
	AFC	0.73	$7.4 * 10^{-10}$
	SF	0.78	$4.1 * 10^{-11}$
Grass	ACC	0.15	0.011
	AFC	0.19	0.003
	SF	0.43	$2.4 * 10^{-5}$

4   Conclusions

Current day vegetation assessments within railway maintenance are (to a large extent) carried out manually; either through visual inspection onsite or by looking at video clips collected by maintenance trains. This study has investigated the quality and reliability of such manual assessments by taking non-domain experts into account to be able to compare inter-rater assessments. Thirty-five track images under different conditions were acquired for the purpose. Three (non-domain experts) raters' were picked at random from among the academic staff at Dalarna University, Sweden. For each image, the raters' were asked to estimate the cover of woody plants, herbs and grass separately (in %) using methods such as aerial canopy cover (ACC), aerial foliar cover (AFC) and sub-plot frequency (SF). Visual estimates of raters' were recorded and analysis-of-variance (ANOVA) tests on the mean estimates were investigated to see whether if there were disagreements between the raters'. ICC (2, 1) was used to study the differences between the estimates. Seven out of the nine ANOVA tests conducted in this study have demonstrated significant difference in the mean estimates of cover (p < 0.05). See Sects. 3.1, 3.2 and 3.3. However estimation of herbs using ACC and SF methods were found to be non-significant i.e. no difference has been observed between the raters'. The fact that raters' expressed difficulty in differentiating between a pine and a tuft of grass (when seen in a nadir perspective) further explains the fluctuation in cover estimates. In the future it would be interesting to extend the work further by carrying out cross investigations between domain experts and people who are instructed (as in the current article) to carry out the assessment intuitively.

References

1. Hulin, B., Schussler, S.: Measuring vegetation along railway tracks. In: Proceedings of the IEEE Intelligent Transportation Systems Conference, pp. 561–565 (2005)
2. Banverket: Vegetation maintenance manual, Bvh 827.1, Original title in Swedish: Handbok om vegetation' (2000)
3. Banverket: Vegetation maintenance requirements, Bvh 827.2, Original title in Swedish: behovsanalys infor vegetationsreglering' (2001)
4. Banverket: Safety inspections manual, Bvf 807.2, Original title in Swedish: sakerhetsbesiktning av fasta anlaggningar (2005)
5. Nyberg, R.G., Yella, S., Gupta, N., Dougherty, M.: Inter-rater reliability in determining types of vegetation on railway track beds. In: Accepted for Publication in the 3rd International Workshop on Data Quality and Trust in Big Data in Conjunction with the 16th International Conference on Web Information Systems Engineering (WISE), Miami, USA (2015)
6. Yella, S., Nyberg, R.G., Payvar, B., Dougherty, M., Gupta, N.: Machine vision approach for automating vegetation detection on railway tracks. J. Intell. Syst. 22(2) (2013). ISSN 2191-026X
7. Nyberg, R.G., Gupta, N., Yella, S. Dougherty, M.: Detecting plants on railway embankments. J. Softw. Eng. Appl. 6(3B), pp. 8–12 (2013) ISSN online 1945-3124
8. Nyberg, R.G., Gupta, N., Yella, S. Dougherty, M.: Monitoring vegetation on railway embankments: supporting maintenance decisions. In: Proceedings of the 2013 International Conference on Ecology and Transportation, Scottsdale, Arizona, USA (2013)
9. Coulloudon, B., Eshelman, K., Gianola, J., Nea, H.: Sampling vegetation attributes, Interagency Technical Reference BLM/RS/ST- 96/002 + 1730, Bureau of Land Management's National Applied Resource Sciences Center, Bureau of Land management. National Business Center. BC-650B. P.O. Box 25047, Denver, Colorado 80225-0047 (1999)
10. Shrout, P., Fleiss, J.: Intraclass correlations: Uses in assessing rater reliability. Psychol. Bull. 86, 420–428 (1979)

A Comparison of Patent Classifications with Clustering Analysis

Mick Smith[1](✉) and Rajeev Agrawal[2]

[1] School of Technology, North Carolina A&T State University,
1601 E. Market Street, Greensboro, NC 27411, USA
csmithl4@aggies.ncat.edu
[2] Department of Computer Systems Technology, North Carolina A&T State
University, 209 Price Hall, Greensboro, NC 27411, USA
ragrawal@ncat.edu

Abstract. There is an abundance of data and knowledge within any given patent. Through the use of textual mining and machine learning clustering techniques it is possible to discover meaningful associations throughout a corpus of patents. This research demonstrates that such relationships between USPTO patents exist. Through the use of k-means and k-medians clustering, the accuracy of the USPTO classes will be assessed. It will also be demonstrated that a more refined classification process would be beneficial to other areas of analysis and forecasting.

Keywords: Clustering · K-means · K-medians · Patent classification · Machine learning · Text mining

1 Introduction

As the age of big data continues to evolve so does the potential for analytic opportunities within large data sets. One area in particular that may lend itself to ongoing analysis comes in the form of patent analysis. In addition to the quantitative analysis that can be done on the frequency or patterns of patents, there exists massive amounts of knowledge that can be extracted from the textual bodies of each document. It's possible that this knowledge could be used to improve on techniques used for searching and retrieving patents of various classification or content. It is important to improve on this process since the quality of the improper searching may result in unexpected overlapping into another person's intellectual property (IP).

The goal of this research is to determine which clustering method most accurately represents the classification of patents as proposed by the United States Patent and Trademark Office (USPTO). The clustering techniques examined in this paper will be k-means and k-medoids. While this is by no means an exhaustive list of clustering methods, it does offer a good starting point of investigation. Both of these methods are efficient, proven approaches for the clustering of textual documents. The establishment of a good clustering technique will provide a baseline for future research on patent clusters. More specifically, it will provide a proven methodology to be applied in a future quantitative and qualitative data mining analysis.

J. Wang et al. (Eds.): WISE 2015, Part II, LNCS 9419, pp. 400–413, 2015.
DOI: 10.1007/978-3-319-26187-4_38

A lot of benefits could be realized from the development of an efficient mechanism for clustering and classifying patents. For instance, due to the sheer volume of patent data there is an increased chance of copyright infringement. Individuals who desire to submit a patent for a new idea may not be aware that the same concept already exists. Such a violation could be avoided if the patents were correctly classified. Furthermore, if the author of the patent had a mechanism to compare similarity of content to either a single patent or group of patents, then this may result in a reduction of IP infringement. Another benefit comes in the form of business value. New marketing and innovation strategies can be discovered proper classification and analysis of patents [10].

The USPTO attempts to make their current classification as specific as possible. At this time it includes 473 classes and offers even more granularity as there are over 150,000 sub-classifications. Although, such specificity may not be beneficial to the analysis of large number of patents. For instance, if someone wanted to conduct research on all "Green Energy", they could perform a key word search on the USPTO website and return a list of 643 related patents. However, this list may not be inclusive of all patents related to green energy. Another approach may be to drill down into the already defined classes that the USPTO has set forth. Though this method might only yield patents related to a specific subset of green energy. Especially since each patent may have multiple sub-classifications.

To properly encapsulate the scope of a patent search, it is necessary to consider textual and content relationships from one patent document to the next. By extension, it is suggested in this research that more meaningful relationships exist between patents than is indicated by the classification and subsequent sub-classification. It should be mentioned that this study is not fully conclusive and that this research is ongoing.

2 Background

One area in particular that exists as a byproduct of big data is the ability to autonomously retrieve textual information and associate various bodies of text. This need spans anything from a web search conducted through an internet web search engine to in-depth textual analysis and natural language processing. Recent research has used semantic and word analysis to identify hidden relationships in social media comments [12]. Additionally textual mining has been used to analyze and compare patents [2, 8, 9, 18].

2.1 Patent Clustering/Classification Techniques

Currently the USPTO suggests using a seven step searching strategy (www.uspto.gov). Their process recommends that the person performing the search, brainstorm some possible relevant terms, and using those terms recursively over multiple searching trials. However, while this method may return a wide breadth of patents related to the search, it is far from an exhaustive list. Furthermore, it is left up to the subjective aptitude of the searching party.

A language based clustering approach was utilized by Kang et al. [9] to associate patent documents. In their research, log-likelihood term frequency and data smoothing techniques are used to establish a good general information retrieval method. Reference [11] used the k-nearest neighbor algorithm, the maximum entropy model, and support vector machines, to classify Japanese patents. Although, their grouping strategy relied heavily on the existing classifications assigned to the patents.

As proposed in this paper, similar research has been done in the field of patent classification. A back propagation neural network was used by Trappery et al. [16] to group a specific class of patents down to the sub-class level. Chen and Chang [2] also proposed a three phase categorization method by which each patent is classified down to the same level of detail. While their method offers an incredible amount of granularity, it differs from this research in that their goal is to match patents to already existing classifications. As previously mentioned, this paper aims to demonstrate that there might be "hidden" clusters defined by textual clustering methods that offer better classification than the current USPTO system.

2.2   Clustering

The data mining technique of clustering is one that has many benefits and can be utilized in a variety of fields. According to Han and Kamber [6] examples of clustering can be seen in marketing, land use, insurance, city-planning, and earth-quake studies. Clustering is also widely used for text mining, pattern recognition, webpage analysis, and marketing analysis [17]. By grouping and subdividing a collection of documents or keyword it may be possible to discover certain trends that exist in data. This grouping is one of the benefits of using clustering techniques. As it pertains to patent mining and technology forecasting there have been many instances of use in other research [4, 7, 8, 17]. Ruffaldi et al. [15] suggest that by using patent citations, that it is possible to understand a given technology's trajectory. The quality of a clustering result depends on both the similarity measure used by the method and its implementation. The quality of a clustering method is also measured by its ability to discover some or all of the hidden patterns.

2.3   K-Means

K-Means Clustering is a clustering technique that clusters data based on the mean vector distance between data points. It is a recursive algorithm that looks to find the minimal mean distances between pieces of data and then group them together. K-Means clustering often terminates at a local optimum. The global optimum may be found using techniques such as: deterministic annealing and genetic algorithms [6]. According to Chernoff et al. [3] there are various versions of k-means algorithms, depending on the method in which covariance matrices are estimated and the procedure of reclassifying the means. K-means has also been proven to be an effective method for classifying patent documents [2].

There are of course some limitations and weaknesses associated with the k-means clustering method. Han and Kamber [6] offer a list of such weaknesses:

- Applicable only when mean is defined, then what about categorical data?
- Need to specify k, the number of clusters, in advance
- Unable to handle noisy data and outliers
- Not suitable to discover clusters with non-convex shapes

2.4 K-Medoids

The K-Medoids clustering method is similar to the K-Means method in that both attempt to minimize the vector distance between data points. The difference between the two is that instead of finding the nearest mean the K-Medoids looks for the center points as defined by the data and works to improve that data center through the use of an arbitrary distance matrix. A common algorithm used for K-Medoids is PAM (Partitioning Around Medoids). K-Medoids clustering has a computational advantage over K-means clustering. K-Medoids clustering finds the representative objects (medoids) in clusters [8].

2.5 Other Clustering Methods

Sometimes when trying to describe a system it is necessary to use a non-discrete metric. This is even truer when attempting to classify two or more potentially similar groups. Through the use of fuzzy logic it is possible to use varying degrees of membership to measure levels of similarity. The principles of fuzzy logic can be extended to the concept of fuzzy clustering. Goswami and Shishodia [5] have shown that it is possible to use the fuzzy c-means (FCM) clustering algorithm to associate different collections of texts. In their approach they use a bag of words approach to determine the frequency of words in a document. Some important steps in their process include pre-processing the text, feature generation, and feature selection. The large dimensionality of textual collections is also challenging for classification algorithms and can drastically increase running time. Researchers apply dimensionality reduction before running the classification algorithm to alleviate these challenges [1].

Li et al. [13] proposed the use of two different document clustering algorithms, Clustering based on Frequent Word Sequences (CFWS) and Clustering based on Frequent Word Meaning Sequences (CFWMS). Their algorithms focused on the sequence of words as opposed to the bag of words approach. The reason behind this was to cluster based on implied meanings within the document instead of word frequency. In their algorithms, each document is reduced to a compact document by keeping only the frequent words. In the compact document, they kept the sequential occurrence of words untouched to explore the frequent word sequences. By building a Generalized Suffix Tree (GST) for all the compact documents, the frequent word sequences and the documents sharing them are found. Then, frequent word sequences are used to create clusters and describe their content for the user [13]. Ideally a comprehensive metric of text semantic similarity should be based upon the relation between the words in addition to the role played by the various entities involved in the interactions described by each of the two texts. Following this, the semantic similarity of textual components is based upon the similarity of the component words in them [12].

3　Experimentation

This research will utilize various data clustering methodologies to better assess the quality of patent classifications by the USPTO. To achieve the results for this research several applications were used. Figure 1 details the process used to create the desired clusters. All of the patents used in this research were obtained from the UC Berkley Fung Institute (https://github.com/funginstitute/downloads). The UC Berkley patent data was extracted from the USPTO website and converted from XML to a SQLite table structure. The database covers all US patents from 1976 to 2013 and consists of 4,823,407 patents. While it would be ideal to perform clustering on all patents, the sheer size and volume patents makes that unrealistic for a short term analysis. To address this issue a smaller sample of patents was extracted. Queries were run on the SQLite database to extract patents that meet the following characteristics:

- Granted between 1995 and 2013
- Contains the word "Technology" in either the abstract or title
- For each patent, the following information was extracted:
  - Patent title, Abstract, USPTO Class, USPTO Patent Number

These initial queries reduced the number of patents to 9,087. However, this was too large a sample to conduct preliminary analysis on, so further filtration was carried out. A random sample of 449 patents was taken from the reduced set of 9,087 and it consisted of 10 USPTO classes. Table 1 contains the list of classes, descriptions, quantities of patents that are used in this experimentation. The data was saved in .csv format and the titles and abstracts were concatenated to create one "document" per row.

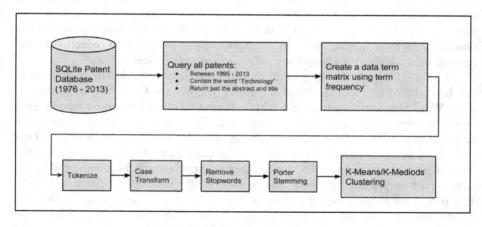

Fig. 1. Workflow for patent clustering

The .csv file was then loaded into RapidMiner so that a term frequency data term matrix could be created. RapidMiner was used to tokenize the text in each row, make all words lower case, remove stop words, and perform Porter stemming. The resulting

Table 1. Technology patent classifications, descriptions, and quantities

USPTO class	Classification description	Quantity
257	Active solid-state devices (e.g., transistors, solid-state diodes)	44
340	Communications: electrical	50
359	Optical: systems and elements	51
365	Static information storage and retrieval	55
370	Multiplex communications	48
435	Chemistry: molecular biology and microbiology	8
439	Electrical connectors	54
514	Drug, bio-affecting and body treating compositions	46
705	Data processing: financial, business practice, management, or cost/price determination	50
707	Data processing: database and file management or data structures	43

data term matrix is then exported to another .csv file. This data term matrix that was consisted of 449 rows and 1,417 columns. R was then used to perform various clustering analyses on the data set.

Two types of clustering was performed in R, k-means and k-medoids. The first objective of our research was to determine if each algorithm could accurately approximate the classifications assigned by the USPTO. For this reason the number of centers (k) was set to 10. Table 2 shows the clustering results for k-means while Table 3 shows the results for k-medoids.

Table 2. K-means cluster results

USPTO Class	Clusters										
	1	2	3	4	5	6	7	8	9	10	
257		39	1	1	2			1			44
340		2	1		33			1	4	9	50
359		7	7		11		19	4	3		51
365		1		39	5			1	1	8	55
370					30			2	6	10	48
435									8		8
439		1	46		3			3	1		54
514						31			15		46
705	11				8				27	4	50
707				1	8			1	19	14	43
Total	11	50	55	41	100	31	19	13	84	45	449

Table 3. K-medoids cluster results

USPTO Class	\multicolumn{10}{c}{Clusters}										
	1	2	3	4	5	6	7	8	9	10	
257	34	2	3	2	2	1					44
340	3	1	17	1		1	12	15			50
359	7	32	5	5	1				1		51
365			3	1	40			10		1	55
370		1	19	2	4		5	17			48
435				1	2				5		8
439	2	1	2	3	1	42	2		1		54
514			1		1		1		43		46
705			7		2		2	22	4	13	50
707			7	1	2		2	29	1	1	43
Total	46	37	64	16	55	44	24	93	55	15	449

A second objective of this research was to determine if the clustering carried out by the USPTO was accurate. In other words, for the group of patents in this study, what is the optimal number of groupings? To address this a few different approaches were used, first a sum of squared error scree plot was created for the k-means clustering. This graph can be seen in Fig. 2, where the objective is to observe an abrupt bend in the curve which would indicate a suggested number of clustering centers. From observing the plot in Fig. 2, there doesn't appear to be a clear point where the curve bends.

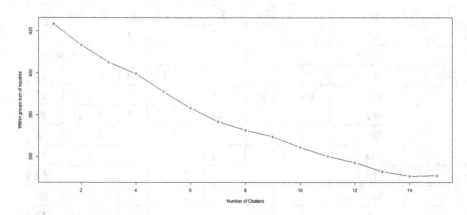

Fig. 2. K-means SSE Scree plot

Table 4. Gap statistics

	k-means gap	k-means SE	k-medoids gap	k-medoids SE
Cluster 1	0.69245	0.00191	0.69159	0.00193
Cluster 2	0.69303	0.00171	0.69630	0.00188
Cluster 3	0.69861	0.00164	0.69984	0.00170
Cluster 4	0.70298	0.00165	0.70463	0.00197
Cluster 5	0.70837	0.00160	0.71212	0.00179
Cluster 6	0.71341	0.00166	0.71853	0.00179
Cluster 7	0.71763	0.00177	0.72389	0.00156
Cluster 8	0.72012	0.00157	0.72818	0.00161
Cluster 9	0.72537	0.00164	0.73156	0.00165
Cluster 10	0.72801	0.00151	0.73414	0.00147
Cluster 11	0.73048	0.00166	0.73712	0.00156
Cluster 12	0.73310	0.00150	0.74055	0.00182
Cluster 13	0.73534	0.00161	0.74380	0.00178
Cluster 14	0.73809	0.00165	0.74538	0.00195
Cluster 15	0.73913	0.00175	0.74710	0.00205

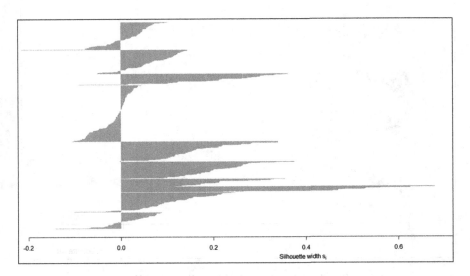

Fig. 3. Silhouette plot of k-means (k = 10)

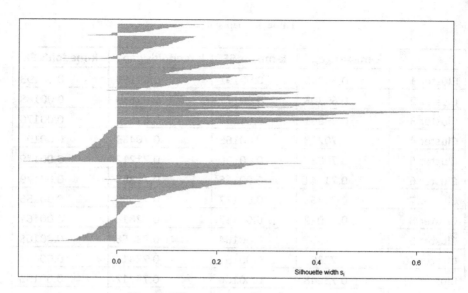

Fig. 4. Silhouette plot of k-means (k = 14)

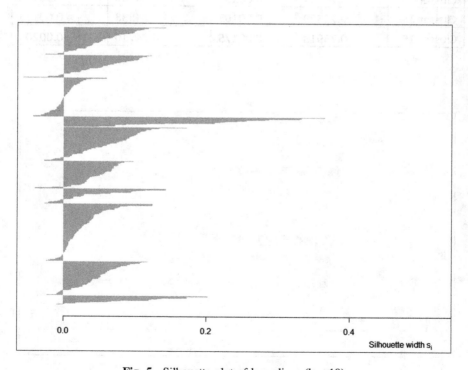

Fig. 5. Silhouette plot of k-medians (k = 10)

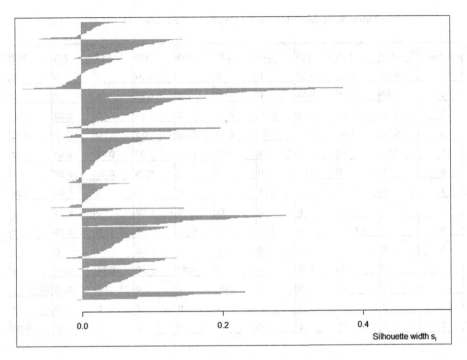

Fig. 6. Silhouette plot of k-medians (k = 14)

Another approach to finding the right number of clusters is to use a gap statistic. Table 4 shows the gap statistic and standard error for both the k-means and k-medoids results. As the highlighted row indicates, the gap statistic suggests 14 clusters for both k-means and k-medoids. It should be noted that the gap statistic identifies the smallest value of k such that $f(k)$ is not more than one standard error away from the first local minimum [14].

Unfortunately, the silhouette plots (Figs. 3, 4, 5 and 6) for each clustering approach don't reveal much of an association between the generated clusters and the data points assigned to them. In the graphs, the clusters are listed in order from top to bottom with each cluster element's silhouette score displayed. The scores can range between -1 and 1, scores closer to 1 are desirable as they demonstrate a stronger association. Low or negative scores may also indicate either too many or too few clusters. However, as it has been stated, this research is in the preliminary stages, and further tweaking and adjustments of the parameters of each algorithm should produce stronger results.

What may be of more interest is to notice that when the number of clusters increased to 14, as suggested by the gap statistic, the strength of the silhouette scores and plots improved. This may be initial support to one of the running arguments in this paper. The classification of patents might be done in a more efficient manner if the context of each patent is assessed analytically instead of subjectively.

To further support these findings the data was also transformed using Principal Component Analysis (PCA). One of the known problems with the clustering of vectors

Table 5. K-Medoids cluster results by percentage (k = 14)

Cluster	257	340	359	365	370	435	439	514	705	707
1	0.14	0.10	0.67	0	0	0	0.07	0	0	0
2	0.80	0.03	0.12	0	0	0	0.03	0	0	0
3	0.22	0.26	0.08	0.08	0.12	0	0.14	0	0.02	0.06
4	0.13	0.06	0.26	0.06	0.13	0.06	0.2	0	0	0.06
5	0.04	0.02	0	0.77	0.04	0.04	0.02	0.02	0.02	0.02
6	0	0.73	0.06	0	0	0	0.06	0.06	0.06	0
7	0	0.10	0.02	0.13	0.24	0	0	0	0.20	0.28
8	0	0.27	0	0	0.35	0	0.02	0	0.22	0.12
9	0	0.08	0	0.23	0.38	0.07	0	0	0.23	0
10	0	0	0.94	0	0	0	0.06	0	0	0
11	0	0	0.02	0	0	0.07	0	0.86	0.03	0
12	0	0	0	0	0.06	0	0	0	0.33	0.61
13	0	0	0	0	0	0	1	0	0	0
14	0	0	0	0	0	0	0	0	0.92	0.08

from a data term matrix is that there may be a lot of sparseness within the matrix. This may reduce the quality of the clustering results. To overcome this problem PCA can be applied and reduce possibility of problems due to dimensionality issues. When applied to this data the resulting gap statistic suggested 14 clusters as well.

```
While column j in classification matrix M has no associ-
ated row
  For each row i in classification matrix M
    x = max(m_i1, m_i2,... m_ij,)
    class assignment is given to the column j associated
with x
    if column j is already assigned to a row then
      Next
    else
      assign class value for column j to row i
    end if
  Next
End While
```

Fig. 7. Pseudo code for assigning USPTO classes to clusters.

Table 6. Patent cluster classifications based on USPTO classes

Cluster	USPTO class	Initial quantity	Cluster quantity	USPTO classification description	Terms
1	n/a	n/a	28	n/a	Optical, light, surface, includes, system, member, plane
2	257	44	31	Active solid-state devices (e.g., transistors, solid-state diodes)	Layer, substrate, region, formed, device, semiconductor, dielectric
3	n/a	n/a	49	n/a	Device, semiconductor, devices, system, signal, control, network
4	435	8	15	Chemistry: molecular biology and microbiology	Sub, signal, surface, conductive, lens, film, light
5	365	55	48	Static information storage and retrieval	Memory, voltage, cell, line, device, bit, cells
6	340	50	15	Communications: electrical	Battery, circuit, rfid, communication, power, message, electrical
7	n/a	n/a	74	n/a	Data, system, method, memory, includes, plurality, device
8	n/a	n/a	40	n/a	Information, network, user, method, system, based, data
9	370	48	13	Multiplex communications	Signal, sequence, reference, using, method, value, assets
10	359	51	17	Optical: systems and elements	Lens, group, side, power, positive, sub, zoom
11	514	46	51	Drug, bio-affecting and body treating compositions	Methods, invention, thereof, treatment, use, pharmaceutical, relates
12	707	43	18	Data processing: database and file management or data structures	Search, content, query, results, method, web, user

(Continued)

Table 6. (*Continued*)

Cluster	USPTO class	Initial quantity	Cluster quantity	USPTO classification description	Terms
13	439	54	37	Electrical connectors	Connector, includes, portion, housing, terminal, body, contact
14	705	50	13	Data processing: financial, business practice, management, or cost/price determination	Order, system, orders, trading, method, deposit, received

One of the challenges of this experimentation is that the cluster classification was not carried through to the final results. While this makes it impossible to construct a true confusion matrix, it allows for interpretation beyond the initial USPTO classification. Still a method for associating the generated clusters to the initial patent groups is good for visual and contextual reference. Especially in the case of 14 clusters. Observing what patents were deemed to be outside the scope of the suggested USPTO class is an important step toward the creation of an alternative classification scheme.

The assignment of USPTO classes back to the created clusters starts with the transformation of the cluster results table to indicate the quantity of patents in each cluster as a percentage. This is shown in Table 5. To assign a cluster to a Classification, the steps outlined with the pseudo code in Fig. 7 were followed. The highlighted cells in Table 5 indicate the assigned class.

After associating a USPTO class to a cluster, the seven most frequent terms from each cluster were extracted. As it can be seen in Table 6, the frequent terms do seem to coincide with the USPTO classification descriptions. The terms listed are in order of occurrence rank within the cluster. Also, clusters 1, 3, 7, and 8 were not assigned a class and may warrant a new classification.

4 Conclusion

In this paper a subset of technology patents were clustered using k-means and k-medians clustering. The goal was to determine how accurate these methods could be and if the classifications suggested by the USPTO were sufficient for further patent analysis. The quality of the classification scheme is extremely important to those looking to delve deeper into trends and themes that are available in a patent corpus. It is almost a certainty that the quality of more advanced textual mining and natural language processing of patents would be improved with a less subjective grouping.

This research demonstrated that it was possible to create clusters of patents based on the frequency of terms within each patent. The initial assessments certainly suggest that it might be possible to improve on the accuracy of the clustering methods by adjusting and testing different parameters. However, since many variations have already been checked, the enhancements may be minimal. In line with the second goal of this research, the results showed that it is possible to generate new clusters and classifications from existing ones. It may be possible to categorize each new cluster better with Topic Modeling techniques such as Latent Dirichlet allocation. This will be the focus of future related research.

References

1. Blake, C.: Text mining. Ann. Rev. Inf. Sci. Technol. **45**(1), 121–155 (2011)
2. Chen, Y.-L., Chang, Y.-C.: A three-phase method for patent classification. Inf. Process. Manage. **48**, 1017–1030 (2012)
3. Chernoff, H., Gillick, L.S., Hartigan, J.A.: k-Means algorithms. Encycl. Stat. Sci. **6**, 3858–3859 (2006)
4. Chou, L.-Y.: Knowledge discovery through bibiometrics and data mining: an example on marketing ethics. Int. J. Organ. Innov. **3**, 106–139 (2011)
5. Goswami, S., Shishodia, M.S.: A fuzzy based approach to text mining and document clustering. Int. J. Data Min. Knowl. Manage. Proc. **3**(3), 43–52 (2013)
6. Han, J., Kamber, M., Pei, J.: Data Mining: Concepts and Techniques. Elsevier, Waltham (2012)
7. Hsu, C.C., Huang, Y.-P., Chang, K.-W.: Extended Naïve Bayes classifier for mixed data. Expert Syst. Appl. **35**, 1080–1083 (2008)
8. Jun, S., Park, S.S., Jang, D.S.: Technology forecasting using matrix mapping and patent clustering. Ind. Manage. Data Syst. **112**, 786–807 (2011)
9. Kang, I.-S., Na, S.-H., Kim, J., Lee, J.-H.: Cluster based patent retrieval. Inf. Process. Manage. **43**, 1173–1182 (2007)
10. Kasravi, K., Risov, M.: Patent mining - discovery of business value from patent repositories. In: Proceedings of the Fortieth Annual Hawaii International Conference on System Sciences, Waikoloa, Hawaii, USA (2007)
11. Kim, J.-H., Choi, K.-S.: Patent document categorization based on semantic structural information. Inf. Process. Manage. **43**, 1200–1215 (2007)
12. Karmakar, S., Zhu, Y.: Mining collaboration through textual semantic interpretation. In: 2011 11th International Conference on Hybrid Intelligent Systems (HIS), pp. 728–733 (2011)
13. Li, Y., Chung, S.M., Holt, J.D.: Text document clustering based on frequent word meaning sequences. Data Knowl. Eng. **64**, 381–404 (2008)
14. Maechler, M.: "Finding Groups in Data": Cluster Analysis Extended Rousseeuw et al, Package "Cluster" (R Documentation). https://cran.r-project.org/web/packages/cluster/cluster.pdf. Accessed 21 July 2015
15. Ruffaldi, E., Sani, E., Bergamasco, M.: Visualizing perspectives and trends in robotics based on patent mining. In: 2010 IEEE International Conference on Robotics and Automation, Anchorage, Alaska, USA 3–8 May 2010
16. Trappery, A.J.C., Hsu, F.-C., Trappery, C.V., Lin, C.-I.: Development of a patent document classification and search platform using a back-propagation network. Expert Syst. Appl. **31**, 755–765 (2006)
17. Trappey, C.V., Wu, H.-Y., Taghaboni-Dutta, F., Trappey, A.J.C.: Using patent data for technology forecasting: China RFID patent analysis. Adv. Eng. Inform. **25**, 53–64 (2011)
18. Tseng, Y.H., Lin, C.J., Lin, Y.I.: Text mining techniques for patent analysis. Inf. Process. Manage. **43**, 1216–1247 (2007)

Combination of Evaluation Methods for Assessing the Quality of Service for Express Delivery Industry

Qing Lou[1], Shao-zhong Zhang[1,2], and William Wei Song[2(✉)]

[1] Institute of Electronics and Information,
Zhejiang Wanli University, Ningbo, Zhejiang, China
[2] Information Systems and Business Intelligence,
Dalarna University, SE-791 88 Borlänge, Sweden
{sza,wso}@du.se

Abstract. In the view of the current service development of the express delivery industry and the data quality problem experienced thereof, we consider to construct an index-based evaluation system for the service quality for the express delivery industry through the analysis of market investigation and data analysis. This system applies analytic hierarchy process (AHP), to survey expert's options and obtain the index weights. The analytical evaluation of service quality for the specific express delivery company is conducted with the fuzzy comprehensive evaluation method. A service satisfaction degree for the express delivery company is generated to improve the overall service performance. Through evaluating results of the solution to the problems in the quality of service, this paper aims at establishing a guideline to improve their service quality for express delivery enterprises. This research aims at the development of a novel method for service quality evaluation in the area of the fast growing businesses of express delivery enterprises.

Keywords: Express delivery enterprise · Evaluation of service quality · Index system · Analytic hierarchy process

1 Introduction

With the rapid development of web based e-commerce technologies, including e-suppliers, logistics and e-distribution, many fast and convenient delivery businesses such as "N deliveries per day" have emerged and progressed. Management of courier companies and their service capabilities have received vast attentions from publics and academics. Many express companies are gradually turning from competition of delivery prices to enhancement of services.

Many researchers and developers have proposed and developed usable and effective theories and methods to evaluate the performance and qualities of logistics enterprises. Saunders and Jones [7] put forward an information system based performance evaluation model, which mainly considers an evaluation of logistics enterprise at the organization level, the management level, and the system function performance level. Several performance evaluation index system and methods were proposed for

© Springer International Publishing Switzerland 2015
J. Wang et al. (Eds.): WISE 2015, Part II, LNCS 9419, pp. 414–425, 2015.
DOI: 10.1007/978-3-319-26187-4_39

information platform of logistics and applying the entropy method and analytic hierarchical process (AHP) method to evaluate the index system [3, 4, 12]. It gave the construction of China express delivery service quality system and fuzzy comprehensive evaluation [8]. Fuzzy evaluation based on SERVQUAL model used for online shopping delivery service quality [9]. The researchers further construct a combination approach of Copeland evaluation method to provide a useful reference for the development of information platform. And in the before, A comprehensive evaluation model was constructed. It based on a gray correlation analysis and entropy weight method, which reduces the evaluation defects from personal factors and hence increases the accuracy of evaluation methods [13]. Yan et al. propose a two-stage method composed of the methods of analysis of network programming (ANP) and goal-oriented programming (GP) [10], which provides support of comparative study of express delivery companies and thus makes an accurate selection of the candidate express companies based on their assignment proportions of businesses.

The SERVQUAL scale was derived from the works examine the meaning of service quality and define service quality and illuminated the dimensions along which consumers perceive and evaluate service quality [5]. In terms of the service quality evaluation, an enhanced SERVQUAL (service quality) evaluation model was proposed in [11], which can effectively evaluate the quality of delivery services. The SERVQUAL model has been widely recognized by many marketing experts in the world. It is considered the most typical method for evaluating the quality of services [1, 2, 6, 15]. Based on the SERVQUAL model, it proposed a method to extend the "security" dimension with one more reference, and based on the analysis of the exploratory factor analysis and confirmatory factor to make it more suitable for the assessment of the quality of Chinese express delivery services [14]. However, due to the secondary indexing factors removed from the process of building the service quality index system, it causes the result of information loss.

From the afore-mentioned, we have observed that very few good research methods have been developed for performance/satisfaction evaluation analysis for express delivery market for the fast growing demand in China. It is necessary for the courier companies to improve their service quality and customer satisfaction before they are driven out of the delivery service market. In this paper, we propose an innovative evaluation model for both quantitative and qualitative analysis of performance evaluation of delivery enterprises. The weight coefficients are selected based on the closeness to reality. The model can not only evaluate the service quality of express delivery enterprises, but also improve the service quality as well as the customer satisfaction.

The rest of this paper is organized as follows. In Sect. 2, we discuss the current status of the service quality problems in express delivery businesses, focusing on challenges caused by the fast growing and problems occurring delivery service quality. In Sect. 3, we propose our method for service quality evaluation, by analyzing the process of evaluation, extending the existing analytic hierarchy process (AHP), and incorporating fuzzy comprehensive synthesis. Section 4 aims to test our method through an establishment of performance evaluation index model and its implemented system, with which we calculate the weights for the entire evaluation model and perform the fuzzy evaluation. In the final section, Sect. 5, we conclude the paper by summarizing our improvement of the AHP method and pointing out how we will further explore the evaluation of express delivery services in a real business environment.

2 Express Service Quality Status

2.1 Fast Development of Express Business

With the growing prosperity of e-commerce businesses, the emerging express logistics industry starts booming and its various functions gradually emerge in many sectors of all kinds, including people's daily life, governments' activities, and public activities.

According to the State Post Bureau, as of 2013 the total amount of postal services reached RMB 267.9 billion yuan, accounting for nearly 9.2 billion couriers. The courier has been approaching towards superpower, with an increase of 60 % over last year, the daily processing maximum reaching more than 65 million items, and the express income 142.9 billion yuan, with an increase of nearly 36% over last year [16].

In the first quarter of 2014, the business volume of express delivery service enterprises totaled 2.6 billion items, an increase of 51.9 % over the same period of last year, and the business revenue totaled 41.35 billion yuan, an increase of 45.6 % of the same period of last year [17]. The express processing volume and business revenue have been rising at a high rate.

2.2 Express Service Quality Status

While the fast-growing of the express delivery businesses, problem comes as well, for example, growing number of customer complaints. According to the general statistics of fast delivery services, the overall rate of complaints has increased by nearly 80 % compared to last year, and among others, the customers mainly complain the commodity damage and the delivery delays. The key problems include the shipment delay, commodity damage, poor delivery services, express item loss, illegal charges, and sign-first-inspection-later. Of these problems, delivery delays account for 52.1 %, unsatisfactory services for 24 %, and item loss for 15.7 % [18].

Apparently, there are many problems existing in the entire procedure of express delivery businesses, from shipment, delivery, payment, and post-delivery services, which has been reflected in the high complaint rate. Probably, the rating of dissatisfaction lies in the topmost and however, it involves in many different aspects (characteristics), such as delivery delay, hidden charges, and careless delivery. In order to follow a healthy development path for the express delivery businesses in the fast growing e-commerce, mobile-commerce, social-commerce industries, an investigation of improvement of the customer service satisfaction in the express delivery business is indispensable and therefore it is necessary to study how to evaluate the performance of express delivery services and propose an advanced evaluation model and method.

3   Service Quality Evaluation Method

3.1   Express Service Quality Evaluation Process

For an express delivery enterprise, an evaluation process of the service quality of express delivery business mainly consists of the following steps.

Step 1 Determination of evaluation subject/agent. It is critical to determine the evaluation subject/agent, which will provide an evaluation to quality of services delivered by express logistics enterprises. The level of knowledge and industry experience, possessed by the subject/agent, is heavily related and crucial to the accurate degree of the evaluation results and thus the success of the evaluation.

Step 2 Seeking for information for evaluation. In order to establish the evaluation goals, all relevant information about the evaluation methods, the procedure of the delivery, and the delivery enterprise, should be properly collected, both from the literature and the actual work, sorted and classified.

Step 3 Establishment of an evaluation index system. The creation of the evaluation index system aims at effectively associating an express enterprise with the participants involved in the procedure of evaluation.

Step 4 Distribution of weights. In the evaluation index system for delivery service quality, the status and function of each index may not be necessarily the same. In other words, some of the functions may have stronger influence on the evaluation than the others. Therefore an appropriate weight is considered to give to each index based on its importance to the evaluation of express satisfaction.

Step 5 Comprehensive evaluation. Using the weight coefficients and the index data gained from the above steps, the service satisfaction degree of an express enterprise is analyzed and the evaluation value is calculated.

Step 6 Analysis of evaluation results. Based on the service quality evaluation results of general express logistics enterprises and the actual situation of a particular express enterprise, a certain evaluation method is selected and applied to judge the comprehensive evaluation value and draw a reasonable conclusion.

3.2   Service Quality Evaluation Method

Analytic Hierarchy Process. The analytic hierarchy process (AHP) is an effective method for the quantitative analysis of the problems in the process of quantitative analysis, which is a combination of qualitative and quantitative analysis model [3].

The steps of the analytic hierarchy process are described as follows [4].

Step 1 Establishing a hierarchical structure. When using the analytic hierarchy process, the problem is divided into several levels, and a multi-level structure analysis model is established based on the relationship between the levels of the factors. See Fig. 1.

Step 2 Constructing judgment matrix. After the establishment of the hierarchical layout, the relationships between the upper and lower factors are established. In order

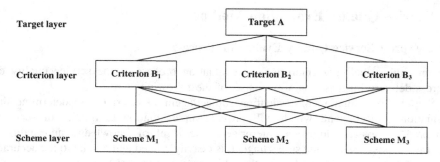

Fig. 1. Multilevel structure analysis model

Table 1. The proportion of scale

B_{ij} assignment	Meaning	B_{ij} assignment	Meaning
1	i and j are equally important		
3	i is more important than j	1/3	i is slightly less important than j
5	i is significantly more important than j	1/5	i was significantly less than j
7	i is strongly more important than j	1/7	i is less important than j
9	i is extremely more important than j	1/9	i is extremely less important than j
2, 4, 6, 8	between{1, 3, 5, 7, 9}	1/2, 1/4, 1/6, 1/8	Between {1,1/3,1/5,1/7,1/9}

to make a quantitative decision, it is essential to determine the proportion (scale) of $1 \sim 9$ by the relative importance of the matrix. Details are shown in Table 1.

Step 3 Hierarchical single sorting consistency checking. The feature vector W is obtained from the matrix A in terms of the maximum eigenvalue λ_{max}, and is normalized in order to obtain a target at the same level compared to the corresponding target level.

The consistency check algorithm is described as follows. Firstly, we calculate the consistency index CI, namely $CI = (\lambda_{max}-m)/(m-1)$, where m is the number of the order number of the judgement matrix. Secondly, we compute the second consistency index RI, denoting the mean random consistency index, to produce the consistency ratio CR, $CR = CI/RI$. Finally, we reckon the consistency of the judgment matrix A. When $\lambda_{max} = m$, $CI = 0$, it means the consistency of the judgment matrix A is entirely consistent. The bigger the CI value and the zero deviation, the worse the consistency is. The values of the judgment matrix RI of the order 1–10 are given in Table 2.

When $CR < 0.10$, it is considered that the judgment matrix A has the same degree of satisfaction. If $CR > 0.1$, we need to adjust the judgment matrix A to achieve the consistency of satisfaction.

Table 2. The *RI* value of judgment matrix

Order	1	2	3	4	5
RI	0	0	0.58	0.9	1.12
order	6	7	8	9	10
RI	1.24	1.32	1.41	1.45	1.49

Step 4 Total ranking and consistency checking. After calculating the weight of each layer and the consistency check, we can get all the indicators from the same level for the relative importance of a level of total sorting weight. This process is to calculate the layers of satisfaction evaluation system, from high to low layer by layer, using the linear weighting method, and in terms of the overall objectives to sort the plans based on their weights to give the pros and cons of each plan.

Step 5 Make a decision. According to the results of the total rankings, the decision is made in terms of the objectives.

Fuzzy Comprehensive Evaluation Method. The fuzzy comprehensive evaluation method is a kind of quantitative evaluation method for the evaluation object. It can provide an accurate decision-making basis for the evaluation method [9]. The parameters and the procedure are described as follows.

(1) A set of factors U: $U=\{u_1, u_2, ..., u_n\}$.
(2) A set of evaluation V: $V=\{v_1, v_2, ..., v_m\}$.
(3) Making single factor evaluation.
(4) Establishing an evaluation matrix.
(5) Determining the weight vector.
(6) Making fuzzy synthesis.
(7) Normalizing the vectors of fuzzy comprehensive evaluation.

For the last two steps, we need to select a weighted average operator, represented by "*". Both the weight vector W and the fuzzy comprehensive evaluation matrix R are integrated to form a fuzzy comprehensive evaluation result vector S, $S = W * R$. According to the evaluation grades of subordinate degree of evaluation object to make decisions, for each plan we normalize the vectors and the criterion for decision making is to prioritize the highest level of evaluation.

4 Express Service Quality Evaluation Model

4.1 Establishment of Performance Evaluation Index System

Evaluation of service quality scale is very impartment, which includes 5 dimensions and 22 indicators in the SERVQUAL model [5]. Based on the above mentioned problems in the express delivery industry and express satisfaction survey, we propose an evaluation of service quality scale, which includes 6 dimensions (i.e. delivery services, security, responsiveness, business level, service level, tangibility) and 18 indicators. The specific indicators are shown in Table 3. The table is divided into two

Table 3. Evaluation index system

	Level indicators	Secondary indicators
Express delivery service quality evaluation system A	Delivery B1 service	the way and the speed of take and send express
		delivery timeliness
		order fulfillment
	Security B2	express without loss
		express without damage
		compensation is reasonable and clear
	Responsiveness B3	timely door-to-door service
		the service hot line open
		timeliness of the complaint resolved
	Business B4 level	scale and credit facilities
		delivery network coverage
		product diversification
	Service B5 level	staff professional and communication skills
		the attitude of take and send express
		the ability to handle complaints
		order tracking query ability
	Tangibility B6	cost reasonable and transparent
		the delivery man dressed in decent appearance

parts: the first level indicators and the second level indicators, and the second level indicators are the expansion of the first level indicators. The selection of these indicators is based on the survey and analysis of the status quote and characteristics of the quality of express delivery services, and can be used as an effective basis for evaluating the quality of express delivery services.

4.2   Calculation of Weights

Here is the combination of the analytic hierarchy process (AHP) and the expert opinion method to calculate the weight of the distribution [8]. A part of these data was obtained through questionnaires. The data set collated in the paper includes a service delivery quality table with 24 indicators. The original data come from the biggest e-commerce company Taobao[1] in China. The time range of soliciting the data is from 30th Oct. to

---

[1] www.taobao.com.

19th Nov. in 2008. We also adopted the research sources (mainly the experiment data) from literature, process quality, corporate image data, as well as other statistical results.

For the index exceeding three dimensions of the secondary indexes, we summarize the expert opinion and use the analytic hierarchy process (AHP) to determine the weights together with the consideration of the express industry survey.

On the basis of the actual investigation and the summary of the expert opinions, we allocate the weights and determine the judgment matrix A as follows:

$$A = \begin{bmatrix} 1 & 1/3 & 3 & 4 & 1/2 & 2 \\ 2 & 1 & 4 & 5 & 3 & 3 \\ 1/3 & 1/4 & 1 & 2 & 1/3 & 1/3 \\ 1/4 & 1/5 & 1/2 & 1 & 1/3 & 1/3 \\ 2 & 1/3 & 4 & 3 & 1 & 3 \\ 1/2 & 1/4 & 3 & 1/3 & 1/3 & 1 \end{bmatrix}$$

The result shows that the largest eigenvalue of the matrix A is 6.2546. We get the result of consistency index 0.05092 by using the formula $CI = (\lambda_{max}-m)/(m-1)$. The order of the judgment matrix A is 6, and the corresponding $RI = 1.24$ is obtained by looking up the Table 2. Using the formula $CR = CI/RI = 0.05092/1.24 = 0.04106$, the result $CR < 0.1$ shows that the judgment matrix A has a consistent satisfaction. The feature vector of the maximum eigenvalue of the matrix is calculated and normalized. The A level of the index weight is W = (0.1953, 0.3603, 0.0693, 0.0502, 0.2318, 0.0932).

Now let us determine the judgment matrix of the B1 level:

$$B1 = \begin{bmatrix} 1 & 1/2 & 1/3 \\ 2 & 1 & 1/2 \\ 3 & 2 & 1 \end{bmatrix}$$

By calculating the matrix, the largest eigenvalue $\lambda_{max} = 3.0092$, $CR = 0.0079 < 0.1$. It has a consistence satisfaction too, with the B1 level index weight W1 = (0.1634, 0.2970, 0.5396).

For the judgment matrix of B2 level, we have:

$$B2 = \begin{bmatrix} 1 & 3 & 2 \\ 1/3 & 1 & 1/2 \\ 1/2 & 2 & 1 \end{bmatrix}$$

Its maximum eigenvalue is $\lambda_{max} = 3.0092$, $CR = 0.0079 < 0.1$. It has also a consistence satisfaction (Note that the maximum eigenvalue λ_{max} is calculated based on the matrixes B1 and B2 respectively). The result is the same due to the same the dimension of judging matrixes. Similarly, the same value is obtained for CR. Its B2 level index weight is W2 = (0.5396, 0.1634, 0.2970).

For the judgment matrix of B3 level, we have:

$$B3 = \begin{bmatrix} 1 & 3 & 2 \\ 1/3 & 1 & 1/3 \\ 1/2 & 3 & 1 \end{bmatrix}$$

Its largest eigenvalue $\lambda_{max} = 3.0536$, and $CR = 0.0462 < 0.1$. It has a consistence satisfaction with the B3 level index weight W3 = (0.5278, 0.1396, 0.3325).

The judgment matrix of B4 level is:

$$B4 = \begin{bmatrix} 1 & 1/5 & 1/4 \\ 5 & 1 & 2 \\ 4 & 1/2 & 1 \end{bmatrix}$$

It has the largest eigenvalue $\lambda_{max} = 3.0246$, and $CR = 0.0212 < 0.1$. It has a satisfactory consistency, with the B4 level index weight W4 = (0.0974, 0.5695, 0.3331).

The judgment matrix of B5 level is:

$$B5 = \begin{bmatrix} 1 & 1/3 & 1/3 & 1/2 \\ 3 & 1 & 2 & 3 \\ 3 & 1/2 & 1 & 2 \\ 2 & 1/3 & 1/2 & 1 \end{bmatrix}$$

and its largest eigenvalue $\lambda_{max} = 4.0710$, and $CR = 0.0263 < 0.1$. It has a satisfactory consistency and its B4 level index weight W5 = (0.1059, 0.4476, 0.2829, 0.1636).

For the B6 level of the index weight determination, we establish the index weight W6 = (0.816, 0.184) using the method of expert evaluation and statistical summary as discussed earlier.

The weights of each layer are listed in Table 4.

In the selection of the 18 customer satisfaction index factors, these factors, delivery timeliness, order fulfillment, price reasonability, items loss, reasonable compensation, the attitude of send-and-take express have the highest weight indicators. However, the facilities, product diversification, and order tracking query have relatively low weights, which is consistent to the actual survey we received from the express delivery enterprises.

4.3 Fuzzy Evaluation

A fuzzy analysis method is used to evaluate the degree of satisfaction of three express delivery enterprises.

The steps of the fuzzy evaluation method are described as follows.

Determine the evaluation factors. This is a two-level evaluation method. The corresponding evaluation factors should belong to two different levels.

The first level contains B1 = {the way and the speed of take-and-send express, delivery timeliness, order fulfillment}, B2 = {no item loss, no damage, the compensation is reasonable and clear}, B3 = {timely door-to-door service, service hot line

Table 4. The value of index weights

Target layer	Level indicators weights	Secondary indicators weights	
		The weight relative to level indicators	The weight relative to target layer
W	W1 0.1953	0.1634	0.0319
		0.2970	0.0580
		0.5396	0.1054
	W2 0.3603	0.5396	0.1944
		0.1634	0.0589
		0.2970	0.1070
	W3 0.0693	0.5278	0.0366
		0.1396	0.0097
		0.3325	0.0230
	W4 0.0502	0.0974	0.0049
		0.5695	0.0286
		0.3331	0.0167
	W5 0.2318	0.1059	0.0245
		0.4476	0.1038
		0.2829	0.0656
		0.1636	0.0379
	W6 0.0932	0.816	0.0761
		0.184	0.0171

open, timeliness of the complaint resolved}, B4 = {scale and credit facilities, delivery network coverage, product diversification}, B5 = {staff professional and communication skills, the attitude of take-and-send express, the capability to handle complaints, order tracking query ability}, and B6 = {the cost is reasonable, transparent, the delivery is appropriate}. The second level has only U = {delivery service, security, responsiveness, business level, service level, tangibility}.

Determine the planning level set. The planning contains four levels "very high", "high", "average", and "low", namely the establishment of the evaluation level set V = {very high, high, average, low}.

Calculation of a single-factor evaluation and building of an evaluation matrix. According to the collected data of research and expert opinions, we evaluate single factors and set up the evaluation matrix R1, R2, R3, R4, R5, and R6.

Establishment of weight vectors with fuzzy synthesis. The analytic hierarchy process (AHP) is applied to obtain index weight for each item W1, W2, W3, W4, W5, and W6. Then we make the fuzzy synthesis as follows:

$$S_i = W_i * R_i, \ 1 <\ = i <\ = 6.$$

The results S1, S2, S3, S4, S5, and S6 form a fuzzy comprehensive evaluation matrix R. The weight vector W and the fuzzy comprehensive evaluation matrix R are combined to provide the fuzzy comprehensive evaluation results vector S, S = W * R.

Normalization process. The normalization process yields S_A = (0.23, 0.31, 0.25, 0.20), S_B = (0.28, 0.28, 0.23, 0.21), and S_C = (0.19, 0.22, 0.29, 0.29).

This decision making process is carried out by using the principle of maximum subordinate degree. By the fuzzy evaluation we find that the overall satisfaction of these three courier companies is not very high, and the results is that the enterprise B has its comprehensive satisfaction higher than that of the enterprise A, which in turn higher than that of the enterprise C. At present, the total score of the express service quality is low and the gap between the customer expectations and the quality of service delivery is large. The express delivery service quality has a lot of space to improve. Enterprises can ameliorate service delivery quality to improve satisfaction according to the different weighting of different dimensions.

5   Conclusion

With the development and enhancement of e-commerce technology, to purchase online becomes more and more popular. This certainly brings a huge development opportunity for express delivery businesses. However, many problems occur in the businesses too, for example, low service quality in delivering items, while it becomes indispensable in our daily life. Therefore, to correctly and efficiently solve these problems in service quality and performance of express delivery businesses is the key to the enhancement of their competition powers.

In this paper, the evaluation index system of express service quality is the combination of quantitative and qualitative analysis, and the weight coefficient is more close to the reality, which makes the evaluation results more accurate. It helps to evaluate the service quality of express enterprise, which makes the enterprise to understand the essence of express service. According to the weight of each index, the enterprise should make a clear direction of service development and find out the method to improve the quality of express delivery service. So is to improve the satisfaction level of express enterprises to enhance their competitiveness. The shortcomings of this paper lie in that the indicators in the evaluation index system constructed is overmuch, the data acquisition caused the limitations to a certain extent.

It is known to us, the subjectivity and arbitrary of measurement of the service quality, particularly in the areas of e-commerce businesses, are the major obstacles to effective and reasonable judgement of service quality and rigorous development of the measure theory and algorithm. We plan to pursue this research direction in our next study.

Acknowledgements. This work was supported by the National Natural Science Foundation of China (Grant 71071145), the National Natural Science Foundation of Zhejiang (Grant Y16G020036), the Social Development Projects of Ningbo (Grant 2012C50045), and the Science and Technology Innovation Team of Ningbo (Grant 2013B82009). This work is completed by the author, Dr. Zhang, during his visit to Dalarna University, Sweden from September 2015.

References

1. Carman, J.M.: Consumer perceptions of service quality: an assessment of the SERVQUAL dimensions. J. Retail. **66**, 33–55 (1990)
2. Cronin, J.J., Taylor, S.A.: Measuring service quality: a reexamination and extension. J. Mark. **56**, 55–68 (1992)
3. Cui, Y.Q.: Research on performance evaluation method of express logistics enterprises (in Chinese). MA Dissertation, Chang'an University, China (2012)
4. Pang, F., Guo, H.L.: Research on the performance evaluation of express enterprise based on AHP (in Chinese). J. Administered Space **2**, 44–45 (2014)
5. Parasuraman, A., Zeithaml, V.A., Berry, L.L.: SERVQUAL: a multiple-item scale for measuring consumer perceptions of service quality. J. Retail. **64**(1), 12–40 (1988)
6. Patrick, A., Karl, J.M., John, E.S.: SERVQUAL revisited: a critical review of service quality. J. Serv. Mark. **10**, 62–81 (1996)
7. Saunders, C.S., Jones, J.W.: Measuring performance of the information systems function. J. Manag. Inf. Syst. **8**, 63–82 (1992)
8. Sun, J.H., Su, Q., Huo, J.Z.: Construction of China express delivery service quality system and fuzzy comprehensive evaluation. J. Ind. Eng. Manag. **4**, 112–116 (2010)
9. Yu, B.Q., Du, G.W.: Fuzzy evaluation based on SERVQUAL model for online shopping delivery service quality. J. Ind. Eng. **4**, 127–132 (2013)
10. Yan, S., Qing, N., Qing, D.Y., Xiao, Y.Y.: Conformation of express delivery service provider based on a two-stage method. In: Proceedings IEEE the 16th International Conference on Industrial Engineering and Engineering Management, pp.1465–1468 (2009)
11. Ye, J.K., Yan, L., Yang, D.H.: Research on the improvement of SERVQUAL evaluation model based on AHP. J. Logistics Eng. Manag. **9**, 78–80 (2010)
12. Zhang, Q.K.: Research on the performance evaluation of third party logistics information platform based on combination evaluation method (in Chinese). MA Dissertation, Wuhan University of Technology, China (2012)
13. Zhang, Y.: Performance evaluation of third party logistics enterprises based on entropy weight and gray correlation. J. China's Circ. Econ. **1**, 19–21 (2008)
14. Zhu, M.H., Miao, S.T., Zhuo, J.: An empirical study on Chinese express industry with SERVQUAL. J. Sci. Technol. Manag. Res. **8**, 38–45 (2011)
15. Zhao, X.D., Bai, C.H., Hui, Y.V.: An empirical assessment and application of SERVQUAL in a mainland Chinese department store. J. Total Qual. Manag. **13**(2), 41–54 (2002)
16. The Report on China Express Delivery Business, Volume of 2014 (August 2015). http://finance.takung-pao.com/
17. The investment analysis and forecast report on Chinese express industry 2014 to 2018 (August 2015). http://bbs.tianya.cn/post-develop-1682706-1.shtml
18. The increased situation of complaints of China's express of 2013. China Industry Research Network (August 2015). http://www.chinairn.com/

Geographical Constraint and Temporal Similarity Modeling for Point-of-Interest Recommendation

Huimin Wu[1], Jie Shao[1(✉)], Hongzhi Yin[2], Heng Tao Shen[2], and Xiaofang Zhou[2]

[1] University of Electronic Science and Technology of China, Chengdu, China
wuhuimin@std.uestc.edu.cn, shaojie@uestc.edu.cn
[2] The University of Queensland, Brisbane, Australia
h.yin1@uq.edu.au, {shenht,zxf}@itee.uq.edu.au

Abstract. People often share their visited Points-of-Interest (PoIs) by "check-ins". On the one hand, human mobility varies with each individual but still implies regularity. Check-ins of an individual tend to localize in a specific geographical range. We propose a novel model to capture personalized geographical constraint of each individual. On the other hand, PoIs reflect requirements of people from different aspects. Usually, places of different functions show different temporal visiting distributions and places of similar function share similar visiting pattern in temporal aspect. Temporal distribution similarity can be used to characterize functional similarity. Based on the findings above, this paper introduces improved collaborative filtering models by jointly taking advantages of geographical constraint and temporal similarity. Experimental results on real data collected from Gowalla and JiePang demonstrate the effectiveness of our models.

Keywords: Recommendation system · Collaborative filtering · Geographical constraint · Temporal similarity

1 Introduction

The popularity of smart mobile devices with positioning technologies triggers the advent of Location-based Social Networks (LBSNs), such as Foursquare, Facebook Place and Yelp, which combine online services and offline activities. In traditional social networks, users build digital social connection. However, in LBSNs, with mobile communication devices now reaching almost every corner of planet earth, users are encouraged to share their location information and extend virtual social connection to real life by sharing their life experiences with "check-ins".

There are two kinds of participants playing important roles in LBSN, namely, *ordinary users* and *business owners*. Ordinary users play the part of consumers

© Springer International Publishing Switzerland 2015
J. Wang et al. (Eds.): WISE 2015, Part II, LNCS 9419, pp. 426–441, 2015.
DOI: 10.1007/978-3-319-26187-4_40

who consume services provided by business owners and LBSN platform developers. Users can choose services that satisfy their requirements or save expenses. Business owners provide information that draws attentions of users and favors profits of business owners themselves. Business owners partnering with LBSN platforms can publish advertisements or discount information. Based on users' historical behaviors, their interests can be explored, which makes it easy for business owners to provide corresponding services.

In this work, we focus on Point-of-Interest (PoI) recommendation in LBSNs. Both of users and business owners can benefit from recommendation systems. For business owners, they can make their services stand out. For users, good services matching their preferences can be provided. The key issue of achieving a successful PoI recommendation is to capture factors that influence users' decisions to visit PoIs and model these factors. Our work is based on exploiting **geographical constraint** and **temporal similarity**.

Geographical constraint of a user influences the possibility of visiting a PoI. It does not make sense to recommend the user a PoI out of her mobility range even though the PoI satisfies her requirements and matches her preferences. There have been several models developed for capturing human mobility pattern. For example, previous work [1,8,11,13] used a power law distribution to model geographical influence. Other work [3,4,14] modeled geographical constraint using Kernel Density Estimation (KDE) method. Gaussian Mixture Model (GMM) can also be adopted when capturing geographical clustering phenomenon [12].

Besides geographical influence, temporal influence should be noted from daily check-in behaviors as well. Human form repetitive behavior patterns, which provides predictability to human mobility, temporal preferences and temporal requirements. Work [1,12] demonstrated that geographical states of some users are influenced by the time factor. Exploring temporal proximity can also contribute to recommendation effectiveness [11,13]. In this work, we examine temporal similarity of PoIs' visiting distributions and use the similarity to characterize functional similarity.

This paper first investigates geographical constraint and temporal similarity separately, and then combines geographical influence with temporal influence in two different ways. The main contributions of our work are:

- We propose a novel geographical pattern model called Short-term Cluster-based Gaussian Mixture Model (SCBGMM) to capture a personalized check-in distribution for each individual.
- We measure functional similarity of PoIs by using temporal distribution. Temporal similarity is capable of explaining characteristics of a PoI.
- We demonstrate the effectiveness of geographical constraint, temporal similarity and two combinations of geographical and temporal influences by incorporating factors mentioned above with a basic model.

The remainder of the paper is organized as follows. We discuss previous studies related to PoI recommendation in Sect. 2. We introduce models based on geographical constraint and temporal similarity in Sects. 3 and 4, respectively. We present two ways to combine geographical and temporal influences in Sect. 5.

In Sect. 6, we verify the effectiveness of our proposal on two real datasets. We conclude our work in Sect. 7.

2   Related Work

Collaborative Filtering. There exist two basic flavors of *collaborative filtering* (CF), *user-based CF* (UCF) and *item-based CF* (ICF). UCF recommends items that users with similar preferences have visited, viewed or purchased. ICF recommends items which are similar to those having been visited, viewed or purchased by same user [5].

In this paper, ICF is used as the basic algorithm. The reasons of this in the context of LBSN are two-fold. First, in UCF, similarity between users is indicated by historical PoI records and is hard to extend to other respect of user behaviors. While in ICF, temporal similarity makes up for lacking of semantic information and provides another angle for evaluating PoI similarity beyond PoI visiting history. Second, in traditional CF application domains (e.g., e-commerce platforms), capturing users' preference is the main purpose. LBSNs bridge the gap between online communications and real life activities, which is an important distinction between LBSN and other virtual applications. In addition to preferences, users' physical and daily life related requirements must be satisfied. This needs more attentions from PoIs' perspective.

Let $N(i; u)$ be a set of PoIs visited by user u, and $N(u; i)$ be a set of users who have visited PoI i. Generally, notations like i, j denote PoIs and u, v denote users. We follow this convention. Given the rate (frequency) r_{uj} of PoI j visited by user u, the recommendation rate of some unvisited PoI is:

$$r_{ui} = \frac{\sum_{j \in N(i;u)} s_{ij} * r_{uj}}{\sum_{j \in N(i:u)} s_{ij}} \tag{1}$$

where s_{ij} represents the similarity of two PoIs. Cosine similarity is one of the most popular measures:

$$s_{ij} = \frac{\sum_{v \in N(u;i) \bigcup N(u;j)} r_{vi} * r_{vj}}{\sqrt{\sum_{j \in N(u;i)} r_{vi}^2} * \sqrt{\sum_{j \in N(u;j)} r_{vj}^2}} \tag{2}$$

Geographical Influences for PoI Recommendation. In previous work, difference approaches have been adopted to describe personalized geographical constraint in an individual's visiting records.

Power law distribution can model the distribution of distances of PoIs [1,11, 13]. It is based on a global observation that the probability of people visiting PoIs decreases with the increase of their distance. Ye et al. [9] proposed a power law distribution of distances between PoIs to estimate check-in probability of an unvisited PoI. Yuan et al. [11,13] modeled the likelihood of a user's check-in at some PoI by using power law distribution of distance between the PoI and previously visited PoI. The distribution of distances of PoIs and current user's "home" location is observed to follow a power law distribution in [1].

KDE is adopted based on a more personalized assumption than power law distribution by modeling PoIs visited by same user. There are some variations when adopting KDE. In some context, KDE was proposed to capture the intuition that every PoI has an influence over nearby PoIs. For example, a famous scenic spot has a higher chance to attract tourists to visit surrounding restaurants or hotels. Capturing the influence spreading phenomenon can more reasonably explain choices of PoIs that users make. Lian et al. [3] proposed kernel density estimation to model the influence areas of a PoI which is fixed to be Gaussian distribution. Besides, KDE can model that the distances of PoIs visited by same user are subject to some kind of distribution, usually Gaussian distribution. A mixture kernel density method which has the ability to model global location data and personal location data is proposed in [4]. Zhang et al. [15] proposed a pilot estimation which is given by a weighed average of distance distribution between current PoI and visited PoIs. KDE is also used to model differences of distances of PoIs. A one-dimensional kernel density estimation to model difference distribution was investigated in [14].

GMM approach can also be used to model check-in distribution. Yuan et al. [12] proposed a probabilistic model based on the intuition that human mobility centers at predefined geographical regions (which is captured by a sampling Gaussian distribution) and influenced by current requirements. GMM describes a user's check-in distribution intuitively and the possibility of a PoI being visited from geographical perspective is represented by GMM value. The GMM describes geographical constraint from another angle.

We choose GMM as our basic model based on the analysis as follows. Power law distribution is often adopted in two aspects - indicating social intimacy [8] and modeling likelihood of users' geographical transfers between PoIs (which is represented by distance distribution of PoIs visited by a same user [9,11,13]). As mentioned above, we adopt ICF as the basic algorithm so that social intimacy has a limited influence. In addition, distance distribution is sensitive to noise spots. KDE is a non-parametric method, but it fails to resist interference of noise spots. GMM needs an assumption about the predefined number of human mobility centers for all users, which can be against the principle of personalization. Based on the global assumption of centers (e.g., "home" and "work"), personalized geographical constraint will take fewer effects. To avoid such global assumption, we use density-based clustering to automatically detect the number of centers to initialize GMM, so that GMM can start from a personalized, reasonable assumption about the number of an individual's mobility centers.

In our model, we adopt a density-based clustering algorithm for each user to capture kernels that the user's check-ins center at, which, at the same time, remove noise spots. However, not all check-in clusters are useful. For example, an individual may move from New York to San Francisco. Clusters in New York that the individual used to visited will be no longer visited. Only short-term clusters should be considered to build the individual's geographical mobility pattern. Thus, Short-term Cluster-based Gaussian Mixture Model (SCBGMM) is built and can be incorporated with ICF for recommendation. Note that we do not make the assumption that a user's check-in distribution is centered at "home" location (such as [1]).

Temporal Influences for PoI Recommendation. Human behaviors show strong periodicity in respect of time [1,11,13]. Temporal influence has been investigated from various aspects. Decisive influence of time on geographical mobility was examined in [1,12]. Cho et al. [1] modeled temporal movement between the "home" state and "work" state based on the Gaussian distribution of each state over the time of the day respectively. Yuan et al. [12] modeled a probabilistic graph, in which, geographical region of a user is decided by time factor. Different from the above studies, we exploit decisive influence of time on clusters to predict which clusters will be visited. Besides temporal influence on geographical information, we use the relationship between temporal similarity and PoIs' functional (categorial) similarity. Yin et al. [10] claimed that PoIs visited at similar time tend to belong to similar category. Ye et al. [7] proved temporal and semantic interaction of PoIs. Different from [7], we use cosine similarity to measure temporal similarity and consider temporal similarity as a factor that influences the possibility of a user's visit to a PoI. In our work, temporal similarity corresponds to current user's requirement rather than short-term preferences such as in [6].

3   Using Geographical Constraint

In our model, we first employ a density-based clustering method in geographical coordinate system to cluster an individual's check-ins. Clusters visited a long time ago are removed and those visited recently are retained. Then, we use remaining clustering centers to initialize parameters of GMM, such as the number of centers and positions of centers. Expectation Maximization (EM) algorithm is employed to learn GMM. Finally, SCBGMM is incorporated with ICF.

3.1   Density-Based Clustering

Yin et al. [10] claimed that PoIs visited by users tend to form several cluster centers and modeled each region with a Gaussian distribution. Different from [10], we assume that PoIs visited by each user are gathering into several centers and capture each region with a density-based clustering algorithm. DBSCAN [2] is capable of clustering a region with higher density than a predefined density value. It detects a kernel based on the principle that if there exist more than an fixed number ($minPts$) of points within an acceptable distance (ϵ), the point is considered as a kernel. Each kernel with neighbors within the acceptable distance forms a cluster. If a neighbor of a kernel is a kernel itself, it will be added into the current cluster with its neighbors. The iteration process continues until no point is added. Traditionally, DBSCAN is employed in two dimensional surface. Here, we extend DBSCAN to the earth surface by replacing Euclidean distance with geodesic distance.

3.2   Decisive Influence of Time

Unlike previous work [1] which uses time to decide which cluster current user to stay, in our work time factor is used to decide which cluster is useless. Firstly, we calculate time span which all check-ins of current user are within. Then check-ins visited after a time spot are marked as recently visited ones. The time spot is set to a value that can split the time span into two intervals with some proportion. Check-ins in the first interval are visited a relatively long time ago, and those in the second interval are visited recently. A cluster without any check-in marked as recently visited one will be removed. Remaining clusters are used to initialize the number of Gaussian components next.

3.3   Gaussian Mixture Model

Given the number of Gaussian model components K, GMM is represented as:

$$p(x) = \sum_{k=1}^{K} \pi_k \mathcal{N}(x|\mu_k, \Sigma_k) \tag{3}$$

where $\mathcal{N}(x|\mu_k, \Sigma_k)$ is k^{th} Gaussian component, π_k is the weight of each Gaussian component, μ_k and Σ_k represent mean and standard deviation of kth Gaussian component. The basic assumption of GMM is that each data point is generated from K Gaussian models jointly so that GMM can provide richer sorts of density models than only one Gaussian model. We use two steps to describe the data point generating process which provides us with a deeper insight into this distribution:

- Choosing one Gaussian model component from K components. The probability of each component being chosen is π_k.
- Picking one data point from the Gaussian model chosen in last step. The process follows a Gaussian distribution parameterized by μ_k and Σ_k.

Suppose N data points are in our dataset. We need to learn parameters mentioned above. EM algorithm is used to fit GMM by maximizing the probability that data points in the dataset are generated from GMM.

3.4   Expectation Maximization for GMM

EM algorithm offers an powerful method to find maximum likelihood solutions for models with latent variables. Parameters of GMM are unknown, and maximum likelihood provides a feasible solution by estimating parameters of the distribution and making the observed dataset fit the distribution most likely. The basic process of EM is: (1) finding the expected value of a maximization function, and (2) maximizing the expectation function.

We will introduce EM for GMM model as an example. For a dataset X, let x_n denote an observed value in X. The likely function of the observations is given by:

$$p(X) = \prod_{n=1}^{N} p(x_n) = \prod_{n=1}^{N} \sum_{k=1}^{K} \pi_k \mathcal{N}(x_n|\mu_k, \Sigma_k) \tag{4}$$

To avoid the underflow of the production above, a log-likelihood function is adopted instead.

$$lnp(X|\pi,\mu,\Sigma) = \Sigma_{n=1}^{N}ln\{\Sigma_{k=1}^{K}\pi_k\mathcal{N}(x_n|\mu_k,\Sigma_k)\} \tag{5}$$

Two-step iteration is implemented:

- **E-Step (Expectation):** Use values of the parameters from last iteration to evaluate the probability of generating an observation from K Gaussian models.

$$\gamma(n,k) = \frac{\pi_k\mathcal{N}(x_n|\mu_k,\Sigma_k)}{\sum_{j=1}^{K}\pi_j\mathcal{N}(x_n|\mu_j,\Sigma_j)} \tag{6}$$

- **M-Step (Maximization):** Re-estimate the distribution parameters to maximize the log-likelihood function.

$$\mu_k = \frac{1}{N_k}\sum_{n=1}^{N}\gamma(n,k)x_n \tag{7}$$

$$\Sigma_k = \frac{1}{N_k}\sum_{n=1}^{N}\gamma(n,k)(x_n - \mu_k)(x_n - \mu_k)^T \tag{8}$$

$$\pi_k = \frac{N_k}{N} \tag{9}$$

We need to reevaluate the log-likelihood function and check for the convergency. The **Expectation** and **Maximization** iteration will not stop until the convergency is achieved.

3.5   Incorporating SCBGMM with ICF

PoIs out of the geographical range described by SCBGMM are less likely to be visited. We propose an algorithm named **ICF-SG**. For each candidate PoI i, we calculate the possibility of i being visited by user u according to Short-term Cluster-based Gaussian Mixture Model (SCBGMM). Those PoIs with lower possibility than a threshold κ will not be recommended. Detailed ICF-SG is presented in Algorithm 1.

3.6   CBGMM

To understand decisive influence of time, we introduce another geographical model slightly different from SCBGMM. GMM is initialized by the clustering result of DBSCAN directly (without filtering clusters that are visited a relatively long time ago). After learning with EM, a simpler model called Cluster-based Gaussian Mixture Model (CBGMM) can be built. We can incorporate CBGMM with ICF in a similar way as Algorithm 1, and we call this algorithm **ICF-LG**. In the experiments, by comparing ICF-SG with ICF-LG, we can verify the effectiveness of decisive influence of time factor in filtering clusters.

Algorithm 1. ICF-SG Algorithm

```
 1: U ← user set;
 2: u ← current user;
 3: C_u ← queue of unvisited PoIs ranked by user's predicted preference calculated by ICF;
 4: R_u ← top-K recommended PoIs;
 5: κ ← threshold to filter;
 6: for each user u ∈ U do
 7:      learn personalized SCBGMM;
 8:      for each unvisited PoI i ∈ C_u do
 9:          if p(i) = ∑_{k=1}^{K} π_{uk}N(i|μ_{uk}, Σ_{uk}) < κ then
10:              continue;
11:          end if
12:          if sizeOf(R_u)|=K then
13:              add i to R_u;
14:          end if
15:      end for
16: end for
```

---

4   Using Temporal Similarity

PoIs satisfy individuals' requirements from various aspects, and normally these requirements are time-related and show periodicity. We propose a way to describe time-relevance and use this characteristic in our model. In this paper, we analyse temporal distribution similarity of PoIs and propose a model to take advantage of temporal similarity. A smoothing approach is discussed as well.

4.1   Characterizing Functional Similarity

People's visits to PoIs are often driven by their requirements. These requirements are different in category and show time-relevance. Here, time-relevance of PoIs does not mean short-term preference [6]. Short-term preference means a user's interest of PoIs with a specific style. For example, someone who is used to go to tidy, quite restaurants, may go to romantic restaurants since his girlfriend who has a long-distance relationship with him comes to visit him. In this work, time-relevance corresponds to an individual's requirements. For example, at mealtime we should pay attention to users' preferences for restaurants, whereas in the afternoon users' preferences for leisure-related spots such as cafes should be concerned.

An intuitive way is to analyse categories of PoIs which are often visited in a time interval and capture the requirements in this time interval. For example, if we analyse categorial information of PoIs often visited in 12 pm or 1 pm in a day, we can find they tend to be restaurants. However, without textual information or categorial annotation in PoI data, we find it hard to analyse users' preferences in PoIs' respect. Previous work [7] mined the relationship between temporal similarity and categorial information. We follow and apply the discovery in our work. We use temporal distribution over 24-hour slots to capture functional characteristic of a PoI. However, considering normalization, we use cosine similarity of temporal distributions to evaluate categorial similarity, which is different from [7].

4.2   Temporal Similarity Evaluation

Let T_{ik} and T_{jk} denote temporal distribution of PoIs i and j in k^{th} hour of a day time, temporal similarity of PoIs i and j is defined as:

$$s_{ij}^t = \frac{\sum_{k=0}^{23} T_{ik}T_{jk}}{\sqrt{\sum_{k=0}^{23} T_{ik}^2} * \sqrt{\sum_{k=0}^{23} T_{jk}^2}} \tag{10}$$

Both of users' time-related requirements and preferences for a PoI should be taken into consideration. PoIs which meet the time-related requirements and satisfy current user's preferences should have high possibilities to be recommended. In traditional ICF (Eq. 1), a user's preference for an unvisited PoI is evaluated by weighted average of the user's preferences for visited, similar PoIs.

4.3   Incorporating Temporal Similarity with ICF

We adopt a linear combination to balance requirement relevance of a PoI and a user' preference for the PoI (Eq. 2), which is called **ICF-T** and defined as follows.

$$
\begin{aligned}
s_{ij}^T &= (1 - \alpha)s_{ij} + \alpha s_{ij}^t \\
&= (1 - \alpha)\frac{\sum_{v \in N(u;i) \cup N(u;j)} r_{vi} * r_{vj}}{\sqrt{\sum_{j \in N(u;i)} r_{vi}^2} * \sqrt{\sum_{j \in N(u;j)} r_{vj}^2}} + \alpha \frac{\sum_{k=0}^{23} T_{ik}T_{jk}}{\sqrt{\sum_{k=0}^{23} T_{ik}^2} * \sqrt{\sum_{k=0}^{23} T_{jk}^2}}
\end{aligned}
\tag{11}
$$

We substitute Eq. 11 into Eq. 1 and get

$$r_{ui}^T = \frac{\sum_{j \in N(i;u)} s_{ij}^T * r_{uj}}{\sum_{j \in N(i;u)} s_{ij}^T} \tag{12}$$

4.4   Smoothing

The idea of smoothing visiting frequency was proposed in [7,11]. Ye et al. [7] applied an empirical window to replace the PoI's visiting frequency in a hour slot with an weighted sum of visiting frequency in current hour, an hour earlier and an hour later. The weight is set to a fixed and empirical value. Yuan et al. [11] evaluated similarity between two time slots by averaging cosine similarity between PoI check-in frequency vectors in the two time slots for all users. In our work, we evaluate similarity between time slots by calculating cosine similarity of all PoIs' check-in frequency vectors in two time slots from a global respective. We use se_h to represent similarity between hour h with an hour earlier and sl_h to represent similarity between hour h with an hour later. We smooth each PoI's number of check-ins in some hour by

$$N(i;u)_h^s = N(i;u)_h + \frac{se_h}{se_h + sl_h}N(i;u)_{(h+23)/24} + \frac{sl_h}{se_h + sl_h}N(i;u)_{(h+1)/24} \tag{13}$$

5   Combination of both Factors

In this section, we will investigate combinations of geographical constraint and temporal similarity factors in two ways, in order to examine effectiveness of these combinations and a mutually reinforcement relationship of them.

Combinations of the two factors can be various for the reason that we set different values for parameter κ in ICF-SG and parameter α in ICF-T, separately. If we set parameter κ to a relatively higher value, PoIs having higher accordance with geographical constraint range will be recommended and vise versa. Parameter α can balance the relative importance between time-relevance of a PoI and current user's preference for the PoI. For simplification and significance, we choose two directions to investigate them in depth. We investigate influences of parameter α on condition that ICF-SG performs best, and influences of parameter κ when best performance of ICF-T is achieved.

5.1   Using Temporal Similarity on ICF-SG

By applying ICF-SG, PoIs which fit a user's preference and are possible to be visited from geographical aspect are recommended. Besides, current user's requirements need to be taken into consideration. When κ equals to a specific value, ICF-SG performs best. Based on ICF-SG, we investigate temporal influences by fixing parameter κ to the value mentioned above and balancing relative importance of time-relevance (by changing the value of α). We call this model as **ICF-GT**. If ICF-GT performs better than ICF-SG, we can conclude that this combination of the two factors is effective and temporal influence can reinforce geographical influence.

5.2   Using Geographical Constraint on ICF-T

ICF-T is capable to capture a user's preference and requirement both. However, only applying ICF-T lacks of filtering power to those PoIs out of geographical mobility range. Geographical constraint needs to be considered. In Eq. 11, we balance the importance of time-relevance and current user's preference using a parameter α. Similar to the method mentioned in Sect. 5.1, we analyse geographical influences through setting parameter α to a fixed value (the value that makes ICF-T achieve best performance) and changing the value of κ. We call this model as **ICF-TG**. If we observe better performances compared with ICF-T, we can prove the effectiveness of this combination of the two factors and the reinforcement of geographical influences to temporal influences.

6   Experiments

6.1   Experimental Setup

Datasets for Evaluation: We conduct our experiments using two datasets crawled from Gowalla [1] and JiePang [3]. The dataset from Gowalla contains

196,591 users and 6,442,890 check-ins from Feb. 2009 to Oct. 2010, and the sample dataset from JiePang (jiepang.com) contains 9,237 users and 1,556,636 check-ins from Mar. 2011 to Mar. 2013. Every check-in record in the test datasets contains *coordinates* and *timestamp* fields. Due to page limit, we mainly report results of experiments on Gowalla dataset. Results on JiePang exhibit similar trend. There are two ways to partition the datasets. Both of them need to divide training set and test set as 7 to 1 in portion. The first way is that we split all check-ins in portion of 7 to 1 in chronological sequence as training set and test set, separately. The second way is to add first 7/8 part according to check-in time of each individual into training set and the others into test set. The first approach cannot guarantee that every user has check-ins in test set. We therefore adopt the second approach so that we can evaluate algorithm effectiveness on every user.

Evaluation Metrics: *Precision, Recall* and *F1* are three most popular measures to evaluate the effectiveness of recommendation algorithms. *Precision* evaluates how many top PoIs recommended are actually visited. *Recall* measures how many PoIs which are visited later are recommended. Empirically, we consider top 7 returned results. Intuitively, stricter constraint condition means higher *Precision* but lower *Recall*, where constraint condition means a standard that we consider a PoI as one that should be recommended. *F1* incorporates both factors of *Precision* and *Recall* and gets a relatively fair measurement.

Evaluated Algorithms: We keep a list of top PoIs for each user in each algorithm tested. We show effects of geographical constraint (Sect. 3), temporal similarity (Sect. 4) and a combination of these factors (Sect. 5) by comparing proposed models with ICF [5]. We set every measurement of ICF as 1, and calculate the ratio of algorithm results of ICF-T, ICF-LG, ICF-SG, ICF-GT, ICF-TG to ICF for each measurement.

- **ICF:** Item-based CF [5] expressed in Eq. 1 predicts user's preference for an unvisited PoI by calculating weighted sum of user's preference for visited PoIs.
- **ICF-2G:** Some of previous studies (e.g., [1]) claimed that an individual's check-ins center at two locations ("home" and "work"). ICF-2G considers two clusters with most check-ins as "home" cluster and "work" cluster so that check-in distribution model for each individual is constituted of two Gaussian components.
- **ICF-LG:** In Sect. 3.6, a long-term check-in distribution is captured for an individual without filtering those clusters which may never be visited. GMM is initialized by clustering results of DBSCAN directly.
- **ICF-SG:** As described in Algorithm 1, ICF-SG filters out those PoIs out of range of personal geographical mobility with decisive influence of time.
- **ICF-T:** In ICF-T expressed by Eq. 12, time-relevance and current user's visiting preferences are combined to evaluate PoIs' similarity. The relative importance of each factor needs to be tuned.
- **ICF-GT, ICF-TG:** ICF-GT focuses on temporal influences on condition that the best performance of ICF-SG is attained, and ICF-TG investigates geographical influences when ICF-T performs best.

6.2 Experimental Results on Gowalla Dataset

Study of Geographical Constraint: We show the influence of κ by comparing *Precision, Recall* and *F1* for different κ. ICF-SG is meaningful when κ is between 0 and 0.9. Specially, when κ is set to 0, the model reduces to basic ICF model. We increase κ from 0 and take 0.1 as step length. The effect of κ is shown in Fig. 1. Overall, ICF-SG outperforms ICF, which proves that geographical constraint takes a promising effect. When κ increases, ICF-SG performs better. ICF-SG performs best when we set κ to 0.9, which confirms that PoIs having high accordance with our model SCBGMM are more likely to be visited.

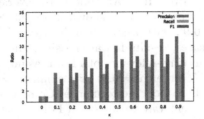

Fig. 1. Performance of ICF-SG

Study of Decisive Influence of Time in Geographical Constraint: We verify the effectiveness of filtering clusters based on time by comparing ICF-SG with ICF-LG. In Fig. 2, we observe clear improvements of ICF-SG in terms of three evaluation metrics and over all values of parameter κ.

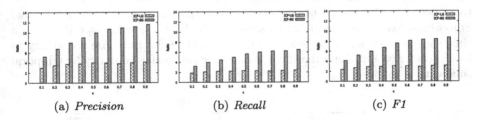

(a) *Precision*   (b) *Recall*   (c) *F1*

Fig. 2. Comparison of performance between ICF-SG and ICF-LG

Study of Number of Gaussian Components: In our model ICF-SG, we do not make any assumption about the number of Gaussian components. However, in model ICF-2G, the number is fixed to 2. By comparison of ICF-SG and ICF-2G showing in Fig. 3, we observe better performance of ICF-SG, and this proves that our model is a more personalized one and can more properly describe individuals' check-in distribution.

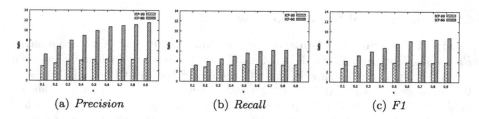

(a) *Precision*   (b) *Recall*   (c) *F1*

Fig. 3. Comparison of performance between ICF-SG and ICF-2G

Study of Temporal Similarity: We tune α to leverage the importance of time-relevance and user's preference as shown in Eq. 11. Same as κ, α is set to be in the interval $[0.0, 0.9]$, with a step length of 0.1. The result is shown as in Fig. 4. When α increases, results fall after rising, and $\alpha = 0.5$ gets the best performance. That means time-relevance of a PoI is as important as current user' preference for the PoI. A large ($\alpha = 0.9$) or a small ($\alpha = 0$) weighted value of temporal similarity leads to poor performance. Compared with previous work in [7], we adopted a different way to smooth a PoI's check-ins distribution over time slots by using temporal similarity rather than a fix window. We calculate similarities between time slots in our dataset. The temporal distribution of each PoI's check-ins is smoothed by Eq. 13. However, smoothed ICF-T takes effect compared with ICF when α is in the interval $[0.1, 0.7]$ but performs worse than ICF-T. As we can see from Fig. 5, the results of smoothed ICF-T show similar tendency with results of original ICF-T when we increase parameter α. However, we only observe slight improvement of smoothed ICF-T compared with original ICF. The reason for the worse performance of smoothed ICF-T (compared with ICF-T) may be that, in our work, temporal distribution of a PoI is used to indicate distinguishing characteristics of PoIs. The number of check-ins in a time slot of a PoI represents an attribute of the PoI's visiting characteristic. Smoothing loses this capability to some extent by sharing the specificity of a time slot with its near time slot. In this way, a time slot may lose the ability to be a special attribute, and the effect of denoting categorial similarity by temporal similarity is reduced.

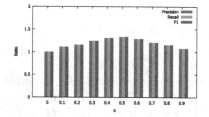

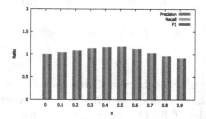

Fig. 4. Performance of ICF-T   **Fig. 5.** Performance of smoothed ICF-GT

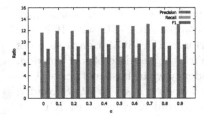

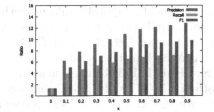

Fig. 6. Performance of ICF-GT   **Fig. 7.** Performance of ICF-TG

Combination of Geographical Constraint and Temporal Similarity: We test the relative importance of time-relevance of PoIs on condition that ICF-SG performs best by fixing parameter κ to 0.9. We set α to best performance value in ICF-T and tune κ to investigate geographical influence. Overall, compared with the best performance of ICF-SG, we observe an improvement of ICF-GT in Fig. 8 for every value of α. Similarly, as shown in Fig. 9 ICF-TG in terms of all values of κ outperforms the best performance of ICF-T, which indicates that with geographical influences, effectiveness of temporal influence is improved. According to the analysis above, we demonstrate mutual reinforcement of geographical and temporal influences and show the superiority of fusing geographical and temporal information.

In the following, we demonstrate the effectiveness of geographical constraint with and without temporal similarity, by comparing ICF-TG with ICF-SG and

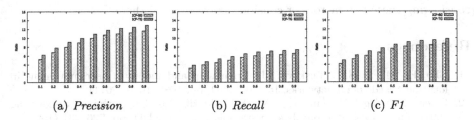

(a) *Precision*   (b) *Recall*   (c) *F1*

Fig. 8. Comparison of performance between ICF-SG and ICF-TG

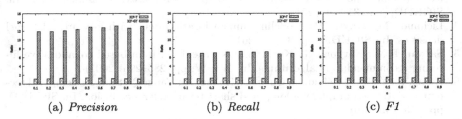

(a) *Precision*   (b) *Recall*   (c) *F1*

Fig. 9. Comparison of performance between ICF-T and ICF-GT

the effectiveness of temporal similarity with and without geographical constraint, by comparing ICF-GT with ICF-T. Firstly, we compare ICF-TG with ICF-SG. Containing temporal similarity, ICF-TG outperforms ICF-SG for each value of κ in terms of each metric, as shown in Fig. 8. With the increase of κ, *Precision*, *Recall* and *F1* of ICF-TG change in a similar tendency with those metrics of ICF-SG. Both of ICF-TG and ICF-SG take best effect when κ is set to 0.9. Secondly, we compare ICF-GT with ICF-T. Involving geographical constraint, ICF-GT outperforms ICF-T clearly for each value of α in all evaluations, as shown in Fig. 9. Variations of three metrics with the increase of parameter α differ a lot between ICF-GT and ICF-T. We can owe this difference to the dominant position of geographical constraint. When comparing ICF-T with ICF-SG, we can draw similar conclusion because the improvement of ICF-SG is more noticeable than that of ICF-T. ICF-GT performs best when α equals to 0.5 in terms of two metrics. In ICF-GT, $\alpha = 0.5$ performs slightly worse than $\alpha = 0.7$.

7   Conclusion

SCBGMM adopts DBSCAN and remove clusters not recently visited to initialize GMM. The proposed ICF-SG based on SCBGMM attains good performance. By comparing ICF-SG with ICF-LG, we prove the effectiveness of filtering based on time. By comparing ICF-SG with ICF-2G, we prove the advantage of varying cluster numbers for individuals. In addition, we take advantage of temporal similarity to describe time-relevance of a PoI and show that time-relevance of a PoI has same importance as preference for the PoI in our model ICF-T. Finally, we combine geographical constraint and temporal similarity to build ICF-GT and ICF-TG, and our experiments show superiority over previous models without considering these two factors or considering only one of them.

References

1. Cho, E., Myers, S.A., Leskovec, J.: Friendship and mobility: user movement in location-based social networks. In: SIGKDD, pp. 1082–1090 (2011)
2. Ester, M., Kriegel, H., Sander, J., Xu, X.: A density-based algorithm for discovering clusters in large spatial databases with noise. In: SIGKDD, pp. 226–231 (1996)
3. Lian, D., Zhao, C., Xie, X., Sun, G., Chen, E., Rui, Y.: GeoMF: joint geographical modeling and matrix factorization for point-of-interest recommendation. In: SIGKDD, pp. 831–840 (2014)
4. Lichman, M., Smyth, P.: Modeling human location data with mixtures of kernel densities. In: SIGKDD, pp. 35–44 (2014)
5. Sarwar, B.M., Karypis, G., Konstan, J.A., Riedl, J.: Item-based collaborative filtering recommendation algorithms. In: WWW, pp. 285–295 (2001)
6. Xiang, L., Yuan, Q., Zhao, S., Chen, L., Zhang, X., Yang, Q., Sun, J.: Temporal recommendation on graphs via long- and short-term preference fusion. In: SIGKDD, pp. 723–732 (2010)
7. Ye, M., Janowicz, K., Mülligann, C., Lee, W.: What you are is when you are: the temporal dimension of feature types in location-based social networks. In: SIGSPATIAL, pp. 102–111 (2011)

8. Ye, M., Yin, P., Lee, W.: Location recommendation for location-based social networks. In: SIGSPATIAL, pp. 458–461 (2010)

9. Ye, M., Yin, P., Lee, W., Lee, D.L.: Exploiting geographical influence for collaborative point-of-interest recommendation. In: SIGIR, pp. 325–334 (2011)

10. Yin, H., Zhou, X., Shao, Y., Wang, H., Sadiq, S.: Joint modeling of user check-in behaviors for point-of-interest recommendation. In: CIKM (2015)

11. Yuan, Q., Cong, G., Ma, Z., Sun, A., Magnenat-Thalmann, N.: Time-aware point-of-interest recommendation. In: SIGIR, pp. 363–372 (2013)

12. Yuan, Q., Cong, G., Ma, Z., Sun, A., Magnenat-Thalmann, N.: Who, where, when and what: discover spatio-temporal topics for twitter users. In: SIGKDD, pp. 605–613 (2013)

13. Yuan, Q., Cong, G., Sun, A.: Graph-based point-of-interest recommendation with geographical and temporal influences. In: CIKM, pp. 659–668 (2014)

14. Zhang, J., Chow, C.: iGSLR: personalized geo-social location recommendation: a kernel density estimation approach. In: SIGSPATIAL, pp. 324–333 (2013)

15. Zhang, J., Chow, C.: GeoSoCa: exploiting geographical, social and categorical correlations for point-of-interest recommendations. In: SIGIR, pp. 443–452 (2015)

A Classification-Based Demand Trend Prediction Model in Cloud Computing

Qifeng Zhou[1], Bin Xia[2], Yexi Jiang[3], Qianmu Li[2(✉)], and Tao Li[3]

[1] Automation Department, Xiamen University, Xiamen, China
zhouqf@xmu.edu.cn
[2] School of Computer Science and Engineering,
Nanjing University of Science and Technology, Nanjing, China
liqianmu@126.com
[3] School of Computer and Information Sciences,
Florida International University, Miami, USA

Abstract. Cloud computing allows dynamic scaling of resources to users as needed. With the increasing demand for cloud service, a challenging problem is how to minimize cloud resource provisioning costs while meeting the user's needs. This issue has been studied via predicting the resource demand in advance. Existing predicting approaches formulate cloud resource provisioning as a regression problem, and aim to achieve the minimal prediction error. However, the resource demand is often time-variant and highly unstable, the regression-based techniques can not achieve a good performance when the demand changes sharply. To cope with this problem, this paper proposes a framework of predicting the sharply changed demand of cloud resource to reduce the VM provisioning cost. In this framework, we first formulate the cloud resource demands prediction as a classification problem and then propose a robust prediction approach by combining Piecewise Linear Representation and Weighted Support Vector Machine techniques. Our proposed method can capture the sharply changed points in the highly unstable resource demand time series and improves the prediction performance while reducing the provisioning costs. Experimental evaluation on the IBM Smart Cloud Enterprise (SCE) trace data demonstrates the effectiveness of our proposed framework.

Keywords: Cloud computing · Capacity planning · Piecewise Linear Representation · Support Vector Machine

1 Introduction

Computing services have become an increasingly popular computing paradigm which provide different styles of services to the cloud resource users with different flavors. Infrastructure as a Service (IaaS), Software as a Service (SaaS), and Platform as a Service (PaaS) are three primary types of cloud computing for both the applications delivered as services over the Internet and the hardware/software systems in the data centers [1].

© Springer International Publishing Switzerland 2015
J. Wang et al. (Eds.): WISE 2015, Part II, LNCS 9419, pp. 442–457, 2015.
DOI: 10.1007/978-3-319-26187-4_41

IaaS cloud is a provision model in which an organization outsources the equipments used to support operations, including storage, hardware, servers, and networking components [1]. In practical application, Iaas is an elastic and economical choice for business IT support. It enables the cloud customers to dynamically request proper amount of virtual machines (VM) based on their business requirements. With the growth of a gigantic number of computing and business server demand, a key issue of IaaS is how to minimize cloud resource provisioning costs while meeting the clients' demands. This is the problem of *effective cloud capacity planning and instant on-demand VM provisioning.*

In general, resource provisioning is challenging due to the pay-as-you-go flexible charging style in IaaS. The amount of resources demand is rarely static, varying as the changes of application number and time. Inefficiency of resource provisioning leads to either over-provisioning or under-provisioning. Over-provisioning may result in idled resources and unnecessary utility costs, while under-provisioning often causes resource shortage and revenue loss. Moreover, initializing a new virtual machine instantly in a cloud is not possible in practice. Therefore, to accomplish effective cloud capacity planning and instant on-demand VM provisioning, application resource needs must be predicted in advance so that the cloud management system can adjust resource allocations in advance.

Capacity planning and instant on-demand VM provisioning problem can be tackled under a unified framework, generally as both problems can be formulated as a generic time series prediction problem [12]. In cloud capacity management, there are two inherent characteristics: **nonlinearity** and **time variability**. The nonlinearity implies that the relationship between the resource demand and its affecting factors is highly nonlinear while the time variability indicates the relationship changes over time. These two characteristics pose a great challenge on effective cloud resource demand prediction.

The existing studies treat the cloud capacity prediction as a regression problem and leverage the state-of-art time series prediction techniques to predict the future capacity of needed resources [8,11]. The Sliding window method [6], Auto Regression (AR) [18], and other methods based on AR such as ARCH(Auto Regressive Conditional Heteroskedasticity) [7], ARMA (Auto Regressive Moving Average) [17] are commonly used techniques to characterize and model observed time series. However, these models are parametric models they only perform well under stable conditions. Artificial Neural Network(ANN) and Support Vector Machine(SVM) regression have also been used to predict the cloud capacity resource demand. These methods decrease the predictive costs compared with the linear regression [9,10].

However, the existing methods can not achieve good performance on resource demand predicting due to the following reasons: (1) The imbalanced demand distribution, dynamic changing requests, and continuous customer turnover make the resource demand highly non-linear and time-varying. Therefore, it is difficult to predict the exact quantity of demand. (2) In practice, the predicting costs are mainly occurred in the cases of sudden changes. However, it is difficult

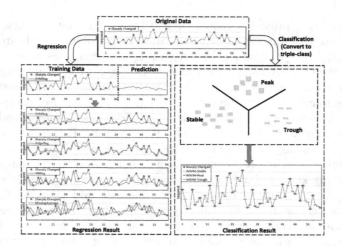

Fig. 1. The illustration of two kinds of cloud resource demand prediction. The left panel includes four commonly used regression-based methods, from top to bottom are Moving average, Nearest neighbour regression, Ridge regression, and SVM regression respectively. The right panel is our proposed PLR-WSVM classification technique. The red line represents real resource demand time series, blue square points are sharply changed demand points, and green line represents the fitting line using different regression methods (Color figure online).

for regression-based methods to capture these changes. (3) Traditional regression cost measures are all symmetric measures [5], but cloud capacity planning is cost sensitive. The estimation for suddenly increasing or decreasing resource demand (noted as peak points and trough points) has different consequences.

In this paper, we propose a framework to address the aforementioned challenges in effective resource provisioning. Our goal is to predict whether the future demand is suddenly changing instead of predicting the actual quantity of demand. First, we formulate the cloud resource demand prediction as a weighted three-class classification problem (peak points, trough points, and stable points). Then, we combine Piecewise Linear Representation (PLR) and Weighted SVM to predict the suddenly changed demand. In addition, we set different weights according to the change rate of the demand, in which the weight reflects the relative importance of each change point.

The main contributions of this paper are describe in Fig. 1. Four commonly used regression-based cloud resource demand prediction methods and their prediction performance on a real world cloud environment are described in the left panel of Fig. 1. Our proposed classification-based method is described in the right panel of Fig. 1. As shown in this figure, commonly used unsupervised regression-based method MA, Nearest Neighbour Regression, Ridge Regression, and SVM Regression cannot capture the suddenly changed points of the resource demand time series. However, our method can identify most of the suddenly changed points and provide a good prediction for all three kinds of points.

The rest of this paper is organized as follows. Section 2 analyses the characteristic of cloud capacity planning problem and briefly introduces PLR and SVM. Section 3 describes the framework of PLR-WSVM. Section 4 presents the experimental results. Finally, Sect. 5 summarizes the paper.

2 Background

To meet the practical demand and reduce the provisioning cost, this paper incorporates PLR and Weighted SVM(WSVM) to predict the change of future cloud resource requirements. PLR is used to extract the peak and trough points, and WSVM is used to model the relationship between the inflection points and the impact factors. We choose these two methods for the following reasons: (1) PLR is simple and the joint points between adjacent segments generated by PLR indicate the change of trends [13–15]. (2) SVM has the excellent generational ability as well as all solutions of SVM model are globally optimal [2, 16].

2.1 Cloud Capacity Planning

Highly Unstable. Effective cloud capacity planning aim to prepare the resources properly. However, unstable customer constituents and the freestyle of resource acquisition/releasing make the cloud resource demand highly unstable. Figure 2 shows the change of the overall customer number over time. As is shown, the total number of customers is continually increasing. Therefore, even the request behaviors of old customers keep stable, the overall request still changes over time. Figure 3 illustrates the request history of three frequently requested customers. We can see that three time series share no common property with each other. As a results, the distributions of the resource demands is highly unstable.

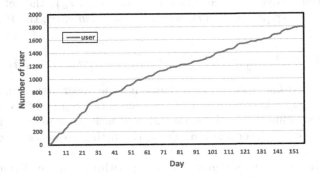

Fig. 2. The change of total cloud service customer over time.

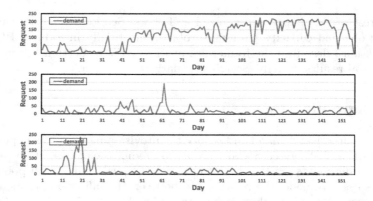

Fig. 3. Time series of resource demands of three frequently requested customers.

Cost Sensitive. Traditional regression-based prediction cost functions such as mean average error (MAE), lease square error (LSE), and mean absolute percentage error (MAPE) are all symmetric measures. In cloud demand prediction, over- and under- prediction will cause different costs, therefore, a symmetric measure is not appropriate for model the asymmetric cost. In this paper, the cloud resource demand prediction is considered as a multi-class classification problem, and we incorporate the different prediction costs as the weights for the samples of different classes. Generally, the weights of peak and trough points should be larger than those of stable points because the predicting costs are increasing when the demand changes suddenly.

2.2   Time Series and PLR

Time series is an ordered set of elements, the element consists of sample values and sample time. Given a time series $T = \{x_1, x_2, ...x_n\}$, the set of segment points is $T_i = (x_{i_1}, x_{i_2}, ..., x_{i_m})$, $(x_{i_1} = x_1, x_{i_m} = x_n, m < n)$, the PLR of T can be described by

$$T_L = \{L_1(x_{i_1}, x_{i_2}), L_2(x_{i_2}, x_{i_3}), ..., L_{m-1}(x_{i_{m-1}}, x_{i_m})\}, \tag{1}$$

where the function $L_{m-1}(x_{i_{m-1}}, x_{i_m})$ represents the linear fitting function at the interval $[x_{i_{m-1}}, x_{i_m}]$. Because the PLR of time series represents a sequence by connecting several linear functions, the value of each point in every interval can be obtained by linear interpolation. Then, the fitting sequence is expressed as $T_i' = (x_1', x_2', ...x_n')$.

Most of the time series segmentation algorithms can be divided into the following three types [13]:

- **Sliding Windows:** A segment is grown until it exceeds some error bound.
- **Top-down:** The time series is recursively partitioned until some stopping criteria are met.

- **Bottom-up:** Starting from the finest possible approximation, segments are merged until some stopping criteria are met. There are two classical ways to find the approximation line [13]:
  - **linear interpolation:** The approximation line for the subsequence $T[a, b]$ is simply the line connecting t_a and t_b.
  - **linear regression:** The approximation line for the subsequence $T[a, b]$ is taken to be the best fitting line in the least squares sense.

2.3 SVM

The main idea of SVM is to generate a classification hyper-plane that separates two classes of data with the maximum margin [2,16,19]. The standard SVM model is as follows:

$$\min_{w,b,\xi_i} \frac{1}{2} \parallel w \parallel^2 + C \sum_{i=1}^{l} \xi_i,$$
$$s.t. \quad y_i(\langle w, \phi(x_i)\rangle x_i + b) \geq 1 - \xi_i,$$
$$\xi_i \geq 0, \quad i = 1, 2, ...l, \tag{2}$$

where $x_i \in R^n$ and $y_i \in \{-1, 1\}$ are respectively the training sample and the corresponding class label, ϕ is a nonlinear map from the original space to a high dimensional feature space, w is the normal vector of hyper-plane in the feature space, b is a bias value, ξ_i is the slack variable, $\langle \cdot, \cdot \rangle$ denotes the inner product of two vectors, and C is a penalty coefficient to balance the training accuracy and generalization ability. The dual form of model (2) is:

$$\min_{\alpha} \frac{1}{2} \sum_{i=1}^{l} \sum_{i=1}^{l} \alpha_i \alpha_j y_i y_j \langle x_i, x_j \rangle - \sum_{i=1}^{l} \alpha_i$$
$$s.t. \quad \sum_{i=1}^{l} y_i \alpha_i = 0$$
$$0 \leq \alpha_i \leq C, \quad i = 1, 2, ...l, \tag{3}$$

The model (3) is an linear SVM method, and it can be easily generalized to non-linear decision rules by replacing the inner products $\langle x_i, x_j \rangle$ with a kernel function $k(x_i, x_j)$. When each training sample has a weight, the standard SVM can be extended to weighted SVM (WSVM) [4], the model (3) is transformed to

$$\min_{\alpha} \frac{1}{2} \sum_{i=1}^{l} \sum_{i=1}^{l} \alpha_i \alpha_j y_i y_j \langle x_i, x_j \rangle - \sum_{i=1}^{l} \alpha_i$$
$$s.t. \quad \sum_{i=1}^{l} y_i \alpha_i = 0$$
$$0 \leq \alpha_i \leq C\mu_i, \quad i = 1, 2, ...l, \tag{4}$$

where $\mu_i(i = 1, ..., l)$ represents the weight of instance x_i. The decision function for WSVM is the same as the standard SVM.

3   The Method

As discussed earlier, a considerable amount of research has been conducted to predict the change of cloud resource demand using regression-based technique. However, due to the characteristics of nonlinearity and time variability, regression-base prediction can only do well in short-term demand prediction. In addition, the predictive errors are usually high when the demand changed suddenly. In this paper, we formulate the cloud demand planning as a classification problem and predict the sharply changed demand.

In this section, we describe the proposed classification-based method named PLR-WSVM. To reduce the time-varying characteristic of resource demand, the whole historic demand dataset is first divided into overlapping training-testing sets. Then, PLR is used to capture the suddenly changed points to form the training dataset, and the weights of the changed points are also assigned according to the changing trend. Finally, WSVM is adopted to build the prediction model.

3.1   The Data Partition

In order to reduce the time-varying feature while maintaining the order of time in time-series data analysis, the whole dataset is often divided into overlapping training-validation-testing sets [3,15]. Suppose the size of whole dataset is m, and the size of each training set and testing set are m_1 and m_2 respectively. Then the whole dataset will be divided into p overlapping training-testing sets:

$$p = \lceil \frac{m - m_1}{m_2} \rceil, \tag{5}$$

where $\lceil x \rceil$ denotes the minimal positive integer that is not less than x.

3.2   Generating the Suddenly Changed Points by PLR

After partitioning the time series dataset into overlapping training-testing sets, PLR is used to automatically generate the suddenly changed demand points. In this work, the top-down algorithm is selected to segment the cloud demand time series and the linear interpolation is adopted to generate the approximation line. The objective segmentation is to produce the best representation such that the maximum error for any segment does not exceed the given threshold δ. The detailed process of PLR is described in Algorithm 1.

Figure 4 presents some examples of using PLR to generate possible suddenly changed points in a period of 120 days. The first subfigure shows the original time series while the rest of the subfigures are generated using different threshold values in PLR. As observed in Fig. 4, the higher the threshold value, the smaller the number of segments generated. For a threshold value of 1.0, there are roughly 65 abrupt changed points while there are only 23 abrupt changed points for a threshold value of 8.0. Each segmentation represents a local peak or trough, and these extremes are transformed into resource demand suddenly changed points.

Algorithm 1. PLR

Input:
  δ: the threshold to decide the point is smooth or not;
  Reqs: the sequence of requests;
Output:
  Label: the type of each point;
1: **for** *index* in *Reqs* (without the first and last point) **do**
2:   **if** $Reqs[index] < Reqs[index + 1]$ **then**
3:     Set $Label[index]$ as pit_{prep}.
4:   **else if** $Reqs[index] == Reqs[index + 1]$ and $Reqs[index] == 0$ **then**
5:     Set $Label[index]$ as $trough_{prep}$.
6:   **else**
7:     Set $Label[index]$ as $peak_{prep}$.
8:   **end if**
9: **end for**
10: Connect the first and last point in *Reqs* with a straight line, and figure out the point P which is farthest from the line. Record the maximum distance as D.
11: **while** $D \geq \delta$ **do**
12:   Update the label of P in *Label* as *trough* or *peak* when the label of P is $trough_{prep}$ or $peak_{prep}$.
13:   Connect the adjacent unstable points (including the first and last even they are treated as stable points) with straight lines, and figure out the points $P_1, P_2, ..., P_n$ which are farthest from the lines in each segmentation. Record the each maximum distance as $D_1, D_2, ..., D_n$.
14: **end while**
15: Update all $trough_{prep}$ and $peak_{prep}$ points in *Label* as 'smooth'.
16: Return *Label*.

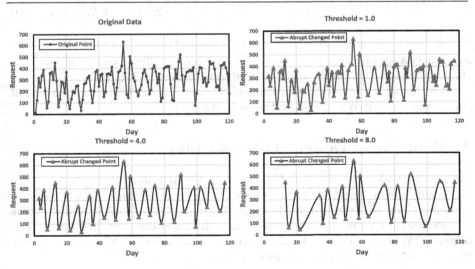

Fig. 4. The possible abrupt changed points generated by PLR

3.3   Constructing Prediction Model by WSVM

We divide all the samples x_i into three classes: peak points(demand suddenly increased points), trough points (demand suddenly decreased points) and stable points(demand changes a little), labeled as 1, 2 and 3, respectively. Furthermore, cloud capacity planning is a cost sensitive problem, the weights of different class instances should be different. According to the cost caused by the error, in model (4), we set

$$\mu_i = \begin{cases} 1 + \alpha & if\ y_i = 1 \\ 1 + \beta & if\ y_i = 2 \\ 1 & if\ y_i = 3 \end{cases} \tag{6}$$

where y_i is the label of x_i, $\alpha = \lambda \cdot \beta$, $\lambda \geq 1$ is a parameter to adjust the cost between peak and trough points.

Then a three-class weighted classification problem can be constructed for each onerlapping training-testing set:

$$T^{(i)} = T^{(i,tr)} \cup T^{(i,ts)}, i = 1, 2, ...p, \tag{7}$$

where

$$T^{(i,tr)} = \{(x_t^{(i,tr)}, y_t^{(i,tr)}, \mu_t^{(i,tr)}) \mid x_t^{(i,tr)} \in R^n, \tag{8}$$

$$y_t^{(i,tr)} \in \{1, 2, 3\}, \mu_t^{(i,tr)} \geq 1, t = 1, 2, ..., m_1\}, \tag{9}$$

and

$$T^{(i,ts)} = \{(x_t^{(i,ts)}, y_t^{(i,ts)}, \mid x_t^{(i,ts)} \in R^n, y_t^{(i,ts)} \in \{1, 2, 3\}, t = 1, 2, ..., m_2\}, \tag{10}$$

denote the training set and testing set respectively, $x_t^{(i,tr)}$ is training sample, $x_t^{(i,ts)}$ is testing sample, $y_t^{(i,tr)}$ and $y_t^{(i,ts)}$ are corresponding class label. $\mu_t^{(i,tr)}$ is the weight of the training sample computed according to Eq. (6).

WSVM is used to model this three-class classification problem. The overall framework of PLR-WSVM is illustrated in Algorithm 2.

4   Experiment Design and Evaluation

We use the real VM trace log of IBM Smart Colud Enterprise to evaluate the effectiveness of our method. The trace data we obtained records the VM requests for more than 4 months (from March 2011 to July 2011), and it contains tens of thousands of request records with more than 100 different VM types. The original trace data include 21 features such as Customer ID, VM type, Requset Start Time, Requset End Time, and etc [12].

Algorithm 2. PLR-WSVM

Input:
    X: Cloud resource demand time series;
    $\delta, r_1, r_2, \alpha, \beta$: the modeling parameters;
Output:
    The testing accuracy, the decision of next day's request (the type of each point);
1: Normalizing the dataset X by $\tilde{x}_i = \frac{x_i - x_{min}}{x_{max} - x_{min}}$;
2: Computing p, the number of partitions according to (5);
3: set $i=1$;
4: **while** $i \leq p$ **do**
5:    Selecting the ith training set and testing set from X;
6:    Generating the three-class sample points by Algorithm 1;
7:    Setting the weights of each instance in the ith training set according to (6);
8:    Training a three-class WSVM model from the ith training set according to (4);
9:    Predicting the labels on ith test set.
10:   Set $i=i+1$;
11: **end while**
12: Computing the test accuracy;
13: Return *Label*;

4.1   Data Preprocessing

There are two data preprocessing steps before the raw data recorded by SCE are used in modeling and prediction:

– **Feature Selection:** The raw data contain some request fields that are used during prediction and also contain some noise during temporal pattern mining. Therefore, not all these twenty-one features of a request record are useful. In this work, we only selected two original features, VM Type (which illustrates the type of VM the customer requests), Company (which include the information of the customer send the request) and four statistics features obtained from history data as the feature subset. The details of features involved in our experiments are described in Table 1.
– **Time Series Aggregation Granularity Selection:** The raw trace data are recorded per second. Aggregate these time series by different granularities would have different levels of information and difficulty for prediction. A too fine granularity would make the value on each timestamp lack statistical significance, however, too large granularity would loss some useful information. Figure 5 shows the different cloud capacity provisioning time series aggregated by hour, day and week, respectively. We can see that the coarser the granularity, the larger the provisioning amount in each time slot. Since the lifetime of a VM is usually longer than hours, aggregate the records by hour is not suitable in practice. On the contrary, if we aggregate the time series by week, the cloud required to prepare the most VMs for each time slot. In this case, the small prediction deviation will result in a large cost [12]. In order to produce the enough modeling data while maintaining the statistical significance of raw time series, in this work, we use the daily time series in our system.

Table 1. The description of features

Field	Description
request1Part	The number of request in current time period
requestAvg	The average of request counts in fixed period recently
requestVar	The variance of request in fixed period recently
requestLastWeek	The number of request in the same time period last week
requestSubject	The subjects of request currently (e.g., types of VM)
requestCompany	The companies of request currently

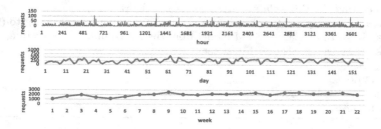

Fig. 5. Time series aggregation granularity selection.

4.2   Experiment Results

We compare the proposed PLR-WSVM method with several commonly used regression-based methods. The detail of methods are described in Table 2.

Regression-Based Methods vs. Real Time Series. Figure 6 displays the original capacity change time series in three different periods and the fitting results using different regression-based methods. From top to bottom, the time series are: (1) Time series predicted by Moving Average; (2) Time series predicted by Nearest Neighbour regression; (3) Time series predicted by Ridge regression; (4) Time series predicted by SVM regression. From left to right, the unstability of three parts of time series is generally increased (i.e., from low to high).

In Fig. 6, we can see that all the regression-based methods can predict well only when the real time series are smooth. With the increasing of unstability

Table 2. The description of methods

Method name	Description
Moving Average	Naive Predictor
Nearest Neighbour Regression	Linear Regression
Ridge Regression	Non-linear Regression
SVM Regression	Non-linear Regression with RBF kernel

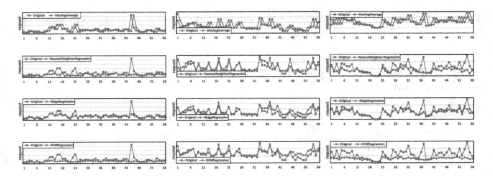

Fig. 6. Regression-based prediction results of three parts of time series.

(sharply changed demand), the fitting error is also increasing. These regression-based methods can give a good prediction in the average sense. For those sharply changed points, regression-based methods cannot predicting well. Among these regression techniques, Moving Average shows a most similar changing tendency with original time series, but its predicting curves have a time delay, limiting its applicability in practice. Ridge Regression has the better performance than Nearest Neighbour regression and SVM regression techniques, but it also cannot predict the sharply changed points well in unstable time series.

PLR-WSVM vs. Regression-Based Methods. The goal of this paper is to study a model predicting the suddenly changed points of cloud resource demand. PLR-WSVM as a classification-based technique can give the results directly. However, regression-based methods must do the following two steps: fitting the original data and setting a threshold to decide whether the next demand is a suddenly changed point or not. It is difficult to set the decision threshold because the resource demand varying from one moment to another.

Table 3 shows the performance of PLR-WSVM compared with other regression-based methods. The experimental setup is using one week time series to predict the demand of next day. To ensure the comparability, the predicting results of regression-based methods are transformed into three classes by comparing the relative change of resource demand with a proper threshold. The transform rule is defined as:

$$label(x_t) = \begin{cases} 1 & if\ c(x_t) \geq\ \theta \\ 2 & if\ c(x_t) \leq -\theta \\ 3 & if\ |c(x_t)| < \theta \end{cases} \qquad (11)$$

where $c(x_t)$ represents change rate of resource demand between the current regression value and the previous one, defined as

$$c(x_t) = \frac{R(x_t) - R(x_{t-1})}{R(x_{t-1})}, \qquad (12)$$

where $R(x_t)$ and $R(x_{t-1})$ indicate the current and previous regression value. θ is the threshold to transform the relative change rate into a label.

Table 3. The comparison of predicted changed point between PLR-WSVM and other regression-based methods.

Method	Threshold	Recall accuracy of Peak(%)	Recall accuracy of Trough(%)
SVMReg	0.5	7.2	4.5
	1.0	0.8	0.0
RidgeReg	0.5	20.7	5.6
	1.0	3.8	0.0
NNReg	0.5	11.2	7.1
	1.0	2.0	0.0
MovingAverage	0.5	0.0	0.0
	1.0	0.2	0.0
PLR-WSVM	-	37.42	37.91

We compared the range of threshold θ from 0.1 to 1.5, Table 3 shows the experiment results. From this table, we can see that PLR-WSVM has the best performance in predicting the troughs and peaks of cloud resource demand while maintain a comparable overall accuracy. Regression-based techniques are very sensitive to the thresholds and have poor performance in predicting suddenly changed points.

We also compare the long term prediction performance of different methods (e.g., using one month time series to predict resource demand of next week). Figure 7 illustrates the prediction results. Compared with short term predicting results, we can see that the predict performance of PLR-WSVM is increased when the period has been extended from one day to one week. The performance of other regression-based methods decreased greatly, especially on trough points. The experimental results indicate that PLR-WSVM has more robust long term prediction ability.

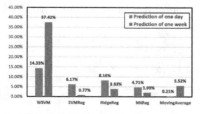
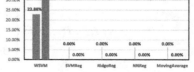

(a) Prediction results for peaks   (b) Prediction results for troughs

Fig. 7. The performance compare on different time span. The left figure is the prediction results of peaks and the right figure is the prediction results of troughs.

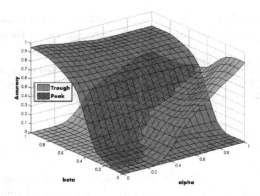

Fig. 8. The effect of turning α and β for capacity prediction.

In addition, PLR-WSVM can balance the cost among different classes of points in cloud resource demand. According to Eq. (6), we can set different weights to trough and peak points using $\alpha = \lambda\beta$. In cloud predicting, since the cost of under-prediction is large than over-prediction, we set $\lambda \geq 1$.

Figure 8 shows the trough and peak points prediction with varying α and β. We can see that the prediction accuracies of trough and peak points have different variation trends. In practice, cloud providers can tradeoff these two costs according the real demand to minimize the total cost.

5   Conclusion

The prediction of cloud resource demands is a very challenging task due to the time-variant and highly unstable characteristics. Traditional regression-based techniques cannot achieve good prediction performance, especially when resource demand changed sharply. In this paper, we discuss the cloud capacity planning problem from a new perspective: predicting the sharply increased and decreased resource demand. Thus the service vendors can cope with the abrupt changed cloud resource demand in advance and improve the quality of cloud service.

We transform the cloud capacity planning problem into a classification problem and use PLR-WSVM to predict the trough and peak points. In particular, PLR is used to generate the training samples from the original resource demand time series, and then WSVM is used to model the prediction of sharply changed demand. Unlike regression-based techniques, our method formulates the cloud resource demand into a three-class classification problem and it does not need to determine the threshold of trough and peak points. Furthermore, WSVM can assign the different weights for peak and trough points to minimize the total provisioning costs.

Experimental results on the trace data of IBM Smart Cloud Enterprise demonstrate the effectiveness of our proposed method. Compared with regression-based techniques, our proposed method achieves more accurate and robust prediction performance on suddenly changed cloud resource demand.

Acknowledgement. This work is supported by Natural Science Foundation of China under Grant No. 61503313 and the Jiangsu Key Laboratory of Image and Video Understanding for Social Safety (Nanjing University of Science and Technology), Grant No. 30920140122007.

References

1. Armbrust, M., Fox, A., Griffith, R., Joseph, A.D., Katz, R., Konwinski, A., Lee, G., Patterson, D., Rabkin, A., Stoica, I., et al.: A view of cloud computing. Commun. ACM **53**(4), 50–58 (2010)
2. Burges, C.J.: A tutorial on support vector machines for pattern recognition. Data Min. Knowl. Discov. **2**(2), 121–167 (1998)
3. Cao, L.J., Tay, F.E.H.: Support vector machine with adaptive parameters in financial time series forecasting. IEEE Trans. Neural Netw. **14**(6), 1506–1518 (2003)
4. Chang, C.C., Lin, C.J.: LIBSVM: a library for support vector machines. ACM Trans. Intell. Syst. Technol. (TIST) **2**(3), 27 (2011)
5. Chatfield, C.: The Analysis of Time Series: An Introduction. CRC Press, Boca Raton (2013)
6. Dietterich, T.G.: Machine learning for sequential data: a review. In: Caelli, T.M., Amin, A., Duin, R.P.W., Kamel, M.S., de Ridder, D. (eds.) SPR 2002 and SSPR 2002. LNCS, vol. 2396, pp. 15–30. Springer, Heidelberg (2002)
7. Engle, R.F.: Autoregressive conditional heteroscedasticity with estimates of the variance of United Kingdom inflation. Econometrica J. Econometric Soc. **50**(4), 987–1007 (1982)
8. Hamilton, J.D.: Time Series Analysis, vol. 2. Princeton University Press, Princeton (1994)
9. Iqbal, W., Dailey, M.N., Carrera, D.: Black-box approach to capacity identification for multi-tier applications hosted on virtualized platforms. In: 2011 International Conference on Cloud and Service Computing (CSC), pp. 111–117. IEEE (2011)
10. Islam, S., Keung, J., Lee, K., Liu, A.: Empirical prediction models for adaptive resource provisioning in the cloud. Future Gener. Comput. Syst. **28**(1), 155–162 (2012)
11. Jiang, Y., Perng, C.S., Li, T., Chang, R.: ASAP: a self-adaptive prediction system for instant cloud resource demand provisioning. In: 2011 IEEE 11th International Conference on Data Mining (ICDM), pp. 1104–1109. IEEE (2011)
12. Jiang, Y., Perng, C.S., Li, T., Chang, R.N.: Cloud analytics for capacity planning and instant vm provisioning. IEEE Trans. Netw. Serv. Manage. **10**(3), 312–325 (2013)
13. Keogh, E., Chu, S., Hart, D., Pazzani, M.: An online algorithm for segmenting time series. In: Proceedings IEEE International Conference on Data Mining, ICDM 2001, pp. 289–296. IEEE (2001)
14. Keogh, E., Chu, S., Hart, D., Pazzani, M.: Segmenting time series: a survey and novel approach. Data Min. Time Ser. Databases **57**, 1–22 (2004)
15. Luo, L., Chen, X.: Integrating piecewise linear representation and weighted support vector machine for stock trading signal prediction. Appl. Soft. Comput. **13**(2), 806–816 (2013)

16. Vapnik, V.N., Vapnik, V.: Statistical Learning Theory, vol. 1. Wiley, New York (1998)
17. Whitle, P.: Hypothesis Testing in Time Series Analysis. Almqvist & Wiksells, Uppsala (1951)
18. Yule, G.U.: On a method of investigating periodicities in disturbed series, with special reference to Wolfer's sunspot numbers. Philos. Trans. Roy. Soc. Lond. Ser. A **226**, 267–298 (1927). Containing Papers of a Mathematical or Physical Character
19. Zhou, Q., Hong, W., Shao, G., Cai, W.: A new SVM-RFE approach towards ranking problem. In: IEEE International Conference on Intelligent Computing and Intelligent Systems, ICIS 2009, vol. 4, pp. 270–273. IEEE (2009)

Query Monitoring and Analysis for Database Privacy - A Security Automata Model Approach

Anand Kumar[1(✉)], Jay Ligatti[2], and Yi-Cheng Tu[2]

[1] Teradata Corporation, San Diego, CA 92127, USA
anand.kumar@teradata.com
[2] Department of Computer Science and Engineering,
University of South Florida, Tampa, FL 33620, USA
{ytu,ligatti}@cse.usf.edu

Abstract. Privacy and usage restriction issues are important when valuable data are exchanged or acquired by different organizations. Standard access control mechanisms either restrict or completely grant access to valuable data. On the other hand, data obfuscation limits the overall usability and may result in loss of total value. There are no standard policy enforcement mechanisms for data acquired through mutual and copyright agreements. In practice, many different types of policies can be enforced in protecting data privacy. Hence there is the need for an unified framework that encapsulates multiple suites of policies to protect the data.

We present our vision of an architecture named security automata model (SAM) to enforce privacy-preserving policies and usage restrictions. SAM analyzes the input queries and their outputs to enforce various policies, liberating data owners from the burden of monitoring data access. SAM allows administrators to specify various policies and enforces them to monitor queries and control the data access. Our goal is to address the problems of data usage control and protection through privacy policies that can be defined, enforced, and integrated with the existing access control mechanisms using SAM. In this paper, we lay out the theoretical foundation of SAM, which is based on an automata named Mandatory Result Automata. We also discuss the major challenges of implementing SAM in a real-world database environment as well as ideas to meet such challenges.

Keywords: Automata · Access control · Differential privacy · Security

1 Introduction

Data is often the most valuable asset to its owner or person whose information is captured in it. Certain information is private and should not be disclosed in any form. Various laws and mutual agreements between parties require the organizations to set strict policies in disclosing certain information. For example, laws

---

Anand Kumar—This work was done at University of South Florida.

J. Wang et al. (Eds.): WISE 2015, Part II, LNCS 9419, pp. 458–472, 2015.
DOI: 10.1007/978-3-319-26187-4_42

such as *Health Insurance Portability and Accountability Act (HIPAA)* mandate the organizations not to disclose personally identifiable information under any circumstances. In the past, enormous efforts have been made to control data access, optimize queries, and process queries efficiently. Little efforts are put to monitor queries to protect against privacy violations occurring through improper data usage [18].

On the other hand, there is the need to find out inherent patterns in the stored data for targeted advertising, marketing, and other business purposes. It is achieved through statistical analysis of the stored data, often done by external analysts. Whether the analysis is done internally or externally, it should follow the aforementioned rules in leaking private information. Thus, privacy-preserving data analysis/mining has become an active field of research and practice. A common practice is to anonymize the data [9] before disclosing to third party analysts. Unfortunately, the anonymous data sets can be combined with information from other sources to extract the private information. In one such example of attacks, users in an anonymized Netflix prize data were identified by using an IMDB movie rating data set [15].

Data protected by mutual agreements and copyright laws needs to be monitored for usages that may affect the businesses. For example, hobbies information of Facebook users should not be combined with Google maps location services. Even if the business has access to both data sets, it should not allow its employees or users to combine them. It may result in violation of privacy as well as affect the data owners financially. Therefore, monitoring data usage becomes an important task for organizations [18].

A number of privacy-preserving policies have been defined and enforced in current database management systems (DBMS) through access control mechanisms (ACM). The main limitation of these ACMs is that they only enforce "black and white" policies; the user is either granted access to the information or completely denied. In the presence of such mechanisms it is challenging to allow third-party analysts to access the database because the ACMs would have to restrict access to private information. Effective analysis is possible if some form of data aggregation, from sources who have private information, is made accessible, without revealing the private information. It is also important to monitor the accesses made by legitimate users. For example, a valid user may access, out of curiosity, important information about important personalities (or celebrities). Such unnecessary accesses should be monitored.

Our vision is to introduce an architecture named security automata model (SAM), shown in Fig. 1, to enforce privacy policies and restrict data usage. An important feature of the SAM architecture is that **we can enforce many types of database privacy policies within a single enforcement architecture.** The SAM model basically enforces ACMs and other declared privacy policies on the data and query results. SAM is a query monitor based on previous work in the area of software security by Ligatti and Reddy, namely, an automata called Mandatory Result Automata (MRA) [11]. All the actions taken by the DBMS for processing input queries and then returning their results are inspected before presenting the results to the user. This inspection feature of SAM is very useful in monitoring authorized users accessing private information out of curiosity. The results that may generate or help infer confidential information are substituted or perturbed by the monitor through differential privacy (DP) [5].

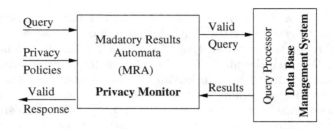

Fig. 1. Architecture: Basic monitor model

SAM works on the basic aggregates supported by state-of-the-art DBMS, while controlling access through ACMs. The data owner can specify policies on these aggregates rather than writing their own interfaces to protect the privacy. This eliminates need to inspect, monitor, and control the programs written by external analysts. The analysts don't have to have any knowledge of special parameters and accuracy requirements, to access the data. The major contributions of this work are:

1. A new privacy enforcement architecture parameterized by various privacy policies.
2. Utilizing SAM to monitor data usage and preserve privacy in databases.
3. Formal model of the architecture using mandatory results automata (MRA).
4. Proof that security mechanisms modeled in our framework can be constructed to enforce any of a large class of privacy policies.

Details of the problem studied and a brief summary of related work are presented in Sect. 2 and the privacy architecture is discussed in Sect. 3. A formal model to enforce policies and challenges involved are discussed in Sect. 4. We conclude our paper in Sect. 5 with some details of possible future work.

2   Problem Description and Related Work

SAM is designed for protection of data stored in relational database management systems (RDBMS). The RDBMS would incorporate SAM as an additional layer to monitor the data usage and preserve privacy. We envision the following three problems to be addressed using SAM.

1. **Monitoring Users:** Access to database from both unauthorized and authorized users must be monitored. The problem of restricting unauthorized users

falls into the realm of ACMs and hence we do not cover it in our work. Some authorized users may access private data out of curiosity. Such accesses must be monitored and appropriate action be taken to preserve privacy. Thus, SAM should automatically audit authorized users accessing data that should not be accessed during normal operations.

2. **Monitoring Data Usage:** Often business agreements and terms of usage restrict the way in which data is used. For example, the use of maps with location data could be restricted. The users of such data must be monitored and combining one data set with other should not be allowed. In such cases, prohibited operations on data are not allowed.

3. **Enforcing Policies:** Various privacy and data usage policies are expressed as predicates that are enforced by the mechanism implemented in SAM. The database administrator declares the policies to be enforced in a query language. Then, SAM enforces ACMs and declared privacy policies to protect the data.

Even though various policies are enforced on every query submitted by a user, there is need to monitor the queries based on history of data accesses. A sequence of relevant queries may give out private information. Therefore, SAM should enforce policies by looking at the history of queries submitted to the database. It becomes more challenging when multiple users collude to access private data. We look at the architecture of SAM to address these problems.

2.1   Related Work

The problem of security in databases is well studied in the past three decades. The security mechanisms of the past are categorized into four classes [1,14]: Conceptual Models, Query Restriction, Output Perturbation, and Data perturbation. A detailed survey is presented in [1]. In the recent past an extensible platform for privacy-preserving data analysis has been studied [12]. There have also been some systems built to enforce privacy-preserving policies [13,16]. These systems have focused mainly on providing database interfaces for the programs of data analysts. The data owner makes settings (specify parameters like privacy budget etc.) so that the programs don't reveal the private information. It is often hard to inspect the bugs and possible privacy breaches in programs written by external agents.

As there are different approaches to preserving privacy in databases, each approach has its own advantages and disadvantages. In our privacy architecture, we have presented a monitoring mechanism that we believe combines the best features of these approaches. It is capable of restricting the queries during normal operation and perturbing the results when it is absolutely necessary. The data owner can define all the privacy policies and input them to the monitor. The privacy model enforces the policies while monitoring external programs for breach in the data privacy. To the best of our knowledge, ours is the first framework for enforcing many types of database privacy policies within a single enforcement architecture.

3   The SAM Architecture

Any database query can be expressed as a relational algebra expression (or **expression** in short) on the values of database table attributes. The set of records R_C whose values satisfy the selection condition C in a given expression is called the **query set**. The basic aggregates used as examples in this model are: averages (AVG), sums (SUM), counts (COUNT), maximums (MAX), and minimums (MIN). The AVG query, and many other statistics, can be computed from the results of these basic aggregate queries. However, the model can be extended to support any type of complex statistical queries.

In this section, we present the basic privacy architecture for databases, using security automata. The runtime policy-enforcement mechanism works by monitoring the queries submitted to the database and the results returned. Runtime mechanisms are quite popular in monitoring database access-control policies, firewalls, operating systems, spam filters, web browsers, etc. We propose a privacy-monitor architecture for DBMS, based on the mandatory results automata (MRA) [11] to enforce the privacy policies in databases, as shown in Fig. 2. The monitor is in-lined with the DBMS, interposing on the input queries and output results. The monitor enforces policies by observing all the input queries and their results. It transforms all queries and results to ensure that the queries processed by the DBMS and the results to be returned to the users are valid. A **valid action** is actually a query or result that satisfies the desired policies of the DBMS. Some of the policies that are enforced by this architecture are discussed in Sect. 4.

Monitor Input Queries: Every input query goes through the *access control* module. Database access is granted only if the user submitting the query has authorization. Otherwise, the query is rejected. Once authorized, the query is inspected to see if the number of entities involved does not exceed the limit set by the database administrator/owner. It is one of the methods to restrict data access that is possible through a sequence of queries [4]. A history of entities and their attributes is maintained to check the overlapping entities in the input queries. A new entry is made in the log of a *history tracker* whenever a valid input query passes through the *entity check* module.

The aggregate queries are modified into their basic components, before submitting to the DBMS, by the *query rewrite* module. For example, an AVG query is split into COUNT and SUM queries. The results of these component queries are monitored to protect privacy.

Monitor Query Results: The results of component queries are inspected by different modules to detect possible situations in which the private information could be revealed. For example, the result of COUNT component of the AVG query is inspected by the *set size restricter*. This module restricts queries from disclosing results with very few or just one tuple [7]. A valid result is inspected by the *query tracker* module for presence of any tracker. A tracker is a query with auxiliary conditions padded to the original conditions that are invalid (or not allowed by

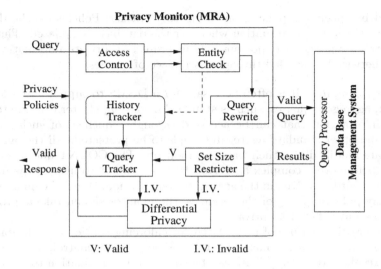

Fig. 2. Architecture: Detailed MRA

the DBMS) [3]. A valid response from the query tracker returns results to the user. If the query set size restricter or the query tracker detect any possibility of revealing private information, the *differential privacy* module perturbs the results. Output perturbation [1,14] is necessary to make sure that nothing is inferred by the user when the results are delivered. We enforce this requirement through differential privacy [5,12].

Definition 1. Differential privacy [5]: *Any randomized algorithm f is said to provide ε-differential privacy if for all data set instances* $G, H \in D^n$ *differing in at most one record, and any set of possible outputs* $S \subseteq Range(f)$,

$$\Pr[f(G) \in S] \le e^\epsilon \Pr[f(H) \in S]$$

where D is the domain of data records, and e is Euler's number (2.71...).

The parameter ϵ is called the *privacy budget* and is a basic requirement for enforcing differential privacy. The accuracy of the results is inversely proportional to the budget value, and hence the privacy guarantee. The database owner can fix the budget requirement through policy predicates that are enforced by SAM. Thus, the analysts do not have to specify any budget or accuracy requirements. This gives the data owner power to control the accuracy precisely to protect the private information.

The history tracker records information about the attributes that are answered in the past queries. The overlap of attributes and results in a sequence of queries is computed and stored for future queries [4]. Whenever a query result violates tracker policy, the monitor applies differential privacy, making it a valid query and protecting the information. The size of the stored history can be managed by specifying policies on the history tracker. Policies can be defined to discard the stored information whenever the database is updated. Thus, the history tracker plays an important role in protecting private information that could otherwise be accessed through a sequence of queries.

Complex Aggregates: It is often necessary for a DBMS to support user-defined, complex, aggregate queries, such as standard deviation, cluster center, etc. The private information that can be accessed through sequences of such complex queries and other standard aggregates needs to be protected. All the aggregate queries are modified into their basic components. The COUNT query is always inserted before every complex aggregate by the monitor, to inspect the number of unique records involved in the aggregate computation. When the number does not satisfy policies of any of the modules, the monitor should take appropriate action to apply differential privacy.

An interesting feature of using MRA for enforcing privacy is their ability to support many policies, such as differential privacy, access control, mechanism to sanitize results, etc. [11]. Policies can be specified in combination to strengthen monitoring process. For example, set-size-restricter and differential-privacy modules, shown in Fig. 2 can work together to decide when to perturb the results. It is also possible to plug-in new modules in this architecture to enforce complex policies, enabling users to tune the level of control. We look into details of the formal model in the following section.

4   Policy Enforcement

The power of SAM is in enforcing various privacy and data use policies. It has various modules that can be utilized to enforce policies effectively. We have identified three classes of policies that are necessary to protect privacy and data usage: query control, result control, and access control.

Query control policies identify various entities and attributes involved in the queries and check for violations of predefined conditions. Number of entities present in the queries can be monitored to identify any privacy violations. Restrictions related to allowed aggregates and joins can also be enforced.

Result control policies control the amount of data accessed by restricting the number of records returned in the query results. A malicious user may issue sequence of queries to access as much data as possible. However, SAM's *history tracker* monitors such activities and denies access. Often it is important to allow full access to the data for processing complex aggregates. In such cases, to preserver privacy, differential privacy is applied to obfuscate the results.

Access control policies are enforced by any standard RDBMS. However, in order to monitor authorized users accessing private data (e.g. celebrity), additional information is required. SAM can store such additional information in a SAM-specific private database and use it to monitor such activities.

It can be noticed that the requirements of data use restrictions fall into all of the above mentioned policy categories. Therefore, the problem of monitoring data use is challenging. Any new policies that can be enforced by SAM fall into one of the above mentioned categories. It is also important that there is need for a query language to define such policies. An example policy P1 to dis-allow entities containing user location and hobbies is shown below.

```
P1: RESTRICT ENTITIES UserLocation, UserHobbies
P2: CHECK HISTORY ( SELECT Entities, Attributes
                    FROM History )
    WHERE Query.Entities IN History.Entities AND
          Query.Attributes IN History.Attributes
```

Policy P2 checks the history of queries and returned results to check if any privacy violations have occurred. Specifying policies in a user friendly language is one of the challenges in enforcing data use restrictions and preserving privacy. Size of history could pose performance issues in efficient query evaluation. In Sect. 4.4 we discuss few possible solutions to overcome these difficulties.

4.1   Formal Model

The DBMS can be abstractly defined in terms of the queries it can process (actions) and the possible results of those queries (results). Any request from users of the DBMS is an aggregate query to be executed. The response to the queries are results computed by the system. We represent the set of actions on a database using the metavariable A, and results to these actions using R. An action can be a query submitted to DBMS. Both sets A and R are nonempty, possibly countably infinite, and $R \cap A = \emptyset$. An event in the DBMS is a query or a result, and it is denoted using E, where $E = A \cup R$.

The sequence of submitting queries to the DBMS and obtaining their results is called an **execution**. Each execution x is defined as a sequence of events with event set E. The set of all finite-length executions possible is denoted by E^*. A session is modeled as an execution. If the session terminates then the execution is finite; otherwise the execution is infinite. Only one infinite-length execution is allowed per user in our system. We denote the empty execution as ε. When an execution x' follows another execution x we denote it by concatenation, $x; x'$. When x is a prefix of x' it is written as $x \preceq x'$. Throughout this paper we abbreviate the formula $\forall x' \in E^* : x' \preceq x \Rightarrow F$ as $\forall x' \preceq x : F$. Finally, we notate the final query in execution x as Q_x.

An MRA M is a quadruple denoted as (E, S, q_0, δ), where E is the event set over which M operates, S is the set of possible states (finite or countably infinite) of M, q_0 is the initial state, and δ is a transition function of the form $\delta : S \times E \to S \times E$. An MRA in its current state takes an event, either an input query or result of a query, and transitions to new state and produces an output. The output can be a valid query to be executed by the query processor, or the result to be presented to the user (Fig. 2). We write $M \Downarrow X$ when M **produces** an execution $X \subseteq E^*$ for input events that match the sequence of input events in X. Section 4.2 will define enforcement as requiring an MRA to produce exactly those executions that the desired policy allows.

MRA treat all actions as synchronous i.e., they finish processing the input action and return results before accepting the next input for processing. Their ability to transform the results of actions is novel and crucial for enforcing policies. This behavior is essential in DBMS and matches our requirement of enforcing privacy-preserving policies.

4.2   Privacy Policies

The SAM is able to enforce any policies that can be defined on the DBMS. In this section we look into different policies that are to be enforced on the queries. The important aspect of the MRA-based architecture is that the MRA can adapt itself to the changing queries over time by enforcing appropriate policies. The policies listed here are well studied in the literature [1, 2, 12], but have never been enforced collectively for protecting privacy.

A **policy** is a predicate defined over executions. A session (execution) $X \in E^*$ is said to satisfy policy P iff $P(X)$.[1]

With the definitions of policies and MRA producing executions we are now ready to define what it means for an MRA to enforce a policy. Definition 2 says that an MRA M enforces a policy P exactly when the set of executions M produces equals the set of execution P allows.

Definition 2. *An MRA M on a system with event set E **enforces** policy P iff*

$$\forall x \in E^* : (M \Downarrow x) \Longleftrightarrow P(x)$$

We next describe different policies that can be enforced by MRA to protect privacy.

Query and Result Control Policies: In this category of privacy policies, the results returned by aggregates depend on the types of attributes involved in the queries. If the attributes involved carry private information, M should enforce the control policies. We examine broad categories of policies on aggregates that M is able to enforce to protect privacy.

Definition 3. *Any execution $x \in E^*$ is said to satisfy the **query-set-size** restriction policy iff there exists a predicate P_s such that,*

$$P_s(x) \Longleftrightarrow \forall x' \preceq x : K \le |R_{x'}| \le N - K$$

where $R_{x'}$ is the query-set (result) of query $Q_{x'}$, N is the number of records in the database, and the value of K is restricted by $0 \le K \le N/2$.

---

[1] Technically, these policies are called "properties" in the literature on formal security models [10, 11, 17].

Query-set-size policies restrict the size of the query set returned by the query processor, to control the information disclosed to users [7]. Such policies can prevent results with very few or just one tuple. The set-size-restricter module of SAM enforces this property. However, when two queries are executed consecutively to get two query-sets G and H respectively, the privacy policy can be violated, as the set difference $|G - H| = 1$. Hence, additional measures should be taken to protect private data.

Definition 4. *Any execution* $x \in E^*$ *is said to satisfy the* **query-set-overlap** *restriction policy iff there exists a predicate* P_o *such that,*

$$P_o(x) \Longleftrightarrow \forall x' \preceq x, \ \forall x'' \preceq x : \left(Q_{x'} \neq Q_{x''} \Longrightarrow |M_{x'} \cap M_{x''}| \leq O \right)$$

where $M_{x'}$ *is the set of entities involved in* $Q_{x'}$ *(and similarly* $M_{x''}$ *for* $Q_{x''}$*) and* O *is the restricted number of overlapping entities.*

In this policy, the number of entities that overlap in successive queries are restricted [4]. When a sequence of queries is executed, the policy is either satisfied or not based on the size of the overlap. Given a minimum query set size K and the number of entity overlaps O allowed in successive queries, the minimum number of queries required to compromise is $1 + (K - 1)O$. Our monitor M can be made to look for this number to secure the data. The entity-check module of SAM is capable of enforcing this policy.

A restricted query result can be calculated with the help of two kinds of trackers [3]. One, using a general tracker when the query-set-size $K \leq N/4$. A policy to preserve privacy in the presence of a tracker can be enforced with the help of set-size restricter in M by changing the set-size conditions in Definition 3 to

$$P_s(x) \Longleftrightarrow \forall x' \preceq x : 2K \not\leq |R_{x'}| \not\leq N - 2K$$

The second type of tracker, called a double tracker, can reveal private information when the query-set-size $K \leq N/3$. This can be prevented with the help of the query tracker in M enforcing the desired tracker policy.

Definition 5. *Any execution* $x \in E^*$ *is said to satisfy the* **tracker** *policy iff there exists a predicate* P_t *such that,*

$$P_t(x) \Longleftrightarrow \forall x' \preceq x, \ \forall x'' \preceq x :$$
$$\left(Q_{x'} \neq Q_{x''} \Longrightarrow \begin{pmatrix} C_{x'} \cap C_{x''} = \emptyset \\ \vee \ K \not\leq |R_{x'}| \not\leq N - 2K \\ \vee \ 2K \not\leq |R_{x''}| \not\leq N - K \end{pmatrix} \right)$$

where $C_{x'}$ *and* $C_{x''}$ *are the selection conditions of the queries* $Q_{x'}$ *and* $Q_{x''}$ *respectively,* N *is the number of records in the database, and the value of* K *is restricted by* $0 \leq K \leq N/3$.

These trackers speculate the privacy compromise by computing all possible combinations of the table attributes that may be queried in the future to get answers to restricted queries. The "query tracker" and "history tracker" modules of M are responsible for enforcing tracker policies. Any algorithm to track quires can be plugged-in.

Differential Privacy Policies: The data owner can restrict the access and accuracy of user queries to the DBMS using differential privacy policies. The idea behind this mechanism is to allow user queries to execute on the actual data, but the results are perturbed before being delivered to the user. The perturbation is a function (differential privacy) $f(R_x)$ on the results R_x of any query $x \in E^*$. The important goal of perturbation is to prevent disclosing the correct value(s) of the result(s). Perturbation may include rounding up (or down) of the resulting values or adding (or subtracting) some pseudo-random numbers [7].

Definition 6. *Any execution* $x \in E^*$ *is said to satisfy the **differential privacy** policy iff there exists a function* f *satisfying Definition 1 such that,*

$$P_d(x) \Longleftrightarrow \forall x' \preceq x : R_{x'} = f(R_{x'}^{orig})$$

where $R_{x'}$ *is the query-set (result) of* $Q_{x'}$ *and* $R_{x'}^{orig}$ *is the original result the DBMS returned for* $Q_{x'}$.

The use of output perturbation satisfies the differential privacy policy, as the actual results are never disclosed to the users. The differential privacy function f perturbs results of each query differently, so that inference of private information using multiple queries is impossible.

The advantages of having differential privacy enforcement in SAM is that the analysts do not have to have knowledge about the budget requirements and accuracy needs. The data owner can set exact accuracy and budget requirements to restrict access to private data. Thus, analysts do not have to invoke special application program interfaces to access data through differential privacy mechanisms.

Access-Control Policies: There has been a tremendous amount of work done in enforcing access control policies. Current DBMSs allow users to access different parts of the data base by enforcing ACMs. The SAM can also enforce these policies, as it is based on MRA, which are capable of enforcing arbitrary access-control policies by monitoring and building a history of entire sessions.

The SAM can take advantage of existing ACM in the DBMS to enforce access-control policies. The underlying MRA M has to request permission from the DBMS before submitting its executions. The request is actually a verification of permission of the current query. When the DBMS grants permission, the MRA can take the action of the input query. This removes the burden of implementing ACMs in SAM. The existing DBMS also does not require any changes.

4.3 Model Properties

In the previous text, we discussed different policies that can be enforced on database queries. Now we need an enforcement mechanism that SAM can follow to preserve privacy. In this section, we show how to enforce the privacy policies defined above by constructing MRA and proving that these MRA can enforce the policies. Given any query-set-size, query-set-overlap, query-tracker, or differential-privacy policy P, Theorem 1 shows that an MRA M can enforce P and Theorem 1's proof shows how to construct such an M.

Theorem 1. *For all policies P such that P is a query-set-size policy, query-set-overlap policy, query-tracker policy, or differential-privacy policy on a system with event set E, there exists an MRA M such that M enforces P.*

Proof. Given any such policy P, we show how to construct an MRA $M = (E, S, q_0, \delta)$ that enforces P on all executions $x \in E^*$. Note that δ is a partial function when δ is undefined on a given input, the MRA cannot transition, so the system effectively halts.

If P is a query-set-size restriction policy parameterized by N and K as in Definition 3, let $M = (E, \{0, 1\}, 0, \delta)$, where δ is:

$$\delta(q, e) = \begin{cases} (0, e) & \text{if } q = 0, \ e \neq query \\ (1, e) & \text{if } q = 0, \ e = query \\ (0, e) & \text{if } q = 1, \ K \leq |e| \leq N - K \end{cases}$$

M enforces P because both produce/allow exactly those executions in which all query results r satisfy the $K \leq |r| \leq N - K$ constraint. Hence, the sets of executions produced by M and allowed by P are equal.

If P is a query-set-overlap restriction policy parameterized by O as in Definition 4, let $M = (E, E^*, \varepsilon, \delta)$, where δ is:

$$\delta(q, e) = \begin{cases} (q, e) & \text{if } e \neq query \\ (q; e, e) & \text{if } e = query, \ q = e_1; e_2; \dots; e_n \text{ and} \\ & \forall i \in \{1, \dots, n\} \ : \ |M_e \cap M_{e_i}| \leq O \end{cases}$$

M enforces P because both produce/allow exactly those executions in which all queries only contain entity sets that never overlap previous queries' entity sets in more than O elements. Hence, the sets of executions produced by M and allowed by P are equal.

If P is a query-tracker policy parameterized by N and K as in Definition 5, let $M = (E, E^*, \varepsilon, \delta)$, where δ is:

$$\delta(q, e) = \begin{cases} (q, e) & \text{if } e \neq query \\ (q; e, e) & \text{if } e = query, \ q = e_1; e_2; \dots; e_n \text{ and} \\ & \forall i \in \{1, \dots, n\} \ : \ (C_e \cap C_{e_i} = \emptyset \text{ or} \\ & K \not\leq |R_e| \not\leq N - 2K \text{ or} \\ & 2K \not\leq |R_{e_i}| \not\leq N - K) \end{cases}$$

M enforces P because both produce/allow exactly those executions in which all queries Q satisfy at least one of the following three constraints: (1) Q has no projection-condition attributes in common with previous queries, (2) $K \not\leq |R_e| \not\leq N - 2K$, or (3) all the previous query results r satisfy the constraint $2K \not\leq |r| \not\leq N - K$. Hence, the sets of executions produced by M and allowed by P are equal.

If P is a differential-privacy policy parameterized by perturbation function f as in Definition 6, let $M = (E, \{0, 1\}, 0, \delta)$, where δ is:

$$\delta(q, e) = \begin{cases} (0, e) & \text{if } q = 0, \ e \neq query \\ (1, e) & \text{if } q = 0, \ e = query \\ (0, f(e)) & \text{if } q = 1 \end{cases}$$

M enforces P because both produce/allow exactly those executions in which all query results are perturbed by the perturbation function f. Hence, the sets of executions produced by M and allowed by P are equal.

Hence, an MRA can enforce all query-set-size restriction, query-set-overlap restriction, query tracker, and differential-privacy polices.

4.4   Challenges

The MRA based monitor can secure the database by enforcing the policies defined above. In this section we discuss some of the challenges in enforcing these policies with MRA, and possible solutions.

Infinite-Length Executions: An infinite-length execution is one that never terminates. For simplicity, the definitions of privacy-preserving policies in Sect. 4.2 only consider finite-length executions. However, these definitions can be generalized to infinite-length executions straightforwardly by placing the same sorts of query-result constraints on every query and result in every execution (including those of infinite-length). All the modules of SAM are capable of monitoring sequences of infinite-length queries, without requiring any changes to their functionalities.

Colluded Attacks: As explained before, MRA is a synchronous automata allowing only one input event and producing a result for that event before processing the next. Thus, concurrent events can't be processed by the MRA. This leads to a possible attack in which multiple users collude and share their results. Since the MRA assumes one infinite-length execution per user, it can protect the private information from only one user, not colluded users.

A possible solution is to assume the queries from all users as events of a single infinite-length execution. Queries from all users can be serialized, possibly based on arrival timestamps. Thus, completely eliminating the possibility of compromise from colluding users. Another direction to eliminate the risk from colluded attacks is taking advantage of differential privacy parameters. The database administrators can specify different types of noise to be added to different users, so the users can't extract any private information by colluding. There could be some limitations on the number of different types of noises that can be generated at run time to a large number of users.

User-Defined Aggregates: It is often necessary to allow the analysts to write their own programs to analyze the data. The data owner can specify differential privacy policies on such programs. The MRA can enforce these policies directly without requiring any changes to the underlying architecture. When different types of aggregates are supported, specifying proper policies becomes a challenging task. The policies should be specified such that the privacy is preserved. It is analogous to specifying access control policies in standard databases.

Performance: Maintaining history of queries and results could impact overall performance of SAM. Logging during query evaluation and log-access for policy enforcement could overload SAM, impacting overall performance. Enforcing policies online as well as offline could alleviate the performance problems. Online setting utilizes only recent past of the history, while offline setting audits to check if any policy violations have occurred due to recent queries. High performance computing strategies can also be applied to utilize large portion of the history for faster processing. The problem of leveraging such strategies is another interesting research challenge.

History tracking is an interesting feature of SAM, which can leverage lineage of queries to preserve privacy and restrict data usage. Query lineage can be utilized for data usage audits [6, 8]. It becomes an effective tool to monitor curious (authorized) users.

5   Conclusions and Future Work

We have presented our vision of a privacy architecture, called SAM, which can enforce various privacy-preserving policies and restrict data usage in databases. MRA are constructed to enforce query, result, and access control policies. An interesting feature of SAM is that it allows accurate query results as long as the privacy is preserved. It can be set up to begin enforcing differential-privacy policies when none of the other policies are satisfied. The data owners can specify policies and input to SAM instead of writing their own database interfaces to preserve privacy. This also alleviates the need for analysts to specify any special parameters to access the database.

The MRA construction method can be followed to enforce the policies defined in this work. We believe that the MRA can be an independent component of the database, without affecting other database modules. Much has to be done to refine the SAM design and ensure effective and efficient implementation of the architecture, as discussed in Sect. 4.4. Here we want to emphasize the very interesting research topic of studying the possibility of utilizing these database modules to improve the performance. SAM can be implemented as part of the DBMS to monitor the results of different database operators. Monitoring operators inside DBMS would eliminate the need for the query-rewrite module in SAM, as the partial results can be monitored even before computing the final result. We believe that the formal architecture presented in this paper is capable of enforcing a wide range of database policies.

Acknowledgements. This project is supported by a grant (No. R01GM086707) from the National Institutes of Health (NIH), USA.

References

1. Adam, N.R., Worthmann, J.C.: Security-control methods for statistical databases: a comparative study. ACM Comput. Surv. **21**(4), 515–556 (1989)
2. Agrawal, R., Srikant, R., Thomas, D.: Privacy preserving OLAP. In: Proceedings of the International Conference on Management of Data, SIGMOD, pp. 251–262 (2005)
3. Denning, D.E., Schlörer, J.: A fast procedure for finding a tracker in a statistical database. ACM Trans. Database Syst. **5**(1), 88–102 (1980)
4. Dobkin, D., Jones, A.K., Lipton, R.J.: Secure databases: protection against user influence. ACM Trans. Database Syst. **4**(1), 97–106 (1979)
5. Dwork, C.: Differential privacy: a survey of results. In: Agrawal, M., Du, D.-Z., Duan, Z., Li, A. (eds.) TAMC 2008. LNCS, vol. 4978, pp. 1–19. Springer, Heidelberg (2008)
6. Fabbri, D., LeFevre, K.: Explanation-based auditing. Proc. VLDB Endow. **5**(1), 1–12 (2011)
7. Fellegi, I.P., Phillips, J.J.: Statistical confidentiality: some theory and application to data dissemination. Am. Econ. Soc. Measures **3**(2), 101–112 (1974)
8. Hasan, R., Winslett, M.: Efficient audit-based compliance for relational data retention. In: Symposium on Information, Computer and Communications Security, pp. 238–248 (2011)
9. Kushida, C., Nichols, D., Jadrnicek, R., Miller, R., Walsh, J., Griffin, K.: Strategies for de-identification and anonymization of electronic health record data for use in multicenter research studies. Med. Care **50**, S82–S101 (2012)
10. Ligatti, J., Bauer, L., Walker, D.: Run-time enforcement of nonsafety policies. ACM Trans. Inf. Syst. Secur. **12**(3), 1–41 (2009)
11. Ligatti, J., Reddy, S.: A theory of runtime enforcement, with results. In: Proceedings of the 15th European Conference on Research in Computer Security, pp. 87–100 (2010)
12. McSherry, F.D.: Privacy integrated queries: an extensible platform for privacy-preserving data analysis. In: Proceedings of the International Conference on Management of Data, SIGMOD, pp. 19–30 (2009)
13. Mohan, P., Thakurta, A., Shi, E., Song, D., Culler, D.: Gupt: privacy preserving data analysis made easy. In: Proceedings of the International Conference on Management of Data, SIGMOD, pp. 349–360 (2012)
14. Muralidhar, K., Batra, D., Kirs, P.J.: Accessibility, security, and accuracy in statistical databases: the case for the multiplicative fixed data perturbation approach. Manage. Sci. **41**(9), 1549–1564 (1995)
15. Narayanan, A., Shmatikov, V.: Robust de-anonymization of large sparse datasets. In: Proceedings of the Symposium on Security and Privacy, S&P, pp. 111–125 (2008)
16. Roy, I., Setty, S.T.V., Kilzer, A., Shmatikov, V., Witchel, E.: Airavat: security and privacy for mapreduce. In: Proceedings of the Conference on Networked Systems Design and Implementation, NSDI, p. 20 (2010)
17. Schneider, F.B.: Enforceable security policies. ACM Trans. Inf. Syst. Secur. **3**(1), 30–50 (2000)
18. Upadhyaya, P., Anderson, N.R., Balazinska, M., Howe, B., Kaushik, R., Ramamurthy, R., Suciu, D.: Stop that query! the need for managing data use. In: Conference on Innovative Data Systems Research (2013)

Correlation-Based Deep Learning for Multimedia Semantic Concept Detection

Hsin-Yu Ha$^{(\boxtimes)}$, Yimin Yang, Samira Pouyanfar,
Haiman Tian, and Shu-Ching Chen

School of Computing and Information Sciences,
Florida International University, Miami, FL 33199, USA
{hha001,yyang010,spouy001,htian005,chens}@cs.fiu.edu

Abstract. Nowadays, concept detection from multimedia data is considered as an emerging topic due to its applicability to various applications in both academia and industry. However, there are some inevitable challenges including the high volume and variety of multimedia data as well as its skewed distribution. To cope with these challenges, in this paper, a novel framework is proposed to integrate two correlation-based methods, Feature-Correlation Maximum Spanning Tree (FC-MST) and Negative-based Sampling (NS), with a well-known deep learning algorithm called Convolutional Neural Network (CNN). First, FC-MST is introduced to select the most relevant low-level features, which are extracted from multiple modalities, and to decide the input layer dimension of the CNN. Second, NS is adopted to improve the batch sampling in the CNN. Using NUS-WIDE image data set as a web-based application, the experimental results demonstrate the effectiveness of the proposed framework for semantic concept detection, comparing to other well-known classifiers.

Keywords: Deep learning · Feature selection · Sampling · Semantic concept detection · Web-based multimedia data

1 Introduction

In recent decades, the number of multimedia data transferred via the Internet increases rapidly in every minute. Multimedia data, which refers to data consisting of various media types like text, audio, video, as well as animation, is rich in semantics. To bridge the semantic gap between the low-level features and high-level concepts, it introduces several interesting research topics like, data representations, model fusion, imbalanced data issue, reduction of feature dimensions, etc.

Because of the explosive growth of multimedia data, the complexity rises exponentially with linearly increasing dimensions of the data, which poses a great challenge to multimedia data analysis, especially semantic concept detection. Due to this fact, it draws multimedia society's attention to identify useful feature subsets, reduce the feature dimensions, and utilize all the features extracted from

© Springer International Publishing Switzerland 2015
J. Wang et al. (Eds.): WISE 2015, Part II, LNCS 9419, pp. 473–487, 2015.
DOI: 10.1007/978-3-319-26187-4_43

different modalities. Many researchers develop feature selection methods based on different perspectives and methodologies. For example, whether the label information is fully explored [1–5], whether a learning algorithm is included in the method [7–11], etc. However, most feature selection methods are applied on data with one single modality. Recently, a Feature-Correlation Maximum Spanning Tree (FC-MST) [12] method has been proposed for exploring feature correlations among multiple modalities to better identify the effective feature subset.

On the other hand, the imbalanced data set is another major challenge while dealing with real world multimedia data. An imbalanced data set is defined by two classes, i.e., positive class and negative class, where the size of positive data is way smaller than the size of negative one. When training a classification model with unevenly distributed data, the model tends to classify data instances into the class with a larger data size. To resolve the issue, two types of sampling methods are widely applied, i.e., oversampling and undersampling. Oversampling methods are proposed to duplicate the positive instances to balance the data distribution. However, the computation time will increase accordingly. Undersampling methods are also widely studied to remove the negative instances to make the data set be evenly distributed. Unlike most undersampling methods, which remove the negative instances without specific criteria, Negative-based Sampling (NS) [13] is proposed to identify the negative representative instances and keeps them in the later training process.

Recently, applying deep learning methods to analyze composite data, like videos and images, has become an emerging research topic. Deep learning is a concept originally derived from artificial neural networks and it has been widely applied to model high-level abstraction from complex data. Among different deep learning methods, the Convolutional Neural Network (CNN) [14] is well established and it demonstrates the strength in many difficult tasks like audio recognition, facial expression recognition, content-based image retrieval, etc. The capability of CNN in dealing with complex data motivates us to incorporate it for multimedia analysis. Specifically, the advantages of CNN are two folds. First, CNN is composed of hierarchical layers, where the features are thoroughly trained in a bottom-up manner. Second, CNN is a biologically-derived Multi-Layer-Perceptron (MLP) [15], thus it optimizes the classification results using the gradient of a loss function with respect to all the weights in the network.

In this paper, an integrated framework is proposed to solve the semantic concept detection problem by applying two correlation-based methods, e.g., FC-MST and NS, on refining the CNN's architecture. FC-MST aims to obtain the effective features by removing other irrelevant or redundant features and it is further applied on deciding the dimension of the CNN's input layer. NS is introduced to solve the data imbalance problem and it is proposed to better refine the CNN's batch assigning process.

The rest of this paper is organized as follows. Section 2 provides related work on training the deep learning models for multimedia data analysis. A detailed description of the proposed framework is presented in Sect. 3. The experiment dataset and the experimental results are discussed in Sect. 4. Lastly, the paper is concluded in Sect. 5 with the summarization.

2 Related Work

With the enormous growth of data such as audio, text, image, and video, multimedia semantic concept detection has become a challenging topic in current digital age [16–18]. Deep learning, a new and powerful branch of machine learning, plays a significant role in multimedia analysis [19–21], especially for the big data applications, due to its deep and complex structure utilizing a large number of hidden layers and parameters to extract high-level semantic concepts in data.

To date, various deep learning frameworks have been applied in multimedia analysis, including Caffe [22], Theano [23], Cuda-convnet [24], to name a few. Deep convolutional networks proposed by Krizhevsky et al. [25] were inspired by the traditional neural networks such as MLP. By applying a GPU implementation of a convolutional neural network on the subsets of Imagenet dataset in the ILSVRC-2010 and ILSVRC-2012 competitions [26], Krizhevsky et al. achieved the best results and reduced the top-5 test error by 10.9 % compared with the second winner. A Deep Convolutional Activation Feature (Decaf) [27], the direct precursor of Caffe, was used to extract the features from an unlabeled or inadequately labeled dataset by improving the convolutional network proposed by Krizhevsky et al. Decaf learns the features with high generalization and representation to extract the semantic information using simple linear classifiers such as Support Vector Machine (SVM) and logistic Regression (LR).

Although deep convolutional networks have attracted significant interests within multimedia and machine learning applications, generating features from scratch and the duplication of previous results are tedious tasks, which may take weeks or months. For this purpose, Caffe, a Convolutional Architecture for Fast Feature Embedding, was later proposed by Jia et al. [22], which not only includes modifiable deep learning algorithms, but also collects several pre-trained reference models. One such reference model is Region with CNN features (R-CNN) [28], which extracts features from region proposals to detect semantic concepts from very large datasets. R-CNN includes three main modules. The first module extracts category-independent regions (instead of original images) used as the inputs of the second module called feature extractor. For feature extraction and fine-tuning, a large CNN is pre-trained using the Caffe library. Finally, in the third module, the linear SVM is applied to classify the objects. Based on the evaluation results on one specific task called PASCAL VOC, CNN features carry more information compared to the conventional methods' extracted simple HOG-based features [29].

Many researchers recently utilize a pre-trained reference model to improve the results and to reduce the computational time. Snoek et al. [30] retrained a deep network, which was trained on ImageNet datasets. The input of the deep network is raw image pixels and the outputs are scores for each concept. These scores are later fused with those generated from another concept detection framework, which uses a mixture of low level features and a linear SVM for concept detection. The overall combination framework achieves the best performance results for 9 different concepts in the Semantic Indexing (SIN) task of TRECVID 2014 [31]. Ngiam et al. [32] developed a multimodal deep learning framework for feature learning using a Restricted Boltzmann Machines (RBMs). To combine information from raw video frames with audio waveforms, a bimodal deep autoencoder is proposed, which is greedily trained by separate pre-trained models for each modality. In this model, there is a deep hidden layer, which models the relationship between audio and video modalities and learns the higher order correlation among them.

Based on the successful results acquired by deep learning techniques, an important question arises: whether deep networks are the solution for multimedia feature analysis or not. Wan et al. [33] addressed this question for Content-Based Image Retrieval (CBIR). In particular, CNN is investigated for the CBIR feature representation under the following schemes: (1) Direct feature representation using a pre-trained deep model; (2) Refining the features by similarity learning; and (3) Refining the features by model retraining using reference models such as ImageNet, which shows the promising results on the Caltec256 dataset. However, the extracted features from deep networks may not capture better semantic information compared with conventional low-level features.

More recent research in multimedia deep learning has addressed challenges such as feature extraction/selection and dimension reduction, where the input is raw pixel values. Specifically, CNN is widely used as a successful feature extractor in various multimedia tasks. However, it is still unknown how it can perform as a classifier for semantic detection tasks.

We address the aforementioned challenges by bridging the gap between semantic detection and a deep learning algorithm using general features including low-level visual and audio features as well as textual information, instead of fixed pixel values of the original images. FC-MST, a novel feature extraction method, is proposed to remove irrelevant features and automatically decide the input layer dimension. Furthermore, NS is utilized to handle the imbalanced datasets. Finally, by leveraging FC-MST and NS in the CNN structure, not only the important and relevant features are fed to the network and the data imbalance issue is solved, but also the computational time and memory usage are significantly reduced.

3   Proposed Framework

As shown in Fig. 1, the proposed framework starts from collecting the data derived from different data types, such as images, videos, and texts. Each modality requires the corresponding pre-processing step. For instance, shot boundary detection and key frame detection are applied to obtain the basic video elements, e.g., shots and keyframes, respectively. Then, low-level visual features and audio features can be extracted from them. For the image data, visual features can be directly extracted from each instance and possibly combined with the corresponding textual information including tags, title, description, etc. For the text

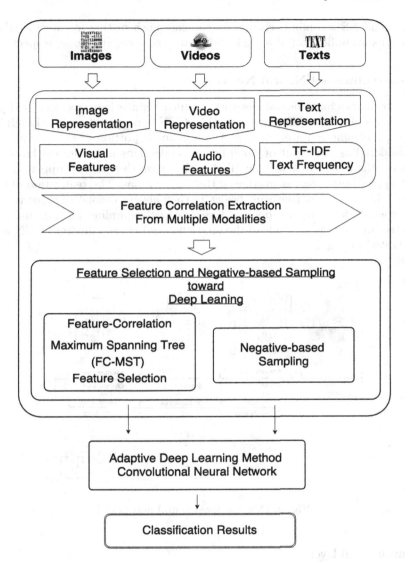

Fig. 1. Overview of the proposed framework

data, it is usually represented by its frequency or TF-IDF [34] values. Once all the features are extracted and are integrated into one, the proposed FC-MST method is adopted to select useful features and decide the dimension of the input layer. On the other hand, NS is carried out to enhance the batch instance selection for every feature map in each iteration process. Hence, the architecture of the original CNN is automatically adjusted based on the FC-MST's feature selection and NS sampling scheme. At the end, each testing instance is labeled as 1 or 0 as an indication of a positive instance or a negative one, respectively.

3.1 Convolutional Neural Network

CNNs are hierarchical neural networks, which reduce learning complexity by sharing the weights in different layers [14]. CNN is proposed with only minimal data preprocessing requirements, and only a small portion of the original data are considered as the input of small neuron collections in the lowest layer. The obtained salient features will be tiled with an overlap to the upper layer in order to get a better representation of the observations. The realization of CNN may vary in the layers. However, basically they always consist of three types of layers: convolutional layers, pooling layers (or sub-sampling layers), and fully-connected layers. One example of the relationships between different CNN layers is illustrated in Fig. 2.

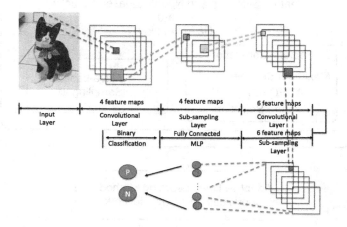

Fig. 2. Convolutional neural network

1. Convolutional layer
   There are many feature maps (representation of neurons) in each convolutional layer. Each map takes the inputs from the previous layer with the same weight W and repeatedly applies the tensor function to the entire valid region. In other words, the convolution of the previous layer's input x is fulfilled with a linear filter, where the weight for the kth feature map is indicated as W^k and the corresponding bias is indicated as b_k. Then, the filtered results are applied to a non-linear activation function f. For example, if we denote the kth feature map for the given layer as h^k, the feature map is obtained as follows.

$$h^k = f((W^k * x) + b_k). \tag{1}$$

The weights can be considered as the learnable kernels, which might be different in each feature map. In order to compute the pre-nonlinearity input to some unit x, the contributions from the previous layer need to be summed up and weighted by the filter components.

2. Pooling layer (Sub-sampling layer)
   Pooling layers usually come after the convolutional layers to reduce the dimensionality of the intermediate representations as shown in Fig. 2. It takes feature maps from the convolutional layer into non-overlapping blocks and subsamples them to produce a single output from each sub-region. Max-pooling is the most well-known pooling method, which takes the maximum value of each block [14, 35], and it is used in the proposed framework. It is worth nothing that this type of layer does not learn by itself. The main purpose of such layer is to increase the spatial abstractness and to reduce the computation for the later layers.

3. Fully-connected MLP layer
   After several convolutional layers and pooling layers, the high-level reasoning in the neural network is done via one fully connected MLP layer. It takes all the feature maps at the previous layer as the input to be processed by a traditional MLP, which includes the hidden layer and the logistic regression process. At the end, one score is generated per instance for the classification. For a binary classification CNN model as depicted in Fig. 2, each instance is either classified as positive or negative class based on the generated score.

Convolutional neural network processes ordered data in an architecturally different way, which transparently shares the weights. This model has been shown to work well for a number of tasks, especially for object recognition [36] and it has become popular recently on multimedia data analysis [22].

3.2 FC-MST Method in Deciding Input Layer Dimension

CNN is a biologically-evolving version of MLP and it is originally implemented for tasks like MNIST digit classification or facial recognition. Though different implementations might have its own unique CNN's architecture, such as different numbers of filtering masks, sizes of the pooling layers, etc., most of them take the original image as the input and process the image as $Height \times Width$ pixel values. Here, the low-level features are selected by the proposed FC-MST and are deployed as the context of CNN's input layer.

FC-MST is proposed in [12], which aims to obtain the effective features by removing both redundant and irrelevant features. The methodology utilizes two correlations listed as follows.

– The correlation among features across multiple modalities;
– The correlation between each feature towards the target positive concept.

Given the revealed correlation, the proposed FC-MST is able to distinguish the effective features from others and greatly reduces the feature dimension. It

---

Algorithm 1. How to decide the dimension of CNN's input Layer by FC-MST

---

    input : The given training data set D with feature set as
                $TDF = F_1, F_2, ..., F_M$, along with the class label C
    output: SF: A set of selected features, which indicates the dimension of
               CNN's input layer $size_H$ and $size_W$

1 $ISF \longleftarrow FCMST(TDF)$;
2 **if** $Num_{ISF} \bmod 6 = 0$ **then**
3 | $size_H = 6$;
4 | $size_W = Num_{ISF}/6$;
5 **else**
6 | $Num_{ISF} = Num_{ISF} - (Num_{ISF} \bmod 6)$;
      | /* Num_{ISF} represents the number of features in ISF */
7 | $Num_{DF} = Num_{ISF} \bmod 6$;
      | /* Num_{DF} represents the number of features which are going to
      | be removed from ISF */
8 | $size_H = 6$;
9 | $size_W = Num_{ISF}/6$;
10 $SF \longleftarrow RemoveNumDF(ISF)$;
11 **return** $SF, size_H, size_W$

---

motivates us to apply FC-MST onto the input layer of the convolutional neural network. Hence, only the important features are considered in the process and the computation time can be greatly reduced. The process is depicted in Algorithm 1. All features from multiple modalities are combined into one unified feature set indicated as TDF. ISF represents the initial selected features after applying FC-MST on the original data set TDF (as described in Algorithm 1, line 1). Next, the number of selected features is checked on two conditions: whether it is a prime number and whether it can be divided by number 6. The checking process is described in Algorithm 1, from line 2 to line 9. The conditions are set because the dimension of the input layer needs to be completely divided by the dimension of the feature map in every convolutional layer, e.g., 2×2. Num_{DF} is obtained by getting the remainder of Num_{ISF} divided by 6. Then, Num_{DF} features are removed based on their correlation towards the positive concept and the deletion operation is performed on the least correlated features (as described in Algorithm 1, line 10). At the end, the selected feature set SF along with the decided dimension of the input layer, e.g., $size_H$ and $size_W$, are returned.

3.3 Negative-Based Sampling in Deciding Batch Sampling Process

The data imbalance problem has been one of the major challenges when classifying a multimedia data set. When the data size of the major class is way larger than that of the minor's, it usually results in poor classification performance. The problem becomes worse when applying the deep learning methods, such as CNN, on the skewed data set. The reason is because most of the deep learning

---

Algorithm 2. Negative-based CNN batch sampling process

---

input : The given training data set D is composed of positive set P and negative set N.

1   **while** *Iterating in Pooling Layer or Convolutional Layer* **do**
2 | $Num_P \longleftarrow |P|$;
3 | $Num_N \longleftarrow |N|$;
4 | $Num_D \longleftarrow |D|$;
5 | $BatchSize = Num_D/100$;
6 | $NF \longleftarrow FCMST(D)$;
7 | **for** *all training negative instances* $I_i, i = 1, ..., Num_N$ **do**
8 | | $NegRank(I_i) = MCA_{NF}(I_i)$;
9 | **for** *Each batch* $B_j, j = 1, ..., 100$ **do**
10 | | $B_j \longleftarrow \emptyset$;
11 | | **if** $Num_P > 1/2BatchSize$ **then**
12 | | | $B_j \longleftarrow$ randomly pick $1/2BatchSize$ from P;
13 | | **else**
14 | | | $B_j \longleftarrow P$;
15 | | $BP_j \longleftarrow |B_j|$;
16 | | $BN_j \longleftarrow (BatchSize - BP_j)$;
17 | | $B_j \longleftarrow$ select BN_j instances with higher Negative Ranking Score from the first $j^{th} BatchSize$ of instances;
18 | Continuing in training CNN model;

---

methods, including CNN, start the training process by assigning instances into different batches and each batch might contain no positive instance but all negative instances due to this uneven distribution. Assigning random instances into each batch is not able to resolve the data imbalance problem and it could result in poor classification results.

To tackle this challenge, "the NS method", which is published in [13], is adopted to improve the CNN batch sampling process as shown in Algorithm 2. As long as the training process is still within either the pooling or convolutional layer, the same negative-based CNN batch sampling process is applied (as described in Algorithm 2, line 1). At the beginning, the number of positive set, negative set, and the combined data set, are obtained and represented as Num_P, Num_N, and Num_D, respectively. The number of instances in each batch is set to be $1/100$ of the total number of instances Num_D. A set of features NF are selected based on the negative-based FC-MST method, which are highly correlated with the target negative concept (as described in Algorithm 2, line 2–6). All the negative instances are looped through to generate the corresponding negative-based ranking score. The negative ranking score is generated by the method called Multiple Correspondence Analysis (MCA) [37,38] using the above-selected features NF. The higher the score is, the more negative-representative the instance is (as described in Algorithm 2, line 7–8). For each batch, it starts with an empty set and then is assigned with either the whole positive set P or the half batch size of the positive instances (as described in

Algorithm 2, line 9–17). The last step in this batch sampling process is to obtain the subtraction of $BatchSize$ and the current numbers of the assigned positive and negative instances are denoted as BP_j and BN_j, respectively. From the $j^{th} BatchSize$ number of instances, the first BN_j instances with higher negative ranking scores are selected into batch B_j. The same process is applied and looped through all the batches.

4 Experiment

4.1 NUS-WIDE Dataset

The proposed framework is validated using the well-known multimedia data set called NUS-WIDE [39]. It is a web image data set downloaded from Flickr website including six types of low-level features. The lite version, which contains 27,807 training images and 27,808 testing images, is conducted in this experiment. The data set contains relatively low Positive to Negative Ratios for all 81 concepts, which is depicted in Fig. 3.

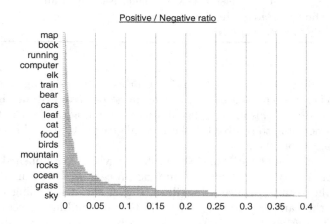

Fig. 3. Positive and Negative Ratios of NUSWIDE Lite 81 concepts

4.2 Experiment Setup and Evaluation

The proposed framework is compared with two well-known classifiers, e.g., K-Nearest Neighbors (KNN) and SVM. It is also compared to MCA-TR-ARC [40], which is applied on the NUSWIDE data set to remove the noisy tags and combine the ranking scores from both tag-based and content-based models. In addition, a sensitivity analysis is conducted to justify which component contributes the most in enhancing the classification results.

Table 1. Average Precision (AP) of the proposed method and other classifiers

Method	Average Precision (AP)
KNN	9.87%
SVM	11.23%
CNN	10.41%
MCA-TR-ARC	33%
Proposed Method	35.61%

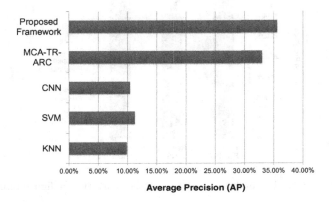

Fig. 4. Average Precision comparing with other methods

4.3   Results

The Average Precision (AP) of NUS-WIDE's 81 concepts for 4 different classifiers and the proposed framework is shown in Table 1. KNN performs the worst with an AP value of 9.87%, which shows that a huge amount of unselected features and the data imbalance issue actually result in very poor classification performance. The same issue affects both SVM and CNN. SVM produces an AP value of 11.24%, which is 1.37% higher when compared to KNN, because it is able to better separate the positive instances from the negative ones. With regard to CNN, it is not able to reach a better performance because how it assigns instances into batches does not resolve the data imbalanced issue. However, CNN has the ability of iterating the training process until it reaches the optimal results, and thus it is able to obtain slightly higher AP values against KNN. MCA-TR-ARC produces a relatively much higher AP value compared to others because of two reasons. First, it applies MCA to remove the noisy tag information. Second, it explores the correlation between the tag-based model and the content-based model, and fuses the ranking scores into one. Finally, the proposed framework, which combines two correlation-based methods, can

Table 2. Sensitivity Analysis (SA) in evaluating contribution for each component

Method	Average Precision	Dropped Performance
The Proposed Work	35.61 %	—
Remove FC-MST	23.85 %	11.76 %
Remove NS	19.39 %	16.22 %
Remove Both	10.41 %	25.20 %

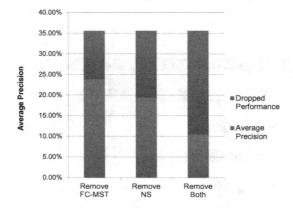

Fig. 5. Sensitivity analysis on the proposed work (Color figure online)

outperform all the other classifiers in the NUS-WIDE dataset. Figure 4 also visually depicts the aforementioned classification results.

A sensitivity analysis is further performed to better analyze the contribution for each component. In Table 2, the first column is the AP values performed by the proposed framework, which includes both FC-MST and NS, and it is able to reach 35.61 %. If FC-MST is removed from the proposed framework, then the AP value dropped by 11.76 %. On the other hand, if NS is removed from the proposed framework, the performance dropped even more. The results indicate that identifying useful features can efficiently increase the average precision, but better assigning the instances into each training batch plays a much important role. Figure 5 highlights the dropped performance in color red when removing different components. The rightmost bar, which is indicated as "Remove Both", represents the performance of the original CNN.

5   Conclusion

In this paper, an integrated framework is proposed to adopt two correlation-based methods, e.g., FC-MST and NS, in adjusting the architecture of one well-known deep learning method called CNN. First, FC-MST is proposed to identify effective features and decide the dimension of CNN's input layer instead of using fixed pixel values of the original images. The features are selected based on their correlation towards the target positive class. Second, NS is proposed specifically to cope with the imbalanced data sets, which usually results in poor classification performance due to its uneven distribution. The problem is worse when the original CNN randomly assigns data instances into each batch. Thus, NS is adopted to alleviate the problem. The experiment shows this proposed integrated framework is able to outperform other well-known classifiers and each correlation-based method can independently contribute to enhance the results.

Acknowledgment. This research was supported in part by the U.S. Department of Homeland Security under grant Award Number 2010-ST-062-000039, the U.S. Department of Homeland Security's VACCINE Center under Award Number 2009-ST-061-CI0001, NSF HRD-0833093, CNS-1126619, and CNS-1461926.

References

1. Zhu, Q., et al.: Feature selection using correlation and reliability based scoring metric for video semantic detection. In: 2010 IEEE Fourth International Conference on Semantic Computing (ICSC) (2010)
2. Shyu, M.-L., et al.: Network intrusion detection through adaptive sub-eigenspace modeling in multiagent systems. ACM Trans. Auton. Adapt. Syst. (TAAS) **2**(3), 9 (2007)
3. Shyu, M.-L., et al.: Image database retrieval utilizing affinity relationships. In: Proceedings of the 1st ACM International Workshop on Multimedia Databases (2003)
4. Shyu, M.-L., et al.: Mining user access behavior on the WWW. In: Proceedings of the IEEE International Conference on Systems, Man, and Cybernetics, pp. 1717–1722 (2001)
5. Shyu, M.-L., et al.: Generalized affinity-based association rule mining for multimedia database queries. Knowl. Inf. Syst. (KAIS) **3**, 319–337 (2001)
6. Ha, H.-Y., et al.: Content-based multimedia retrieval using feature correlation clustering and fusion. Int. J. Multimedia Data Eng. Manage. (IJMDEM) **4**(5), 46–64 (2013)
7. Li, X., et al.: An effective content-based visual image retrieval system. In: Proceedings of the 26th IEEE Computer Society International Computer Software and Applications Conference (COMPSAC) (2002)
8. Huang, X., et al.: User concept pattern discovery using relevance feedback and multiple instance learning for content-based image retrieval. In: Proceedings of the Third International Workshop on Multimedia Data Mining (MDM/KDD), in conjunction with the 8th ACM SIGKDD International Conference on Knowledge Discovery and Data Mining (2002)
9. Chen, S.-C., et al.: Augmented transition networks as video browsing models for multimedia databases and multimedia information systems. In: Proceedings of the 11th IEEE International Conference on Tools with Artificial Intelligence (ICTAI), pp. 175–182 (1999)
10. Chen, S.-C., et al.: Identifying overlapped objects for video indexing and modeling in multimedia database systems. Int. J. Artif. Intell. Tools **10**(4), 715–734 (2001)

11. Chen, X., et al.: A latent semantic indexing based method for solving multiple instance learning problem in region-based image retrieval. In: Proceedings of the IEEE International Symposium on Multimedia (ISM), pp. 37–44 (2005)
12. Ha, H.-Y., Chen, S.-C., Chen, M.: FC-MST: feature correlation maximum spanning tree for multimedia concept classification. In: IEEE International Conference on Semantic Computing (ICSC) (2015)
13. Ha, H.-Y., Chen, S.-C., Shyu, M.-L.: Negative-based sampling for multimedia retrieval. In: The 16th IEEE International Conference on Information Reuse and Integration (IRI) (2015)
14. LeCun, Y., et al.: Gradient-based learning applied to document recognition. Proc. IEEE **86**(11), 2278–2324 (1998)
15. Ruck, D.W., et al.: The multilayer perceptron as an approximation to a Bayes optimal discriminant function. IEEE Trans. Neural Netw. **1**(4), 296–298 (1990)
16. Yang, J., Yan, R., Hauptmann, A.G.: Cross-domain video concept detection using adaptive svms. In: Proceedings of the 15th ACM International Conference on Multimedia (2007)
17. Meng, T., Shyu, M.-L.: Leveraging concept association network for multimedia rare concept mining and retrieval. In: IEEE International Conference on Multimedia and Expo (ICME) (2012)
18. Ballan, L., et al.: Event detection and recognition for semantic annotation of video. Multimedia Tools Appl. **51**(1), 279–302 (2011)
19. Mobahi, H., Collobert, R., Weston, J.: Deep learning from temporal coherence in video. In: Proceedings of the 26th ACM Annual International Conference on Machine Learning (2009)
20. Zou, W., et al.: Deep learning of invariant features via simulated fixations in video. In: Advances in Neural Information Processing Systems (2012)
21. Yang, Y., Shah, M.: Complex events detection using data-driven concepts. In: Fitzgibbon, A., Lazebnik, S., Perona, P., Sato, Y., Schmid, C. (eds.) ECCV 2012, Part III. LNCS, vol. 7574, pp. 722–735. Springer, Heidelberg (2012)
22. Jia, Y., et al.: Caffe: convolutional architecture for fast feature embedding. In: Proceedings of the ACM International Conference on Multimedia (2014)
23. Bastien, F., et al.: Theano: new features and speed improvements. arXiv preprint arXiv:1211.5590 (2012)
24. Krizhevsky, A.: Cuda-convnet (2012). https://code.google.com/p/cuda-convnet/
25. Krizhevsky, A., Sutskever, I., Hinton, G.E.: Imagenet classification with deep convolutional neural networks. In: Advances in Neural Information Processing Systems (2012)
26. Berg, A., Deng, J., Fei-Fei, L.: Large scale visual recognition challenge 2010 (2010). www.imagenet.org/challenges
27. Donahue, J., et al.: Decaf: a deep convolutional activation feature for generic visual recognition. arXiv preprint arXiv:1310.1531 (2013)
28. Girshick, R., et al.: Rich feature hierarchies for accurate object detection and semantic segmentation. In: IEEE Conference on Computer Vision and Pattern Recognition (CVPR) (2014)
29. Felzenszwalb, P.F., et al.: Object detection with discriminatively trained part-based models. IEEE Trans. Pattern Anal. Mach. Intell. **32**(9), 1627–1645 (2010)
30. Snoek, C.G.M., et al.: MediaMill at TRECVID 2013: searching concepts, objects, instances and events in video. In: NIST TRECVID Workshop (2013)
31. Over, P., et al.: TRECVID 2010: an overview of the goals, tasks, data, evaluation mechanisms, and metrics (2011)

32. Ngiam, J., et al.: Multimodal deep learning. In: Proceedings of the 28th International Conference on Machine Learning (ICML) (2011)
33. Wan, J., et al.: Deep learning for content-based image retrieval: a comprehensive study. In: Proceedings of the ACM International Conference on Multimedia (2014)
34. Salton, G., Buckley, C.: Term-weighting approaches in automatic text retrieval. Inf. Process. Manage. **24**(5), 513–523 (1988)
35. Serre, T., et al.: Robust object recognition with cortex-like mechanisms. IEEE Trans. Pattern Anal. Mach. Intell. **29**(3), 411–426 (2007)
36. McCann, S., Reesman, J.: Object detection using convolutional neural networks
37. Lin, L., et al.: Weighted subspace filtering and ranking algorithms for video concept retrieval. IEEE MultiMedia **18**(3), 32–43 (2011)
38. Yang, Y., Chen, S.-C., Shyu, M.-L.: Temporal multiple correspondence analysis for big data mining in soccer videos. In: The First IEEE International Conference on Multimedia Big Data (BigMM) (2015)
39. Chua, T.-S., et al.: NUS-WIDE: a real-world web image database from National University of Singapore. In: Proceedings of the ACM International Conference on Image and Video Retrieval (2009)
40. Chen, C., et al.: Web media semantic concept retrieval via tag removal and model fusion. ACM Trans. Intell. Syst. Technol. (TIST) **4**(4), 61 (2013)

Social Network Privacy: Issues and Measurement

Isabel Casas, Jose Hurtado[✉], and Xingquan Zhu

Department of Computer and Electrical Engineering and Computer Science,
Florida Atlantic University, Boca Raton, FL 33431, USA
{icasas1,jhurtad2,xzhu3}@fau.edu

Abstract. Social networks are becoming pervasive in todays world. Millions of people worldwide are involved in different form of online networking, with Facebook being one of the most popular sites. Online networks allow individuals to connect with friends and family, and share their private information. One of the reasons for the popularity of virtual communities is the perception of benefits received from the community. However, problems with privacy and security of the users information may also occur, especially when members are not aware of the risks of posting sensitive information on a social network. Members of social networking sites could become victims of identity theft, physical or online stalking and embarrassment as a consequence of malicious manipulation of their profiles data. Although networking sites often provide features for privacy settings, a high percentage of users neither know nor change their privacy preferences. This situation brings to consideration about many important aspects of social network privacy, such as what are the privacy issues in social networks? what are common privacy threats or risks in social networks? how privacy can be measured in a meaningful way? and how to empower users with knowledge to make correct decisions when selecting privacy settings? The goal of this paper is twofold. First, we discuss potential risks and attacks of social network site users privacy. Second, we present the measurement and quantification of the social privacy, along with solutions for privacy protection.

1 Introduction

With the arrival of the Internet, online social networking has transitioned from being used by selected user groups to mass adoption. While personal information is occasionally made available via the Internet, these sites further promote the sharing of personal related content. Boyd and Ellison presented in [21] a definition of online social networks: "web-based services that allow individuals to (1) construct a public or semi-public profile within a bounded system, (2) articulate a list of other users with whom they share a connection, and (3) view and traverse their list of connections and those made by others within the system". With their commercial success and rapid growth in participation, online networking has diversified its usage across a myriad of different websites. In [2], Acquisty

© Springer International Publishing Switzerland 2015
J. Wang et al. (Eds.): WISE 2015, Part II, LNCS 9419, pp. 488–502, 2015.
DOI: 10.1007/978-3-319-26187-4_44

and Gross presented a classification of sites as follows: common interests, dating, business, pets, photos, face-to-face facilitation and friends.

Even though each social network site has its own unique concept and themes, most of them require users to create a representation of themselves or profile with the purpose of interacting with other users. Because user profiles may contain private or sensitive information, such as home address, school, personal preference etc., the protection of private data has become increasingly important. Although the visibility of a profile can be fine-tuned by users, this feature varies from site to site, and information is often completely visible by default settings for many networks, such as Facebook. To aggravate the situation, online networking companies often have nontransparent ways of handling users data since they intend to maximize their revenue by targeted advertising or other channels.

The advancement of online social networking has brought changes and new perspectives to numerous already established concepts and ideas. In 1890, Warren and Brandeis [22] created what is considered the first United States publication advocating for the right of privacy, with privacy being simply defined as: "the right to be let alone". On his book Privacy and Freedom [23], Westin defines privacy as the right "to control, edit, manage, and delete information about them[selves] and decide when, how, and to what extent information is communicated to others". While these definitions have been accepted for many years as synonym of privacy, their limitations become clear when applied to cyberspace, particularly in online social networking environments.

Indeed, privacy has brought new challenges to the academic community and much research has been done in the subject [25–29]. Still, privacy understood in the online situation is a rather elusive term; users should have the right to be left alone but also, as presented in [30], should have the right to be "left in secret". The link between the usage of social networks and privacy is not easy to understand. On one hand, some users would like to share their personal information only with friends or family members but not strangers [2,31]. On the other hand, some users are happy to publicize personal related content with total strangers. In [48], the author suggests to look at privacy from different perspectives, including governmental policies, citizen rights, and consumers protection. In the case of online social networks, its consumers have the right and must know, what information is being collected, by whom and how it will be used. In any case, when information is published, users expose themselves to threats that range from identity theft, embarrassment, stalking to hiring discrimination. Furthermore, personal data from numerous users from a social network can be profitable to other users and third parties.

Due to many concerns raised regarding privacy issues in social networks, research has been made in developing efficient ways of measuring and evaluating privacy. Some works have proposed brand new models to approach privacy in online social networks in order to address flaws of existing models [32]. Following a different approach, some researches propose smart ways of using the existent privacy settings from popular networking sites such as Facebook. Applications that analyze current privacy preferences and recommend more privacy-aware ones have also been developed [11,12]. The authors in [3,4] constructed a privacy index, where the calculation of the privacy index incorporates sensitivities and visibilities of known attributes in a users profile. In this paper, we examine two aspects of online privacy. First, we make an analysis on privacy issues in online social networks as well as attacks against users privacy. Next, we review the literature to find what has been studied in the subject of privacy measurement in online social networks. While our findings show that many works have been done regarding privacy issues in social networks, very few works, however, have addressed the measurement of the privacy, which leaves plenty of room for improvement.

A brief view of the social network entities, and their roles and relationships from data perspectives is shown in Fig. 1.

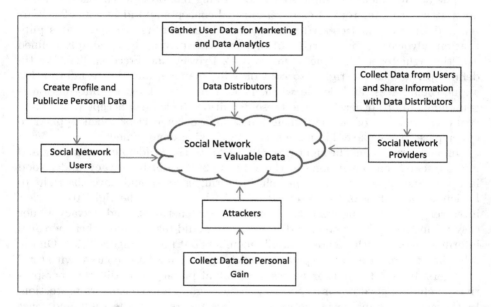

Fig. 1. Social network entities, and their roles and relationships from data perspectives

2   Privacy Issues in Social Networks

With the advancement of the internet, private information has been made available more than ever and online social networks have played a major role in this situation. Privacy implications are determined by how identifiable private information can be derived from provided data, its recipients and its uses [2]. Numerous discussions have been previously made on privacy in social networks [30,33–35].

Researchers have identified two main methods of private information gathering: data leakage due to privacy disclosure and data leakage based on attack techniques [30]. Yang *et al.*, [24] further separated these two main methods into four categories: (1) individual items (profile picture, home address etc.) are available to attackers, (2) aggregation of items of the same person collected though different social sites, (3) inference of hidden attributes from public attributes and, (4) de-anonymizing datasets. In [24], the authors modeled two types of attackers: tireless attacker and resourceful attacker. Resourceful attackers have enough means such as storage and knowledge to gather information from social networks and create his/her own database or digital dossier. Tireless attackers, although lack material means, are willing to dedicate plenty of time and energy. These types of attackers are usually knowledgeable in information retrieval techniques. Information can be collected by crawling the web to obtain personal data, particularly from social networks or mining datasets published by networking sites.

In the following, we will review privacy disclosure issues in social networks and major means of risks and attacks. A brief summary of the privacy disclosure issues, misusage, and common private information gathering methods is shown in Table 1.

2.1   Privacy Disclosure Issues

When creating a profile in a social network, especially the one like Facebook where users are encouraged to provide real data, personal information is at risk. As the study in [2] reveals, users are happy to share personal information: more than 90 % of users uploaded a clear and identifiable picture to their profile, more than 50 % had their current address and almost 40 % had their phone numbers. Users might want certain information to be seen by a close circle, but not by strangers or vice versa. They might also want people who know them well, being kept away from their information. Current privacy models in social networks have been criticized because there is a disconnection between users believed privacy settings and what is in fact happening to their data [32]. Researchers have also concluded that privacy policies and privacy settings are hard to understand for the general social networks members. As a consequence, privacy protection choices are oftentimes poorly selected.

Plenty of controversy has generated the fact that employers began requesting social network credentials from their applicants or current employees. Several states including California, Illinois, Maryland, Michigan, Missouri, New York, Minnesota, South Carolina and Washington have passed legislations protecting the rights of the employees to keep privacy on some parts of their online profile [47]. Some schools have also been involved in similar discussions for trying to control students social networks accounts. A simple joke from a friend posting and tagging an awkward picture could leave social media users exposed. Not only credentials have been requested by employers but also deletion of social media accounts and addition of human resources director as "friend" [43]. Even though social networking has radically changed the definition of privacy, with more personal information being made public, situations like this undermined social sites members privacy expectations.

2.2   Privacy Attack Issues

All this profusion of information could potentially be a target for privacy breach type of attacks. These attacks can come from three main sources: users' so called friends (other users within the network), third party applications, and service providers.

Re-identification and De-anonymization. Anonymization is one of the common techniques for user privacy protection. The idea of anonymizaiton is to strip the data from all attributes that might identify an individual. For example, demographic information, names or social security. This process of anonymization has often been considered a synonym of privacy and it is usually referred to as personally identifiable information removal [44].

When collected data can be linked to a person or an individual, it is called reidentification. To anonymize data, a social network is assumed to be a graph and transformations are applied to achieve privacy preservation. Assuming a social network to be a graph has many advantages but also has some disadvantages. For example, a graph behavior can be modeled by using analytical tools. Link prediction is one of the tools that can be used for modeling and prediction of graphs evolution over time. Combining link prediction with de-anonymization algorithms could be a channel to re-identification.

Anonymised information from social networks is being shared with advertisers and other businesses which expose users' data to re-identification. Several de-anonymization techniques have been recently discussed in the literature [44–46]. An attacker could create several accounts in a social network which share a link pattern among them and connect them to target users. It would be very easy for the attacker to identify his/her accounts and the target account after anonymization. A study presented in [46] showed that knowing someones group membership in a social network is enough to identify that person. Furthermore, the authors in [44] presented a class of de-anonymization algorithms that with minimum background information, an attacker is able to identify a persons record in an anonymized dataset with a high accuracy. These algorithms include three main parts: a scoring function to measure how well a record from a non-anonymized dataset (background knowledge from social network users) matches a record from the anonymized dataset; a matching criterion which is an algorithm to decide based on the results of the scoring function, if there is a match between records; and record selection that selects a "best guess" record if necessary [44].

Phishing. Phishing is a well-known form of social engineering. These attacks involve a perpetrator impersonating a trusted party with the goal of obtaining users private information such as passwords, credit card numbers, and social security numbers. These attacks have generalized ways of luring victims into accepting, for example, a request from a popular banking business, but generally almost no information is known about the receiver. However, when phishing is combined with elements of context, it reaches incredible high success; as it has been studied, it becomes four times more effective [40]. With the massive amounts of data available from social networks, context aware phishing can be dangerously increased. Users leave behind trails of information such as likes on Facebook, stories and videos posted and tweets. A smart fisher can exploit all these elements to increase the yield of an attack.

Is My "friend" a Threat? In social networks, communication between friends is facilitated. Unless a profile is completely open for everyone, only friends of the user have access to view his/her personal information. However, in [2], Acquisti and Gross noted how the word friend has a different meaning in the online and offline context. While offline relationships are extremely diverse in terms of how close a relation is perceived to be, we only have simplistic binary relations online: friends or not. As a consequence, a friend in a social network could perfectly be someone who we would not consider friends offline. Therefore, befriending users in online social networks could open a door for stealing information.

Privacy attacks coming from friends have been presented in [36]. The authors present what is called same-site and cross-site profile cloning. In a same-site profile cloning attack, the perpetrator creates a duplication of a users profile within the same network with the goal of befriending the victims friends. Because the request comes from a known person, the victims contacts are likely to accept and expose their personal information to the perpetrator. A more vicious attack is cross-site cloning because it raises less suspicion. The perpetrator knows a user and his friends from network A and creates a duplicate profile in network B where the user is not registered. Friendship requests are then sent to the users friends who also have a profile in network B. User awareness is crucial to avoid these attacks.

Malicious Third Party Applications. Third party applications provide online social network users with additional functionality, for example, games and horoscope. They are extremely popular and most users take advantage of them. While these applications are built using the social network API and oftentimes reside on the platform, a different company develops them thus; they are considered untrusted [39]. Careless and naive users are perfect victims for malicious third party applications; a well-known example is the Facebook worm "Secret Crush" [37]. Targeted users received a message saying that someone had a crush on them; to reveal the identity of the crush, users had to forward the invitation to five of their contacts and install an application called "Crush Calculator" that was in fact spyware. Meanwhile, there are applications that access users public and private attributes to perform their intended functionality; a restaurant recommender application must have users current location or a horoscope application must have users birth date. Oftentimes, users are unaware of the usage of their private information by third party applications.

In recent news [38], Facebook decided to take action on the way third-party apps published stories to the News Feed, without explicit action from the users.

According to the Facebook website: "Weve also heard that people often feel surprised or confused by stories that are shared without taking an explicit action".

Social Networking Service Providers. When users upload their information to a social network, they are trusting the company (or service providers) to protect their privacy. However, it is known that online networking business profit from sharing their members information [30] therefore, there is a fine balance between securing members privacy and distributing their data to advertisers. Meanwhile, poor software engineering practices offer hackers the opportunity to access private data. As it was published in December 2011 [41], a bug on the "reporting flows" of Facebook caused Mark Zuckerbergs private pictures being leaked.

Table 1. Privacy disclosure and misusage (left column) and common private information gathering methods (right column).

Privacy disclosure	Privacy attacks
Improper use of disclosed information	Re-identification and de-anonymization
Hiring discrimination	Context aware Phishing
School admittance discrimination	Information leakage from friends
	Malicious third-party applications
	Attacks to social network providers

3   Privacy Evaluation and Measurement

With the popularity of online networking sites and the many concerns that have been discussed in regards to privacy issues, one fundamental challenge is *how to measure, evaluate and guarantee privacy and security in online social networks.*

A practical way for privacy evaluation is one of the steps to empower users to a better and robust information protection as well as a powerful method to make users attentive of how their privacy will be exposed, when certain data is posted online or when they make changes to their privacy choices. Two main approaches have been followed to pursue this goal. The first approach takes as the main component the existing privacy options already provided by the networking sites [6]. The second methodology aims to create an index that is in indication of good or bad privacy [3–5]. Both approaches have its advantages and disadvantages. A benefit of fine tuning existing privacy settings is that, since they are already part of the networking sites, with some assistance offered to users, they could become part of their everyday practice. However, it has been demonstrated that users tend to not change default settings [2,7]. In addition, recent surveys suggest that users often have no knowledge about these settings or have a strong perception of complexity that in many cases is justified. On the other hand, methods for privacy index creation go further than suggesting ways of better hiding information by taking into account relationships between actors and attributes as well as predictive power of combinations of attributes. However, several of the magnitudes used for calculations and formulations come from very subjective areas such as users perception of an attributes influence on privacy and are not very accessible to the general users population.

3.1 Privacy Setting Recommendation

Recommending well informed and carefully selected privacy settings for online social networks, has been proposed by researchers. For example, as of March 2013, Facebook reported 1.11 billion users whose profile information is set to be shareable/accessible with friends, friends of friends, and very often, the rest of the public. By default, privacy settings of users accounts on Facebook are open to everyone searches, inside and out of Facebook. Furthermore, research has demonstrated that not only users in general are more likely to leave default settings as is [7] but also, that a small number actually change default privacy preferences on Facebook [2].

One of the first known intents of aiding users with privacy choices was Reclaim.org. The company offered an open source application which worked as a scanner of Facebook members privacy settings [9]. After downloading the application, it would evaluate the accounts privacy selections showing what settings had been securely configured as well as what information was available to the public; it would also make recommendations to enhance privacy selections. The Green Safe is an application that was created to keep Facebook members data out of Facebooks control [10]. Users data is imported into the application, and subsequently, deleted from Facebook. Friends are still able to access the information but it will be hidden from third party applications and partner sites. The Green Safe mines users profiles for targeted advertising, yet the privacy policy ensures that the company will not "share, trade or sell your information with anyone".

3.2 Machine Learning Techniques for Setting Recommendation

An interesting idea that has not been extensively researched is the usage of recommender systems. This type of systems would recommend privacy choices by establishing links between members that have been found to share similar privacy preferences.

Specifically, the work made in [6] uses machine learning as the basis for a recommender system. The authors in [6] created a training dataset which includes attributes from users profile (for example, name, work experience and time zone), interests (such as communities and groups), privacy settings on photo albums and privacy settings on posts.

Surveys conducted by Westin [13] have assisted researches in classifying online network members into three groups based on their privacy concerns: High and Fundamentalist, Medium and Pragmatic, Low and Unconcerned. Fundamentalists refer unwillingness to provide data on websites and go to extremes to avoid revealing any type of personal information. Pragmatics are described as willing to share personal information if they find it being beneficial. Unconcerned users have no problem revealing personal information upon request as well as not having concerns with their privacy. By analyzing users attitude towards sharing their photo albums and posts, a model can be trained to classify members as one of the above mentioned categories (Fundamentalist, Pragmatic, and Unconcerned). This categorization process creates another attribute called privacy_category which is the class label given to a user based on their privacy choices. A standard decision tree was used to infer the privacy_category of the users selected for the training dataset. When a new user joins a social network (Facebook was used as the platform for the study) a k-nearest neighbor classifier would determine the privacy_category for the user. Based on the characteristics of the predicted class, the application recommends which attributes should be disclosed and which ones should be hidden. An application with this type of feature would provide an improvement to privacy settings rather than leaving them completely open which is the default.

3.3   PrivAware and Privometer

In 2010, Facebook provided a software environment intended for third-party developers to create their own applications that access Facebook members information. Numerous applications have been developed using this platform, such as games, information sharing, social causes promotion and privacy protection. Two interesting examples of these applications designed for privacy protection are PrivAware [11] and Privometer [12].

PrivAware [11] was designed based on a basic principle that information should be protected from escaping its intended boundaries. Therefore, PrivAware assists users quantifying privacy risks associated to friend relationships in Facebook. Specifically, PrivAware deals with the attributes inference problem. It has been shown by studies [14–16] that even if a social network member is cautious towards privacy, certain attributes can be inferred based on the values of those of his/her friends; for example political view and affiliations.

PrivAware methodology to the inference reduction problem is: given a set of friends and a privacy requirement represented as the maximum acceptable number of predictable attributes, find the maximum set of friends that fulfills the privacy requirement. PrivAware applies a basic algorithm for attribute inference: given an attribute, the algorithm finds its most popular value among the users friends; this value is assigned to the user if the number of friends sharing this attribute surpasses a previously selected threshold. PrivAware then gives users the choice of either delete unsafe friends, partition friends into groups and apply access control (set risky group to invisible) to each group or "contaminate" his/her friends network with users who do not share common attributes with the target user. The second approach is more desirable since users are not likely to remove friends particularly those who are more similar to them or pollute their network and create confusion among desirable contacts. From the results of collecting data from PrivAware, the authors showed that 59.5 % of the time attributes were correctly inferred based on users social contacts. Also, their heuristic approach to friends removal or grouping for privacy preservation was 19 less than the baseline (removing contacts at random).

Following a similar principle as PrivAware, but with more extended functionality, Privometer [12] is presented as a privacy protection tool that measures the extent of information revelation in a user profile and suggests self-sanitation activities to control the amount of leakage. At the time Privometer was presented, it was the "first functional prototype of a privacy measuring tool to be implemented on a social network" [12]. It provides users with an insight on how a malicious application that could be installed in a contacts profile can access beyond public profile information. Privometer works under the assumption that a third-party application installed in a contacts profile runs an attribute inference algorithm using one of the most popular inference models. Privometer determines the amount of information leakage using some renowned inference models to identify the one that causes the most damage to users privacy. The final leakage value is derived from combine probability of inference. Users are then presented with a graphical measurement of their privacy, a ranking of friends based on their individual contributions to information revelation and actions for remediation. The actions suggested by Privometer are in addition to the privacy settings already provided by Facebook and range from requesting a contact to hide specific attributes to deleting a contact.

3.4   Privacy Index

Quantifying and measuring privacy can be very challenging, mainly because the definition of privacy is very subjective and each individual might have a different opinion about this concept. In 2013, Yong Wang *et al.*, [3–5] proposed to use privacy index (PIDX) to quantify privacy. For the authors, privacy can be assessed based on three metrics: known attributes, their sensitivities, and visibilities. Furthermore, they consider that a combination of attributes may also compromise users privacy; combinations of attributes are called virtual attributes. Based on these assumptions, three privacy measurement functions are discussed and evaluated: weighted privacy measurement function, maximum privacy measurement function, and composite privacy measurement function. Three privacy indexes were further created on the privacy measurement functions: weighted privacy index (w-PIDX), maximum privacy index (m-PIDX) and composite privacy index (c-PIDX).

To reflect the sensitivity of an attribute, a privacy impact factor is assigned to each of them. For full privacy disclosure the value of the impact factor is 1 and it is calculated as a ratio of its privacy impact to full privacy disclosure. A larger numbers indicate that the information is more sensitive. Probabilities are also used to describe attributes visibility. An unknown attribute would have visibility of 0, for a known attribute it is 1 and values between 0 and 1 represent partial disclosure.

Let (L, S, V) represent the set of actors' complete attribute list, their privacy impact and visibilities, respectively.

$$L = (a_1, a_2, \ldots, a_n) \qquad S = (s_1, s_2, \ldots, s_n) \tag{1}$$

$$V = (p_1, p_2, \ldots, p_n) \tag{2}$$

For the purpose of this experiment, $p_i = 1 \ (1 \leq i \leq n)$

Let (L_k, S_k, V_k) represent and actors complete attribute list, their privacy impact and visibilities.

$$L_k = (a\prime_1, a\prime_2, \ldots, a\prime_m) \qquad S_k = (s\prime_1, s\prime_2, \ldots, s\prime_m) \tag{3}$$

$$V_k = (p\prime_1, p\prime_2, \ldots, p\prime_m) \tag{4}$$

- Weighted Privacy Measurement Function and Weighted Privacy Index (w-PIDX)

$$f_w(L_k, S_k, V_k) = p\prime_1 + p\prime_2 + \ldots + p\prime_m = \sum_{i=1}^{m} p\prime_j s\prime_j \tag{5}$$

$$w - PIDX = \frac{f_w(L_k, S_k, V_k)}{f_w(L, S, V)} * 100 = \frac{\sum_{i=1}^{m} p\prime_j s\prime_j}{\sum_{i=1}^{n} s\prime_j} * 100 \tag{6}$$

- Maximum Privacy Measurement Function and Maximum Privacy Index (m-PIDX)

$$f_m(L, S, V) = max(p\prime_1 s\prime_1 + p\prime_2 s\prime_2 + \ldots + p\prime_m s\prime_m) = \sum_{i=1}^{m} p\prime_j s\prime_j \tag{7}$$

$$w - PIDX = f_m(L_k, S_k, V_k) * 100 = max(p\prime_1 s\prime_1 + p\prime_2 s\prime_2 + \ldots + p\prime_m s\prime_m) * 100 \tag{8}$$

- Composite Privacy Measurement Function and Composite Privacy Index (c-PIDX)

$$f_c(L, S, V) = f_m(L_k, S_k, V_k) + (1 - f_m(L_k, S_k, V_k)) = \frac{f_w(L_k, S_k, V_k)}{f_w(L, S, V)} \tag{9}$$

$$w - PIDX = f_c(L_k, S_k, V_k) * 100 \tag{10}$$

As shown, w-PIDX is a centrality measure useful for measuring attribute incremental changes, although it is not useful for privacy ranking. However, this is not the case with m-PIDX, which is good to measure privacy relative value but not incremental changes. Based on their experiments, the authors selected c-PIDX as the most accurate and complete measure of privacy for a social network actor.

Previous works in privacy indexes [17, 18] are mainly based on the item response theory (IRT). Although IRT is a powerful tool, it is designed based on three assumptions: users are independent, items are independent, and users and items are independent. In reality, these assumptions do not apply very well to real-world social networks. The work in [18] goes to the extent of assuming independence between attributes without considering attributes underlined relationships. This represents a problem as it has been confirmed by [19, 20] that combination of attributes may harm users privacy since it can lead to the inference of unknown attributes.

A brief summary of the privacy measurement and evaluation is shown in Table 2.

Table 2. Privacy measurement and evaluation

Measurement or Evaluation	Detail
Privacy settings recommendation	PrivAware, Privometer
	Machine learning techniques to recommend attributes based on similar users height
Privacy index	Weighted Privacy Index (w-PIDX)
	Maximum Privacy Index (m-PIDX)
	composite privacy index (c-PIDX)

4   Conclusion

In this paper, we discussed two important topics, privacy issues, and privacy measurement and evaluation, in online social networks. Large amounts of personal related content have become accessible over the internet via online social networks. While users enjoy connecting with friends and family, concerns over privacy are becoming an increasing important factor. Advances in information retrieval and data analytics provide adversaries with almost unlimited access to the plentiful personal information on the web. We have seen different types of attacks against users privacy, and how with a small portion of information, attackers are able to connect users online persona, to the real life individual. Due to all these concerns, security and privacy must be quantified and evaluated. This paper has shown studies aimed to measure privacy. Some applications have been built that can be integrated with networking sites such as Facebook with the goal of assisting users to evaluate their current privacy choices and recommend settings for maximum protection. An equally important question about the privacy concerns is the measurement and the quantification of the privacy, and we have discussed three privacy measurement in the paper. Our research shows that not much has been implemented and passed along to users. Although social networking companies offer some form of privacy settings, there is a great interest in sharing users data for profit. As a result, it becomes users' responsibility to be knowledgeable and be aware of every decision they make when networking online.

References

1. Bilge, M., Strufe, T., Balzarotti, D., Kirda, E.: All your contacts are belong to us: automated identity theft attacks on social netorking. In: Proceedings of the 18th International World Wide Web Conference, Madrid, Spain. ACM (2009)
2. Gross, R., Acquisti, A.: Information revelation and privacy in online social networks. In: Proceedings of ACM Workshop on Privacy in the Electronic Society, pp. 71–80, November 2005
3. Kumar, R.N., Wang, Y.: SONET: a SOcial NETwork model for privacy monitoring and ranking. In: Proceedings of the 2nd International Workshop on Network Forensics, Security and Privacy (2013)
4. Wang, Y., Nepali, R.K.: Privacy measurement for social network actor model. In: Proceedings of the International Conference on Social Computing, pp. 659–664 (2013)
5. Wang, Y., Nepali, R.K., Nikolai, J.: Social network privacy measurement and simulation. In: Proceedings of the International Conference on Computing, Networking and Communications (ICNC) pp. 802–806 (2014). doi:10.1109/ICCNC.2014.6785440
6. Ghazinour, K., Matwin, S., Sokolova, M.: Monitoring and recommending privacy settings in social networks. In: Proceedings of the Joint EDBT/ICDT 2013 Workshops, pp. 164–168. ACM (2013)
7. Mackay, W.: Triggers and barriers to customizing software. In: Proceedings of CHI 1991, pp. 153–160. ACM Press (1991)
8. Duffany, J., Galban, O.: Hacking Facebook Privacy and Security. Polytechnic Univ. of Puerto Rico San Juan (2012)
9. Shimel, A.: Reclaim your privacy from facebook. PCWorld. Network World, 19 May 2010
10. Perez, S.: New App Helps Keep Facebook's Hands Off Your Data. Readwrite. N.p., 10 May 2010
11. Becker, J.L.: Measuring privacy risk in online social networks. ProQuest, UMI Dissertations Publishing (2009)
12. Talukder, N., Ouzzani, M., Elmagarmid, A., Elmeleegy, H., Yakout, M.: Privometer: privacy protection in social networks. In: 2010 IEEE 26th International Conference on Data Engineering Workshops (ICDEW), Long Beach, CA (2010)
13. Kumaraguru, P., Cranor, L.F.: Privacy indexes: a survey of westins studies. Technical report CMU-ISRI-5-138, Carnegie Mellon University, CMU, Pittsburgh, PA, USA, December 2005
14. Staddon, J., Golle, P., Zimny, B.: Web-based inference detection. In: SS 2007: Proceedings of 16th USENIX Security Symposium on USENIX Security Symposium, pp. 1–16. USENIX Association, Berkeley (2007)
15. Macskassy, S.A., Provost, F.: Classification in networked data: a toolkit and a univariate case study. J. Mach. Learn. Res. **8**, 935–983 (2007)
16. Neville, J., Jensen, D.: Leveraging relational autocorrelation with latent group models. In: Proceedings of International Workshop on Multirelational Mining, pp. 49–55 (2005)
17. Maximilien, E.M., Grandison, T., Sun, T., Richardson, D., Guo, S., Liu, K.: Privacy-as-a-Service?: models, algorithms, and results on the facebook platform. In: Web 2.0 Security and Privacy Workshop (2009)
18. Liu, K.U.N.: A framework for computing the privacy scores of users in online social networks. ACM Trans. Knowl. Disc. **5**(1), 1–30 (2010)

19. Sweeney, L.: Uniqueness of simple demographics in the U. S. population. Data privacy Lab white paper series LIDAP-WP4 (2000)
20. Golle, P.: Revisiting the uniqueness of simple demographics in the US population. In: Proceedings of the 5th ACM Workshop on Privacy in Electronic Society. ACM (2006)
21. Boyd, D., Ellison, N.: Social network sites: definition, history, and scholarship. J. Comput. Med. Commun. **13**(1), 210–230 (2008)
22. Warren, S., Brandeis, L.: The right to privacy. Harvard Law Rev. **4**(5), 193–220 (1890)
23. Westin, A.: Privacy and Freedom. Athenaeum, New York (1967)
24. Yang, Y., Lutes, J., Li, F., Luo, B., Liu, P.: 26 Stalking online: On user privacy in social networks. Paper presented at the 37-48 (2012). doi:10.1145/2133601.2133607
25. Backstrom, L., Dwork, C., Kleinberg, J.: Wherefore art thou r3579x?: anonymized social networks, hidden patterns, and structural steganography. In: Proceedings of ACM International Conference on World Wide Web, pp. 181–190 (2007)
26. Hay, M., Miklau, G., Jensen, D., Towsley, D., Weis, P.: Resisting structural re-identification in anonymized social networks. Proc. of VLDB Endow. **1**(1), 102–114 (2008)
27. He, J., Chu, W.W.: Protecting private information in online social networks. In: Chen, H., Yang, C.C. (eds.) Intelligence and Security Informatics, vol. 135, pp. 249–273. Springer, Heidelberg (2008)
28. He, J., Chu, W.W., Liu, Z.V.: Inferring privacy information from social networks. In: Mehrotra, S., Zeng, D.D., Chen, H., Thuraisingham, B., Wang, F.-Y. (eds.) ISI 2006. LNCS, vol. 3975, pp. 154–165. Springer, Heidelberg (2006)
29. Liu, K., Terzi, E.: Towards identity anonymization on graphs. In: Proceedings of the 2008 ACM SIGMOD, pp. 93–106 (2008)
30. Chen, X., Michael, K.: Privacy issues and solutions in social network sites. IEEE Technol. Soc. Mag. **31**(4), 43–53 (2012). doi:10.1109/MTS.2012.2225674
31. Austin, L.: Privacy and the question of technology. Law Philos. **22**, 119–166 (2003)
32. Tierney, M., Subramanian, L.: Realizing privacy by definition in social networks. In: Proceedings of 5th Asia-Pacific Workshop on Systems (APSys 2014) (2014)
33. Chen, X., Shi, S.: A literature review of privacy research on social network sites. In: Proceedings of International Conference on Multimedia Information Networking and Security (MINES), vol. 1, pp. 93–97, November 2009
34. Joshi, P., Kuo, C.-C.: Security and privacy in online social networks: a survey. In: 2011 IEEE International Conference on Multimedia and Expo (ICME), pp. 1–6, July 2011
35. Zhang, C., Sun, J., Zhu, X., Fang, Y.: Privacy and security for online social networks: challenges and opportunities. IEEE Netw. **24**(4), 13–18 (2010)
36. Bilge, L., et al.: All your contacts are belong to us: automated identity theft attacks on social networks. In: Proceedings of the 18th International Conference World Wide Web (WWW 2009), pp. 551–560. ACM Press (2009)
37. Mansfield-Devine, S.: Anti-social networking: exploiting the trusting environment of Web 2.0. Netw. Secur. **11**, 4–7 (2008)
38. King, R.: Facebook dials back on third-party app shares, 27 May 2014. http://www.zdnet.com (retrieved)
39. Gao, H., Hu, J., Huang, T., Wang, J., Chen, Y.: Security issues in online social networks. IEEE Internet Comput. **15**(4), 56–63 (2011). doi:10.1109/MIC.2011.50
40. Jagatic, T.N., et al.: Social phishing. Commun. ACM **50**(10), 94–100 (2007)

41. Duell, M.: Mark Zuckerbergs private Facebook photos revealed: Security glitch allows web expert to access billionaires personal pictures. The Daily Mail (MailOnline), December 2011. http://www.dailymail.co.uk/news/article-2070749/Facebook-security-glitch-reveals-Mark-Zuckerbergs-private-photos.html

42. Valdes, M., McFarland, S.: Employers Ask Job Seekers for Facebook Passwords. Associated Press, 20 March, 2012

43. Vijayan, J.: New laws keep employers out of worker social media accounts. Computer World, 4 January 2013. http://www.computerworld.com/article/2505609/data-privacy/ill-bans-firms-from-asking-workers-job-seekers-for-social-media-info.html

44. Backstrom, L., Dwork, C., Kleinberg, J.: Wherefore Art Thou r3579x?: anonymized social networks, hidden patterns, and structural steganography. In: Proceedings of the 16th International Conference on World Wide Web (WWW 2007), pp. 181–190. ACM Press (2007)

45. Narayanan, A., Shmatikov, V.: De-anonymizing social networks. In: Proceedings of the 20th IEEE Symposium on Security and Privacy (SP 2009), pp. 173–187. IEEE CS Press (2009)

46. Wondracek, G., et al.: A practical attack to de-anonymize social network users. In: Proceedings of IEEE Symposium on Security and Privacy (SP 2010), pp. 223–238. IEEE CS Press (2010)

47. Vijayan, J.: Ill. Bans firms from asking workers, job seekers for social media info. PCWorld. ComputerWorld, 7 August 2012

48. Barnes, S.B.: A privacy paradox: social networking in the United States. First Monday, 4 September 2006. http://firstmonday.org

Building Secure Web Systems Architectures Using Security Patterns

Eduardo B. Fernandez

Department of Computer Science and Engineering, Florida Atlantic University,
Boca Raton FL, 33431, USA
ed@cse.fau.edu
http://www.cse.fau.edu/~ed

Abstract. Software patterns are encapsulated solutions to recurrent problems in a context. Patterns combine experience and good practices to develop basic models that can be used to build new systems, to evaluate existing systems, and as a communication medium for designers. *Security patterns* provide guidelines for secure system requirements, design, and evaluation. We consider their structure, show a variety of them, and illustrate their use in the construction of secure web-based systems. These patterns include among others Authentication, Authorization/Access Control, Firewalls, Secure Broker, Web Services Security, and Cloud Security patterns. We have built a catalog of over 100 security patterns. We complement these patterns with *misuse patterns*, which describe how an attack is performed from the point of view of the attacker, we show XSS as an example. We integrate patterns in the form of *security reference architectures* that represent complete systems. Reference architectures have not been used much in security and we explore their possibilities. We show how to apply these patterns through a secure system development methodology. We introduce patterns in a conceptual way, relating them to their purposes and to the functional parts of the architecture. Example architectures include a browser and a cloud computing system. The use of patterns can provide a holistic view of security, which is a fundamental principle to build secure systems. Patterns can be applied throughout the software lifecycle and provide an ideal communication tool for the builders of the system. They are also useful to record design decisions. The patterns and reference architectures are shown using UML models and examples are taken from my book: "*Security patterns in practice: Building secure architectures using software patterns*", Wiley Series on Software Design Patterns, 2013. The talk also includes some recent patterns, e.g. Network Function Virtualization. Security is a fundamental quality for any web system. Most proposed solutions are ad hoc or partial; regretfully security must be holistic and systematic. Patterns provide the basis for systematic and holistic approaches and are becoming more important every day. Attendees will be able to understand the idea behind security patterns and security reference architectures, get acquainted with some of them, and use them to build architectures for secure systems.

© Springer International Publishing Switzerland 2015
J. Wang et al. (Eds.): WISE 2015, Part II, LNCS 9419, p. 503, 2015.
DOI: 10.1007/978-3-319-26187-4

The WordNet Database:
Form, Function, and Use

Mark A. Finlayson

Assistant Professor of Computer Science, School of Information and Computing
Sciences, Florida International University, 11200 S.W. 8th Street,
ECS Room 362, Miami FL, 33199, USA
markaf@fiu.edu

Wordnet[1] is a large lexical database of the English language. Like a regular dictionary, it indexes base form words (such as the word *run*) to meanings (e.g., "move fast by using one's feet" as well as "a score in baseball"). Unlike a regular dictionary, it encodes significant amounts of additional information about the interrelationships of word meanings and lexical forms. Perhaps most helpfully, it marks what words are almost exactly synonymous, and so can be used as a thesaurus in addition to a dictionary. Beyond this, however, Wordnet encodes a number of other relationships, such as the fact that an *animal* (synonymous with *animate being*, *creature*, or *fauna*) is a type of *organism*, which is in turn a type *living thing*. This is called the semantic relationship of type-subtype, and Wordnet encodes semantic and lexical relationships between its entries such as type-subtype, part-whole, substance-whole, member-set, domain-topic, antonymy, derivationally related forms, among others. In addition to this rich repository of language meaning, Wordnet is further notable for its size, containing over 155,000 base wordforms, 117,000 meanings, and 188,000 relationships beyond synonymy, including over 46,000 lexical relationships and 142,000 semantic relationships.

Wordnet can be of great use to any application that has to interact with natural language text. In this tutorial, we will first learn about the form of the Wordnet database: the core concepts, what kinds of relationships are encoded in the database, and some caveats about the database contents. We will also examine a small selection of tasks enabled by each type of information encoded in the database. These tasks are provided only as a sample of potential applications, as the range of uses is limited only by one's imagination. Tasks we will learn about include low-level NLP tasks such as lemmatization or root finding (given the inflected form "running" return the root "run", or given the irregular form "is" return the root "be"), all the way up to conceptual processing tasks such as determining that *cats* and *dogs* are more similar to one another than to *turtles*, *plants*, or *cars*.

In addition to the form and utility of the database, we will learn how to interact with the database programmatically. We will first review ways of loading Wordnet into common databases such as MySQL, Sqlite, PostgresSQL, and the like, such that it can be. After this we will examine how to interface with the database directly within a Java programming language environment, focusing on the library the MIT Java Wordnet Interface (JWI)[2]. JWI is small, extremely fast, easy to use, and provides API access to all available Wordnet database information.

---

[1] http://wordnet.princeton.edu/

[2] http://projects.csail.mit.edu/jwi

© Springer International Publishing Switzerland 2015
J. Wang et al. (Eds.): WISE 2015, Part II, LNCS 9419, p. 504, 2015.
DOI: 10.1007/978-3-319-26187-4

Author Index